Lecture Notes in Earth Sciences 86

Springer
Berlin
Heidelberg
New York
Barcelona
Hong Kong
London
Milan
Paris
Singapore
Tokyo

Henry V. Lyatsky   Vadim B. Lyatsky †

The Cordilleran Miogeosyncline in North America

Geologic Evolution and Tectonic Nature

With 78 Figures

Authors

Dr. Henry V. Lyatsky
Lyatsky Geoscience Research & Consulting Ltd.
4827 Nipawin CR. N.W.
Calgary, Alberta, Canada T2K 2H8
E-mail: lyatskyh@cadvision.com

Dr. Vadim B. Lyatsky†

"For all Lecture Notes in Earth Sciences published till now please see final pages of the book"

Cataloging-in-Publication data applied for

Die Deutsche Bibliothek - CIP-Einheitsaufnahme

Lyatsky, Henry V.:
The Cordilleran miogeosyncline in North America : geologic evolution and tectonic nature / Henry V. Lyatsky ; Vadim B. Lyatsky. - Berlin ; Heidelberg ; New York ; Barcelona ; Hong Kong ; London ; Milan ; Paris ; Singapore ; Tokyo : Springer, 1999
(Lecture notes in earth sciences ; 86)
ISBN 3-540-66197-2

ISSN 0930-0317
ISBN 3-540-66197-2 Springer-Verlag Berlin Heidelberg New York

Typesetting: Camera ready by author
SPIN: 10736077 32/3142-543210 - Printed on acid-free paper

Foreword

The Earth's rock-made outer shell, the lithosphere, contains entities of two main types, continental and oceanic. Cratons and mobile megabelts are lower-rank lateral lithospheric tectonic units into which continental lithospheric masses are subdivided. Mobile megabelts develop when large zones of continental crust and lithosphere are tectonically mobilized and deeply reworked; the reworked continental lithosphere could have previously been cratonic. Many of the mobile megabelts in the world lie in peripheral regions of continents; others are located far within continental interiors. A megabelt, being a first-order continental lateral tectonic unit, consists of subordinate major units: orogenic zones and median massifs. Median massifs are less tectonically mobilized and reworked than orogenic zones, and they retain more of their ancient (commonly, ex-cratonic) characteristics.

Yet, despite the ancient continental inheritance, it is conventional for modern tectonists to regard mobile megabelts not from the vantage point of their parent cratons but from distant oceans genetically unrelated to the continent. The underlying false assumptions are that continents are essentially inert, incapable of self-development, and that any tectonism in them must be induced externally - from the sub-lithospheric mantle below or from oceanic plates to the side. In reality, continental lithosphere has its own radioactive energy sources, and its tectonic development is mostly an expression of indigenous, internal processes.

The ocean-based, "Poseidonian" approach to continental tectonics is illogical at its core: instead of relying on direct observations of rocks in land areas where these rocks are exposed, it puts too much emphasis on speculative, assumption-based reconstructions of distant plate motions in the past and on the assumed effects of those supposed motions on continental areas. Plate reconstructions are based not on the observed geology of continental mobile megabelts and cratons but on model-driven interpretations of geophysical anomalies in oceanic regions far away. The global approach has in the last several decades usefully extended the vision of tectonists, which was previously restricted to directly accessible continental regions. Unfortunately, a casualty of this model-based kind of globalization has been the traditional practice of basing tectonic conclusions on the observed, sampled and analyzed rocks and the mapped rock-body relationships. But it is observational geological information that provides the ultimate means to test abstract models, and this fact-based approach is adopted in the present book.

Being generally unable to detect high-angle discontinuities, seismic reflection profiles may give the viewer a false, partial impression that only low-angle discontinuities exist in continental crust. The rock nature and origin of these seismically reflective low-angle crustal discontinuities are unknown, and they could be related to a multitude of possible causes: original discontinuities in the protolith, subsequent metamorphic fronts and rheologic boundaries, ductile and brittle shearing, intrusive igneous bodies, and so on. In the absence of direct geological observations, the inherent non-uniqueness of geophysical interpretations precludes definitive conclusions about the rock composition and structure of deep crust. Even in the exposed and drilled uppermost part of the crust, every dry oil or water well, and every failed mining or geotechnical project represents a failure of prediction.

Superdeep wells drilled into the upper continental crust in Europe, Russia and North America have presented many surprises quite unanticipated by prior geophysical predictions, including very unexpected findings about sources of seismic reflections and potential-field anomalies, as well as position of the brittle-ductile boundary and thermal and hydrologic conditions at depth. Not even such sparse geologic constraints are available for the middle and lower crust. The arbitrary and model-driven assumption that seismic-reflection geometries deep in the crust mostly represent thrusting leads to an incorrect perception of the continental crust as necessarily a stack of thrust sheets. Meanwhile, the block structure of the crust goes unnoticed.

Tectonics synthesizes the factual findings from field mapping and many specialized geological disciplines - petrology, geochemistry, paleontology, and so on. From a vast amount of diverse information, a tectonist selects that which permits to discern the phenomena induced by endogenic processes in the lithosphere, and combines a broad range of facts and data into an internally consistent concept. This productive, practical approach is utilized in the present study of the Cordilleran mobile megabelt, which lies on the western flank of the North American continent.

Steep, deep-seated, long-lived crustal weakness zones are well known to have existed in the western part of the North American craton since the Late Archean or Early Proterozoic. Trending mostly NE-SW, these ancient crustal weakness zones divide the craton into distinct large blocks with dissimilar crystalline-crust and volcano-sedimentary-cover properties. Some of these zones predate cratonization (which is known to have occurred in the Early Proterozoic in western Canada). Many more steep brittle faults, variously following and cutting across the ancient ductile structures, were created in the upper crust after the end of cratonization. Rejuvenation of old weakness zones and fractures in various crustal tectonic regimes and conditions influenced block movements and the formation of volcano-sedimentary basins on the craton in the Proterozoic and Phanerozoic. Across the Cordilleran tectonic grain, these ancient weakness zones continue from the craton all across the Late Proterozoic-Phanerozoic Cordilleran mobile megabelt, where geological and geophysical evidence shows they influenced many tectonic manifestations (depositional, magmatic, metamorphic, structural) throughout the megabelt's lifespan.

Wide westward extent of ancient pre-Cordilleran craton(s) is suggested by other evidence as well. Tectonically reworked Precambrian rocks of various Archean and Proterozoic ages are recognized increasingly commonly all over the Canadian Cordillera, while the paleontological, paleomagnetic and structural evidence for exoticism of assumed accreted terranes is continually revised or eliminated. The oldest known incidence of the "Cordilleran" NNW-SSE tectonic trend is a ~1,760-Ma geochemical anomaly in the cratonic basement of the Alberta Platform, but thereafter orogenic activity shifted westward. Two-sided volcano-sedimentary and sedimentary basins in the eastern Cordillera - Middle Proterozoic, Late Proterozoic, late Paleozoic, Mesozoic - are known from sedimentological studies to have received sediments from land sources on the east and west. The Middle Proterozoic Belt-Purcell Basin, though very deep, was evidently intracratonic and non-orogenic; the ancient cratonic area to the west of it is conventionally called the Western craton. The NNW-SSE Cordilleran tectonic trend was not firmly established till the onset of Windermere rifting at ~780 Ma. The Late Proterozoic Prophet-Ishbel trough was probably a foredeep to the well-known Antler orogen, whose continuation from the U.S. Cordillera lay in the Canadian Cordilleran regions to the west.

In the cratonic Phanerozoic Alberta Basin just to the east of the Canadian Cordillera, many Paleozoic platformal sedimentary units fail to exhibit a wedge-like thickening to the west. Occasional westward continuation of cratonic depositional settings far into the Cordilleran interior is indicated by the occurrences of Paleozoic carbonate-platform remnants similar to the coeval carbonate platforms in the cratonic Alberta Basin. Contrary to a common assumption, at no time did the eastern Canadian Cordillera contain an Atlantic-type continental margin.

Geochemical evidence of reworked, ex-cratonic basement rocks is well known from the metamorphic core complexes in the Canadian Cordilleran interior. Highly metamorphosed former supracrustal rocks are known in these complexes as well.

To achieve their high metamorphic grades, in the late Mesozoic and early Cenozoic these rocks were first lowered to great crustal depths of 20-30 km, and then rapidly returned to the surface. Mapped tectonic manifestations indicate that several intense orogenic episodes had occurred in that region in the Middle Jurassic and mid-Cretaceous, each ending with decompression and crustal extension. The Late Cretaceous-Early Tertiary Laramian orogenic cycle, with its own extensional pulse in the end, was only one in this long series. This multi-cycle tectonic history rules out the commonly assumed scenario with a passive continental margin in the eastern Cordillera before the mid-Mesozoic; compression and exotic-terrane accretion and stacking from then till the Early Tertiary; and extension thereafter.

The cratonward-vergent Laramian Rocky Mountain fold-and-thrust belt on the eastern flank of the Cordilleran mobile megabelt has long been known to be a thin-skinned, rootless thrust stack consisting of supracrustal rocks. It evidently does not involve the cratonic basement, which in that region is not tectonically reworked. Although the Rocky Mountain fold-and-thrust belt is conventionally assigned to the Cordillera, the evident Cordilleran tectonic reworking of the ancient basement begins near this belt's western boundary, in the Cordilleran interior. Rather than the eastern Laramian deformation front, it is that interior line that should be considered the modern western edge of the North American craton.

West of the Rocky Mountain fold-and-thrust belt lies the eastern Cordilleran miogeosynclinal Omineca orogenic zone, characterized by strong crustal reworking, metamorphism and deformation. These manifestations were comparatively slight in the more-rigid crustal blocks of the Intermontane-Belt and Yukon-Tanana median massifs farther west: their volcano-sedimentary cover is preserved largely intact, with its upper parts metamorphosed only slightly or not at all; block faulting is common. The magmatism there was largely teleorogenic, rooted in adjacent orogenic regions (mostly, the eugeosynclinal Coast Belt to the west). Blocks of rigid, semi-reworked ex-cratonic Precambrian crust probably lie beneath the exposed Late Proterozoic and Phanerozoic volcano-sedimentary cover of the median massifs. Presence of older continental basement in the median massifs is suggested by radiometric inheritance, xenolith evidence, seismic-velocity structure of the crust, and long-time rigidity of these crustal blocks that prevented a more profound orogenic reworking. Transcurrent, ex-cratonic, ancient crustal weakness zones are expressed particularly strongly in the median massifs.

Rock evidence indicates that local rifts, Mediterranean- or Red-Sea-type deep marine basins and even ephemeral minor subduction zones might nonetheless have existed in the Canadian Cordillera at different times and in different localities. Yet, because crustal blocks with a shared ancient crustal ancestry in the Cordilleran interior evolved throughout the Phanerozoic side by side, roughly *in situ*, the oft-postulated big intra-Cordilleran former oceans are unlikely. Tectonic zones and belts of the Cordilleran mobile megabelt developed essentially *in situ*, but the median massifs retain more of their ancient cratonic crustal inheritance than do the orogenic zones on both sides of these massifs.

Outward-verging Mesozoic and Cenozoic fold-and-thrust zones of various sizes in the Cordilleran interior bilaterally flank the miogeosynclinal and eugeosynclinal orogenic belts. Like the huge Rocky Mountain fold-and-thrust belt on the east side of the Cordilleran mobile megabelt, they are probably rootless. These shallow deformation zones on the orogens' flanks obscure the deep crustal boundaries of the orogenic and median-massif belts.

Seismic evidence that the Cordilleran crust is a stack of exotic-terrane thrust sheets is likewise unconvincing. Geophysical anomalies carry information about the spatial distribution of some specific physical properties of rocks, but they say nothing about the discontinuities' geologic nature, genesis or age. High-angle crustal discontinuities are normally missed in seismic reflection images altogether, and the geologic origin of low-angle reflections is unknown. Correlation of seismic events is complicated by these events' unknown nature, common lack of continuity and character variations, and gaps in the data. In a strongly deformed and magmatized region, many off-line arrivals and other forms of coherent noise contaminate the seismic data at short and long traveltimes. The technical need during data acquisition for road access in the mountains required the available seismic reflection profiles to be shot largely along passable valleys, which tend to follow steep Cordilleran and transcurrent crustal faults; as a result, some of the seismic lines were shot not across but along large-scale structures.

Postulations of trans-Cordilleran crustal detachments rely on assumption-based correlations of selected, disconnected low-angle seismic events and their assignment to faults that at the surface are known to be variously low-angle or steep. Many of the supposedly trans-crustal detachment faults are correlatable only with reflections that dissipate in the mid-crust. The use of discontinuous "floating" events to justify these faults' deeper continuity is unfounded, and it contradicts the results of refraction seismic surveys. On the other hand, important steep fault systems that cannot easily be correlated with big low-angle reflections are overlooked in the model-driven tectonic analysis.

Precambrian craton(s) continued far to the west in pre-Cordilleran time. Today, their semi-reworked remnants are found in the Cordilleran median massifs, which are surrounded by orogenic zones where the crustal tectonic reworking was much greater. The main differences between tectonic zones in the Cordilleran interior lie not in their assumed exotic-terrane composition but in the degrees of orogenic reworking of the indigenous North American continental lithosphere.

ACKNOWLEDGMENTS

This book has arisen from our decades-long experience of practical research and exploration in western Canada and elsewhere. Our work has benefited from many informative and helpful discussions with our colleagues in the oil and mining industries, at the Universities of Calgary and British Columbia, and at the Vancouver, Calgary and Ottawa offices of the Geological Survey of Canada. These colleagues have brought to our attention many relevant facts, data and scientific ideas. Particular thanks go to the scientific staff at the Geological Survey of Canada office in Vancouver, who initially introduced the first author to the Cordilleran geology and whose interest, support and encouragement are deeply appreciated. Our colleagues at Lithoprobe have also been helpful with geophysical data and current tectonic ideas. Responsibility for scientific conclusions presented here, however, rests with the authors alone.

Garth Keyte and Four West Consultants (Calgary) kindly provided printing, drafting and reproduction facilities. Gerald M. Friedman (Northeastern Science Foundation and City University of New York) looked after the scientific review of the manuscript.

Sadly, the second author, my father, passed away as this manuscript was near completion, and he did not see its publication. This book, therefore, is in his memory.

Henry Lyatsky

TABLE OF CONTENTS

1 - INTRODUCTION

2 - GOVERNING PRINCIPLES IN REGIONAL TECTONIC EVOLUTION

3 - ROCK UNITS IN REGIONAL TECTONIC ANALYSIS

4 - CONCEPTUAL FUNDAMENTALS OF REGIONAL TECTONIC ANALYSIS ON CONTINENTS

5 - VIEW OF THE CORDILLERAN MOBILE MEGABELT'S EVOLUTION FROM THE CRATON

6 - OMINECA OROGENIC BELT AS TECTONOTYPE OF THE EASTERN CORDILLERAN MIOGEOSYNCLINE

7 - BROAD LOOK AT GEODYNAMICAL MECHANISMS OF CRUSTAL RESTRUCTURING IN THE CORDILLERAN OROGENS AND MEDIAN MASSIFS

8 - LESSONS FROM CORDILLERAN GEOLOGY FOR THE METHODOLOGY OF REGIONAL TECTONIC ANALYSIS OF MOBILE MEGABELTS

9 - PRACTICAL UTILITY OF ROCK EVIDENCE AND GEOPHYSICAL STUDIES IN RESTORING REGIONAL TECTONIC HISTORY IN MOBILE MEGABELTS

LIST OF TABLES AND FIGURES

1 - INTRODUCTION

Region of study

Cordillera is a Spanish word for a big, linear, high-standing piece of land. In particular, it denotes ranges and ridges in mountain chains - and in the Americas, Spanish conquistadors used it to describe the Andes. The name Cordillera also spread into western North America, where it is now conventionally applied to the entire mountain region west of the Great Plains. On its western side, the Cordillera of North America continues to the very edge of the continent. The Great Plains occupy the continental Interior from the Gulf of Mexico to the Canadian Arctic. The North American Cordillera, in the modern designation, runs N-S for thousands of kilometers from Central America to Alaska. Its width varies from just ~200 km in Central America to ~1,500 km in the western U.S. Much of the southern part of the Cordillera, in the southwestern U.S. and northern Mexico, has a peculiar physiography of alternating basins and ranges. The broad Colorado Plateau separates this province from the more continuous mountains to the north. The Columbia Plateau, in northwestern U.S., is also extensive. Covered by voluminous basaltic flows, it clearly segments the central Cordillera. North of it lies the most ordered, linear segment of this mountain region, trending NNW-SSE through Canada and containing an interior upland zone of subdued relief. This zone, called Intermontane Belt, separates two main, rugged mountain belts to the west and east. In central Alaska, the Cordillera swings to the east (Figs. 1-10; Table I; the reader is encouraged to examine these and following diagrams before reading further, to familiarize himself/herself with the location and place names mentioned in the text).

Based on its main physiographic forms, the Cordillera in Canada and adjacent parts of the U.S. is conventionally subdivided into five subparallel physiographic belts (from east to west): Rocky Mountains, with an imbricated structure and many high limestone cliffs; Omineca, of rugged and high mountains with a block-like fabric; Intermontane, of gentle plateaus cut by river valleys; Coast, containing the highest and most rugged mountains of the Canadian Cordillera, with big glaciers at high elevations and with fiords cutting across this entire belt; and Insular, consisting of a chain of islands separated by wide waterways and narrow fiords, whose backbone is comparatively low mountains. This book concerns itself with the Cordillera north of the Colorado and especially Columbia plateaus, as far as the Yukon-Tanana lowlands in central Alaska, and from the Foothills of the Rocky Mountains to the boundary of the Intermontane and Coast belts. Most of this geographic region lies in Canada, and abundant information is available in the public domain to make the descriptions in this book.

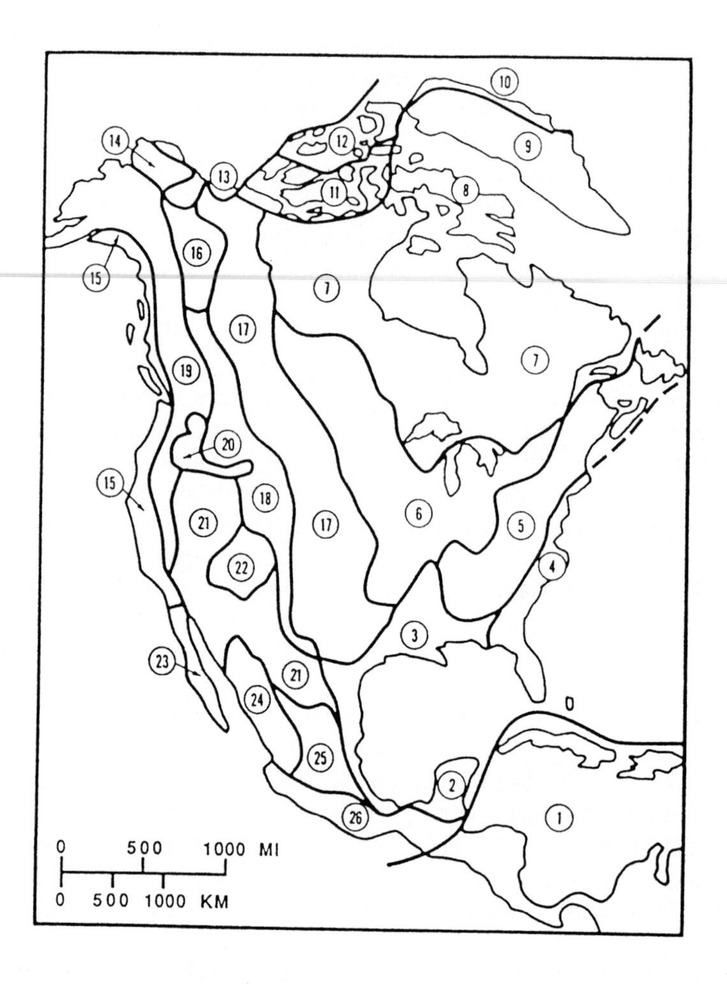

Figure 1. Main geomorphologic provinces of the North American continent and vicinity (modified from Bally et al., 1989): 1 - Central America and Caribbean; 2 - Yucatan platform; 3 - Gulf of Mexico coastal plain; 4 - Atlantic coastal plain; 5 - Appalachian mountains and plateaus; 6 - Central lowland; 7 - Canadian Shield; 8 - Baffin Island; 9 - Greenland ice cap; 10 - West and North Greenland mountains and fiords; 11 - Arctic lowland; 12 - Innuitian; 13 - Arctic coastal plain; 14 - Brooks Range; 15 - Pacific Rim and Pacific Coast Ranges; 16 - Mackenzie; 17 - Great Plains; 18 - Rocky Mountains; 19 - Interior plateaus and mountains; 20 - Columbia Plateau; 21 - Basin and Range; 22 - Colorado Plateau; 23 - Baja California; 24 - Sierra Madre Occidental; 25 - Sierra Madre Oriental; 26 - Sierra Madre del Sur.

Figure 2. Tectonic index map of North America, showing main tectonic provinces (simplified from Bally et al., 1989). Abbreviations of tectonic provinces: BR - Basin and Range; BRO - Brooks Range; F.B. - fold belt; MA - Marathon uplift; MK - Mackenzie Mountains; OU - Ouachita Mountains.

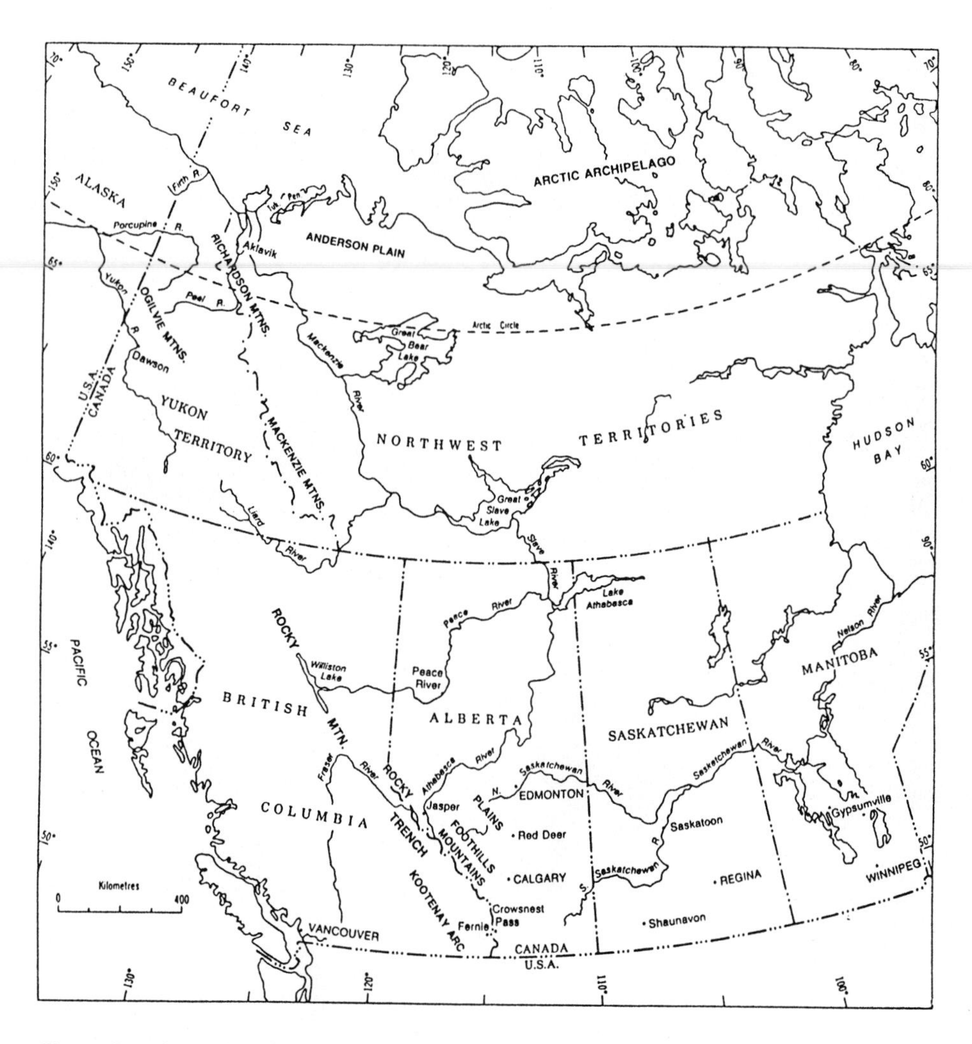

Figure 3a. General geographic index map of western Canada (modified from Poulton et al., 1993), showing major geographic features. To become familiar with the region of study discussed in this book, and particularly with the place names and localities mentioned in the text, the reader is encouraged to review this and subsequent diagrams before continuing with the reading. Virtually all the main localities and geographic names used in this book are shown in Figs. 3 and 4.

Figure 3b. Detailed geographic index map of western Canada, including the Cordillera and the adjacent western parts of the North American craton (simplified from *Canada Gazetteer Atlas*, 1980), showing main rivers, lakes and cities. Physiographic zones in the Canadian Cordillera are shown in more detail in Fig. 4.

Figure 4a. Physiographic zones and regions of the Canadian Cordillera (from Mathews, 1986).

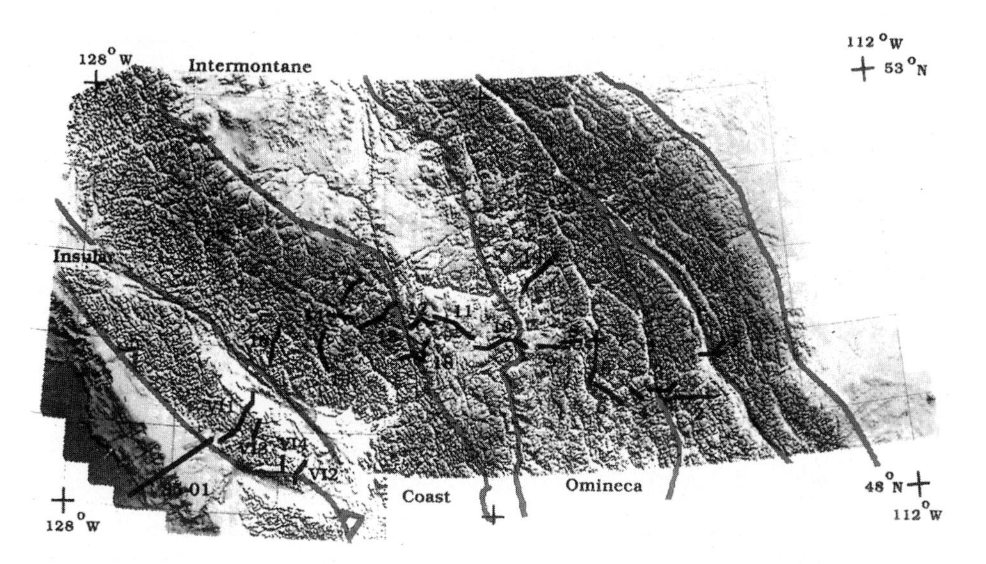

Figure 4b. Shaded-relief topography of the southern Canadian Cordillera, with an overlay of major Cordilleran geologic belts (Figs. 6, 7a). Numbered heavy lines mark the Lithoprobe seismic reflection profiles (modified from Cook et al., 1995).

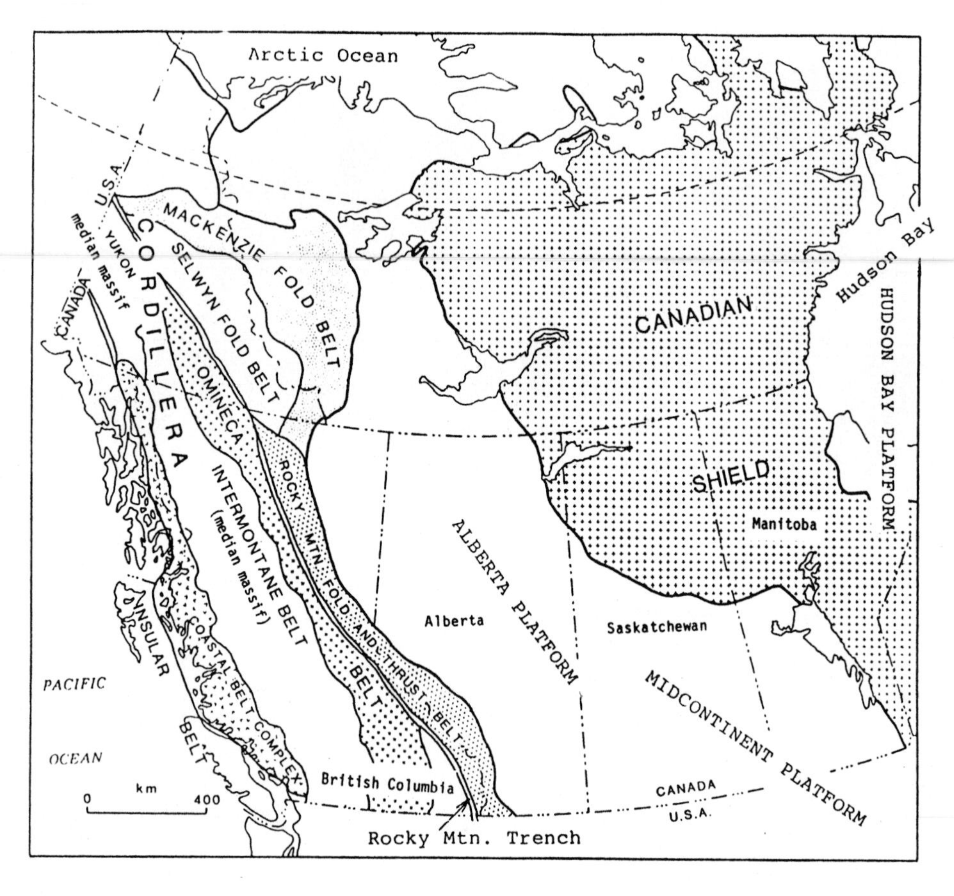

Figure 5. Principal present-day geologic zoning of western Canada (modified from Douglas et al., 1970 and Richards, 1989).

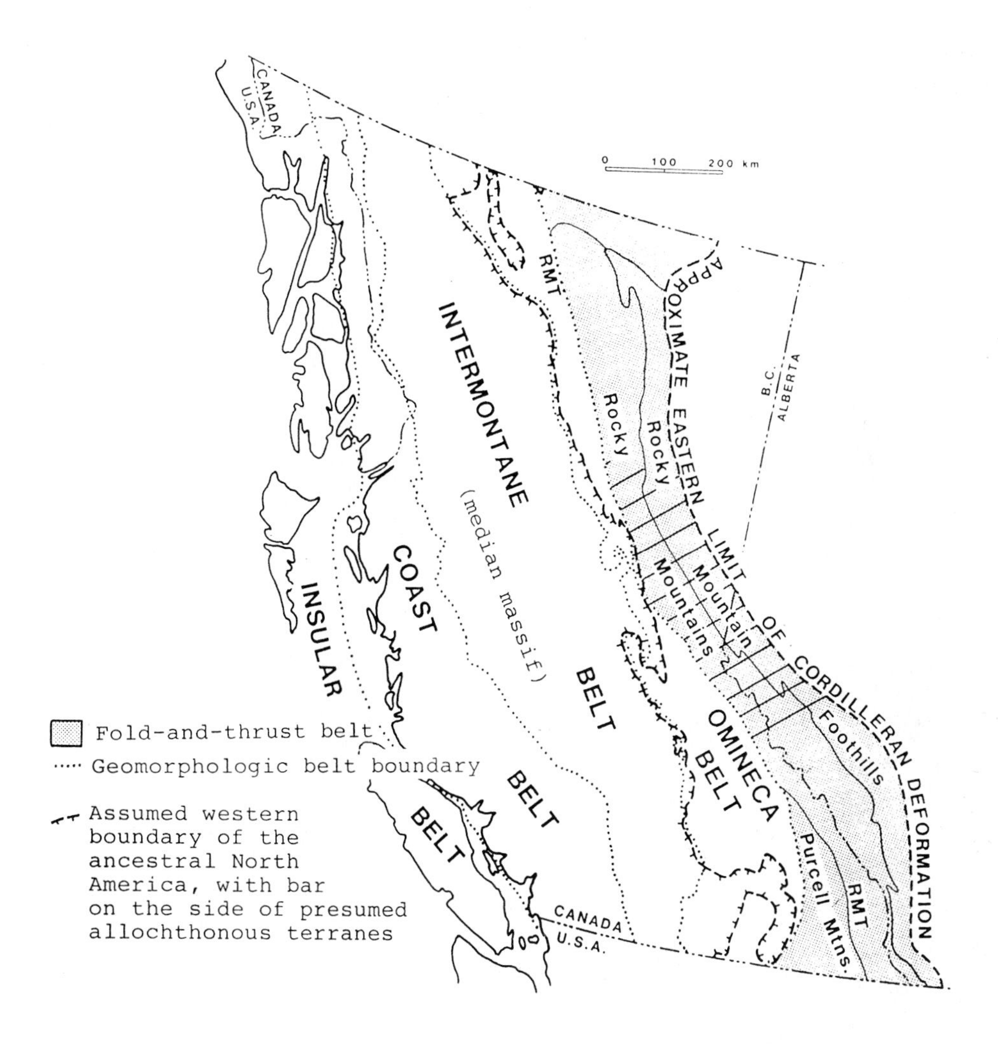

Figure 6. Main tectonic belts in the Canadian Cordillera (modified from McMechan and Thompson, 1989). The Rocky Mountain fold-and-thrust belt contains three main segments, from south to north: thrust-dominated, transitional (ruled pattern) and fold-dominated; the amount of shortening decreases from south to north. The Purcell Anticlinorium was assigned by these authors to the Rocky Mountain fold-and-thrust belt. This book, however, follows the more-common practice to assign it to the Omineca orogenic belt (Fig. 7a) because, unlike the Rocky Mountain Belt, the anticlinorium area was orogenized more profoundly than just by rootless, thin-skinned Laramian thrusting. RMT - Rocky Mountain Trench.

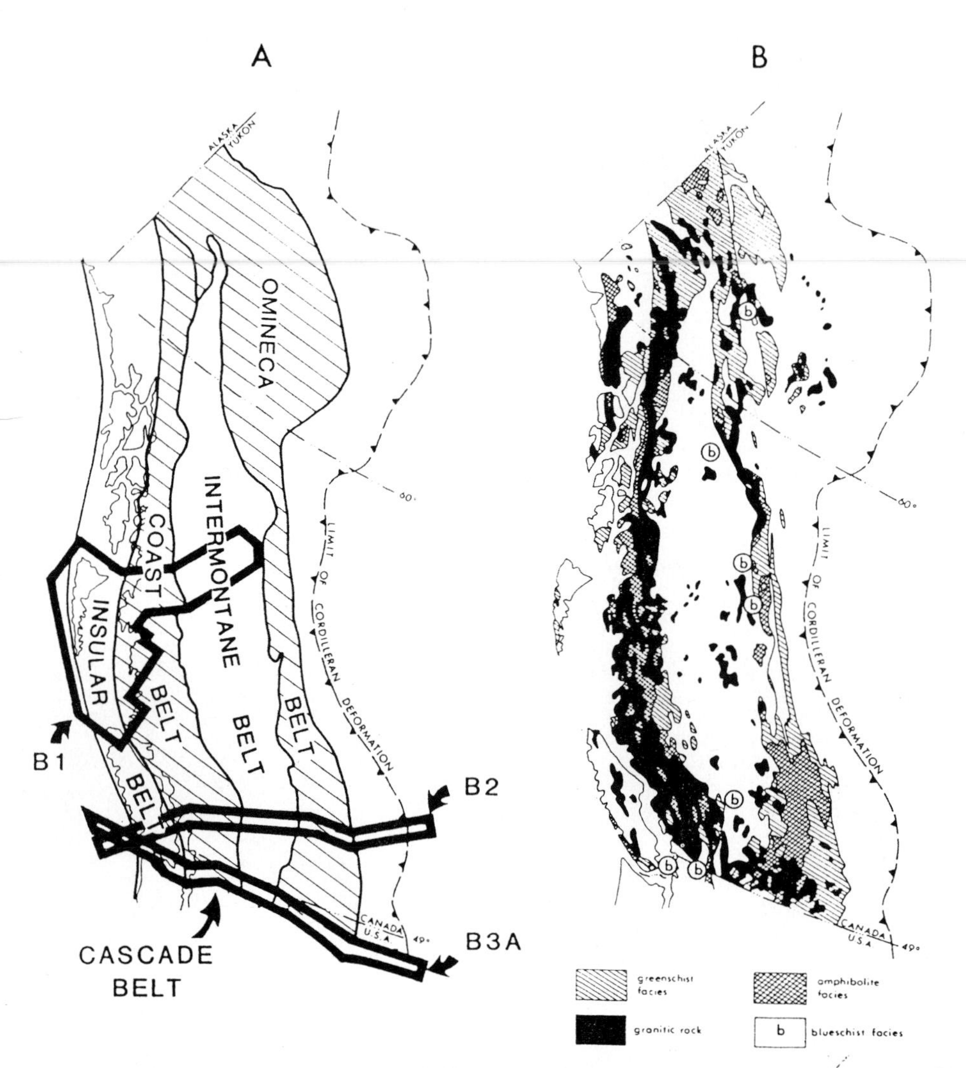

Figure 7. Five main conventionally defined geologic belts of the Canadian Cordillera (**a**) and simplified metamorphic map of the Canadian Cordillera (**b**), after Monger et al. (1994). *DNAG* (*Decade of North American Geology*) transects B1, B2 and B3A in Fig. 7a correspond, approximately, to the COCORP (B3A), southern Canadian Cordillera Lithoprobe (B2) and ACCRETE (1) crustal transects across the Cordillera. Most of the attention in this book is given to the B2 (Lithoprobe) area, but comparisons with the other transects are made commonly. The Purcell Anticlinorium is included into the Omineca Belt. Rocks in the unpatterned areas in Fig. 7b lack significant metamorphism.

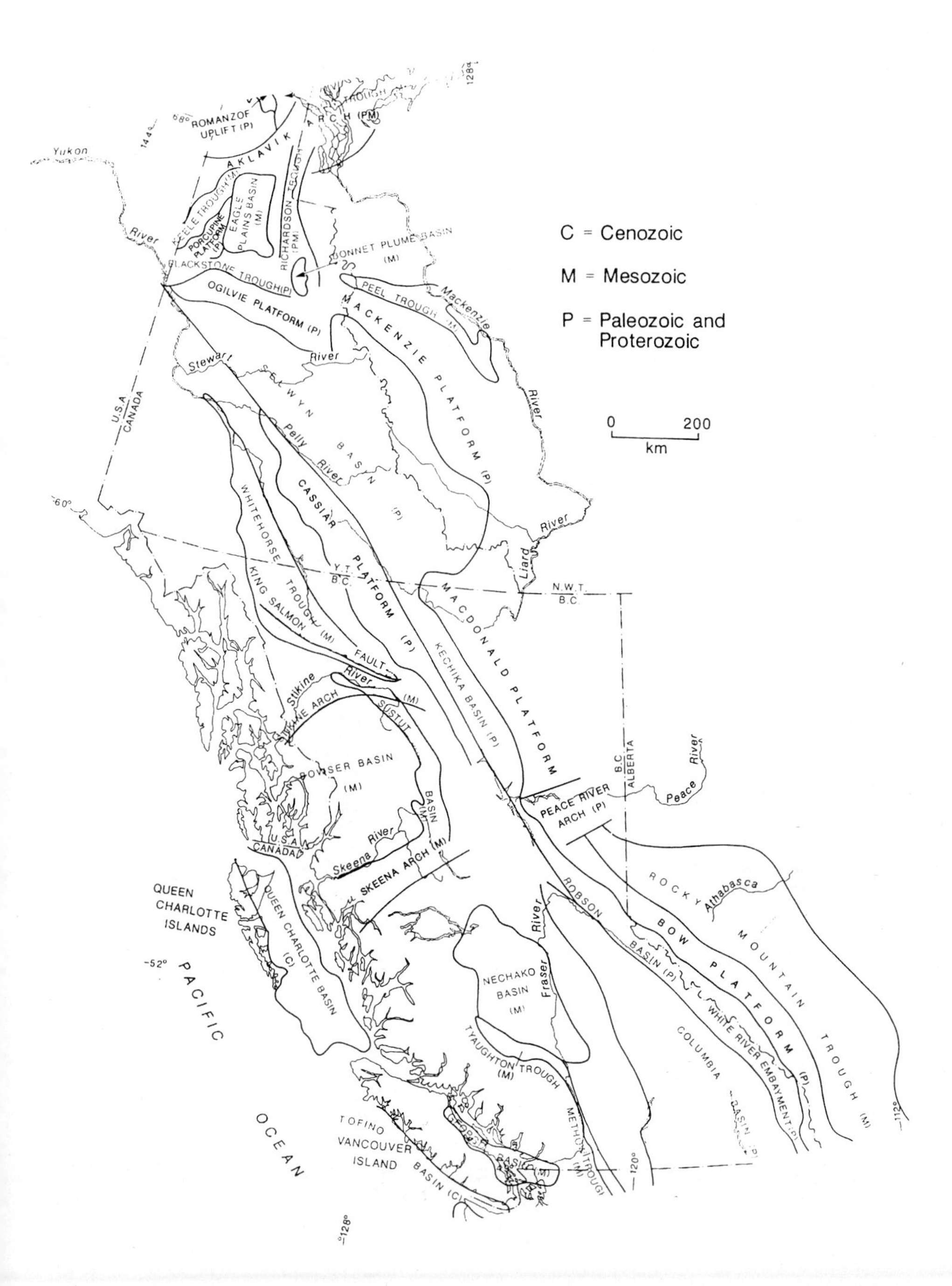

Figure 8. Major tectonic features of the Canadian Cordillera discussed in this book (simplified from Gabrielse et al., 1991a).

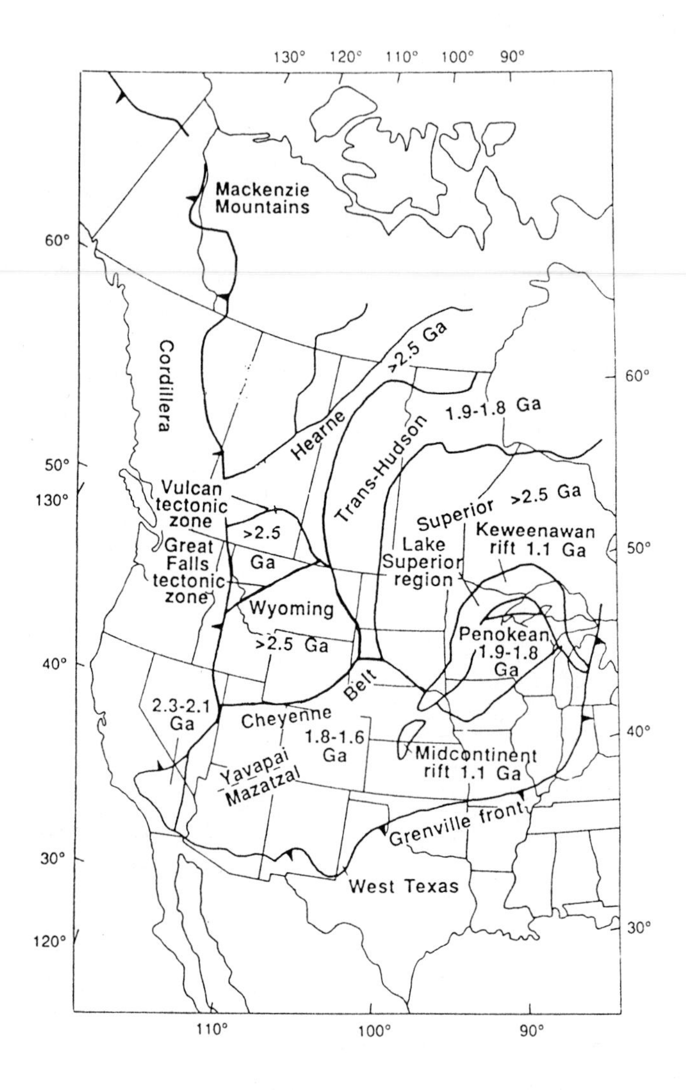

Figure 9. Precambrian tectonic zonation of the central and western part of the North American craton, according to Link et al. (1993, with modifications). An update of this scheme has been proposed by Lyatsky et al. (1999).

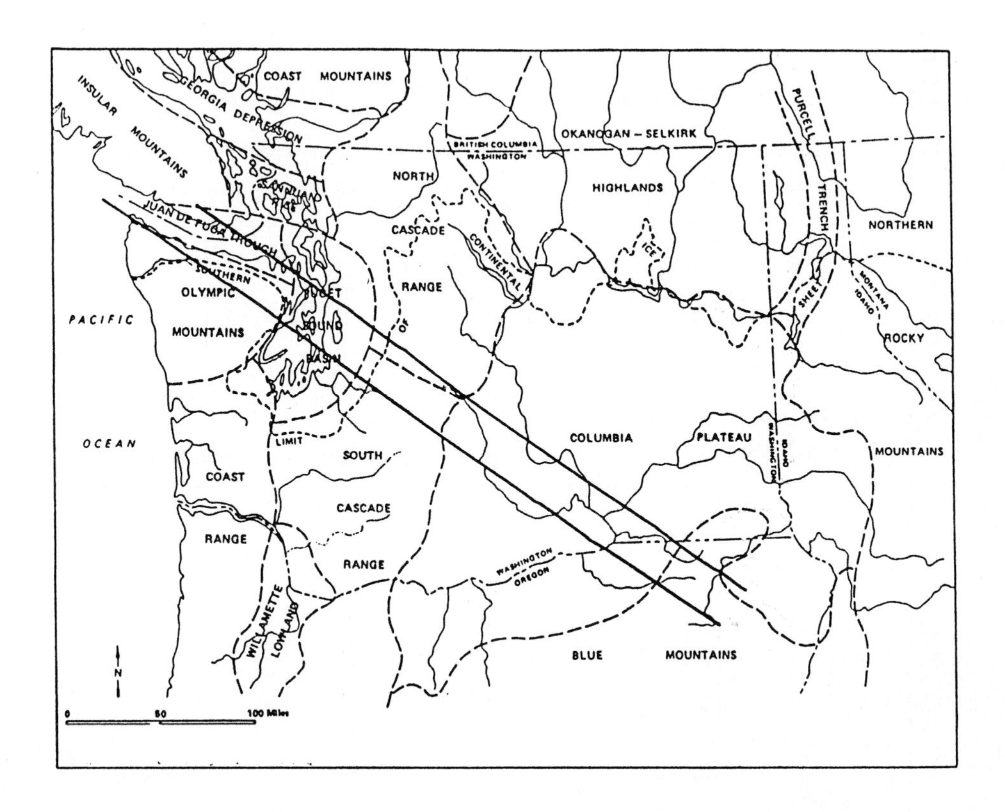

Figure 10. Geologic provinces in the northern U.S. and southernmost Canadian Cordillera (modified from Galster et al., 1989). The heavy double line marks the Olympic-Wallowa zone of crustal weakness (Lyatsky, 1996).

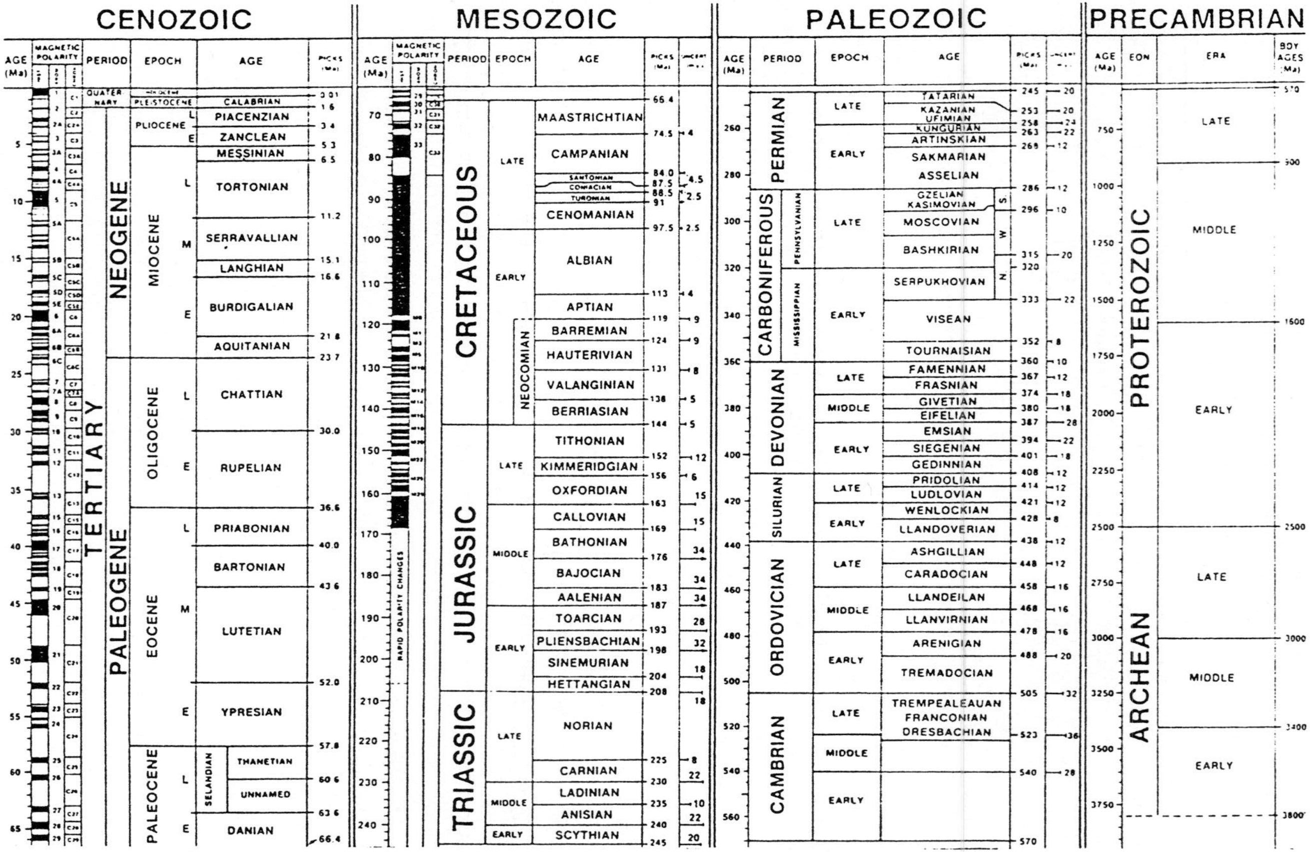

CENOZOIC

Period		Epoch		Age	Picks (Ma)
Quaternary		Holocene			0.01
		Pleistocene		Calabrian	1.6
Tertiary	Neogene	Pliocene	L	Piacenzian	3.4
			E	Zanclean	5.3
		Miocene	L	Messinian	6.5
				Tortonian	11.2
			M	Serravallian	15.1
				Langhian	16.6
			E	Burdigalian	21.8
				Aquitanian	23.7
	Paleogene	Oligocene	L	Chattian	30.0
			E	Rupelian	36.6
		Eocene	L	Priabonian	40.0
			M	Bartonian	43.6
				Lutetian	52.0
			E	Ypresian	57.8
		Paleocene	L	Thanetian (Selandian)	60.6
				Unnamed (Selandian)	63.6
			E	Danian	66.4

MESOZOIC

Period	Epoch	Age	Picks (Ma)	Uncert. (m.y.)
Cretaceous	Late	Maastrichtian	74.5	4
		Campanian	84.0	4.5
		Santonian	87.5	
		Coniacian	88.5	
		Turonian	91	2.5
		Cenomanian	97.5	2.5
	Early	Albian	113	4
		Aptian	119	9
	Neocomian	Barremian	124	9
		Hauterivian	131	8
		Valanginian	138	5
		Berriasian	144	5
Jurassic	Late	Tithonian	152	12
		Kimmeridgian	156	6
		Oxfordian	163	15
	Middle	Callovian	169	15
		Bathonian	176	34
		Bajocian	183	34
		Aalenian	187	34
	Early	Toarcian	193	28
		Pliensbachian	198	32
		Sinemurian	204	18
		Hettangian	208	18
Triassic	Late	Norian	225	8
		Carnian	230	22
	Middle	Ladinian	235	10
		Anisian	240	22
	Early	Scythian	245	20

Magnetic polarity: 70–170 Ma, anomalies 29–33, C29–C34; 170–210 Ma, rapid polarity changes.

PALEOZOIC

Period		Epoch	Age	Picks (Ma)	Uncert. (m.y.)
Permian		Late	Tatarian	245	20
			Kazanian	253	20
			Ufimian	258	24
			Kungurian	263	22
		Early	Artinskian	268	12
			Sakmarian		
			Asselian	286	12
Carboniferous	Pennsylvanian	Late	Gzelian (S)		
			Kasimovian (S)	296	10
			Moscovian (W)		
			Bashkirian (W)	315	20
	Mississippian	Early	Serpukhovian (N)	320	
			Visean	333	22
			Tournaisian	352	8
Devonian		Late	Famennian	360	10
			Frasnian	367	12
		Middle	Givetian	374	18
			Eifelian	380	18
		Early	Emsian	387	28
			Siegenian	394	22
			Gedinnian	401	18
Silurian		Late	Pridolian	408	12
			Ludlovian	414	12
		Early	Wenlockian	421	12
			Llandoverian	428	8
Ordovician		Late	Ashgillian	438	12
			Caradocian	448	12
		Middle	Llandeilan	458	16
			Llanvirnian	468	16
		Early	Arenigian	478	16
			Tremadocian	488	20
Cambrian		Late	Trempealeauan	505	32
			Franconian		
			Dresbachian		
		Middle		523	36
		Early		540	28
				570	

PRECAMBRIAN

Eon	Era	Bdy Ages (Ma)
		570
Proterozoic	Late	900
	Middle	1600
	Early	2500
Archean	Late	3000
	Middle	3400
	Early	3800

Table I. Geologic time scale, adopted by the Geological Society of America (1983) an used in this book: Precambrian and Phanerozoic.

Scope and main topics of the study

This book is about tectonic evolution of the eastern Cordillera of North America - i.e. about its evolution due to interacting processes in the crust and subcrustal lithosphere. Geology deals with all processes of formation and transformation of rocks (which are mineral aggregates) and rock-made bodies. These processes can be deep-rooted (*endogenic*) or surficial (*exogenic*). The main forces driving the tectonics of lithosphere are induced by gravitation, which pulls things down, or by heat, which causes things to expand and rise. In this book, whose focus is tectonics, exogenic development of the Cordillera is considered only to the extent that it illustrates endogenic forces.

In a continental tectonic region, gravity and thermal forces are indigenous, and they are usually the defining factors in the region's evolution. But, being part of larger systems, a region also feels the external influences from its neighbors, creating complexities that may be difficult to decipher. In a marginal continental region like the Cordillera, the most obvious source of external influences is plate motions and plate interactions at the continental margin. Subduction of oceanic lithosphere beneath the continental plates, where it occurs, can considerably distort the indigenous force pattern, although subduction at a continental margin does not eliminate the indigenous crust-mantle processes in a large, Cordillera-size region.

The Cordillera is an orogenic tectonic province, very non-uniform across and along its trend. The outside influences are the strongest in its western, peri-Pacific belts, and the weakest close to the North American craton. South of the big transcurrent Olympic-Wallowa zone of crustal weakness, close to the Canada-U.S. border, ongoing subduction of the oceanic Juan de Fuca plate is indicated by a variety of manifestations. North of it, no active subduction is occurring. As a result, north of the Olympic-Wallowa Zone, indigenous tectonic processes of the Cordillera can be studied without a modern external overprint of active subduction processes.

Orogeny and *orogenic* vs. *mountain-building* are notions from different fields of science, and the overlap between them is no more than partial. Etymologically and historically, the term *orogen* was initially invented to denote areas contrasting to the plains, where mountains were created and still stand. Elongated orogenic zones were supposed to be big furrows in continents, infilled with sediments and volcanic strata, whose axial belts were later magmatized, folded and inverted. These furrows were called *geosynclines* (i.e. big geo-features with inward-inclined flanks), and the geosyncline concept had been an important first approximation in the tectonic understanding of the genesis of continental mountain zones.

This 19th-century meaning of these terms has long been obsolete, and in the 20th century *geosyncline* and *orogen* have been modernized. Geosynclines are no longer presumed to be synformal, and orogens are known to be much more than just makers of mountains. Both these terms are applied to continental-scale tectonic *mobile megabelts* which rim the vast stable *cratons*. These megabelts are surface expressions of deep lithospheric zones of increased mobility and rock transformation. Called simplistically orogenic, these thermo-tectonic transformations affect the entire crust and upper mantle. Tectono-metamorphic and tectono-magmatic reworking accompanies the tectonic deformation of the crust, which is exhibited in folding of stratal rocks, faulting and warping. Large continental depressions and arches form commonly at the early stages of megabelt development. Deep downwarps and grabens make room in which large orogenic volcano-sedimentary basins form. Large-amplitude upwarps and horsts make landmasses, which are attacked by exogenic processes of erosion and denudation. Rivers and glaciers carve valleys, canyons and fiords, creating a rugged mountainous relief which marks mature orogenic zones following the upwarping and inversion of the mobile-megabelt crust.

Continental-scale mobile megabelts may be classified in a variety of ways: based on their age, intensity of tectonism, abundance of certain rocks types, and so on. A very important criterion is presence of a pre-existing basement. The crustal affinity of this basement may be oceanic (e.g., in westernmost California) or, more commonly, continental (e.g., Archean-Paleoproterozoic cratonic remnants in British Columbia). Tectonic reworking in orogenic zones may cause the old basement's deterioration or complete disappearance. Its semi-reworked preserved blocks in a megabelt are recognized in median massifs, around which the orogenic zones are forced to swing. Folds in orogenic zones commonly verge towards the median-massif buttresses.

In the Cordillera, such a buttress has been recognized from geologic mapping since the 1950s and 1960s to continue from Idaho and Washington, through British Columbia and Yukon into east-central Alaska. Like similar massifs in Europe, it was termed *Zwischengebirge* (German for "intermontane area"). The Intermontane physiographic belt is an expression of a series of slightly folded and moderately uplifted semi-reworked basement blocks. To the west lies the Coast Belt orogenic zone, stretching from the North Cascade Mountains in Washington through British Columbia and southeastern Alaska, which corresponds only roughly to the physiographic Coast Belt. East of the Intermontane median-massif belt, the Omineca orogenic zone runs from western Montana to Yukon, where its linearity is disrupted by the eastward-protruding Mackenzie Mountains orogen. Between the uppermost Missouri River in Montana and the Liard River in northeastern British Columbia, parallel to the Omineca orogen runs the Rocky Mountain Belt, whose appearance resulted not from deep lithospheric reworking but from eastward displacement of a surficial (~10 km thick) thrust stack. The tectonic nature, evolution and resulting modern characteristics of the Intermontane median-massif belt, Omineca orogen and Rocky Mountain fold-and-thrust belt are the main subject of this book.

Objectives and overall approach

The first comprehensive and fairly detailed geologic summary of this part of the Cordillera was included in a monumental Canada-wide synopsis compiled by the Geological Survey of Canada in the late 1960s (Douglas, ed., 1970), though a less-detailed summary had been produced a decade before (White, 1959). The entire Canadian Cordillera was denoted *Cordilleran Orogen* or *Cordilleran Geosyncline*, as at that time these two notions were not well differentiated. In the description of Douglas et al. (1970, p. 367), "the western part of the Cordilleran Orogen embodies the parts of the Cordilleran Geosyncline that received eugeosynclinal and epieugeosynclinal assemblages". The eugeosynclinal rock successions in question comprise huge volcanic and sedimentary assemblages and contain some ultramafic complexes. The younger epieugeosynclinal successions are dominated by marine and non-marine sedimentary rocks in syn-orogenic and successor basins, and their deformation is less intense than in their eugeosynclinal predecessors. On top of these rocks lie post-orogenic, continental clastic and volcanic rocks, whose deformation is only slight. The eastern part of the Cordilleran Orogen, by contrast, "...embraces the parts of Cordilleran Geosyncline that generally received miogeosynclinal and exogeosynclinal sediments and that underwent surficial or *décollement* deformation with little or no metamorphism, volcanism, or plutonism during the orogenic phases" (Douglas et al., 1970, p. 366-367).

Inconsistently, though, these authors (op. cit., p. 368) added that "the eastern or Omineca Crystalline Belt includes extensive areas of metamorphic rocks that form the core zone of an alpine-type orogen, the marginal zone of which is the Eastern Cordilleran Orogen". To the west, the Zwischengebirge is dominated by disconnected horsts and by two transverse arches running across the main NNW-SSE orogenic grain of the Canadian Cordillera.

This first overview of the entire Cordilleran geology, though imperfect, had an advantage of not being influenced by the rising ideas about the defining role of subduction and other plate-margin processes in continental orogenic evolution (those were in their infancy: e.g., Dewey and Bird, 1970). But the revision came soon, as a new generation of geologists advanced plate-tectonics-oriented hypotheses of Cordilleran evolution (Monger et al., 1972). In the new vision, as outlined by Wheeler and Gabrielse (1972), tectonic environments were distinguished strongly: miogeosynclinal (continental-margin), collisional (between plates of continental and oceanic crust) and island-arc (oceanic, or Indonesian-type, or Andean-type). The Cordilleran evolution and structure were subdivided into two fundamentally different parts: in the west, the subduction-induced Pacific Orogen, including the Intermontane median massif(s); in the east, the Columbian Orogen, embracing the assumed Paleozoic continental shelf-slope-rise system or continental-terrace wedge. This revision was nothing short of revolutionary. It included the newly fashionable ideas about free drift of continents, recycling of lithospheric slabs into the mantle by subduction, and of plate-margin interactions and oceanic-plate restructuring as the main driving influences on continental tectonism (Atwater, 1970; Coney, 1970; Dietz and Sproll, 1970). It was the age when the U.S. National Academy of Sciences (1980, p. 6; italics in the original) recommended firmly that "*the revolution in the earth sciences that resulted from the study of the ocean crust should spread to the continents*".

For all the benefits of expanding the geologic scope, this approach distracted the geologists from their traditional studies by direct observation of rocks through mapping supplemented with geophysical and geochemical techniques. The aim of the traditional approach is to picture a geologic area as a three-dimensional continuum made of rocks, before its history and evolution can be understood. Contrary to this practice, many new workers came to believe that a basic tectonic framework for on-land geologic studies should be provided by the set of ideas called plate-tectonic theory, derived chiefly from geophysical studies in remote oceanic regions. The general concepts in this set of ideas are as follows: (a) the Earth's outer shell (lithosphere) is rigid to depths ranging from 5 km to 100 km under oceans and from 20 km to 250 km under continents; (b) this outer shell consists of a small number of large rigid plates which move relative to one another and interact at their boundaries; (c) "the lithospheric plates ride as passengers on the convecting [and ductile] asthenosphere" (National Academy of Sciences, 1980, p. 5). From this standpoint, it was recommended that "the study of continents should proceed from an understanding of modern plate-boundary processes to efforts to apply this understanding to the geological record of continents" (op. cit., p. 6).

After two decades of trying to put these recommendation into practice, it is becoming clear that the key to understanding the tectonics of the lithosphere still lies on continents. Facts from oceanic geology are of great importance, but they are supplementary. The general principles of rock-body behavior, evolution and structure are common for all rock bodies, wherever they occur, and rocks on land are much more accessible for observation than at the ocean bottom. As well, unlike the oceanic crust, the continents are known to have a huge variety of rock compositions and structures. This variety and accessibility make the studies on continents a sound foundation for realistic and practical tectonic theories. Only abundant and verifiable *facts*, obtained by *direct observation* of rocks and rock bodies, of their internal structure and mutual interrelationships, can verify the hypotheses and models. Not model-driven but fact-derived conclusions have a chance to stand the test of time.

In this book, we assemble and analyze facts related to the Canadian Cordillera, while paying less attention to many existing speculations (unless they are too influential to be ignored: e.g., concepts of terrane tectonics, externally-induced tectonism, whole-crust thrusting and orogen-scale detachments, etc.). The main objective is to return the Cordilleran tectonic studies to the reliable and productive track where tectonic characteristics are inferred from a critical analysis of the descriptions of rocks and rock bodies. These descriptions, obtained from direct observations, provide the ultimate constraints on tectonic speculations and interpretations of indirect geophysical data.

In the last few decades, with the benefit of extensive mapping and new methods of laboratory analysis of rocks, a huge amount of geological information has become available in many parts of the Cordillera in Canada and the U.S. Good-quality geophysical gravity and magnetic coverage is available in most regions. The Cordillera and the submerged continental margin are imaged in many deep seismic profiles: Lithoprobe in Canada, COCORP in the U.S., as well as released petroleum-industry data. Vast useful factual information is contained in the publications of the Geological Survey of Canada, U.S. Geological Survey, provincial and state agencies, and academic institutions.

All this creates a sound foundation for further consideration of the Cordilleran evolution, without prejudices or biases. Our concepts about different parts of the Cordillera, presented in this book, have been tested and confirmed by our extensive practical experience. Most of our work was done to make predictions about natural resources and environmental hazards. Comparisons of the Cordillera with other geologic regions in North America and Eurasia, also largely based on personal experience, strengthens our conclusions.

2 - GOVERNING PRINCIPLES IN REGIONAL TECTONIC EVOLUTION

Basic issues in determining the tectonic evolution of orogenic regions

Four aspects of rock-based tectonic analysis

Geology, as a practice and a field of knowledge, operates with rocks and rock-made bodies. Rocks are natural aggregates of minerals, which normally take the form of crystals and clasts. They make up geologic bodies that can be as large as the lithosphere or as small as a lamina or a vein. Each body has its own mineral composition, which distinguishes it from other bodies. In monomineralic and polymineralic rocks, crystals, clasts and lower-order rock bodies are in some arrangement. Their organization defines the body's internal structure, which distinguishes the body even if the adjoining bodies have the same mineral composition. In a hierarchy of rock bodies of different orders (from lamina or vein to lithosphere), compositional and structural characteristics together define the bodies' properties, which change at the bodies' boundaries.

Tectonics considers all geologic (rock-made) bodies whose genesis and transformation responded more or less directly to deep-seated processes which occur continually in the lithosphere. Rocks and rock bodies which were formed near the ground surface due to exogenic processes are beyond the scope of tectonics, unless the surficial conditions were created by endogenic forces of gravitational or thermal origin. *Tectono-sedimentary* analysis aims to reconstruct the history of tectonic surface depressions, arches, sediment-provenance areas, etc.

Of the most interest in tectonic analysis are those aspects of rock genesis and transformation which reflect the deep-seated tectonic processes. Almost never can these processes be observed in action or replicated in the laboratory, but observable fossil products of these processes are preserved in abundance in the rock record. Metamorphic and magmatic rocks are the most obvious of these products. The depth at which rocks were produced or altered can be estimated fairly reliably from specific observable rock properties, and metamorphic and igneous rocks provide the most reliable information about the crustal and subcrustal conditions at the time of their formation. Secondary transformations of rocks, including metamorphic overprints, give indications of subsequent tectonic processes in the region - in particular, about the history of rocks' burial, heating, elevation and exposure.

The *tectono-metamorphic* and *tectono-magmatic* aspects of regional geology are thus fundamental components of tectonic analysis. In orogenic regions especially, tectonic hypotheses and geophysical models must be verified by the analysis of these rocks. Without such confirmation, speculative hypotheses and models have little value for geologic (including tectonic) studies, and may be misleading.

A special place belongs to the analysis of *tectono-deformational* styles of rocks. From the time of their genesis, geologic bodies are in some arrangement with respect to one another. Sequentially deposited stratal sedimentary and volcanic rocks, for example, obey the principles of Superposition and (unless the bedrock surface is inclined) Original Horizontality. Intrusive igneous bodies, in contrast, are usually discordant with their country rocks. Secondary deformation can be produced by or accompany the tectono-metamorphic, tectono-magmatic and tectono-sedimentological processes. Non-tectonic deformation is also possible: collapse structures above zones of karsting and salt dissolution, downslope sliding of unconsolidated sediments, deformation of rocks by glaciers, and so on. Igneous plutons and salt and mud diapirs commonly cause very complex country-rock deformation near their flanks and tops.

Metamorphic processes add to the complexities induced by deformation. Typically, their combined effects are seen in Precambrian terrains that during their history passed through very deep crustal levels. In ancient and young orogenic belts affected by large vertical displacements, originally supracrustal volcanic and sedimentary rocks sometimes have metamorphic grades acquired in the conditions of the middle or lower crust. Subsequent uplift and erosional unroofing caused these rocks to be exposed at the surface.

In continental-scale mobile megabelts, orogenic zones of deep rock reworking are squeezed between less-reworked blocks (such as median massifs). Orogenic zones have experienced a complex, multi-phase history of deformation, metamorphism and magmatism, whose cumulative and overlapping effects are hard to decipher. Axes of inversion are commonly flanked by outward-vergent folds and faults. At the boundaries with more-stable blocks, the vergence is away from the orogenic zones and towards the blocks' interiors, and recumbent folds and low-angle thrust sheets are common. Fold-and-thrust belts propagate over undisturbed crystalline basement of rigid blocks, and are thus broader than the whole-crust orogenic zones which induced their appearance. In marginal parts of mobile megabelts facing the very stable cratonic masses, fold-and-thrust sheets between subhorizontal thrusts may be far removed from their orogenic hinterlands. Being orogen-sourced, they are exotic to the areas they overlap. They lack underlying orogenic roots, and may be pushed over areas whose deep crust is not orogenic at all.

Hierarchy of continental lateral tectonic units

As lateral tectonic units of continents, *mobile megabelts* stand as opposites of stable *cratons*. Like cratons, they are units of whole-lithosphere extent (e.g., see in Lyatsky et al., 1999). Within mobile megabelts, whole-lithosphere (or at least whole-crust) lateral tectonic units of a second rank are *median massifs* and *orogenic zones*. Boundaries of these crustal or lithospheric features may not coincide with the edges of mountainous zones. Fold-and-thrust belts, for example, grow outward, away from the deeply orogenized and tectonically reworked zones, and overlap marginal parts of cratons and median massifs. Tectonically, mountain ranges may be produced in a wide variety of ways: by rootless thrusting (Rocky Mountains), by rise of cratonic blocks due to forces induced from mobile-megabelt hinterland (Colorado Plateau), by inherently cratonic upward movements (Boothia Arch in the Canadian Arctic), or as direct result of orogenic processes in mobile megabelts and their subsequent inversion (Omineca Belt in the Paleozoic, Coast Belt in the Tertiary). Presence of mountains is thus not a reliable indicator of tectonic setting.

To interpret tectonic settings correctly, the tectono-deformational aspect of tectonic analysis of a region must always be done cautiously and in conjunction with the tectono-sedimentary, tectono-metamorphic and tectono-magmatic considerations.

Interaction of internal and external forces in the evolution of mobile megabelts and their constituent tectonic zones

Mobile megabelts, as distinct constituent parts of continents, probably first appeared in the Late Archean. By then, proto-continents had already been differentiated in the Earth's perisphere from the primordial material. In that ancient continental crust, maximum tectonic activity was concentrated in broad zones, whereas proto-cratonic parts of the lithosphere remained comparatively more stable and resistant to intense reworking. Preserved fragments of ancient cratons are found on all continents, sometimes incorporated as median massifs into Early and Middle Proterozoic mobile megabelts. Much of the globe's continental crust was already in place by the Proterozoic (e.g., Armstrong, 1991; Rudnick, 1995). As the continental crust evolved and grew stronger, Early and Middle Proterozoic megabelts were differentiated from their coeval cratons more sharply than their Archean predecessors. Deep-seated reworking nonetheless mobilized the previously cratonized crust and lithosphere, and some zones of reworking were expressed at the surface as elongated rifts. As reworking in some of these zones progressed further, these incipient mobile megabelts expanded and became internally differentiated.

Some of the Proterozoic continental rifts died before they had a chance to become full-fledged megabelts. One example is the Midcontinent Rift which cuts across North America from the Great Lakes to the southern U.S. Far less impressive is the recently delineated Kimiwan zone of recrystallization and restructuring recognized from drillhole samples of the crystalline basement in west-central Alberta (Fig. 11; Burwash et al., 1995; Chacko et al., 1995). This zone, dated at ~1,760 Ma, has a NNW-SSE orientation parallel to the much younger Cordilleran mobile megabelt to the west. This orientation is atypical for the Proterozoic Laurentian craton. The Kimiwan anomaly is a superimposed on the pre-Cordilleran crustal fabric, and it seems to be the first known precursor of the Cordilleran megabelt.

A very broad Belt-Purcell volcano-sedimentary basin, now heavily deformed and displaced by the Laramian thrusting that created the Rocky Mountain fold-and-thrust belt, lies on the site of toady's eastern Cordillera straddling the Canada-U.S. border (Figs. 12-13). To the north lies a similar but less deformed Mackenzie Mountains Basin (Aitken, 1993a-c). Both these basins, formed around 1,600-1,500 Ma, still have essentially cratonic geologic characteristics. If Cordilleran mobile-megabelt development was occurring at that time, it must have been happening farther west or between these basins, having shifted away from its original Kimiwan locality. A Middle Neoproterozoic rift-like tectonic feature along the present-day eastern Cordillera between these two basins is well mapped in the western Rocky Mountain and Omineca belts, from latitude ~52°N northward. Known as Windermere Rift (basin), it was initiated at ~780 Ma.

Whenever the Cordilleran mobile megabelt began, its development continues to this day. During this long time, crust in the megabelt has been reworked, metamorphosed, plutonized and deformed many times over in the Late Proterozoic, Paleozoic, Mesozoic and Tertiary (Gabrielse and Yorath, eds., 1991). In such episodes, conventionally called "orogenies", the reworking began at deep crustal and subcrustal levels due to new pulses of thermo-dynamic activity. Each such pulse in the megabelt was concentrated in one or two orogenic zones, while the mobile megabelt's other parts remained largely quiescent except for teleorogenic influences manifested in superimposed faulting, magmatism, etc.

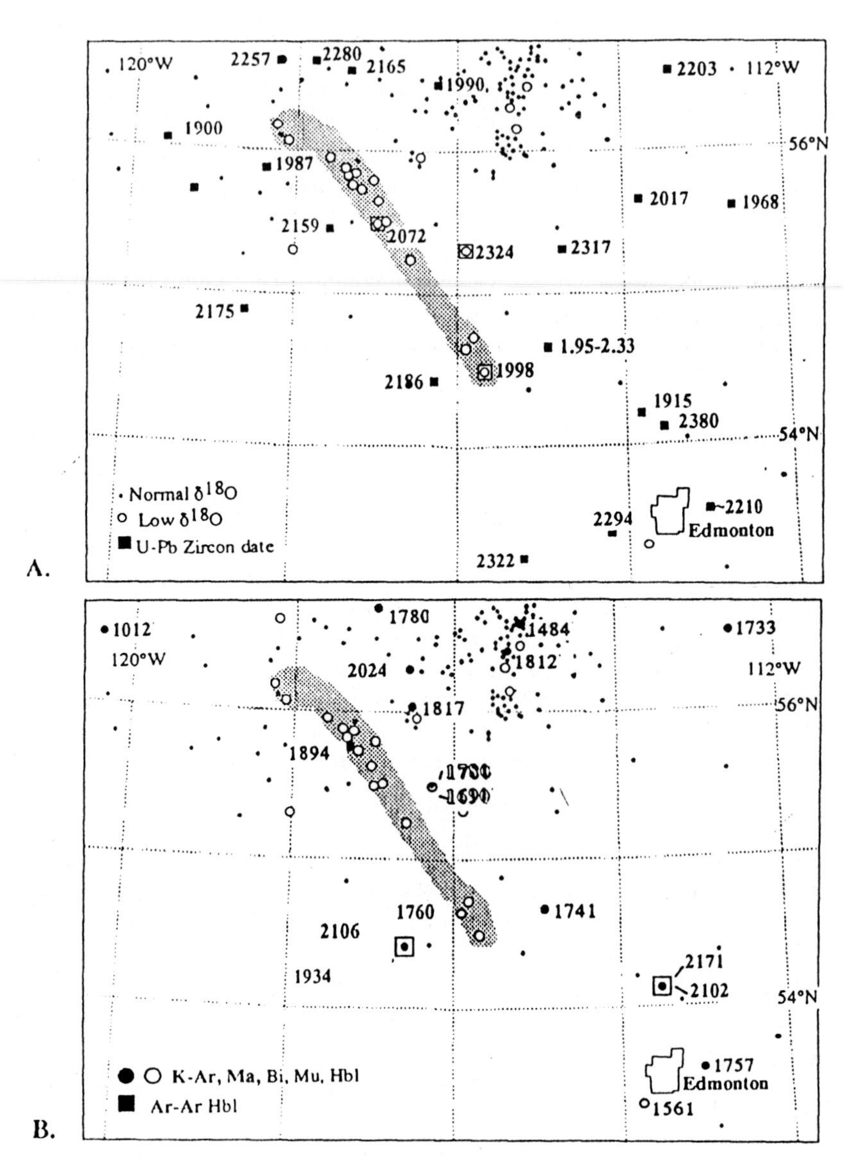

Figure 11. Position of the Kimiwan isotopic anomaly, age ~1,760 Ma, in the cratonic crystalline basement of the Alberta Platform (modified from Chacko et al., 1995). This anomaly seems to have formed by reheating of this zone in the cratonic crust. The NNW-SSE elongation of this zone of crustal reworking makes it the earliest known probable tectonic precursor of the Cordilleran mobile megabelt in western Canada.

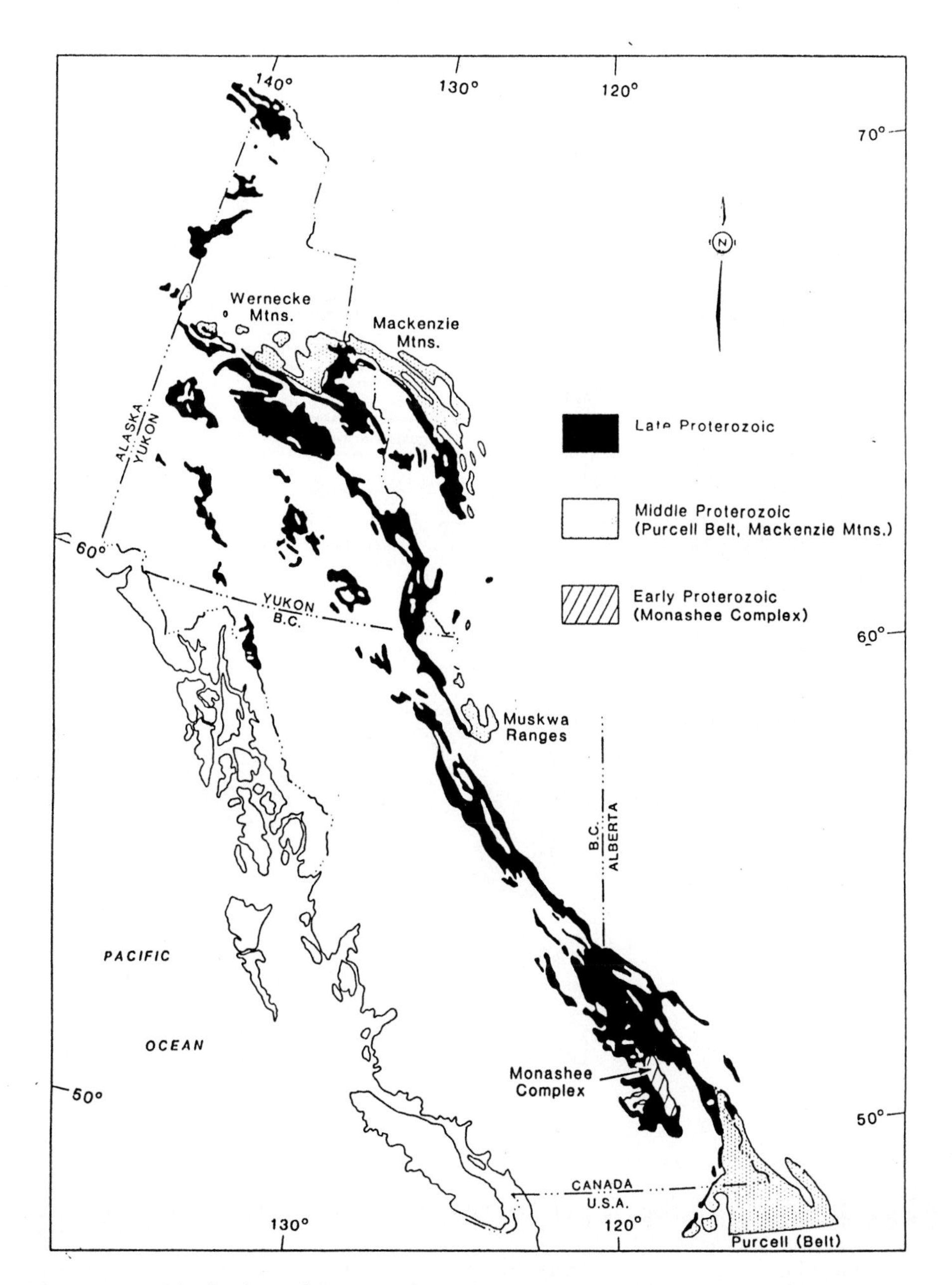

Figure 12. Distribution of Proterozoic rock assemblages (Belt-Purcell, Windermere, their equivalents, and others) in the Canadian Cordillera (modified from Monger, 1989). Distribution of Late Proterozoic Windermere Supergroup rocks is shown in black.

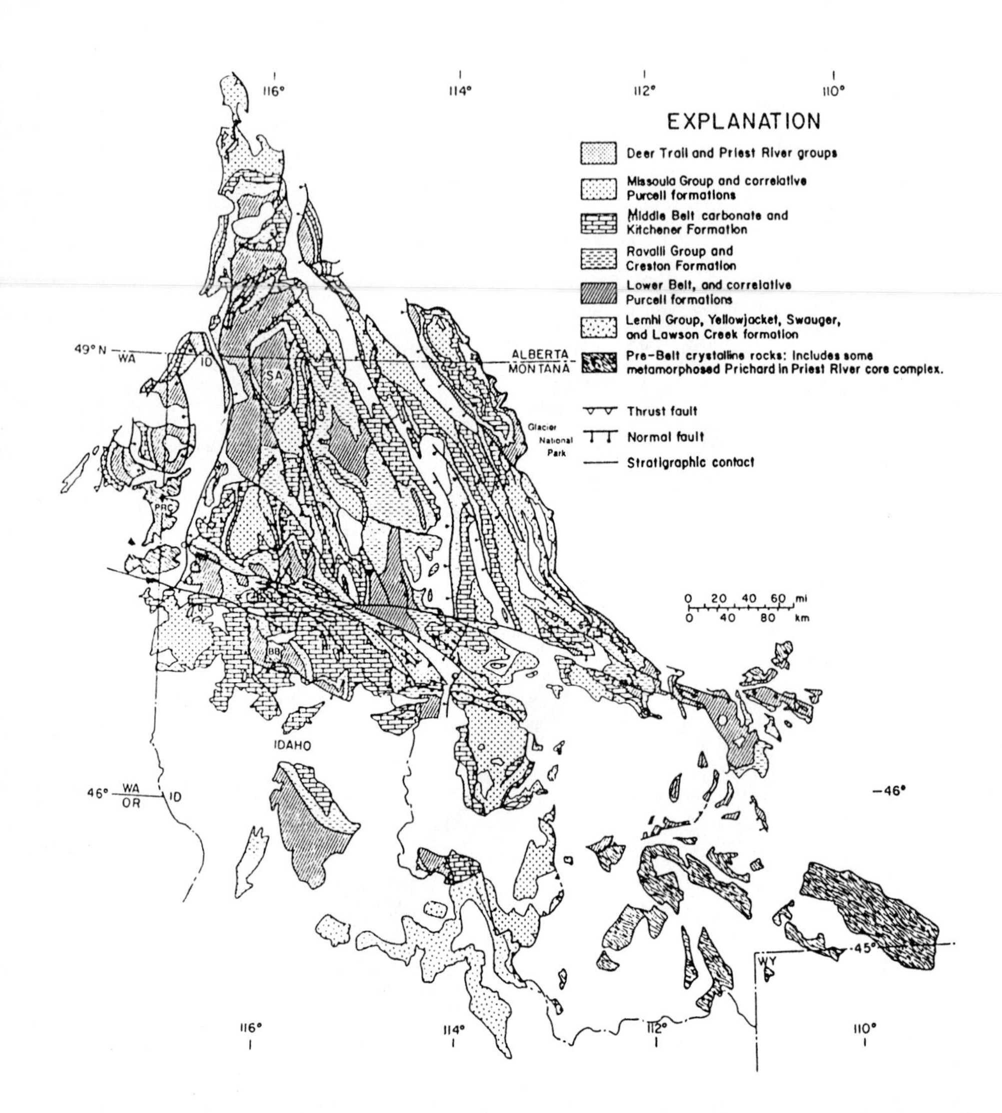

Figure 13. Generalized geologic map of the Middle Proterozoic Belt-Purcell Basin, using the stratigraphic nomenclature accepted in the U.S. where most of this basin lies (modified from Link et al., 1993). BB - Boehls Butte; PRC - Priest River Complex; SA - Sylvanite Anticline.

The main causes of these thermo-dynamic events lay in the lithosphere. Continental lithosphere has radioactive elements and thus its own sources of heat. Besides, additional heat is supplied from the sublithospheric mantle. A sudden increase in heating, from whatever sources, intensifies the reworking of rock bodies in orogenic zones. In such zones, pre-megabelt rocks and bodies may be reworked completely. Elsewhere in the megabelt, the degree of reworking (metamorphism, palingenic and other magmatism, deformation, restructuring) may be lower, depending on the distance from the strongly mobilized orogenic zones. Inter-orogenic median massifs are stronger and less reworked, and they retain older cratonic crustal fragments. As lateral tectonic units of the lithosphere, median massifs bear a mixed heritage of internal, indigenous old features and newer ones imposed in the megabelt.

An important factor in a mobile megabelt's evolution is its position on the continent. Intracontinental megabelts, such as Eurasian, differ in some ways from those which, like the Cordilleran megabelt, lie in marginal continental regions. Both these megabelt types lie on continental crust. Yet, since the appearance of the modern terminological system for description of lithospheric plates, crucial for the understanding of marginal-continental mobile megabelts has been the notion of subduction. In the simple geophysical model developed in the 1960s, subduction occurs in zones where plates meet, and their motion towards each other causes one of them to be overridden by the other. Where continental and oceanic plates converge, the overriding plate is typically continental (overthrusting by the oceanic plate, or obduction, occurs rather seldom). In the 1970s and 1980s, plate convergence and subduction came to be viewed as the most powerful or even only force controlling orogenesis and mountain-building in the continental lithosphere, even in cratonic regions.

These ideas seemed plausible as long as plates were regarded as rigid, stiff lithospheric slabs floating passively as "passengers on the convective asthenosphere" (National Academy of Sciences, 1980, p. 5). Forced towards each other by mantle convection or other causes of plate motion, their edges in the zone of convergence are deformed. Since the plates were assumed to be inert, intraplate deformation also had to be explained by external plate-margin processes, however far the area lay from a plate edge.

At present, many contradictory facts suggest this original model was too simplistic. Commonality of continental and oceanic intraplate deformation indicates that plate rigidity was overestimated. Considerable plicative (about a vertical axis) and disjunctive tectonic deformation has been noted on the southern and northern ends of the Juan de Fuca oceanic plate (Lyatsky, 1996). In the Pacific plate and elsewhere, abundant seismic reflectors in the oceanic lower crust (Eittreim et al., 1994) may be signs of shearing and delamination, and internal thrusts have been described the most convincingly in the Indian Ocean (Chamot-Rooke et al., 1993).

The original model of plate motions has undergone some modifications. In the current literature, other plate-driving mechanisms besides asthenospheric drag are considered as well: trench pull, slab push (e.g., Wilson, 1993). But in reality, great disruptions, folds and warps that could be attributable to such forces not are observed in plates. Dragging due to trench pull, in particular, would induce widespread intraplate extension, which is not observed in sea-floor bathymetric forms or in seismic images. Thus, the question of what causes plate movements remains unresolved.

No clearer is how exactly plates are formed at mid-ocean ridges. In the original rigid-plate model, oceanic lithosphere is created along spreading centers in a systematically, orderly fashion, in parallel belts of successive ages. In a reversing geomagnetic field, formation and spreading of the ocean floor is recorded in magnetic-anomaly bands tens of kilometers wide and thousands of kilometers long. These anomalies reflect remanent magnetization of mafic igneous rocks solidified and cooled in an alternating normal or reverse ambient geomagnetic field (Vine and Matthews, 1963). Based on the unproved assumption about constancy of spreading rates, this succession of anomalies was converted into a time scale (Heirtzler et al., 1968) which has come to be treated as a global reference frame for the timing of plate formation and the paths of their motions (e.g., Kearey and Vine, 1990).

Though this reference frame is still in wide use, many relevant uncertainties are mentioned often as well. Igneous rocks in the oceanic crust are known to be disrupted by many faults and block tilts, and these faults could readily serve as conduits for juvenile magmas whose presence would cause distortions of the original magnetic anomalies. Where secondary faults or dikes run parallel to the original oceanic-crustal fabric, their presence may be hard to recognize. Besides, the basaltic layer in the oceanic upper crust is no longer thought to be sufficient as a source of magnetic anomalies, and other, unknown sources lie in the ill-explored deeper crust (e.g., Ballu et al., 1998; Lawrence et al., 1998; Tivey and Tucholke, 1998). In the absence of abundant data about the structure and exact composition of the mostly-hidden oceanic crust, these pitfalls risk contaminating the age correlation of oceanic magnetic anomalies. By contrast, continental regions are well studied by direct observation. To extrapolate tectonic conclusions obtained in poorly studied oceanic regions to much better studied continents is an inverse of proper logic.

Tectonic influences of plate interactions must, of course, be taken into account, particularly in regions near plate boundaries. The Cordilleran megabelt lies in such a region. However, regional tectonic reconstructions on continents must be derived first and foremost from multi-aspect analysis of observed geologic facts within the region of study.

Some basic tectonic definitions

Tectonics is one of the most widely used words in geology, but it is far from clearly formalized. With the discovery of lithospheric plates (which are themselves still imperfectly defined), the notion of tectonics was greatly transformed. Tectonics, which deals with deep-seated endogenic geologic processes and their results, is an intrinsic part of geology. As a part of geology, tectonics, or geotectonics as it is sometimes called, deals with rocks and rock-made bodies.

Price (1972, p. ix) discussed what he called the "principal structural *or* tectonic provinces" (italics ours). Each tectonic province is "...a discrete area set apart from the others by the relative age and other salient characteristics of orogenic deformation (tectogenesis)... The "tectonic style" of each province is the aggregate expression of a whole array of distinctive but interrelated attributes, each of which provides an indication of some aspects of its tectonic evolution". The listed "tectonic attributes" seemed rather arbitrarily selected, but these authors addressed the most important characteristics of a province: its dimensions and shape; crustal structure; sedimentary, volcanic, intrusive and metamorphic record; peculiarities of orogenic deformation (structural styles) and vertical crustal displacements; patterns of erosion (unconformities) and geomorphologic development; relationship to adjacent provinces.

In a still-practical view of the Canadian Cordilleran province, Wheeler (1970, 1972) distinguished in it the Western and Eastern foldbelts separated by the Intermontane Zone (median massif). "Each orogen or fold belt is composed of a mobile uplifted core zone of granitic and medium- to high-grade metamorphic rocks, flanked by belts in which the tectonic transport has been directed principally away from the core zone" (Wheeler, 1972, p. 2).

Departure from the practicality normal for field geologists became apparent in North America in the 1980s. National Academy of Sciences (1980, p. 3-4) defined tectonics as "...a branch of the earth sciences dealing with the origin, evolution, structure, and internal relations of regional features of the earth's crust. Tectonics is closely related to geodynamics, which is that branch of the earth sciences that deals with the forces and processes of the earth's interior." Though influential, this definition of tectonics is flawed. To list both origin and evolution is tautological, because evolution includes origin (as well as subsequent development). To separate "structure" and "internal relations" is incorrect, because they mean the same thing. What is meant by "regional features of the earth's crust" is not explained, and to restrict the features of tectonic interest to the crust rather than the whole lithosphere is artificial and incomplete. To separate tectonics from lithosphere geodynamics is nonsense, but geodynamics of sub-lithospheric levels is related to tectonics only partly. Tectonics remains an intrinsic part of geology, limited in its scope to the Earth's rock-made shell - lithosphere.

Misconceptions about the definition of tectonics remain common, and have even made their way in the standard *Glossary of Geology* published by the American Geological Institute (Gary et al., eds., 1972; Bates and Jackson, eds., 1980, 1987; Jackson, ed., 1997). The *Glossary* rightly considers tectonics as a branch of geology, but incorrectly reduces tectonics to just its structural aspect: it is argued that tectonics deals with "structural *or deformational* features", and that it is "closely related" to structural geology . "...The distinctions [from structural geology] are blurred, but tectonics generally deals with lager features" of unspecified type (Bates and Jackson, eds., 1987, p. 675; Jackson, ed., 1997, p. 653, italics ours). This reduction to just large-scale structural geology robs tectonics of its rich geologic substance and predictive power.

Like all of geology, tectonics deals with the entire rock-made perisphere of the Earth. As a geological discipline, tectonics has as its object all rock-made bodies created and transformed by endogenic tectonic processes. Each body is studied as a compositional-structural entity, regardless of its position in the hierarchy of endogenically formed and transformed bodies from the smallest tectonites to the entire lithosphere.

The distinction between tectonics and structural geology is very clear. Structural geology is a specialized discipline dealing with deformational reworking of rocks and rock-made bodies. It focuses on phenomena like fissures, faults and folds; huge features such as lithospheric and crustal blocks are beyond its scope. In contrast to this specialized discipline, tectonics is broader, as it studies rocks and rock bodies in four aspects: tectono-sedimentary, tectono-metamorphic, tectono-magmatic and tectono-deformational. The last category includes deformation of all sorts and scales: cracking, fissuring, faulting, folding, warping, etc. Structural geological, sedimentological, petrological and other geological analysis is incorporated into the much larger tectonic analysis, but the latter is not reducible to any of the former.

The *Glossary of Geology* (Bates and Jackson, eds., 1987, p. 675; Jackson, ed., 1997, p. 653) states that tectonics deals with "the broad architecture of the outer part of the earth". This is not a definition at all, because, except for *earth*, none of these words is formalized: the meaning of *broad, architecture, outer part* is unclear. A proper definition is possible only with well-defined terms and words. For example, the *outer part* has no clearly defined boundaries and is not a geological term, whereas *lithosphere* or *crust* do and are.

Another quasi-term in the *Glossary* is *lithotectonics*. It is tautological, because tectonics, being a geological science, deals with rocks (*lithos*) by definition. Tectonics is studied from its manifestation in rocks. Other terms are also formalized incorrectly: *litho-tectonic unit* is a fundamental element of regional tectonic analysis, but it is confusingly defined as "a mixture[?] of lithostratigraphic units[?] resulting from tectonic deformations[?]" (Bates and Jackson, eds., 1987, p. 675).

Geologic units are conceptual, but derived from the observation of the real geologic world. They are used to divide the rock-mass continuum into discrete elements, to be used in a mental system representing the observed geologic reality. In geology and tectonics, it is often convenient to distinguish *lateral rock-made units* such as plates, continental-crust and oceanic-crust masses, continental cratons and mobile megabelts, median massifs and orogenic zones. *Vertical rock-made units* are distinguished as well: lithosphere, crust, mantle, ductile and brittle crustal layers, structural-formational étages, formations, beds and so on. Compositional and structural definition of rock bodies is based on any set of characteristics useful for a particular practical purpose - as, for example, metamorphic zonation is defined from certain mineral assemblages.

In the successions of sedimentary and volcanic stratal rocks, units have traditionally been defined based on their lithologic continuity, fossil assemblages, and so on (Laffitte et al., 1972). Today, other useful criteria are available as well, and units are defined as hydrogeological, geochemical, even paleoclimatic (e.g., Harland, 1992). In regional tectonic analysis, vertical tectono-stratigraphic units are distinguished from the lithostratigraphic and chronostratigraphic successions. A simple way to define them is to look for obvious discontinuities between stratigraphic rocks successions (or sequences), whereas a more sophisticated approach is to look for breaks in tectonically induced characteristics. The sequence-based approach, pursued in North America by Sloss et al. (1949) and Sloss (1963, 1988), is weakened by the difficulty in determining whether any particular unconformity was induced tectonically or non-tectonically (e.g., eustatically or epeirogenically). The other approach, based on looking for tectonic breaks in a rock record, has been less popular because it is cumbersome, involving joint multi-aspect analysis of diverse mineralogical, petrological, structural and other regional information (e.g., Pettijohn, 1960). A prominent North American user of this approach was Weller (1958, 1960).

Tectono-stratigraphic units are not "mixtures" but distinct entities with designated common characteristics. They are not simply assemblages of "lithostratigraphic units", although often they have their typical predominant rock types. Lithostratigraphic and tectono-stratigraphic units are actually parts of different conceptual systems, and should not simply be lumped together. As well, tectono-stratigraphic units are not simply results of "tectonic deformations" but multi-aspect manifestations of regional tectonic reworking of the lithosphere (Lyatsky et al., 1999).

3 - ROCK UNITS IN REGIONAL TECTONIC ANALYSIS

Lateral and vertical crustal units: general classification

Geological analysis requires strong definitions of basic notions. For the purposes of this book, mobile megabelts must be clearly distinguished from cratons, and major constituent lateral tectonic units need to be distinguished within megabelts themselves.

The modern cratons are just the latest in a long line spanning several generations of cratons and continent-wide tectonic restructuring. Cratons are relatively immobile parts of continents, in contrast to mobile megabelts (cp. Kober, 1923). Cratons are defined in relation to megabelts, and each appearance of new megabelts and cratonization of older ones results in a craton of a new generation. New mobile megabelts are formed largely by reworking and mobilization of previously cratonic crust and lithosphere.

A cycle of reworking begins at deep levels in the lithosphere, usually during periodic intensification of continental geothermal dynamics (cp. Lyatsky, 1965; Cooper, 1990). Reworking in a mobilized continental zone propagates upward, and its manifestations in the crust are identified by geologic mapping. The initial near-surface manifestation is commonly rifting. Some rifts die before they have a chance to develop into full-fledged mobile megabelts, but others continue to evolve as reworking spreads laterally and vertically. Shallow expressions of the reworking are elongated volcano-sedimentary basins, chains of volcanoes, faulting, uplift and subsidence of crustal blocks, and so on. The causative processes remain unseen, and can only be inferred from these visible manifestations in exposed rocks. But not just shallow-formed rocks are exposed at the surface: elevated crustal zones and blocks contain unroofed rocks that were formed or modified at mid-crustal and even lower-crustal depths. Within uplifted lateral tectonic units - zones and blocks - vertical tectonic units may be discriminated based on metamorphic zonation due to burial-related regional metamorphism, impregnation with intrusive rocks of different ages, changes of deformation style from one structural-formational level to another, and variations in the geometry and distribution of sedimentary basins in tectonic depressions.

Depending on how vertical tectonic units are defined, their boundaries may be breaks in rock composition and structure, metamorphic fronts, brittle-ductile transitions, structural unconformities, thrusts, etc. Within a lateral tectonic unit, boundaries of vertical tectonic units may represent episodes of restructuring - sharp changes in the distribution of uplifted and subsided areas, block tilts and so on - which are expressed in changes of depositional systems. The stratigraphic aspect of tectonic analysis then becomes important. Stratigraphy deals with stratal bodies which were formed obeying the Principle of Superposition of younger bodies on older ones. Tectonic (structural-formational) units in a regional lithostratigraphic record bear marks of their depositional origin in a normal succession from older units below to younger ones above.

Tectonic stratigraphic boundaries may be sharp or subtle, depending on the nature of the reworking episodes and their specific manifestations in a given area. Megabelts are, compared to cratons, tectonically unstable, and stratigraphic breaks in them have a better chance to represent tectonic events in megabelt evolution. Structural unconformities typically mark such events, and their variations from region to region reflect lateral variations in the tectonic history of a megabelt. Vertical rock units bounded by such tectonic breaks reflecting pulses of restructuring are called *structural-formational (i.e. compositional-structural) étages.*

Depositionally sequential étages, whose collection varies from one lateral crustal unit to another, are best recognized where rock successions are least reworked. An étage contains a set of lithostratigraphic formations, a *formation* being a rock assemblage of distinct lithologies and age span which can be mapped reliably in a particular region. Formations are basic units of stratigraphic grouping of continuous paragenetic rock bodies, clearly separated from rock assemblages below and above by changes in lithology reflecting changes in paleoenvironments induced by causes of endogenic or exogenic nature. Internal lithic homogeneity and continuity are the main defining criteria, although lateral facies transitions may be contained within a formation. Whether or not it has dynamic tectonic causes, a lithostratigraphic formation is a static system of stratal rock bodies, usually subdivided into smaller members. Formations may be combined into bigger and more complex units, groups and supergroups. All these lithostratigraphic units, at different hierarchical levels, reflect certain overall depositional settings. Each such unit is a series of stratal rocks forming a coherent succession, distinct on some set of lithological parameters from the successions above and below. Rocks that were not formed by natural superposition of younger formations on older ones (e.g., most metamorphic complexes) are not subject to stratigraphic division, but they are still divisible into tectonic (structural-formational) étages.

In mobile megabelts, orogenic development and deep lithospheric reworking usually begin before any mountain-building. Deep restructuring of the lithosphere starts long before it acquires any expression at the ground surface, and the initial depressions and rises postdate the first phases of orogenic tectonism. But unlike the deep-seated orogenic precursors, shallower manifestations of tectonism are preserved in the mappable rock record and mark the visible beginning of mobile-megabelt development. These initial visible manifestations are commonly deep volcano-sedimentary basins recognized on top of the ancient, pre-megabelt crystalline basement.

The initial orogenic volcano-sedimentary cover étages develop over crustal zones, commonly rift-like, that are already mobilized to some degree at depth. The contact between the basement and the cover is sharply unconformable. At the very base of these early cover étages, just above the basement, commonly lies a layer of conglomerate. Where recognized, it is a useful marker horizon marking the onset of cover sedimentation. No such markers have been recognized in the Cordillera, but they have been described in other megabelts (Appalachians, Tien Shan, eastern Sayan and others).

The term *orogenic grain* is often applied only to this supra-basement rock complex. This is misleading, as orogenic reworking affects the entire crust and lithosphere, where it overprints the older structure. Reworking begins before the formation of orogenic cover, and continues during the entire lifespan of the mobile megabelt. It affects only the basement at first, but as the cover develops, it too begins to be altered by continuing orogenic reworking.

Structural-formational étages, thrust sheets, and metamorphic and rheologic panels as fundamental tectonic units of the crust

Lateral and vertical tectonic rock units are basic components of the crust. Without them, tectonic analysis is impossible. Initial orogenic rifts, subsequent orogenic zones, and median massifs are lateral units of crustal or lithospheric dimensions. *Median massifs* in mobile megabelts are blocks preserved from complete reworking. *Orogens* and *orogenic zones* are the most-reworked belts. Each lateral tectonic unit has its own general physiographic expression and deep physical (geophysical) properties. Working from the ground surface down, geologists are able to distinguish lateral tectonic units based on their compositional and structural properties. The main, indispensable source of information is field mapping of exposed rocks. More easily than the basement, the post-basement cover is subdivided into vertical structural-formational units, or étages, each representing a particular tectonic setting.

Burial-related metamorphism, common in mobile megabelts, imposes a new zonation on any pre-existing geologic framework, in the basement and in the cover. At high grades, it can obliterate previous geologic boundaries altogether (for this reason, the ancient basement is commonly too homogenized to be divisible into étages). Commonly, as when burial is accompanied by tilting, metamorphic zonation is not coplanar with the previous set of étages, and metamorphic fronts cross the older tectonic boundaries. Contact metamorphism, such as in the aureoles of batholiths, disturbs the original structural-formational framework as well. Due to obliteration of older fabrics at high metamorphic grades, to trace a pre-existing framework from slightly metamorphosed areas to strongly metamorphosed ones is not always possible. Tectono-metamorphic complexes require special attention: whereas information about the original étages in them may be lost, they contain their own information about the younger episodes of tectonic mobilization and restructuring.

Depending on its causes and conditions, metamorphism may be of different kinds - prograde or retrograde, strong or weak, regional or local. Results of metamorphism depend on the rocks' temperature and pressure paths, protolith composition, presence of water. Mechanical results of heating, extending or compressing a rock mass also vary: depending on rock composition and rheological conditions, there may be diffuse creep, local mylonitization, brittle breakage. In the upper crust rocks usually behave brittly, whereas below the brittle-ductile transition the norm is ductile behavior. Flowage of rocks is particularly common under high-grade dynamo-thermal metamorphic conditions. Ductile and brittle shearing at shallower crustal levels might accompany the flowage below. The depth to the brittle-ductile transition varies in mobile megabelts, but may be as little as ~10 km (as in young tectonic platforms over the Variscan megabelt in Europe; Emmermann and Lauterjung, 1997). As a region evolves, the position of brittle-ductile transition may change with time. Older transition zones may be lowered to deeper levels and reworked there as a block subsides. Uplift, on the other hand, permits old transition zones to escape destruction by reworking and be preserved. These fossil brittle-ductile transition zones complicate the regional framework of tectonic units.

Also common in mobile megabelts and megabelt-craton transition areas are sheets of rocks bounded by low-angle brittle thrust faults. These sheet stacks form progressively, as the thrusts prograde away from a core zone of an orogen. To distinguish between these brittle thrusts and low-angle ductile shear zones is often easy, but in some cases the distinction is unclear. Within orogenic zones, inverted and uplifted as much as 30-40 km, subhorizontal shear and rheological panels are easy to confuse with brittle thrust sheets. Yet, properly distinguished, brittle thrusts on orogen flanks and deeper ductile shears in orogen cores may in some cases be parts of common deformational systems with different manifestations in different tectonic zones.

All lateral and vertical tectonic units in a megabelt are, in some fashion, records of a mobile megabelt's tectonic evolution. All the units are rock-made, and distinguished in the observed rock mass. But understanding their evolution requires a time dimension as well. The common reference frame is provided by standard global Geological Time Scale (e.g., Harland et al., 1990). Provincial stratigraphic correlation charts apply this scale to specific regions. Within a region, recognition of rock-made structural-formational étages goes hand in hand with definition of temporal stages of the region's evolution.

An étage is assigned well-defined boundaries that reflect regional episodes of sharp restructuring. In mobile megabelts, such episodes are often full-fledged orogenies accompanied by metamorphism and magmatism. Gentler restructuring episodes are marked by just structural unconformities. The unrecorded time interval between the top of the underlying rock-made étage (which is usually erosional) and the base of the overlying one may be quite long. By convention, such a time gap is considered to be part of the tectonic time stage lasting from the base of the underlying étage to the base of the overlying one. Thus, a tectonic stage includes the age span of its corresponding rock-made étage as well as the gap at that étage's top. Because the rock record is incomplete and the definition of étages is interpretive, tectonic stages are in large measure conceptual. But their definition from étage bases adds objectivity, as these bases are mappable and observed in real rock masses.

Rock bodies contain tectonic features that are records of events during and after these bodies' formation (e.g., syn-depositional and post-depositional faults, folds, etc.). Although tectonic features of different generations may be present in the same rock body, their causative events must be discriminated clearly because they belong to and characterize different tectonic stages. A restructuring episode affects all the previous étages. Confusingly, some rocks of a particular étage, such as igneous plutons, may be emplaced into older rock units - however, they should be assigned not to the étage they are physically contained in but to the one during whose tectonic stage they were emplaced.

Stage-by-stage evolution of mobile megabelts and their lateral and vertical tectonic units may thus be reconstructed from the analysis of genetic links between rock bodies of similar age falling into the same tectonic stage. The largest, first-order stages in an orogenic zone are usually few. Generally, the initial stage is represented by big volcano-sedimentary basins and rift-related mafic-ultramafic complexes; the middle stage by flysch formations and voluminous granitoid complexes; the young stage by successor basins of clastic rocks and small fault-related igneous intrusions. Typically, an orogenic zone is first manifested by the creation of trough-like volcano-sedimentary basins, followed by magmatic activity along these zones of weakness, then by thermally induced uplift and inversion. As reworked rocks pass through deep crustal horizons, they acquire distinctive metamorphic and structural fabrics and minerals reflecting the time and nature of their reworking.

Within these first-order time stages, smaller tectonic stages can usually be distinguished. Tectono-sedimentary, tectono-magmatic, tectono-metamorphic and tectono-deformational phenomena occur episodically, related to pulses that could be specific to an individual orogenic zone or to the entire megabelt. The scale of a tectonic stage or reworking episode can only be reconstructed from regional analysis of the entire megabelt, relying on the fundamental principles of geology and considering all four main types of tectonic manifestations.

Methodologies of tectonic analysis based on physical models of lithospheric plates

In the last three decades, geology has been distracted from its traditional rock-orientedness, as it has begun to rely more on physical interpretation of data obtained by indirect geophysical methods. These interpretations and models are based on specific *a priori* assumptions about the composition and structure of deep crustal levels. For a model to be internally consistent is not enough, and verification by real observational data is needed to check the model's correctness. A lot of assumptions made commonly in geophysical interpretations are now known to be geologically flawed, too simplistic or outright incorrect. Incorrect assumptions are recycled in the model to produce internally consistent but unrealistic solutions (Lyatsky, 1994a; Oreskes et al., 1994), which disregard crucial geological principles established by long experience of direct observation and mapping of real rocks and rock bodies.

Lithospheric plates were originally discovered and delineated by geophysicists, mainly by using magnetic, gravity and seismic techniques in oceanic regions. An unfortunate by-product of this success was a perception that geologic field mapping on land is somehow of secondary importance, as a traditional mapper drowns in small-minded empiricism while overlooking a broad vision afforded by global tectonics. This disregard for geology among some geophysicists is strengthened by an idea that physics is the most fundamental of all sciences, from which other fields such as geology are mere derivatives. These views are epistemologically incorrect.

Excessive optimism about omnipotence of current global tectonic models is expressed often, as, for example: "The plate-tectonics model has had marked success in explaining the first-order features of the earth; it has been particularly successful in explaining the nature and age of the crust in the ocean basins" (Burchfiel, 1980, p. 15). Two decades later, many observations have been reported which the 1960s-vintage concept about the nature of plates is unable to explain. The original plate-tectonics model rested on several pillars, some of which remain more postulated than confirmed. It is indeed very plausible that uneven heat distribution in the Earth's interior causes mantle convection, and that this convection affects the lithosphere. It is also apparent that the lithosphere consists of a small number of plates, which in some fashion move away from spreading centers. These ideas have been reasonably integrated with many old and new observations made possible by ever-improving technology.

The idea of thermal currents in the mantle under the more-solid perisphere has existed for some two centuries, applied to subcrustal and sublithospheric levels. A lively discussion about the possible patterns of such currents began in the 1920s. In the 1950s, compilations of earthquake data showed the earthquakes to be concentrated in distinct global-scale bands, partly coincident with the newly discovered system of mid-ocean ridges. Bands containing the deepest earthquakes, with hypocenters hundreds of kilometers below the surface, were found to rim some continental landmasses and large island archipelagoes. In cross-section, these bands of hypocenters were shown to dip towards the continents or archipelagoes, to a maximum depth of ~600-700 km. These dipping focal zones were often associated with pronounced elongated negative gravity anomalies and deep bathymetric troughs on the ocean floor (as in the Java, Japan, Aleutian, Chile and other regions). Logically, it was suggested that oceanic crust and lithosphere are being sucked or pushed downward along these tectonically active zones.

These tectonic zones were interpreted as zones of subduction, where oceanic lithosphere is consumed and recycled into the mantle - as opposed to spreading centers, where it is created. The global balance of total plate area is maintained. The concept combining plate creation, lateral motion and destruction, improperly called *plate tectonics*, offers a presumed mechanism of plate motions over the globe, but it does not explain these motions' dynamics or even kinematics.

What actually drives plate motions is still unclear, as different ideas have competed for decades without an obvious winner. The original concept (e.g., Morgan, 1968) was that a plate, even though it may contain both oceanic and continental parts, is essentially a rigid slab whose stiffness prevents large-scale internal deformation. Only at plate boundaries is tectonic deformation induced by plate interactions. At divergent boundaries, where plates move apart as oceanic lithosphere is created, normal faulting occurs in rifts that have narrow axial structural valleys. At convergent boundaries, where plates move together head-on or obliquely, subhorizontal compressional or transpressional forces have produced all known foldbelt systems (including those in the orogenic zones on continents).

Erroneously, a new era was proclaimed in the science of geology: "Until the advent of plate-tectonic concepts in the late 1960s, geologists and geophysicists studying the components of ancient orogenic and magmatic terranes on the continents had no actualistic framework within which to comprehend the origins and interrelationships of those components" (Hamilton, 1980, p. 33). Of course, comprehension of tectonics had always been actualistic, regardless of prevailing dogma. As well, the view of the Indonesian islands and adjacent bathymetric plateaus as a modern example of developing orogenic areas predates the plate-tectonic concepts.

Reasons for plate movements are not yet understood. Whether plates are pushed away from their spreading centers, sucked down subduction zones, or dragged along by convecting sublithospheric mantle is still unclear (Wilson, 1993). The existing models have so far failed to solve this "conundrum" (Bercovici, 1993). The competing elegant numerical models have been unable to describe all the observed relevant facts (e.g., Richardson, 1992; Bott, 1993). Ridge push would induce in-plate compressional thrusting and folding, especially near the mid-ocean ridges, on a scale that is not observed. Subduction-zone suction would tend to pull the oceanic plates apart, producing widespread extensional deformation on a scale that is not seen.

For plates to be able to propagate the enormous stresses over thousands of kilometers, as these models would require, the plates would have to be extremely stiff. Such stiffness was assumed in the 1960s when the data base about the oceanic crust was small, but growing modern evidence points to significant intraplate deformation including warping, extensional faulting, thrusting, block tilting and so on (e.g., Chamot-Rooke et al., 1993; Macdonald et al., 1988, 1991, 1996; Ballu et al., 1998; Lawrence et al., 1998; Mitchell et al., 1998). Some workers have ascribed intra-plate compressional stresses to the push from ridges (e.g., Stefanick and Jurdy, 1992), but such effects could also be produced by upwelling and downwelling limbs of thermal convection cells in the mantle (Bott, 1993). Analogously, Forte et al. (1993) suggested such convection-induced forces could cause warping in vast continental regions. Little is known about how mantle convection actually occurs, and it is reasonable to expect cells of different dimensions to affect the overlying lithospheric plates differently.

Some rheological complexities in the Earth's crust and upper mantle

One effect of mantle convection is to change the lithosphere's thermal and rheological conditions and properties. Shearing may develop in boundary zones. Drag at the base of the lithosphere seems probable (though its exact mechanics remains unclear and direct confirmation is impossible; cp. Richardson, 1992 and Bott, 1993). Shearing and flowage of ductile rocks at the base of the crust are consistent with strong seismic reflectivity of the lower crust and the Moho. Lateral migration of ductile material in the continental lower crust has been suggested repeatedly (e.g., Dohr, 1989; Lyatsky, 1993, 1994a).

In the oceanic lower crust in vast regions, abundant seismic reflections have been found to dip at low angles in various directions, stopping at the crust-mantle boundary (Eittreim et al., 1994). Only some of these reflections continue upward through the entire crystalline oceanic crust, and a few penetrate the overlying sedimentary cover (see also White et al., 1990; Morris et al., 1993; Ballu et al., 1998; Lawrence et al., 1998). As far as can be seen in two-dimensional profiles, most these reflections dip towards the spreading centers (which may lie thousands of kilometers away), but some dip away from the centers. Some oceanic-crust reflectors and confirmed faults on the ocean floor are oriented obliquely or perpendicular to mid-ocean ridges. From ocean-floor mapping and seismic profiles, shearing in the oceanic crust seems to have varied in orientation and intensity from place to place and from time to time.

If slip occurs between the crust and upper mantle and between various horizons in the crust, simple extrapolations of geologic properties from one lithospheric level to another may be misleading. Doglioni (1990, 1993) favored an idea of general westward drift of the lithosphere with respect to the underlying asthenosphere. This hypothesis is speculative and at present not verifiable, but structures and rock ages may vary with depth in the oceanic and continental lithosphere in any locality.

Within continental crust, vertical variations in the structural and rheological properties of rocks above and below a detachment level have also been noted (Wegmann, 1930; van Bemmelen, 1967) in various regions. Extensive geological mapping in the Basin and Range province in the U.S. Cordillera revealed a series of normal faults separating alternating asymmetric ridges and basins. Displacements on large-scale listric normal faults made room for basins containing Tertiary sediments. These faults are steep at the surface, but many of them apparently flatten out with depth, and seismic profiles suggested they merge at depth into master detachments. This regional structural system also contains a series of cores made up of high-grade metamorphic rocks complexly (and probably genetically) interconnected with granitic intrusions. The reasonable conclusion was that the entire Basin and Range province was formed by whole-crust tectonic extension, and geophysical surveys show that crustal thickness there is indeed reduced to just ~30 km. At the brittle-ductile transition, shearing is though to have produced a detachment between brittle rocks in the "upper plate" above and ductile rocks in the "lower plate" below, with markedly different structural patterns (see review papers in Pakiser and Mooney, eds., 1989; Beratan, ed., 1996).

Continental upper-crustal stress patterns, measured in drillholes and on the surface, cannot be simply extrapolated down into the lower crust. Deep drilling in crystalline-rock terrains in Russia, Europe and the U.S. has shown thermodynamic conditions to change with depth non-linearly and unexpectedly, with big lateral variations from region to region. Each drillhole negated countless previous speculations about the crustal composition, structure and physical conditions (e.g., Kozlovsky, ed., 1984; Emmermann and Lauterjung, 1997).

In the 1970s and 1980s, active tectonism-inducing forces were assigned to sub-lithospheric mantle and boundary zones of interacting lithospheric plates (e.g., Burchfiel, 1980). Continental-lithosphere masses were treated as essentially inert, responding only passively to external influences. Intracontinental tectonic stresses, deformation and magmatism were ascribed to distant causes from plate-boundary zones or from the mantle below. Convergent plate boundaries, in particular, were thought to be responsible for orogenic structures. Where no plate boundaries were apparent in modern intracontinental settings, ancient ones were postulated to account for orogenic events in the past. Tectonics within lithospheric plates was reduced to just effects of convective currents in the mantle and interactions between plates. The notion of inert, dead continental (and oceanic) lithosphere fits the simple physical models of assumed plate behavior, but geologically it is in contradiction with reality.

Factual departures from the simplistic 1960s-vintage models of lithospheric plates began in the 1970s. Plate thickness was found by geophysical studies to be not uniform but highly variable, reaching ~400 km under continental Archean cratons (Jordan, 1975). Internal deformation was discovered in oceanic plates, indicating their rigidity had been overestimated. Along spreading centers, extensional structural characteristics were found to vary. Some mid-ocean ridge segments were found to coincide exactly with upwelling of hot mantle, but others were not. Bott (1993) calculated that presence or absence of upwelling could alter the stress field in the spreading-center zones by a factor of 2 to 3.

That mantle upwelling does not everywhere coincide with the system of mid-ocean ridges is now accepted widely. Localized plumes, in particular, penetrate and disrupt the assumed system of mantle currents. Heating variations in space and time would affect the lithosphere and crust, inducing lateral variations in the modes of lithosphere formation, reworking and destruction. Spots and zones of increased mantle heating have been shown to coincide with a variety of surface features in both oceans and continents: seamounts, islands, flood-basalt provinces. High elevation of the entire African continent since the Mesozoic has been attributed to the action of a huge mega-plume beneath (e.g., Pavoni, 1993). Existing seismic models of the Earth's deep interior (e.g., Su et al., 1994; Su and Dziewonski, 1995) suggest an overheated mantle in the equatorial Africa and the Pacific, lending support to the ideas about mega-plumes.

Assessment of geological information from oceanic-crust magnetic anomalies

Geochemical, petrological and structural variations in ocean-island and mid-ocean ridge basalts have been noted repeatedly (e.g., Macdonald et al., 1988, 1991, 1996, and references therein). Close dependence of rock magnetization on petrology (e.g., Hailwood, 1989; Opdyke and Channell, 1996) suggests variations in the magnetic properties of the oceanic crust should be considerable. The attribution of the oceanic magnetic anomalies to just the basaltic layer of the oceanic crust (Vine and Matthews, 1963) has also been proven incorrect: regional and local considerations with the benefit of magnetic sea-bottom and satellite surveys require additional contributions from sub-basalt horizons. Yañez and LaBreque (1997) have modeled some sources of oceanic magnetic anomalies to lie in the lower crust and even the upper mantle (see also Tivey and Tucholke, 1998). The relief of the Curie isotherm, beneath which the rocks are unable to hold magnetization, is indeed uneven (Cohen and Achache, 1990), and from Magsat data Yañez and LaBreque (1997) have estimated the bulk magnetization of magnetic-anomaly rock sources in the North Pacific to be at least 50% higher than in the North Atlantic. The sources of oceanic-crust magnetic anomalies, with their remarkable linear character, remain unclear.

Oceanic linear magnetic anomalies form coherent patterns in huge domains hundreds and even thousands of kilometers across, bounded by huge intraplate fracture zones and plate boundaries. Distinctiveness of the oceanic anomaly pattern comes from the regularly alternating, parallel positive and negative anomaly bands usually ~30-40 km wide. The reasons for such an alternating pattern were suggested by the discovery of episodic reversals of the geomagnetic field (see Hailwood, 1989; Opdyke and Channell, 1996). By an ill-understood process, occasionally and abruptly the Earth's geomagnetic field reverses its polarity. These reversals take just several thousand years, and are usually followed by longer intervals during which the polarity remains more stable.

These reversals are imprinted in the remanent magnetization of oceanic-crust rocks created at spreading centers, forming a zebroid pattern of anomalies. Recent basalts bear the present-day polarity (defined as normal), and the modern predominantly-normal anomaly band follows the axial zones of mid-ocean ridges. Ocean-floor basalts become older away from the axial rift valleys of the ridges, and basalts older than ~800,000 years are magnetized predominantly reversely. Farther out lie normally magnetized basalts, and then reversely magnetized ones again, and so on, and the striking age-dependent distribution of magnetic polarities in axial zones of mid-ocean ridges (where ocean-floor basalts lack a sedimentary cover and can be sampled for dating) is observed worldwide. These successions of normal- and reverse-polarity anomaly bands on both sides of mid-ocean ridges tend to be strikingly symmetric. Correlation of age and magnetic-anomaly polarities of sea-bottom basalts along ridges with well-studied magnetized stratal rock successions on land offered a convincing calibration of oceanic magnetic anomalies with the known polarity zonation. Then, from basalt exposures along mid-ocean ridges, the Geomagnetic Polarity Time Scale (GPTS) was extrapolated back in time and laterally into ocean-floor regions where basalts are hidden beneath a sedimentary cover. Magnetic anomalies were numbered sequentially away from the spreading centers, and sparse drillhole data were used to obtain chronostratigraphic information from the ocean-floor sedimentary cover above the basalts. Onshore, paleomagnetic and chronostratigraphic data were used for GPTS calibration (Harland et al., 1990; Berggren et al., 1995).

Detailed studies of oceanic magnetic-anomaly records across mid-ocean ridges began in the 1950s and 1960s. The South Atlantic anomalies were found to be the most obviously symmetric, and relatively little disrupted by transform faults and intraplate deformation. Magnetic-anomaly stripes, in the South Atlantic and elsewhere, looked coherent and ordered for thousands of kilometers, except for a few zones that by comparison were magnetically quiet.

Vine and Matthews (1963) proposed that elongated magnetic anomalies over oceanic crust are a direct reflection of sea-floor spreading (hypothesized previously by Hess, 1962), as anomaly source rocks are formed continually along spreading centers. As the geomagnetic field reversed occasionally, basalts in the oceanic upper crust retained a remanent magnetization imposed by the ambient field at the time of their cooling. This idea was broadly accepted, and has served as a foundation for reconstruction of plate motions (Vine and Wilson, 1965; Pitman and Heirtzler, 1966; Heirtzler et al., 1968; Stock and Molnar, 1988; Atwater, 1989; DeMets et al., 1990; and many others). Without sufficient regard for the real complexities of magnetic-anomaly sourcing in the oceanic crust, this simple explanation has been used for many exciting speculations.

Vine and Matthews (1963) assumed that remanent magnetization of oceanic crust basalts was preserved from the time of basalt extrusion without significant alteration. Modern studies show that, on the contrary, basalts are altered considerably in the ocean water during and after their cooling. In any case, thermoremanent magnetization of basalts along does not account for the entire oceanic magnetic anomaly pattern (e.g., Yañez and LaBrecque, 1997; Tivey and Tucholke, 1998), and the nature of the required deeper anomaly sources is not clear. At the base of the oceanic crust there appears to lie a layer of mafic an ultramafic cumulates, whose distribution, like that of the upper-crust basalts, is probably subhorizontal rather than banded (cp., e.g., Malpas et al., eds., 1990). A plausible source of the striped magnetic field is the horizon of densely packed, steep sheeted dikes beneath the basalts but above the cumulates. The dikes are mafic, and lie shallower than the Curie isotherm.

Few drillholes have penetrated igneous rocks of the oceanic crust, and their locations are far apart. No drillhole has penetrated the sheeted-dike layer to its base, so geological ideas about this layer's composition and structure are rather speculative. The ~50% difference in estimated bulk magnetization between the North Pacific and North Atlantic regions suggest strong regional differences in the magnetic sources. These variations may be due to dissimilar volume and/or magnetization of the source rocks. While the reasons for regional variations in oceanic-crust magnetization remain unknown, Yañez and LaBrecque (1997) examined the magnetic-anomaly domains east and west of the East Pacific seamount chain and attributed them at least partly to "structurally related magnetization anomalies".

Whereas rock bodies in the oceanic upper-crustal sedimentary and basaltic layers are flat-lying, bedded and broad, the sheeted dikes are subvertical and seem like a plausible source for striped magnetic anomalies. Seismic data suggest thickness of the sheeted-dike interval varies considerably, as does thickness of the lower-crustal layer of presumed cumulates. Big thickness variations are apparent in the volcanic and sedimentary layers as well. Nonetheless, the lineated magnetic anomaly pattern is ubiquitous are regular in appearance, regardless of these changes, and the relationship between these anomalies and their unidentified crustal sources remains unclear.

Another simplifying assumption in the initial models of plate motion was that spreading rates had remained constant during much of the Cenozoic and Late Cretaceous (Heirtzler et al., 1968). In the modern literature, however, changes in the rates of plate spreading are acknowledged to occur often. None of this complexity was taken into account in the simplistic initial explanation of the oceanic magnetic anomalies.

The conceptual mechanism where oceanic lithosphere is formed continually at spreading centers in long belts reflected in magnetic anomalies, and moves steadily away from the centers to make room for new lithosphere, no longer explains the observed facts. For example, "...this simple mechanism cannot explain some characteristics of seafloor spreading magnetic anomalies nor the magnetization of dredged oceanic crustal samples" (Yañez and LaBrecque, 1997, p. 7947; also Tivey and Tucholke, 1998). This hypothetical mechanism overlooks many known complexities in magmatism and fault deformation mapped at spreading centers (e.g., Macdonald et al., 1988, 1991, 1996; Smith et al., 1997; Lawrence et al., 1998). It is based on an unproved presumption that oceanic magnetic anomalies reflect the age of unsampled rocks to the very base of the lithosphere, and assumes implausibly that the original lithospheric structure and magnetization formed at spreading centers are preserved to the present day essentially unchanged.

Observations of ocean floor from submersible craft and geophysical surveys show the mid-ocean rift morphology to be similar to that on land, with many variations in the distribution of strain and magmatism along rift zones. Complex fault systems have numerous strands that split and merge, and lavas erupt episodically from fissures and volcanic centers. Many faults in the oceanic crust have been found to run obliquely or perpendicular to the ridge. Due to technical difficulties with sample recovery and laboratory analysis, only a small number of age determinations are available for ocean-floor basalts. Deeper source rocks of magnetic anomalies are not sampled or even seen. Geophysical anomalies reflect variations in specific physical properties in source rocks, but they say nothing about rock age. Besides, not all magnetic anomalies observed today represent the source rocks' original properties acquired at the time of their genesis. In the presence of saline water, geochemical alterations are common. Fractures in the crust do not form in just a single event, but are rejuvenated repeatedly, and new fractures are created due to stress-field fluctuations. Metamorphism is an ever-present likely possibility, given an irregular heat supply from the mantle.

New and rejuvenated fractures can serve as conduits for magma injection long after the crust's genesis. Juvenile dikes that are parallel to the original fabric would contaminate the original magnetic-anomaly pattern, but their presence would be exceptionally difficult to recognize. Besides, many fault-bounded blocks in the oceanic crust are tilted, invalidating the assumptions about the paleohorizontal that are integral to paleomagnetic determinations. Changes in thermo-dynamic conditions result in variations of spreading rates, which are now recognized to have occurred frequently. All in all, the oceanic lithosphere is far more complex, diverse and reworked than the initial models assumed.

Cohen and Achache (1994) presented a possibility that some components of the oceanic magnetization pattern may not be related to lithospheric age at all. Correlation between some magnetic and gravity anomaly domains in the North Pacific suggests some sort of late-stage undulations imposed on lithosphere of different ages (see also geoid anomaly maps of Bowin, 1983). More locally, mean values of Bouguer gravity anomalies are different across the Mid-Atlantic Ridge at equatorial latitudes, causing Ballu et al. (1998) to conclude that deep-seated processes related to the spreading center in that area are "very asymmetric". Large-scale lateral variations in seismic anisotropy of the oceanic lithosphere have been interpreted as products of post-genetic, secondary reworking (Wolfe and Silver, 1998).

Existence in the oceanic crust of new fractures and shears, both steep and low-angle, is now acknowledged widely (e.g., White et al., 1990; Banda et al., 1992; Morris et al., 1993; Mitchell et al., 1998). Unfortunately, seismic reflection methods are designed to detect low-angle but not high-angle discontinuities, and vertical faults are often unseen seismically. Each generation of faults offers potential conduits for magma injection, with unexamined effects on the magnetic anomalies.

Shearing and decoupling at the oceanic crust-mantle transition, discussed above, complicate downward extrapolations of inferred upper-crust ages all the way to the base of lithosphere. Abundant seismic reflections in the lower crust (cp. White et al., 1990; Banda et al., 1992; Eittreim et al., 1994) seem to stop at the Moho level, suggesting a degree of detachment between the crust and the mantle. Magnetic-anomaly sources must lie above the Curie isotherm, i.e. predominantly in the crust - but even if crustal-age determinations from magnetic anomalies were somehow to be confirmed, detachment and decoupling would make it hard to extrapolate them into lithospheric mantle.

Dikes in the oceanic crust are evidently not arranged in a simple, uniform pattern. Sheeted dikes most probably originated in a strongly extensional tectonic regime. In modern mid-ocean-ridge axial valleys and spreading centers, extensional normal faults seem to continue downward 7-8 km, through the entire oceanic crust (e.g., Toomey et al., 1985). But even early surveys of the Mid-Atlantic Ridge showed various fault groups with different orientations. Recently, near this ridge south of the Kane transform fracture zone, Lawrence et al. (1998) reported rotated blocks bounded by normal faults. They also noted two different groups of dikes, with dissimilar orientations and paleomagnetic properties, causing them to question the "assumptions about uniform dike orientation at oceanic spreading centers".

Geologically it is more plausible that accretion to plates of new material at spreading centers occurs not in an immaculately orderly fashion, belt by belt, but rather more patchily. At present, magmatism at mid-ocean ridges occurs occasionally, and in restricted localities. Faulting within oceanic plates also took place repeatedly, making conduits for new magmas.

The assumption that magnetic anomaly stripes correspond simply to a lateral age progression of the entire oceanic lithosphere, belt by belt, is unverified and very speculative. This assumption does not rest on anything like a sufficient knowledge of the geologic nature of the oceanic magnetic field or oceanic-crust rock properties, genesis and alteration.

In the eastern Pacific, for example, reconstructions of plate motions from interpretations of magnetic-anomaly stripes, still produce uncertainties, overlaps and gaps that are not permissible in rigid-plate tectonics (e.g., Stock and Molnar, 1988). Even as the data base becomes more detailed, the methodology of plate reconstructions from magnetic stripes seems to have many unaccounted-for systematic shortcomings. To base a crucial part of plate-kinematic reconstructions on such a shaky foundation, and then to extrapolate these unreliable inferences onto continents, is speculative to the extreme. It has been found to cause big mistakes (Lyatsky, 1996). While traditional, rock-based magnetostratigraphy has been helpful to geologists and tectonists, simply ascribing ages to unsampled oceanic lithosphere based on a nomenclature of magnetic anomalies can be quite misleading. Also misleading would be speculations about the possible effects of hypothesized past plate interactions on continental tectonics.

The needs of practical tectonics (Lyatsky et al., 1999) call for a return to the traditional focus on meticulous studies of observable rocks and rock-made bodies. These observations would serve as a sound basis for regional tectonic conclusions. This sound approach is employed in the forthcoming chapters.

4 - CONCEPTUAL FUNDAMENTALS OF REGIONAL TECTONIC ANALYSIS ON CONTINENTS

Continental and oceanic lithospheric masses as lateral tectonic units

The words *ocean*, *oceanic*, *continent* and *continental* are often used to mean very different things. Their sense may variously be geographical, referring to big bodies of water and continuous lands; geophysical, referring to types of Earth's crust possessing or lacking specific seismic-velocity characteristics; or tectonic, referring to two types lateral lithospheric units with contrasting geologic properties. As tectonic units with multi-aspect geological definitions, oceanic and continental lithospheric masses are bounded by discontinuities where properties of the lithosphere change. These boundaries do not necessarily coincide with the boundaries of lithospheric plates. Systems of lithospheric plates and systems of oceanic vs. continental masses are not the same. Some plates, such as North and South American or African, are composite, incorporating continental and oceanic lithospheric masses. On the other hand, parts of a single oceanic-lithosphere mass (Atlantic, Pacific) may be assigned to different plates.

Oceanic and continental lithospheric masses are distinguished by different fundamental characteristics of their geology and tectonics. Continents are the thickest lithospheric blocks, with crust-mantle properties which are sharply different than those in oceanic blocks. Enriched with felsic material, they are coherently high-standing, demarcated by the lower (outer) continental slopes or, where present, troughs (trenches). But topographic and tectonic boundaries of continents are not everywhere obvious. There is no clear boundary between North and South America, for example, or between North America and Asia. The most apparent are the continent-ocean lithospheric boundaries at the Atlantic and Pacific margins of the conterminous U.S., but in Canada their definition is still unclear (Grant, 1987; Lyatsky, 1996). Still, continental lithospheric masses are very different from the surrounding oceanic ones, having contrasting structure and bulk petrological and geophysical properties.

Continental masses are distinguished by their enrichment in incompatible chemical elements, average andesitic composition, presence of virtually all known rock types, and ages exceeding 4 Ga (Anderson, 1995; Rudnick, 1995). Continental crust is strikingly inhomogeneous in its rock properties vertically and laterally, as indicated by geologic mapping and geophysical surveys. The bases of continental lithosphere and crust correspond to changes in mineral composition and/or physical state or rocks, but the debate about which of these parameters yield the most reliable definition still continues.

Interrelations of the rock-made lithosphere with the sub-lithospheric mantle are important for understanding continental evolution. Galactic and Solar-System influences on the Earth may affect continental evolution as well, but they are outside the scope of tectonics, which deals with rock-made bodies and endogenic processes. Continental lithosphere has its own sources of energy, which gives it a geodynamical system all of its own. Even cratons are actively self-developing (Lyatsky et al., 1999), and it is even more true of mobile megabelts.

The North American continent's cratonic core is still defined loosely, and sometimes this definition is largely physiographical (Canadian Shield vs. Central Lowlands and Great Plains: e.g., in Flint and Skinner, 1974; or flat plains vs. mountains: e.g., in King, 1977). Flint and Skinner (1974) defined the craton as including the Canadian Shield but not the adjacent platforms, whereas King (1969) regarded the Shield as a collection of foldbelts and distinguished only the platforms as cratonic. In Europe, Kober (1923) and Staub (1924) used the term *Kratogen* to describe a tectonic entity, as opposed to *Orogen*, which is strong enough to resist the forces that caused the spectacular Alpinotype deformation (folding, thrusting). They distinguished cratonic regions as stable and rigid, as opposed to the heavily folded Alpine structures. A similar tectonic approach was use by Stille (1924, 1941), who combined shields and platforms into *Kratons*, in contrast to regions of orogenic deformation. Kay (1951), who applied these ideas to North America, Anglicized the original German spelling to *craton*; and so this word entered the mainstream English-language literature. To Stille and his contemporaries, cratons were whole-crust entities, rigid and resistant to orogenization for long periods of time. To Stille (1924, 1941) and Shatsky (1956, 1964), *craton* was a purely tectonic notion, opposite to *geosyncline* or *geosynclinal orogenic belt.*

Two contrasting structural and topographic styles separate continents into principal domains: mobile megabelts, partly coincident with the most mountainous areas; and relatively uniform, low-relief platforms, covered mostly by flat-lying sedimentary rocks. Land accounts for only about one-third of the Earth's surface, but continental masses, defined tectonically based on their whole-lithosphere properties, cover almost half the globe (cp. Cogley, 1984; White, 1988). On several lines of geochemical and geophysical evidence, roots of continents, including North America, continue downward 400 km or more (e.g., Jordan, 1975; Gossler and Kind, 1996; also see review by Lyatsky, 1994a). This is much deeper than the elastic layer (which could be >100 km thick; Anderson, 1995) required to support even the highest isostatically balanced mountains.

Seismic anisotropy in the continental root under the Canadian Shield roughly matches the Late Archean geologic grain mapped in the cratonic Superior province on the surface (Silver and Chan, 1988). The cold, stable and possibly compositionally distinct cratonic roots are characterized by relatively high seismic velocities, especially under shields. Rudnick (1995) argued that at least ~70% of present-day continental masses existed since the end of the Archean. Unlike the oceanic lithospheric masses, continents thus appear to have been rooted deep in the Earth for most of their history.

Recognition of lateral tectonic units within mobile megabelts

Cratons are the most stationary and least mobilized parts of continental-lithosphere masses. Their oldest parts formed in the Archean. Some parts of cratons were later mobilized and reworked in large megabelts. As mobile megabelts died, their crust and lithosphere were again cratonized and incorporated into the next generation of cratons. The time-variant distribution of cratons and mobile megabelts determined the main tectonic and structural peculiarities of continents during different tectonic periods of their history. The modern North American continent has a craton in the middle, flanked by two main mobile megabelts: Appalachian-Ouachita and Cordilleran. On the current evidence, these megabelts originated no earlier than ~1,700 Ma (end of the Paleoproterozoic), and were imposed on the Archean-Early Proterozoic Laurentian craton. Remnants of pre-existing cratons are found in the hinterlands of both these great mobile megabelts (e.g., Crittenden et al., eds., 1980; Rankin et al., 1993).

Median massifs, at their basement level, are semi-reworked remnant pieces of older cratons. In the Cordillera, they are located mostly in the hinterland, between the mountain ranges to the east and west. The Intermontane and Yukon-Tanana median massifs are well-known distinct lateral tectonic units, recognized for decades (Douglas, ed., 1970; Gabrielse and Yorath, eds., 1991; Plafker and Berg, eds., 1994). With incomplete crustal and subcrustal reworking, they are comparatively rigid and provide tectonic settings conducive to the appearance of rectangular volcano-sedimentary basins whose metamorphism is only slight (commonly, to no higher than prehnite-pumpellyite grade). These massifs and basins are deformed much less than the neighboring orogenic zones, and their magmatism is largely teleorogenic.

In the Canadian Cordillera, orogenic belts on the west side of the median massifs in the mobile-megabelt hinterland were previously assigned eugeosynclinal affinities, whereas orogenic belts on the massifs' east side were thought to be miogeosynclinal (cp. Kay, 1951; Wheeler and Gabrielse, coords., 1972). In Stille's classical nomenclature, western belts (Coast and Insular) were produced in a full-fledged geosyncline, while the eastern one (Omineca) formed in a less-developed geosyncline. Stille (1941) combined both these geosyncline types together under a common name *orthogeosyncline*, distinguishing eugeosynclinal belts of abundant volcanism from the near-craton miogeosynclinal belts where volcanism was subdued. In this classification, the Omineca Belt in the eastern Cordillera is a miogeosynclinal orogenic zone.

With the rise of the plate tectonics concepts, most geosyncline terminology was dropped: *eugeosyncline* went almost completely out of use, and *miogeosyncline* was reduced to a less meaningful word *miocline* (in Greek, it means "lesser incline", which is not what these features are). To Stille (1924, 1941) and many Russian (Shatsky, 1956; Beloussov, 1962) and French (Aubouin, 1965) tectonists, eugeosynclines and miogeosynclines were crustal-scale orogenic belts (the Russians also called them *mobile belts*, from which this book's *mobile megabelt* is derived) of similar mountain-building potential but differing by the presence or absence of intrusive mafic magmatism, volume of extrusive magmatism, and position relative to the neighboring craton (cp. Alpine internides and externides of Kober, 1923, 1951). North American tectonists were first to directly equate miogeosynclines (mioclines) to continental slopes, initially at the Atlantic passive continental margin (Dietz, 1961; Dietz and Holden, 1967, 1974). This approach was essentially sedimentological, but it became common. Its weakness is that it largely ignores the continental margins' tectonic nature and history beyond just sediment accumulation and plate interactions.

In the Cordillera, the old regional nomenclature underwent some redefinition in the last three decades. Wheeler (1970) regarded the Intermontane Zone as distinct, separating the Western and Eastern Cordilleran foldbelts (the latter included the Omineca and Rocky Mountain belts). Soon after, he (Wheeler, 1972, p. 2) renamed the Eastern Fold Belt as Columbian Orogen, and the Western Fold Belt became Pacific Orogen. In his description, "each orogen or fold belt is composed of a mobile uplifted core zone of granitic and median- to high-grade metamorphic rocks, flanked by belts in which the tectonic transport has been directed principally away from the core zone". Precambrian basement was supposed to continue from the modern North American craton into the Columbian Orogen's core zone - the Omineca crystalline belt. Indeed, remobilized cratonic-basement rocks were suggested in gneiss domes of the Shuswap Metamorphic Complex (Figs. 14-24; the reader is encouraged to examine these maps before reading further). A similar basement was assumed to lie in the North Cascade Mountains, west of the Intermontane Belt (cp. Misch, 1966).

The Columbian Orogen was thought in the 1970s to contain a miogeosynclinal-platformal or miogeosynclinal sequence, including the 10-12-km-thick Late Proterozoic Windermere Supergroup. The older (Middle Proterozoic), huge Belt-Purcell Basin straddled the Canada-U.S. border, containing granite stocks and orogenic deformation dated at 1,260±50 Ma (Ryan and Blenkinsop, 1971).

Much confusion surrounded the tectonic nature of the easternmost zone within the belt of Hinterland or Intermontane plateaus (Figs. 3, 4, 6, 7; cp. Wheeler, 1970, 1972). This zone was included into the Columbian Orogen, as indeed was most of the Intermontane Belt except for its western parts which were folded and thrusted due to orogenic evolution of the Coast Belt (Wheeler and Gabrielse, coords., 1972, their Fig. 12).

Figure 14. Geologic map of the southern Canadian Cordilleran interior, from the Rocky Mountain fold-and-thrust belt on the east to the Coast Belt on the west (from Wheeler and McFeely, comps., 1991).

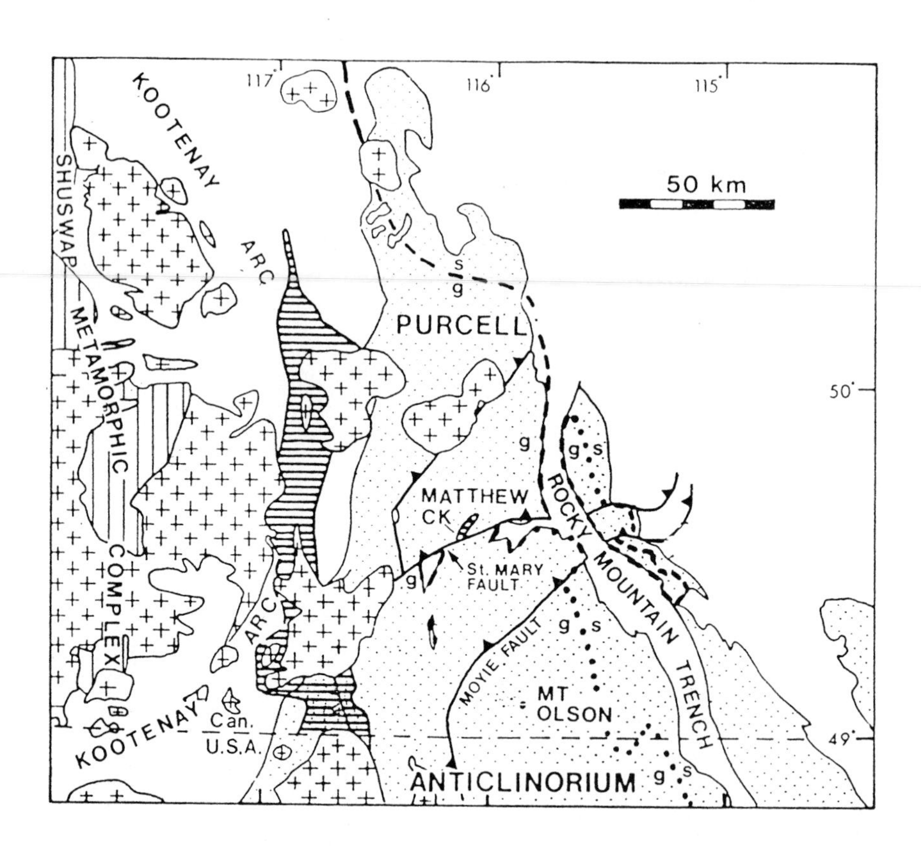

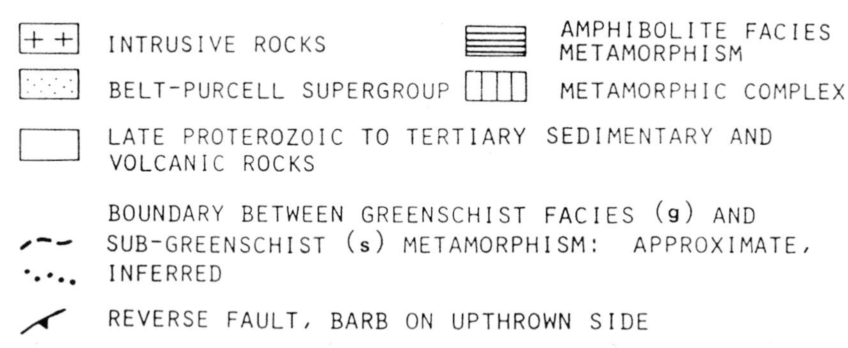

Figure 15. Major geologic features in the Purcell Anticlinorium and adjacent parts of the Kootenay Arc and Shuswap metamorphic complex in the southern part of the eastern Canadian Cordilleran miogeosyncline, as identified by geologic mapping (simplified from McMechan and Price, 1982).

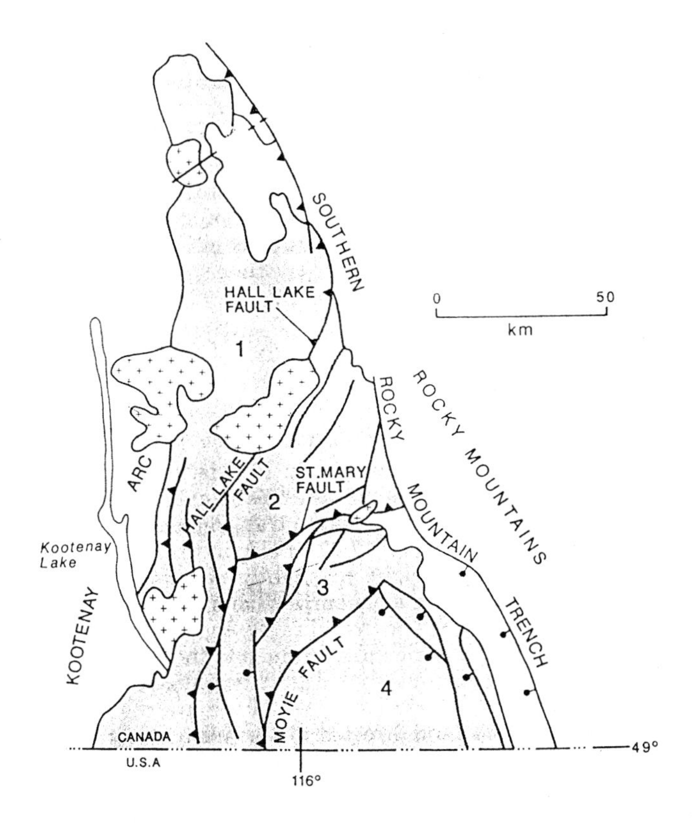

Figure 16. Major faults and plutons in the Purcell Anticlinorium in the southern Canadian Cordilleran miogeosyncline, as identified by geologic mapping (modified from McMechan, 1991). The Purcell Anticlinorium represents the inverted and largely tectonically reworked northern protrusion of the Belt-Purcell Basin.

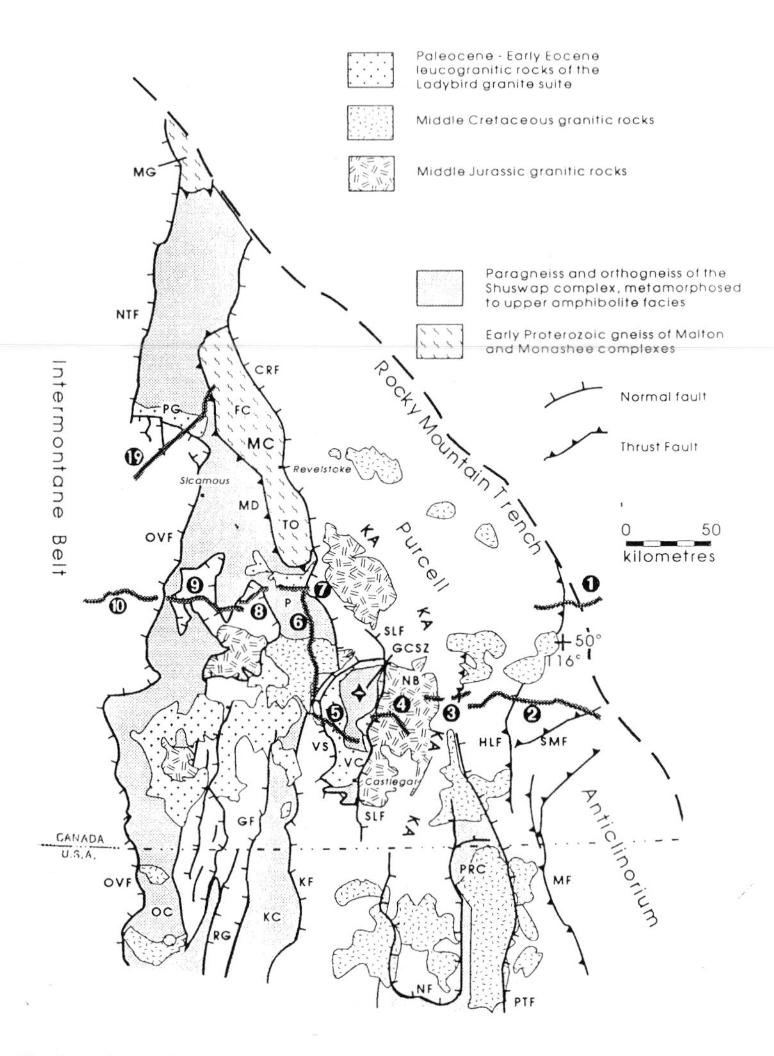

Figure 17. Locations of metamorphic core complexes and major faults in the southern part of the miogeosynclinal Omineca Belt of the Canadian Cordillera, based on geologic mapping (modified from Carr, 1995). KA - Kootenay Arc; NB - Nelson batholith; PG - Pukeashun granite; RG - Republic graben. Fault systems and shear zones: CRF - Columbia River; GCSZ - Gwillim Creek; GF - Granby; HLF - Hall Lake; KF - Kettle; MD - Monashee; MF - Moyie; NF - Newport; NTF - North Thompson; OVF - Okanagan-Eagle River; PTF - Purcell Trench; SLF - Slocan Lake; SMF - St. Mary; VS - Valkyr; WBF - West Bernard. Metamorphic core complexes and their culminations: FC - Frenchman Cap; KC - Kettle-Grand Forks; MC - Monashee; MG - Malton; OC - Okanagan; P - Pinnacles; PRC - Priest River; TO - Thor-Odin; VC - Valhalla. Numbers 1-10 and 19 mark the Lithoprobe seismic reflection profiles.

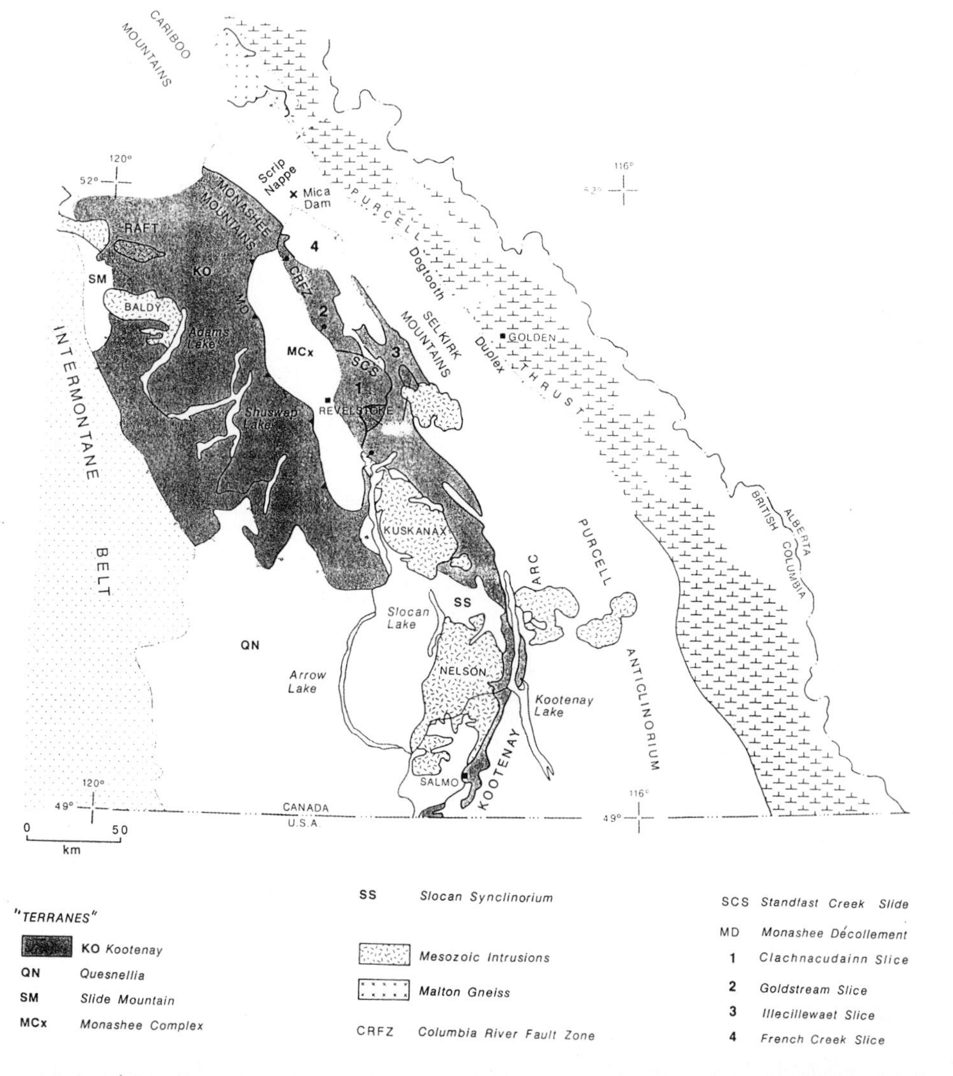

Figure 18. Major plutons, faults and conventionally assumed terranes in the southern part of the eastern Canadian Cordilleran miogeosyncline (simplified from Tempelman-Kluit et al., 1991).

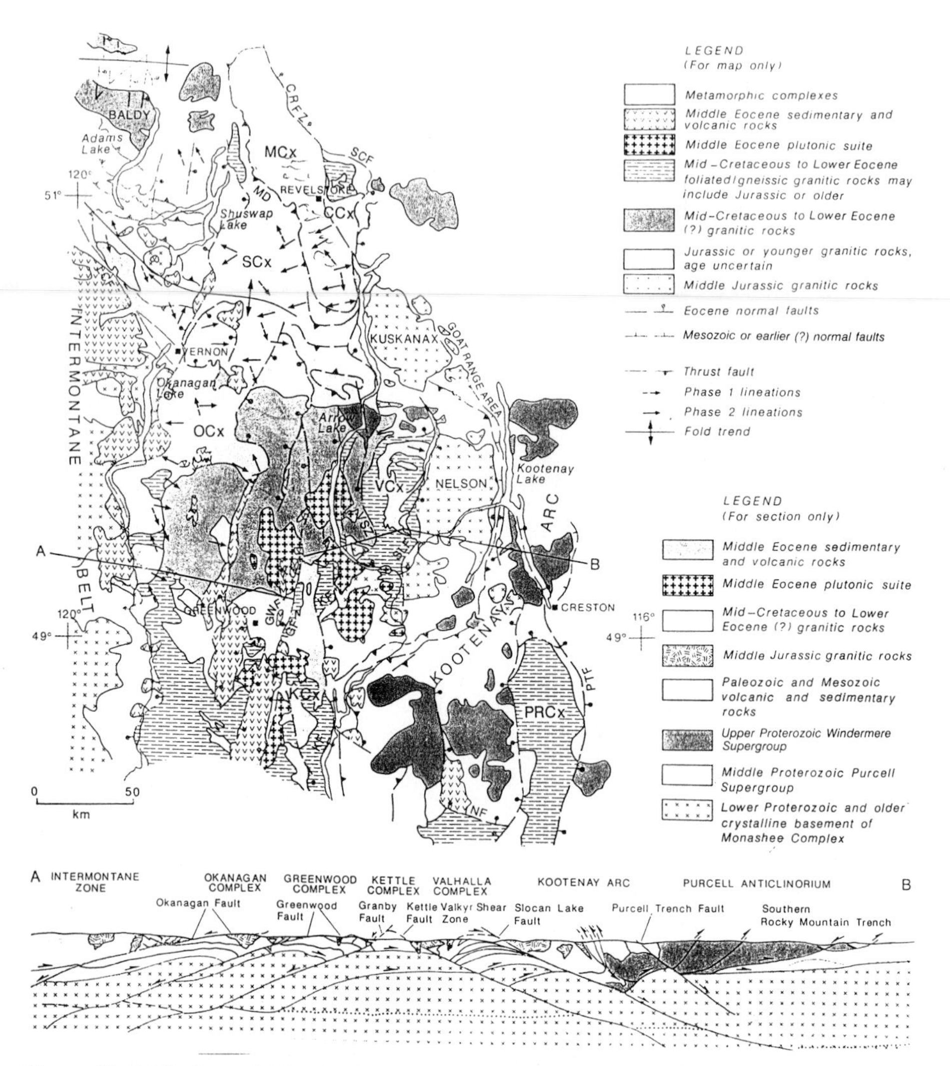

Figure 19. Main faults, folds, plutons and metamorphic core complexes in the southern part of the miogeosynclinal Omineca Belt, as identified by geologic mapping at the surface and assumed in projections deeper into the crust (simplified from Gabrielse, 1991b). The cross-section A-B illustrates the conventional view of the Cordilleran crust as a stack of thrust slices, overlooking high-angle faults. CCx - Clachnacudainn rock assemblage; CRFZ - Columbia River fault; GF - Granby fault; GWF - Greenwood fault; KF - Kettle fault; LCF - Lewis Creek fault; MCx - Monashee complex; MD - Monashee fault; NF - Newport fault; OCx - Okanagan complex; OF - Okanagan fault; PRCx - Priest River complex; PTF - Purcell Trench fault; SCF - Standfast Creek fault; SCx - Shuswap complex; SLF - Slocan Lake fault; VCx - Valhalla complex; VSZ - Valkyr fault.

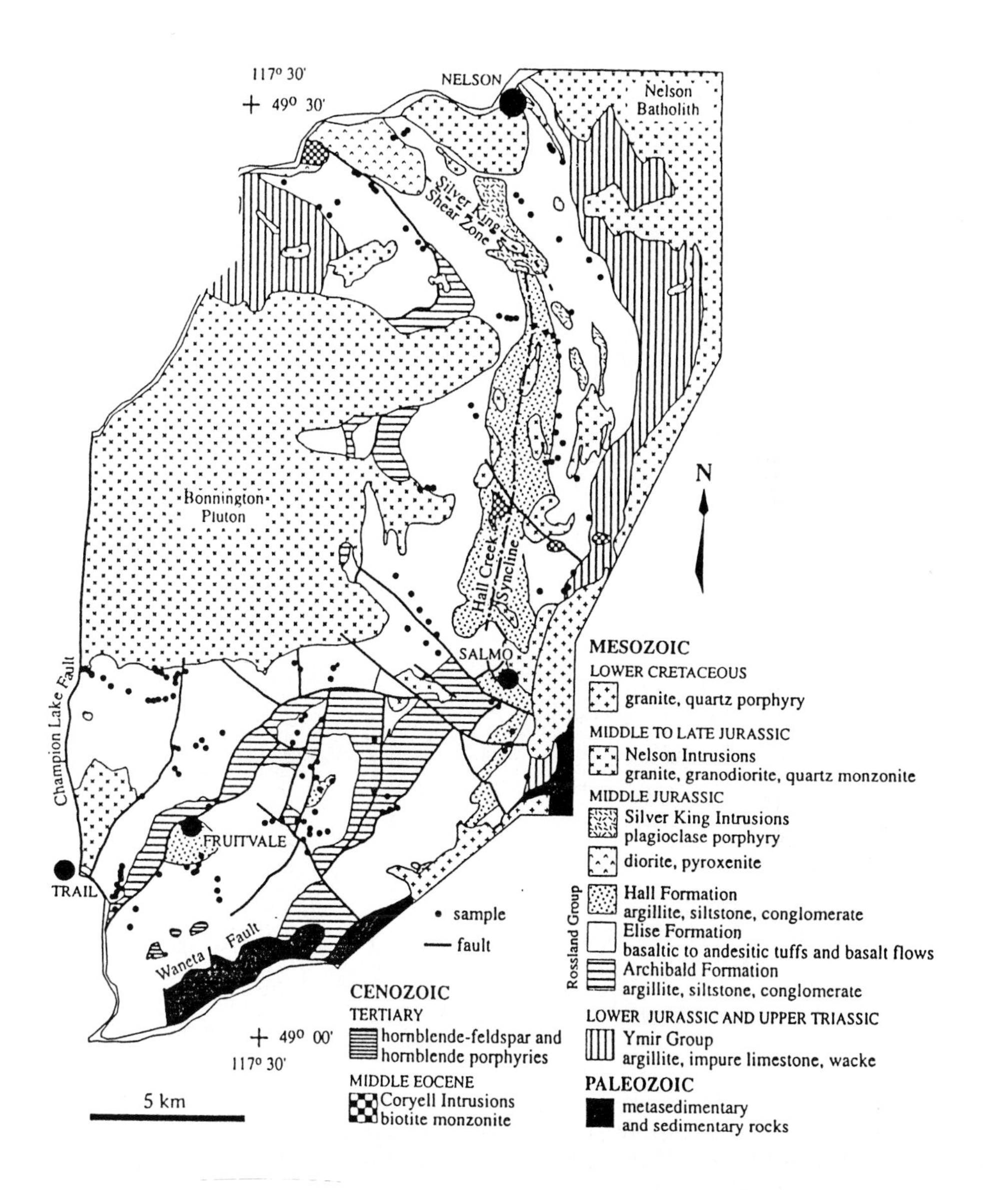

Figure 20. Structures between the Nelson and Bonnington plutons and the Waneta fault, as identified by geologic mapping (modified from Andrew et al., 1991; Powell and Ghent, 1996). Black dots mark rock samples analyzed by Powell and Ghent (1991).

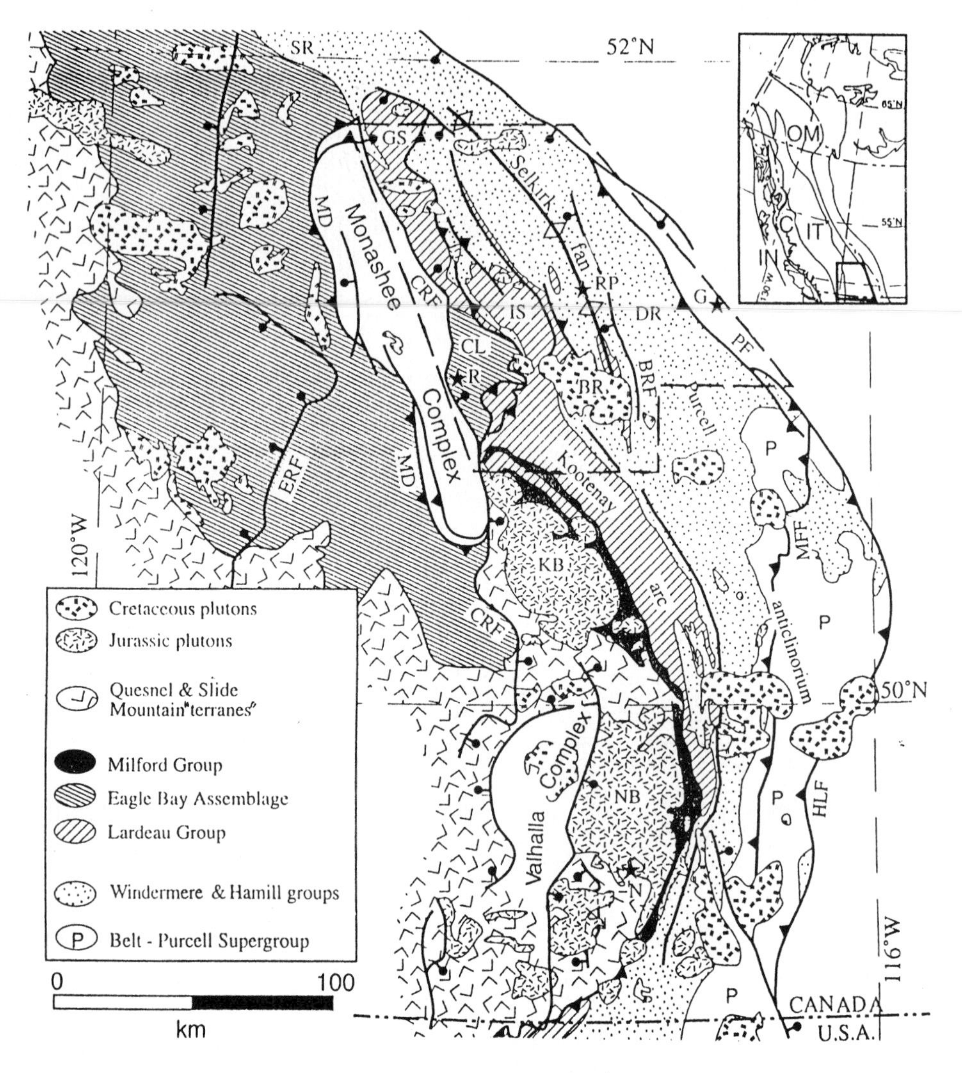

Figure 21. Principal rock assemblages in the southern part of the miogeosynclinal Omineca Belt, as identified by geologic mapping (simplified from Colpron et al., 1998). Area in the polygonal dashed-line box is shown in more detail in Fig. 23. BR - Battle Range pluton; BRF - Beaver River fault; C - orogenic Coast Belt; CL - Clachnacudainn rock assemblage; CRF - Columbia River fault; DR - Dogtooth Range; ERF - Eagle River fault; G - Golden; GS - Goldstream fault-bounded rock unit (block); HLF - Hall Lake fault; I - Intermontane median-massif belt; IN - Insular Belt; IS - Illecillewaet synclinorium; KB - Kuskanax pluton; MD - Monashee fault; MFF - Mount Foster fault; N - Nelson; NB - Nelson pluton; OM - Omineca miogeosynclinal orogenic belt; OF - Purcell Trench fault; R - Revelstoke; RP - Rogers Pass; SP - Scrip Range.

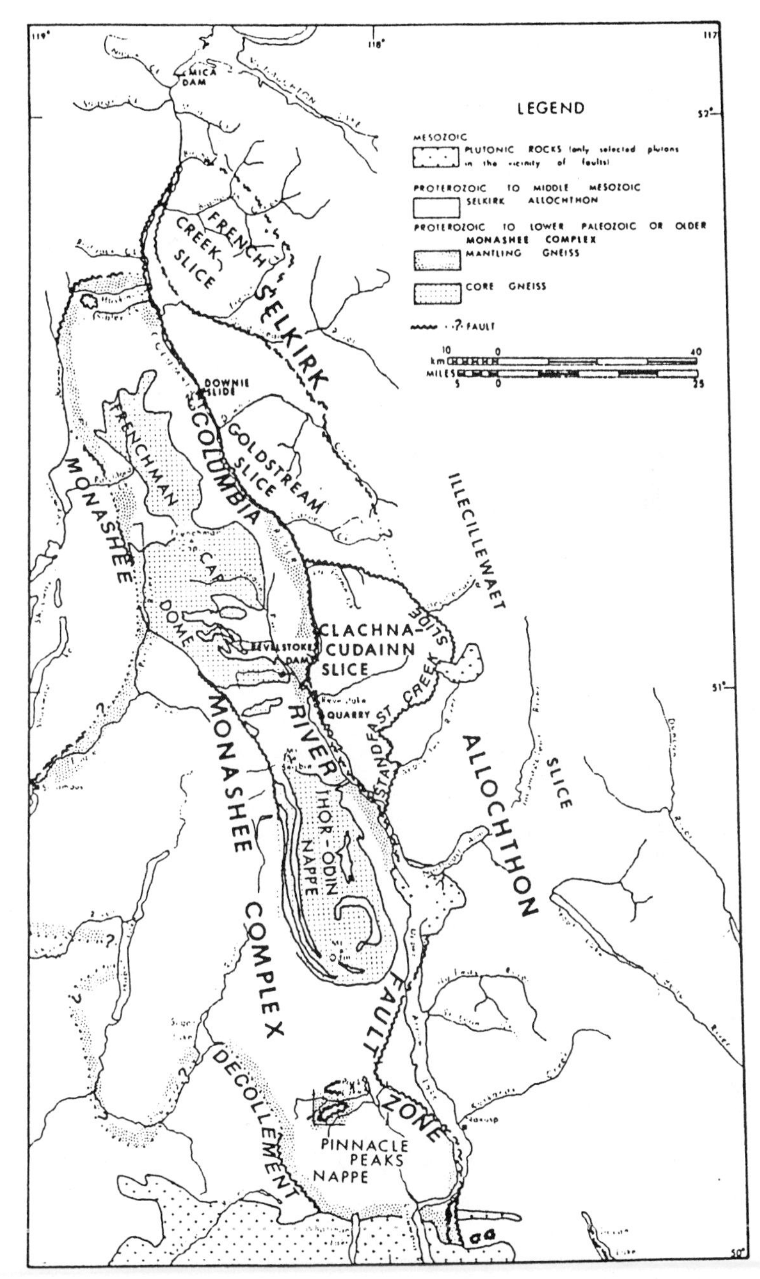

Figure 22. Major fault zones and blocks in the southern part of the Canadian Cordilleran miogeosyncline, as identified by geologic mapping (simplified from Read and Brown, 1981).

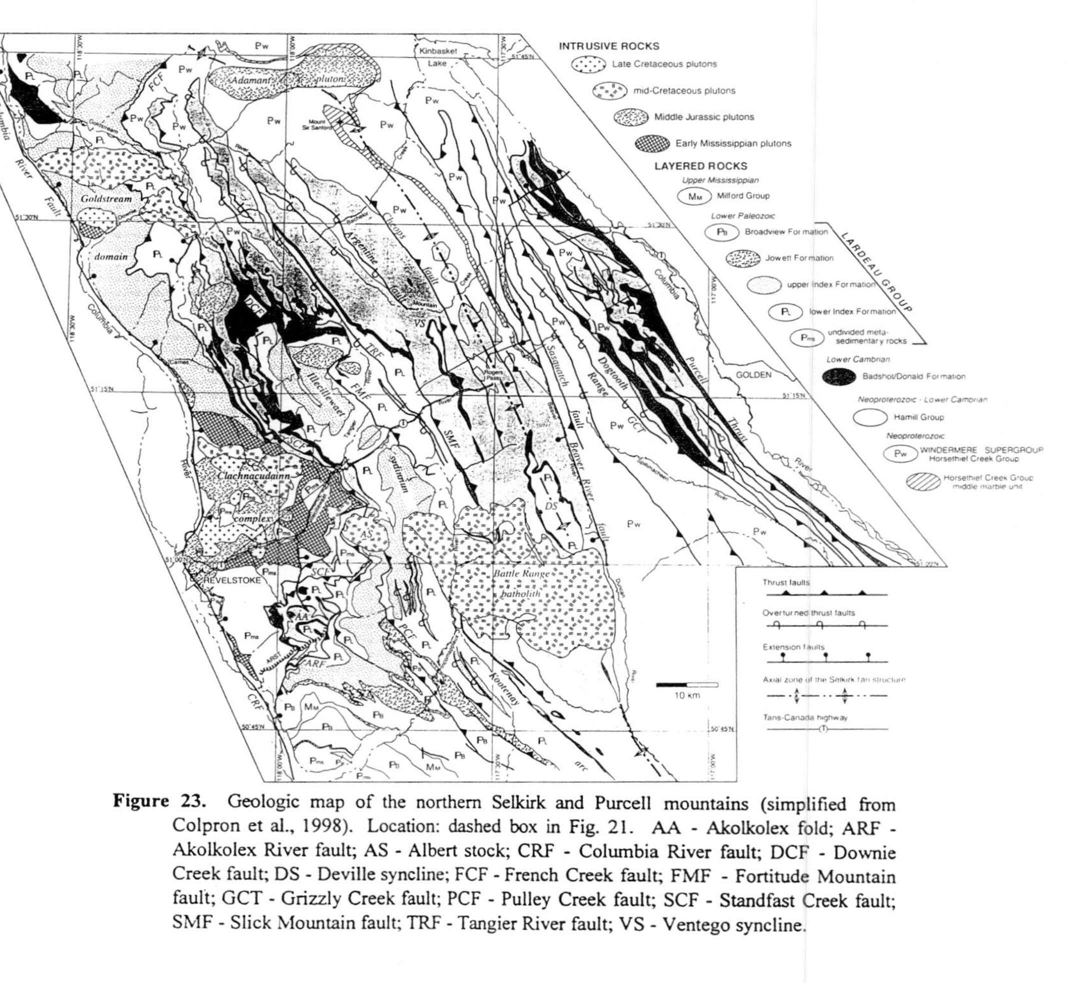

Figure 23. Geologic map of the northern Selkirk and Purcell mountains (simplified from Colpron et al., 1998). Location: dashed box in Fig. 21. AA - Akolkolex fold; ARF - Akolkolex River fault; AS - Albert stock; CRF - Columbia River fault; DCF - Downie Creek fault; DS - Deville syncline; FCF - French Creek fault; FMF - Fortitude Mountain fault; GCT - Grizzly Creek fault; PCF - Pulley Creek fault; SCF - Standfast Creek fault; SMF - Slick Mountain fault; TRF - Tangier River fault; VS - Ventego syncline.

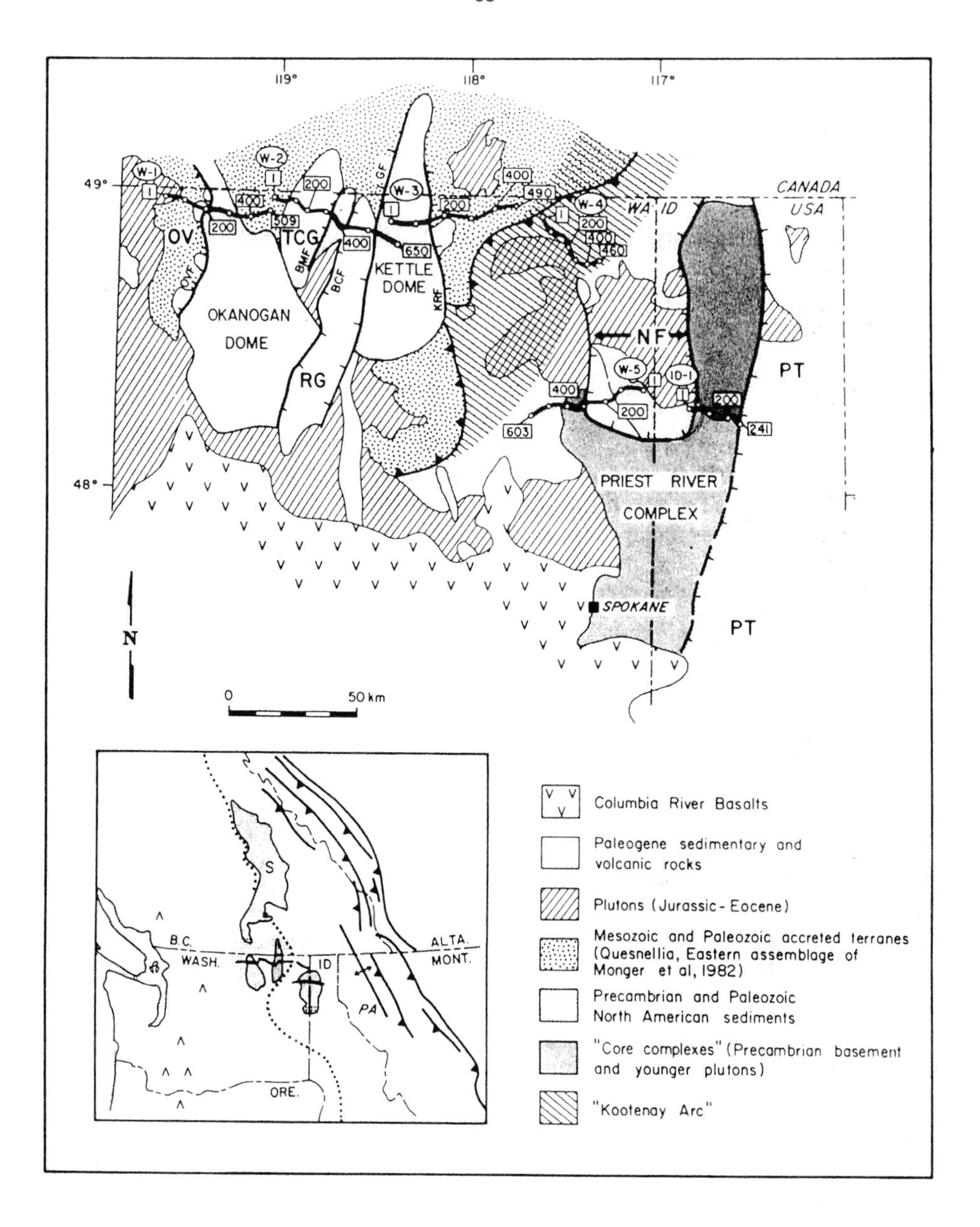

Figure 24. Major faults and metamorphic core complexes in the eastern Cordilleran Omineca miogeosyncline along and just south of the Canada-U.S. border, with conventionally assumed terrane designations and locations of COCORP seismic reflection profiles W-1 to W-5 (simplified from Monger et al., 1994). BCF - Bacon Creek fault; BMF - Bodie Mountain fault; GF - Granby fault; KRF - Kettle fault; NF - Newport fault; OV - Okanagan Valley; OVF - Okanagan Valley fault; PA - Purcell Anticlinorium; PT - Purcell Trench; RG - Republic graben; TCG - Toroda Creek graben. The COCORP seismic reflection data are presented in Fig. 39.

In the early 1980s, the entire Hinterland (Intermontane) Belt was reassigned to the Pacific Orogen (cp. Wheeler, 1970; Monger et al., 1982), based on assumed oceanic-crust affinity of rocks west of the Omineca Belt and the then-growing ideas about far-traveled small blocks ("terranes") delivered to the Cordillera on the back of oceanic lithosphere converging with North America at the continental margin. Western and central Cordillera was supposedly constructed by docking to North America of arriving terranes to form a Cordilleran "Collage" (Davis et al., 1978). Driven by numerical models of plate motions in the northeastern Pacific Ocean, which were based on assigning ages to oceanic magnetic-anomaly lineations (e.g., Atwater, 1970), hypotheses arose about terranes wandering all over the Western Hemisphere and the entire globe, arriving to the Cordillera from South America or Southeast Asia (cp. Hamilton, 1980; Irving and Monger, 1987). These ideas have misguided geologists to a path of loose and groundless speculations.

More usefully, this incorrect methodology inspired a new wave of targeted field mapping, aimed to seek evidence of terrane boundaries and accretion. These boundaries were expected to be faults across which compositional and structural characteristics of rocks change drastically. That search has revealed many new facts incompatible with the pre-conceived terrane hypotheses that inspired it, resulting in better overall knowledge of the geology of the Cordillera.

Vertical tectonic units within crustal blocks

Each crust-mantle block has its own typical rock composition and rock-body relationships. Composition and type of rocks depend on the rocks' genesis and subsequent transformations. Structure of rock bodies changed in the course of a region's evolution, with new restructuring episodes masking or obliterating older fabrics. Reconstruction of these events is the essence of regional tectonic analysis. These reconstructions can be realistic only if the analysis is based on thorough examination of rocks and rock bodies.

Major blocks making up a tectonic region are laterally restricted tectonic units encompassing the entire crust and even lithosphere. In a vertical dimension, they are layered, as rock properties change with depth. Conclusions from gravity data that somewhere in the Earth there must be a level of isostatic compensation, for example, led to the mechanical notion of a strong lithosphere floating on a weaker asthenosphere (Barrell, 1914). Thermal demagnetization of rocks hotter than the Curie temperature is evident from simple laboratory experiments. Other considerations and experiments point to rheological breaks, finite effective thickness, and so on. These vertical variations in the lithosphere arise due to dissimilar physical and chemical rock properties in different layers.

Depositional stratification of rocks results from a complex interplay of physical and chemical processes in various sedimentary settings. Usually, a stratum is a tablet-like body, with its width greatly exceeding its thickness. By convention, not all mapped stratal rock bodies are defined as separate units in their own right. Laminae, for example, are usually considered to be not separate beds but characteristic structural or textural components of bigger rock bodies: small-scale cross-bedding and lamination are usually treated as internal to the layer. Suites, or formations, of bedded rocks - respectively, in the Russian/Soviet (Zhamoida, ed., 1979) and North American (North American Commission on Stratigraphic Nomenclature, 1983) stratigraphic codes - are essentially equivalent notions. In the geological practice in English-speaking countries, *formation* is considered to be the fundamental stratigraphic unit. Sometimes formations are combined into groups and supergroups. The most important characteristic of formations is their mappability, based on their lithologic content distinct from that of underlying and overlying formations. This lithologic criterion gives the definition of formations some independence from the rock ages, though each formation spans a certain time interval.

On a more regional scale, delineation of a formation may be complicated by intraformational lateral facies variations, or by secondary metamorphic overprints of original rock-body characteristics. The grade of contact and burial metamorphism may also vary laterally. Sometimes high-grade metamorphic rock bodies are homogenized into coherent new entities, but in lower-grade rocks the original stratification is often still discernible. Correlations of high-grade and low-grade metamorphosed bodies can be problematic. In metamorphic-rock terrains, the nature of protolith and the original character of formations are sometimes unclear, and pre-metamorphic boundaries of formations and blocks may be obscured. In western Canada, Collerson et al. (1990) have noted that regional metamorphism obliterated many original rock characteristics of the North American cratonic basement even at the relatively high (and young) basement levels. The widely exposed relatively young level is related to the Hudsonian thermo-metamorphic event at ~1,900 Ma. Younger, ~1,760 Ma, non-orogenic granitic bodies comagmatic with rhyolite volcanics, together with sedimentary quartzite, overlie Archean gneisses (e.g., in the Swift Current area in southern Saskatchewan). They form coherent platforms with the Early Proterozoic cover.

The most objective and exact time frame for age correlation of rock units of different nature (stratigraphic and intrusive, sedimentary and metamorphic) is provided by radiometric dating. The most common unit of time in geology is million years. The Geologic Time Scale for the Phanerozoic sets a calendar where the boundaries of periods and their subordinate time units are defined to within 0.1 Ma (Harland et al., 1990).

Stacked tectonic rock units are primarily lithostratigraphic, normally deposited one by one in an upward-younging progression. Each geologic (rock-made) body is tectonic, in that it bears some information about the tectonic regime during and after deposition. But tectonic rock units are those geologic bodies whose boundaries correspond to regional restructuring episodes and changes in tectonic regime.

At the highest level of hierarchy of layered tectonic bodies is the whole lithosphere, which is made up of rocks as we know them from observations at the ground surface. In this sense, the asthenosphere, where the material is in different phases (Anderson, 1995), is not made of rocks and is not a subject of geology and tectonics. If the lithosphere-asthenosphere boundary is a phase transition, its position depends on the chemical composition of material, temperature, pressure, water content and other intraplanetary conditions. Below the lithosphere, these factors are considered by mantle geodynamics and other disciplines, which fall outside geology. As the physical and chemical conditions change, the transition zone may shift up or down, and traces of its former positions may remain in the lower lithosphere. The same is the case with the base of the crust, called the Moho boundary. Discontinuous remnants of ancient Mohos may appear in places in deep seismic images, and disruptions of the modern Moho by faults is recognized commonly in seismic profiles (e.g., Potter et al., 1986).

Shifting subhorizontal brittle-ductile transitions within the crust are now also reported often. Remnants of such transitions are found in the Precambrian craton of North America (Lewry and Stauffer, eds., 1990), as well as in Paleozoic-Mesozoic successions of the Cordilleran mobile megabelt (Simony, 1991; Tempelman-Kluit et al., 1991). An extreme ultra-mobilistic opinion in the 1960s and 1970s briefly held that there is no such thing as vertical tectonics and that all crustal and lithospheric tectonism is driven laterally, but in the face of overwhelming rock evidence this view is now deservedly out of fashion (see review by Friedman, 1998). Vertical displacements of large crustal blocks can now be reconstructed with some precision from paleothermometric and paleobarometric studies (e.g., Parrish, 1995; Colpron et al., 1996, 1998).

Within individual coherent crustal blocks, old brittle-ductile transitions can be raised, lowered, tilted or curved in response to tectonic causes. These transitions may be recorded in seismic profiles as intracrustal reflections. Unfortunately, reflections cutting across the dominant seismic-reflection fabrics are often interpreted as igneous sills, without sufficient consideration of other options. Large, thick, extensive mafic sills have indeed been mapped in many areas in the Belt-Purcell Basin and elsewhere (e.g., Litak and Hauser, 1992). But not all tabular reflection bands are due to sills, and even extensive data analysis often fails to distinguish them from similar-looking reflections due to low-angle faults, zones of shearing, current or fossil metamorphic fronts and zones of brittle-ductile transition, and so on. Only geological information, where available, permits realistic seismic modeling and calibration and is able to provide a conclusive answer.

Continuing development of the theory of mobile megabelts

Before the 1960s, a standard template in tectonics was derived from the geosynclinal theory dating back to the work in the U.S. of Hall (1859) and Dana (1873). In some parts of the world, variants of this theory were regarded as almost absolutely true. But as it spread worldwide from its birthplace in North America, this theory underwent many mutations. For Dana, and especially Haug (1900), a *geosyncline* was a large volcano-sedimentary furrow formed over a pre-existing basement, and the geosynclinal orogenic grain was restricted mostly to that basin (also Aubouin, 1965). The essence of this theory was used broadly by Kay (1951), who regarded all volcano-sedimentary basins as geosynclines of different kinds. This was an oversimplification inadequate for practical tectonics, but it was not so completely false as to merit its current near-total oblivion. Deep-seated processes of crustal reworking do create depressions, including those along crustal weakness zones used by rifts and orogenic furrows, which are filled with volcanic and sedimentary deposits. Such processes are responsible for the formation of many deep basins, potentially serving as a blanket beneath which the temperature increases and metamorphism takes place. It is now well recognized that the orogenic grain in a mature orogenic zone is not merely surficial but involves the whole crust or even lithosphere.

In Europe and Russia, the geosynclinal theory was enhanced by the addition of a complementary theory of cratons, and a single whole-crust geosyncline-craton theory arose (mostly due to Suess, Staub, Stille and Shatsky). In this new application, the term *geosyncline* became a misnomer, but was retained to denote whole-crust or whole-lithosphere entities in mobile megabelts, affected by deep-seated orogenic (but not necessarily mountain-building) processes. The great potential of the modernized geosyncline theory was not fully appreciated in North America, where a negative reaction to Kay's overinterpretation eventually shaped the scientific opinion. Since the 1960s, the attention has been focused on many unexpected facts that came to light, largely through U.S. efforts, from the ocean floor. The geosyncline theory, even in its improved mid-20th-century form, was unable to explain these new facts, and so it was discarded as useless.

The new theory, based on geophysically modeled plates and their motions, soon became a template also widely accepted as uniquely true, absolute and eternal. In the 1960s and 1970s, deductive reasoning was conferred upon it, making this theory a framework for interpreting all geological, geochemical and geophysical facts. In the worst cases, facts have been filtered to select only those which seem to fit the presumably-uncontestable theory.

But the classical plate tectonic theory of the 1960s is also becoming outdated. New geophysical, geochemical and geological information suggests continents have deep roots, >400 km under cratons. Stability of such roots is not accounted for in the modern models of plate motion. Neither is it considered in the tectonic reconstructions that assume an assembly of modern continents from microcontinents and suspect terranes that arrived from unknown parts of the globe (Davis et al., 1978; Bickford, 1988; Hoffman, 1988).

Some modern geologists give the applications of deductive logical thinking a fancier-sounding name “deductive-hypothetical”, as opposed to presumably non-hypothetical, empirical thinking derived from observations. But actually, hypothesizing is involved each time we try to grasp the essence of observed phenomena and especially their causal connections. In geology, in order to reach the likeliest conceptual conclusion, we must consciously give priority to rock-based information about the region under investigation. To make our conclusions practical and to minimize speculation, we must verify and justify them by facts. We need to restrict our logical thinking to the least-speculative methods of analysis and synthesis, combined inextricably.

Each researcher and practitioner works in some temporary intellectual environment of prevailing tectonic ideas about the Earth and its regions. Anyway, each age bears an impact of its dominant philosophical, methodological and specific preferred concepts. This guidance helps us focus our efforts and select the useful information from the ever-growing tide of new data and facts. The danger is that these guiding concepts may begin to drive our thinking to much, causing us to confuse the passing with the absolute. This matters relatively little to a geologist mapping the outcrop, because, whatever the prevailing dogma, the field geologist stands face-to-face with independent facts. But for a tectonist who puts these data together to obtain a conceptual genetic interpretation, a reasonable measure of detachment in his/her thinking is crucial.

5 - VIEW OF THE CORDILLERAN MOBILE MEGABELT'S EVOLUTION FROM THE CRATON

Choice of perspective for undistorted visualization of mobile-megabelt evolution

As in visual art, perspective matters in geological reasoning. The angle from which we describe natural geologic phenomena affects how we perceive them. An incorrectly chosen angle distorts the phenomena's significance, proportions and details. It is usual, for example, to regard the North American continent as a passive element of a lithospheric plate (e.g., Burchfiel, 1980) or as a succession of orogens from the Late Archean onward (Hoffman, 1988). Both these views have some partial validity, but both are distorting. Reduction of continental tectonics to little more than interplay of plates robs tectonics of its very essence, which is to deal with self-developing systems. A continent's evolution and tectonic history are more realistically understood in terms of specifically continental, cratonic and mobile-megabelt, tectonic regimes. Cratonic regimes are succeeded by mobile-megabelt regimes in regions where new megabelts develop, and coexist with these regimes side by side where cratons persist (Lyatsky et al., 1999).

Since the mid-19th century, geologic mapping in the eastern U.S. has revealed that the stratigraphic horizons of consolidated rocks that underlie the Quaternary till in the Mid-continent Plains continue into the mountain regions of the Appalachians, where thickness of these rocks increases greatly. This led to an inference that the tectonic boundary between the Plains and the Appalachian Mountains does not coincide with the boundary apparent in topography. The idea of post-Precambrian, continental-scale volcano-sedimentary furrows or geosynclines that precede mountain-building took hold in North America (Hall, 1859; Dana, 1873) and then elsewhere. It is in these geosynclines that the greatest thickness of sediments and volcanics accumulated, later to be variously plutonized, metamorphosed and deformed (for reviews, see, e.g., Rast, 1969; King, 1977; Friedman et al., 1992). Since the early 20th century, European and Russian workers (Suess, Stille, Karpinski, von Bubnoff, Shatsky) have developed a concept of continental-scale, stable Precambrian cratons underlying vast plains such as those in western Russia, distinct from the mountain-bearing deformed geosynclines (e.g., Spizharsky, 1973; Spizharsky and Stelmak, 1977).

In tectonics as a separate branch of science, flat plains were understood to be physiographic expressions of tectonic platforms, and mountainous areas to be expressions of geosynclinal-orogenic regions, though their topographic and tectonic boundaries do not coincide fully. Shield areas, where Precambrian crystalline rocks are exposed, were recognized to be in direct continuity with the platform basement, being made up of similar crystalline rocks. Shields and platforms together were combined into *cratons*, as opposed to mobile-megabelt provinces. This term was reserved only for shields and platforms with old Precambrian basement, whereas tectonic platforms whose basement is younger (post-Precambrian) were classified separately (Stille, 1941). But subsequent recognition that many modern mobile megabelts had been initiated in the Precambrian, coupled with the knowledge that the basement of cratonic platforms is Precambrian as well, created confusion in distinguishing within continents of cratonic and vs. non-cratonic provinces, orogenic vs. non-orogenic entities, constituent parts of cratons, and so on. With the advent of plate tectonics, the usage of words like *craton, orogen, platform* became even looser than before.

In the North American and other continents, the original cratonic masses were broader than today, but they shrank as their pieces were reworked in orogenized mobile-megabelt zones that developed later, mostly on ex-cratonic continental crust. In North America, cratonized crust of Archean and Early Proterozoic rocks appeared long before the Phanerozoic sedimentary cover. That old, Laurentian craton was a single tectonic entity, standing apart from the Mesoproterozoic Grenville-Appalachian and incipient Cordilleran mobile megabelts. In post-Hudsonian time, remobilization in the western part of the Laurentian craton involved the development of a new, superimposed, typically "Cordilleran" NNW-SSE tectonic trend. In the crystalline basement in Alberta, it is recorded as the Kimiwan Zone of isotopic alteration and extensive restructuring at ~1,750 Ma (Muehlenbachs et al., 1993; Burwash et al., 1995). In the later Proterozoic, the locus of Cordilleran tectonism shifted to the west, and at ~780 Ma it was manifested in the Windermere rift precursor of the Cordillera (Douglas et al., 1970; Wheeler and Gabrielse, coords., 1972; Gabrielse and Yorath, eds., 1991).

New tectonic continental features reworked the ancient crust almost completely in orogens, incorporated semi-reworked crustal pieces in median massifs, and left the remainder of the craton generally undisturbed. Whole-crust boundaries between these new lateral tectonic units were commonly deep, high-angle fault systems. The Grenvillian-Appalachian mobile megabelt was polycyclic, with several distinct orogenies. While the main orogenic megacycles were two, Grenvillian and Appalachian proper, three regional orogenies - Taconic, Acadian, Alleghenian - affected the Appalachian region in pre-Jurassic Phanerozoic time. Grenvillian blocks as well as remnant pieces of pre-existing crust are recognized intermittently in the Appalachians (Rodgers, 1987, 1995; Rankin et al., 1993).

The Cordilleran mobile megabelt, in contrast, has evolved in just one orogenic megacycle which is still continuing at present. Several orogenies took place in it between the Late Proterozoic and the Tertiary. Remnant blocks of the previous craton are preserved in the Cordillera even better than in the Grenville-Appalachian megabelt. Geochemically, cratonic crystalline basement remnants are detected over most of the Cordillera (Reed et al., eds., 1993). Petrologically, Laurentian-craton rocks are preserved in a series of so-called metamorphic-core complexes from Arizona to British Columbia (particularly in the Omineca Belt; Gabrielse and Yorath, eds., 1991; Stott and Aitken, eds., 1993).

Typical continental-crust crystalline rocks of Early to Middle Proterozoic ages are known in many parts of the Cordillera (Reed et al., eds., 1993; Parrish, 1995). No such rocks are yet recognized from direct observations in the Intermontane median-massif belt in Canada, but next to it to the north, in Alaska, the Yukon-Tanana crystalline massif contains Upper Proterozoic to Lower Paleozoic stratified rocks intruded by Devonian granites with zircons dated radiometrically as Proterozoic (Aleinikoff et al., 1986, 1987; Plafker and Berg, eds., 1994). In the Intermontane Belt, the exposed volcano-sedimentary rocks are usually metamorphosed to low grades (prehnite-pumpellyite, mostly) and folded only gently. This points to the presence of an underlying rigid basement. The oldest known rocks in the volcano-sedimentary cover of the Intermontane median-massif belt are Middle Devonian. A Precambrian age of the basement is indicated by the studies of xenoliths and isotopic signatures of igneous rocks, and Precambrian rocks are found in the metamorphic core complexes in the Omineca Belt nearby (Armstrong and Ghosh, 1990; Armstrong et al., 1991; Parrish, 1995).

In the Omineca Belt near the Canada-U.S. border at 49°N, between the Southern Rocky Mountain and Purcell trenches on the east and the Okanagan Lake and Eagle River faults on the west, high-grade metamorphic core complexes have been mapped in a broad area as much as 250 km wide. They are clustered into three elongated, N-S-oriented bands along longitude 118°W (the Kettle core), west of 119°W (the Okanagan core), and between them (the Priest River and Valhalla cores). North of latitude 50°N, these bands merge into a single zone only 75 km or less in width. In the northern part of this single zone, where it narrows to just 20-35 km, outcrops of crystalline basement rocks are mapped discontinuously for a distance of some 300 km, from the Thor-Odin core to the Malton core.

The N-S trend of metamorphic complexes and exposures of pre-megabelt basement rocks is oblique to the general NNW-SSE structural grain of the Cordillera. These N-S trends were superimposed on the Cordilleran structural pattern only in the Tertiary.

The NNW-oriented Southern Rocky Mountain Trench fault, also of Tertiary age, cuts these N-S zones of deep metamorphosed rocks along the boundary of the miogeosynclinal Omineca Belt with the rootless Rocky Mountain fold-and-thrust belt to the east. Tertiary whole-crust extension and uplift of crystalline-basement rocks from 20-30-km depths (Spear and Parrish, 1996) along N-S faults may mark the latest pulse of incipient rifting in the eastern Cordilleran miogeosyncline (the previous two or three rifting episodes there have also been noted; Constenius, 1996; Cecile et al., 1997). The Southern Rocky Mountain Trench fault helped accommodate the large Tertiary tectonic extension, but the elevated high-grade rocks are found on both the eastern and western sides of the Omineca Belt and in places on both sides of the Rocky Mountain Trench (McDonough and Simony, 1988). This indicates that tectonic movements in that region attributed to the Laramian orogeny had a variety of manifestations. Vertical crustal movements on steep faults in the Early Tertiary in places exceeded 30 km (cp. Parrish, 1995; Spear and Parrish, 1996).

Visualizing the tectonic evolution of the Cordilleran mobile megabelt from the vantage points of its previous cratonic background and of the modern craton nearby helps to better understand the megabelt's development.

Demarcation of the boundary between craton and mobile megabelt

For some practical purposes, it is conventional to demarcate modern cratons simply by the obvious outer edges of deformation in fold-and-thrust belts that rim the mobile megabelts. However, because the fold-and-thrust deformation is superficial and rootless, cratonic structural-formational étages could continue into the present-day fold-and-thrust belt and even beyond (as is well known to be the case in the Rocky Mountain Belt). To put the edge of a craton at the deformation edge of marginal fold-and-thrust belts limits the extent of the craton artificially, ignoring the geologic history and deep crustal structures of craton-mobile megabelt transitions. In the age of deep geophysical imaging, this approach is obviously obsolete: exploration for hydrocarbons and minerals requires knowing where true, deep-rooted orogens end and cratonic platforms begin, and the old superficial approach may misguide these exploration efforts.

The craton-megabelt boundary may lie at the back of the fold-and-thrust belt, but there it is hard to locate due to strong thrusting folding (Sloss, ed., 1988; Stott and Aitken, eds., 1993; Lyatsky et al., 1999). Along the western edge of the modern North American craton, there are big contradictions and misconceptions about the tectonic position of the Interior Platform's western boundary in the Rocky Mountains (Fig. 25; see also King, 1977; Brewer et al., 1981; Ricketts, ed., 1989). The Canadian *DNAG* (*Decade of North American Geology*) volume D-1 (Stott and Aitken, eds., 1993, p. 3) contains rather arbitrary suggestions that "the western boundary of the Interior Platform is marked by the Sweetgrass Arch extending from Montana into Alberta", which lies hundreds of kilometers from the fold-and-thrust front, and that the Western Canada Sedimentary Province is "inextricably linked with the histories of the mountain chains that form the outer boundaries" of the Cordillera. Such definitions are too loose to be useful, if not outright misleading: the boundary between the plains and the mountains is not the same as the boundary between the craton and the mobile megabelt. The Sweetgrass arches (Northern, in Alberta, and Southern, in Montana) are not the edge of the modern or previous North American craton (Lyatsky et al., 1999). Other authors put

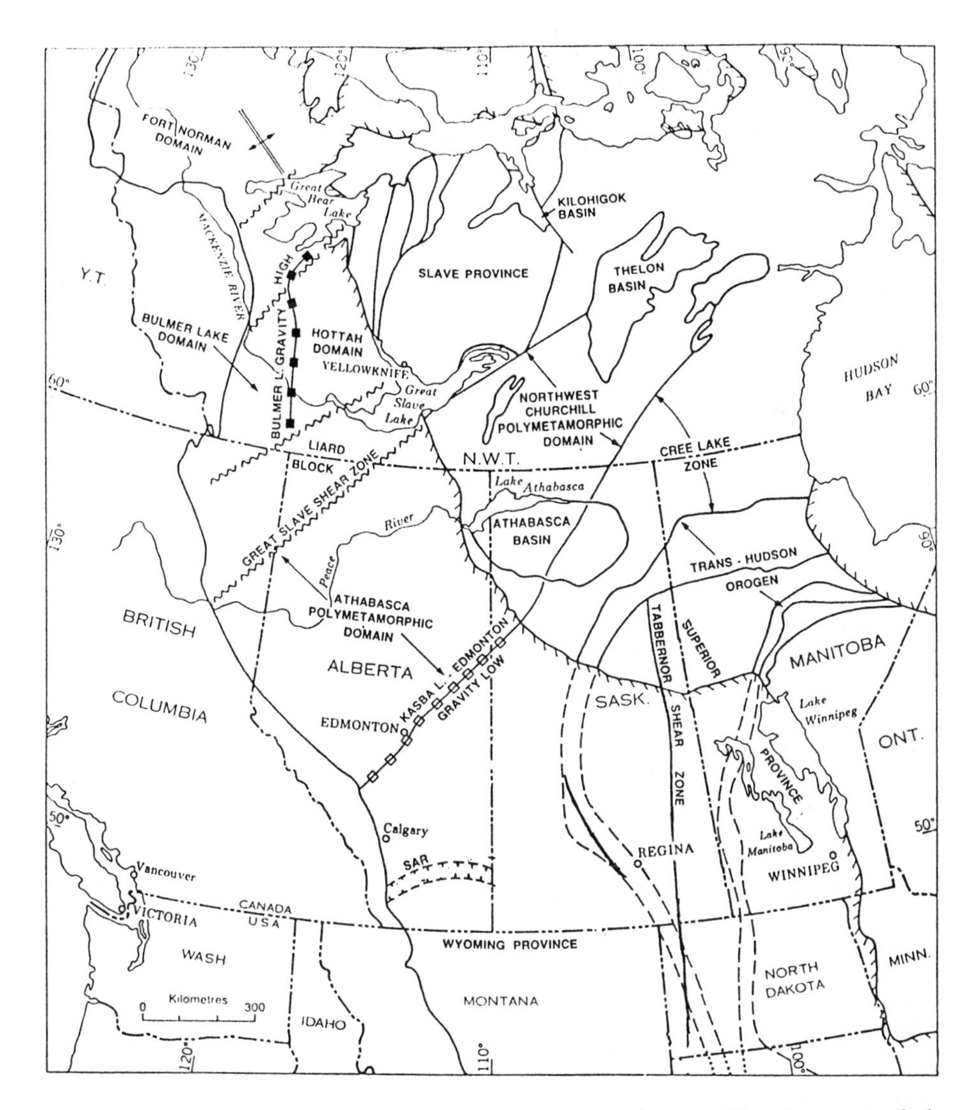

Figure 25. Tectonic provinces in the Precambrian cratonic crystalline basement of the platformal Western Canada Sedimentary Province (simplified from Burwash et al., 1993). SAR - Southern Alberta zone of gravity lows. Several major NE-SW-oriented ancient zones of crustal weakness continue from the Canadian Shield across the Phanerozoic Western Canada Sedimentary Province into the Cordilleran mobile megabelt.

the western boundary of the Western Canada Sedimentary Province at "the structural front of the Rocky mountains" (e.g., Podruski et al., 1988, p. 14), or at the Rocky Mountain Trench far to the west of the deformation front. Cecile et al. (1997) distinguished in the pericratonic region the Alberta Block containing the Bow and Kakwa platforms, the West Alberta Ridge, Peace River Arch, Purcell High, and White River and Robson troughs (Fig. 26). They stated (their p. 59) that "the Alberta Block east of the miogeocline includes the Sweetgrass Arch", but defined it loosely as "a large positive area that runs northeasterly from Montania, through Alberta to Saskatchewan between the Cordilleran Miogeocline and Williston Basin". Price (1994), on the other hand, followed the long-standing tradition of recognizing a single miogeosyncline-platform tectonic domain (cp. Wheeler and Gabrielse, coords., 1972).

Tectonic attribution of the miogeosyncline has also been unclear. In the 1970s, it was customary to contrast the Proterozoic-Paleozoic Omineca Belt with the "shallow-deformation" Rocky Mountain (Foreland) fold-and-thrust belt. The boundary between these belts was traced in the north "slightly east of the Rocky Mountain Trench, and then west of Rocky Mountain Trench along the east side of Cassiar Platform" (Wheeler and Gabrielse, coords., 1972, p. 33). Although the Rocky Mountain Trench is Tertiary, existence there of a deep-seated and long-lived structural break was later confirmed for the entire Paleozoic (Struik, 1987).

The fold-and-thrust belt, east of the Southern Rocky Mountain Trench and eponymous fault, consists of two parts: more-deformed Rocky Mountain ranges and less-deformed Foothills, with deformation decreasing towards the craton. In the Rocky Mountain fold-and-thrust belt, from seismic and drillhole data, only the sedimentary cover of the craton is affected by Late Cretaceous-Early Tertiary Laramian deformation, whereas the cratonic basement in the Foothills and Rocky Mountains is essentially undisturbed (Bally et al., 1966). Stratal rocks of the platformal cover were delaminated from the cratonic basement and translated eastward as thrust sheets. Some rocks of the miogeosynclinal hinterland were also involved in allochthonous displacements (Figs. 27-28).

Tectonic shortening in the fold-and-thrust belt has been estimated from restorations of folds and thrusts variously at 200 km (Price, 1981) to just ~70 km (Jerzykiewicz and Norris, 1992, 1994; McMechan and Thompson, 1993). Some eastward thrust displacements across the Southern Rocky Mountain Trench fault, with shortening of a few tens of kilometers, involved basement-rock slivers from the miogeosynclinal Omineca Belt (McDonough and Simony, 1988, 1989).

Low-angle faults and detached folds in the Rocky Mountain fold-and-thrust belt, lie in the hanging wall of a regional detachment generally recognized as the master thrust-fault zone dipping gently to the west. The hinge line probably runs near the axis of the Pur-

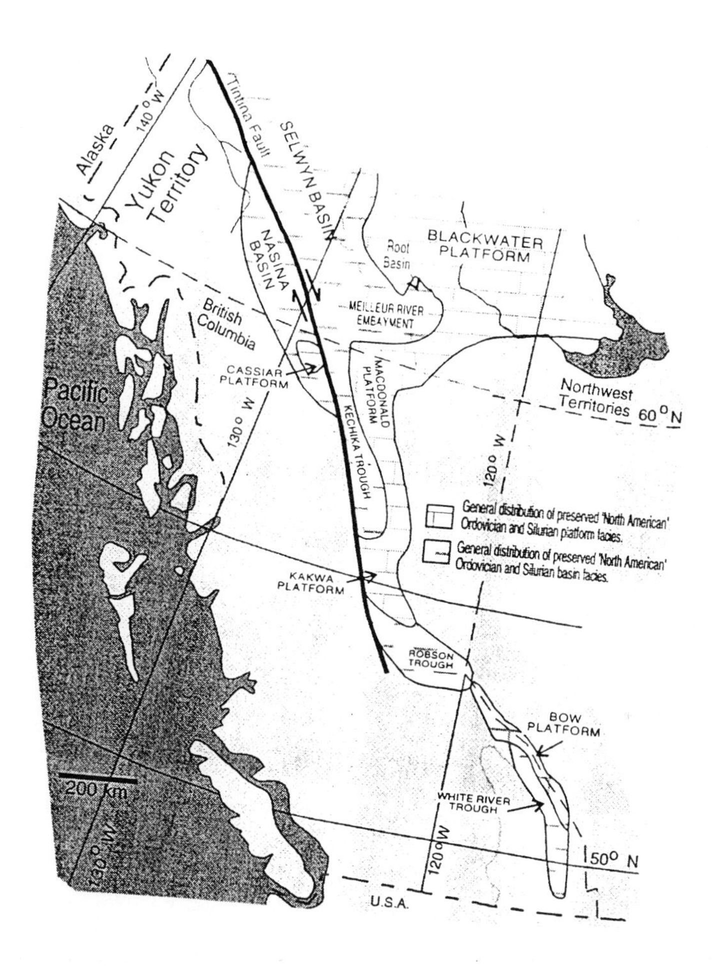

Figure 26. Map of Early Paleozoic carbonate platforms and deeper-water basins in the eastern Cordilleran miogeosyncline and adjacent regions (modified from Cecile et al., 1997).

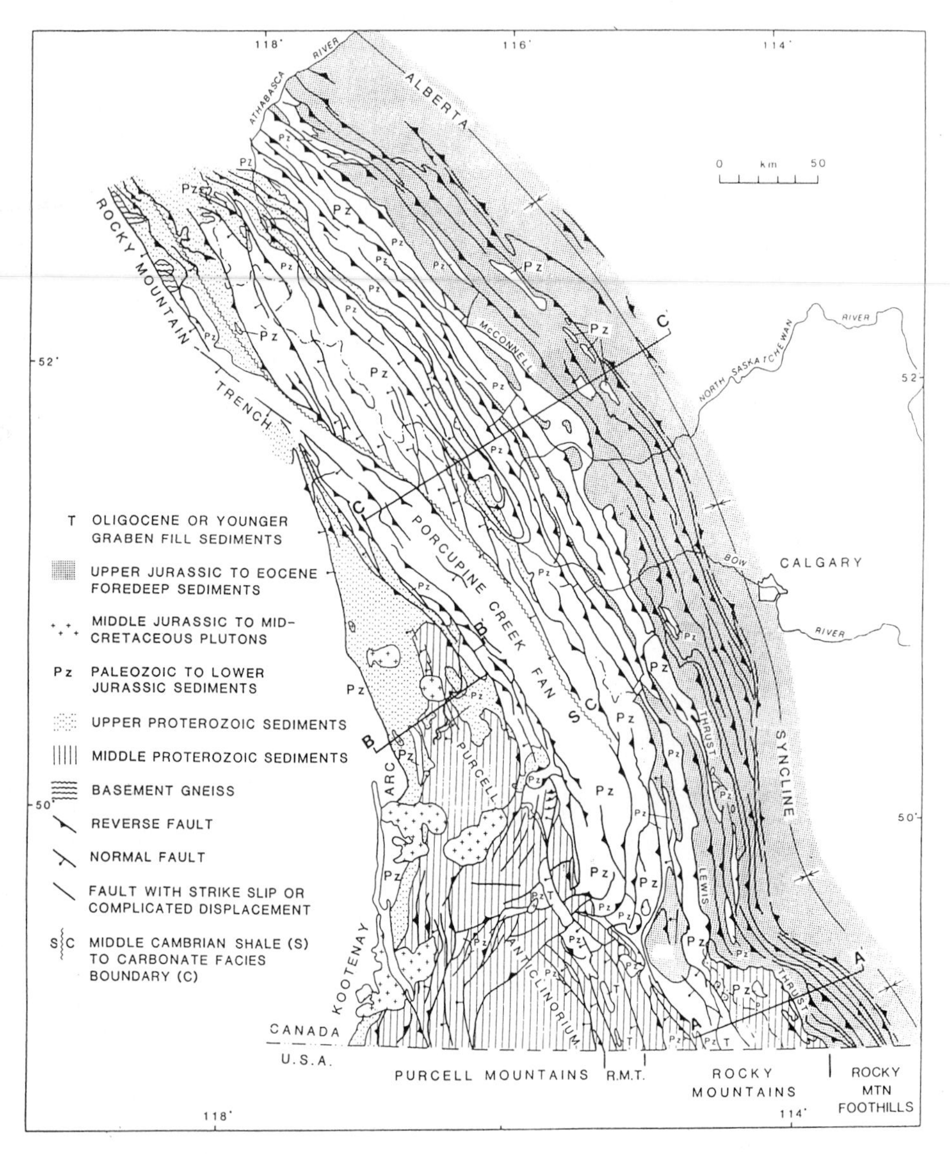

Figure 27. Generalized structural map of the southern Laramian Rocky Mountain fold-and-thrust belt and Purcell Anticlinorium in the southern Canadian Cordillera, Alberta and British Columbia (modified from McMechan and Thompson, 1989). Cross-section C-C' is shown in Fig. 28. RMT - Rocky Mountain Trench.

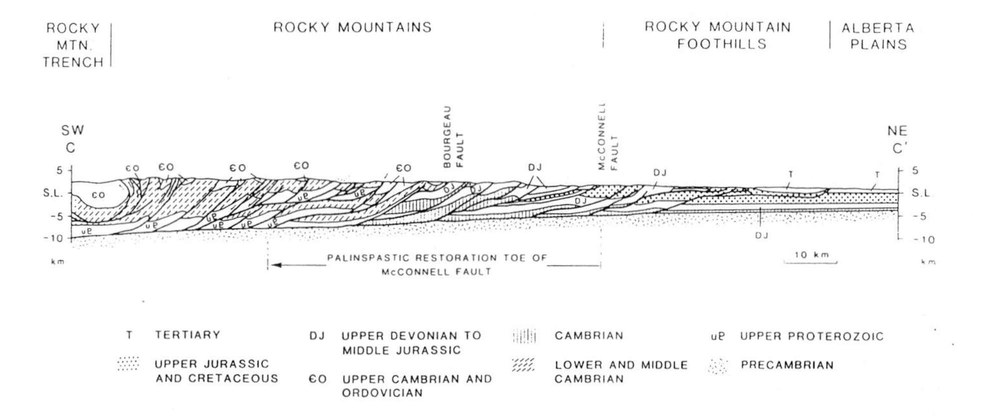

Figure 28. Structural cross-section C-C' of the Rocky Mountain fold-and-thrust belt (modified from McMechan and Thompson, 1989). Location of the cross-section is given in Fig. 27.

cell Anticlinorium. Farther west in the Cordilleran mobile megabelt, crystalline basement rocks with Early Proterozoic radiometric dates are broken into distinct pieces, reworked and partly exposed in granite-metamorphic core complexes.

The Laramian fold-and-thrust belt developed as a stack of rock packages (thrust sheets) which disturb the normal stratigraphic successions (Lawton et al., 1994; Lebel et al., 1996; Fermor, 1999; McMechan, 1999). Each thrust slice was not displaced across the entire width of the belt, but their cumulative displacements were considerable. Miogeosynclinal and cratonic-cover rocks are sometimes hard to distinguish, because many of their stratigraphic units are similar (e.g., the Paleozoic carbonate-platform facies). Because tectonic regimes were occasionally shared across the entire miogeosyncline-platform domain, settings of sedimentation were at times indistinguishable as well. A strong influence on thrust geometry came from the distribution of more-competent vs. less-competent stratal rocks (e.g., carbonates vs. shales) in the succession subjected to thrust deformation, and these lithostratigraphic variations are conducive to the development of glide planes. Where thrust planes jumped upward from one bedding plane to another, they locally steepened and ramps developed. These and other space problems were accommodated by thrust splays and elongated anticlines. Deep high-angle faults are difficult to identify in seismic sections across fold-and-thrust belts, but their presence is often confirmed by field mapping.

The Rocky Mountain Belt is segmented along its length, as is well seen in its formational, structural and physiographic variations. Many segments are distinctly linked to large, continuous, NE-SW-oriented transcurrent faults that dissect the craton as well as the mobile megabelt. These faults were rejuvenated during various tectonic stages.

Segmentation of the Rocky Mountain Belt by major NE-SW-trending zones of crustal weakness was inherited from ancient times. These faults existed even in the Early Proterozoic predecessors of the North American craton (Lyatsky et al., 1999). They are traced into the hinterland of the Cordilleran megabelt, where they control the position of the transverse Skeena and Stikine arches, Bowser Basin and some other prominent features in the Intermontane median-massif belt. This indicates that a pre-modern North American craton extended far into what would become the Cordilleran interior. At present, though, the craton ends where whole-crust mobilization becomes evident, along the boundary of the rootless Rocky Mountain fold-and-thrust belt with the deep-rooted miogeosynclinal Omineca orogenic zone.

Prejudices in existing interpretations of geophysical data

The Canadian Lithoprobe program includes several transects across the Canadian Cordillera from the Interior Platform to the Pacific Ocean, which make it possible to examine the deep structure of the Cordilleran mobile megabelt in more detail than ever before. The Southern Canadian Cordillera Transect, in particular, has provided a large amount of diverse data. Some of the conventional interpretations were published in a special volume of the *Canadian Journal of Earth Sciences* (Cook, ed., 1995). Establishing a pre-defined interpretational framework, the transect's objectives were formulated as follows: to study "the way in which the [tectonic] terranes were attached to North America, as evidenced by the nature of the boundaries between them and North America, and the way in which they were detached from their subducted lithosphere and subsequently stretched during crustal extension constitute the basic objective of the Lithoprobe Southern Canadian Cordilleran Transect" (Cook, 1995a, p. 1483). However, the collected data permit alternatives to this picture. A broader mission was set out for the Lithoprobe program by its director: a multidisciplinary collaboration among geologists, geochemists and geophysicists from the academia, government and industry, aimed at "enhancement of our understanding of the tectonic development of our continent" (Clowes, 1996, p. 109).

Contrasting interpretations of deep seismic data in various parts of the Cordillera in Alaska, Yukon, British Columbia and the conterminous U.S. (cp., among others, Potter et al., 1986; Clowes et al., 1995; Hollister et al., 1996; Miller et al., 1997) underline the need for revision of modern tectonic templates. Viewed from the craton, tectonic history of the Cordilleran mobile megabelt can be elucidated differently than the way continental geology is often visualized from models relying on assumed plate interactions and ages of oceanic magnetic stripes (Atwater, 1970, 1989; Riddihough and Hyndman, 1976, 1991). Model-driven estimates require accretion and subduction of some 13,000 km of oceanic lithosphere at the western North American continental margin since the Jurassic (during the past 185 Ma; see also Monger, 1993). Strangely, all that supposed plate convergence produced only enough terrane accretion to account for a modest westward growth of North America by ~500 km or less. Therein, according to Cook (1995a), lies a "basic" and "outstanding" problem.

In the now-standard reasoning (Cook, 1995a-c), ~13,000-km of oceanic lithosphere was driven towards ancestral North America (by unknown forces) and subducted more or less continuously and consumed into the mantle. Some pieces of these plates were detached from the downgoing lithosphere and attached to the leading edge of the continental mass as tabular slices or wedges 10-15 km thick. The ~500-km-wide span of the Cordillera west of its miogeosyncline (which is assumed to be a Paleozoic-early Mesozoic continental margin of an ancestral North America) is presumed to be a mosaic of randomly accreted, far-traveled terranes of oceanic origin, thrusted over each other and over an underlying wedge of North American cratonic basement (see also Monger et al., 1982, 1994 for details of this scheme). But "the fact that less than 5% of the lithospheric material that converged with western North America over the past 200 Ma has found its way into the rocks of the visible portions of the Cordillera leads to the obvious conclusion that the remaining >95% was detached and subducted, or recycled, into the sub-lithospheric mantle" (Cook, 1995a, p. 1483).

Each new arriving terrane (or composite superterrane, amalgamated previously from constituent smaller terranes) was supposedly pushed over an old continental margin from west to east. The western edge of the North American continental lithosphere, abraded from below by the downgoing oceanic plates, was at the crustal levels continually affected by low-angle thrusting; a result was a stack of slices composing most of the crust in the Cordillera. These thrusts merge at the lower-crustal levels into a single detachment thought to be orogen-scale (Cook and Varsek, 1994; Varsek, 1996). In this scheme, the Cordilleran megabelt has no roots in the mantle.

The standard Lithoprobe tectonic conclusions relied heavily on these premises, overlooking many real geological facts. In fact, no geological or petrological evidence requires this scenario, and many observations contradict it (see, e.g., Ernst, 1988; Woodsworth, 1991; Woodsworth et al., 1991; Conrey et al., 1995). The current interpretations of Cordilleran evolution rest on a great many unconfirmed or demonstrably incorrect assumptions. The eastern Cordilleran miogeosyncline, for example, is supposed to be represented by a one-sided Paleozoic sedimentary basin at an old continental margin of an ancestral North America, analogous to modern continental margins elsewhere (e.g., Monger, 1993). This assumption is negated by mapping in the Omineca Belt which shows the Paleozoic basin to have been two-sided (Aitken, 1993a-c; Henderson et al., 1993; Richards et al., 1993).

Clowes (1996, p. 112) stated that "unroofed metamorphic core complexes preserve older [cratonic] structures". This is not entirely correct: the core complexes only contain some rocks and minerals with ages similar to those in the basement of the Interior Platform of the modern North American craton. This suggests a common basement in both these tectonic provinces, but whereas the Interior Platform's basement was not substantially reworked since Hudsonian time, the miogeosyncline's turbulent tectonic history included several strong reworking episodes which affected the basement rocks. The latest such reworking, as indicated by the studies of metamorphic core complexes, took place as late as in the Early Tertiary Eocene. Thermobarometry (e.g., Spear and Parrish, 1996) indicates that basement rocks of the miogeosynclinal Omineca Belt had originally been at supracrustal levels, were then lowered to crustal depths of ~30 km, and then raised back to the surface.

Overmobilistic speculations about a terrane "collage" of the Cordillera (Davis et al., 1978; Monger, 1993) have been translated into a presumed succession of tectonic units, regimes and deformation episodes (Monger et al., 1994, p. 359) as follows: "(1) A Late Archean/Early Proterozoic (2.8/1.8 Ga) crystalline basement... [exists under] much of the eastern Foreland [Rocky Mountain] Belt. (2) This basement was rifted during at least three periods of extension, in Middle and Late Proterozoic and early Paleozoic times... (3) In the eastern Cordillera, between Early Cambrian and Late Jurassic times, platform-shelf-slope-rise or miogeosynclinal, mainly carbonate-shale deposits accumulated along the passive western margin of ancestral North America. In the western Cordillera, there is a record of plate convergence between oceanic and arc terranes that had uncertain paleogeographic positions relative to the eastern passive margin. (4) The mid-Mesozoic to Recent active continental margin evolved in three stages: (4a) In Middle Jurassic (ca. 170 Ma) to Paleocene (ca. 60 Ma) time, extensive terranes (Insular, Intermontane Superterranes...) composed of intra-oceanic arc and ocean floor rocks of Paleozoic and younger ages were accreted to the ancient continental margin, a process accompanied and followed by stacking of thrust sheets, mainly eastward-verging in the eastern Cordillera and westward-verging in the western Cordillera... (4b) In Eocene (ca. 60 to 40 Ma) time, crustal extension and transtension in the central part of the orogen were associated with widespread calc-alkaline and alkaline magmatic arc activity; in late Eocene time there was local accretion of oceanic basalt and sediment near the present continental margin. (4c) In late Tertiary (40 Ma) to Recent time, magmatic arc and back-arc activity, areally more restricted than that of the early Tertiary interval, was associated with subduction of the downgoing Juan de Fuca plate."

This scenario has in turn led to an implausible postulation that the Late Proterozoic-early Paleozoic continental margin of an ancestral North America is still preserved in the eastern Cordilleran crust as a west-facing crustal ramp (Cook et al., 1991a-b, 1992). Cook (1995b) supposed that this ramp (which he thought is reflected in the pattern of Bouguer gravity anomalies in the miogeosynclinal region), as well as the continental Moho in the Cordillera, survived essentially unchanged during the entire Phanerozoic time. This idea overlooks the orogenic nature of and crustal reworking in miogeosynclines, where severe metamorphism and deformation, and vertical movements on the order of tens of kilometers, fundamentally changed the crustal structure in orogenic zones.

It would thus be against the basic principles of geology to regard the Cordilleran Moho and "the associated basement ramp" as preserved Proterozoic features "not related to either Mesozoic contractional deformation or middle Tertiary extension" (Cook, 1995b, p. 1525). Indeed, just south of the Canada-U.S. border, the Moho in the Cordillera has been interpreted from COCORP deep seismic data to be as young as Cenozoic (Potter et al., 1986).

Two regional Lithoprobe seismic lines (as well as several short ones) have been shot E-W across the southern Canadian Cordillera: lines 7 and 9, with the N-S Line 8 connecting them in the Omineca Belt (see Figs. 29-39, which are intended to be viewed as a set). Reflection seismic data from these lines were interpreted mostly by Cook et al. (1992), Clowes et al. (1995) and Zelt and White (1995), and refraction seismic data by Kanasewich et al. (1994) and Burianyk and Kanasewich (1995, 1997). Though both these techniques depicted the seismic-velocity and acoustic-impedance structure of the crust, the discontinuities they helped identify were not the same, and considerable disa-

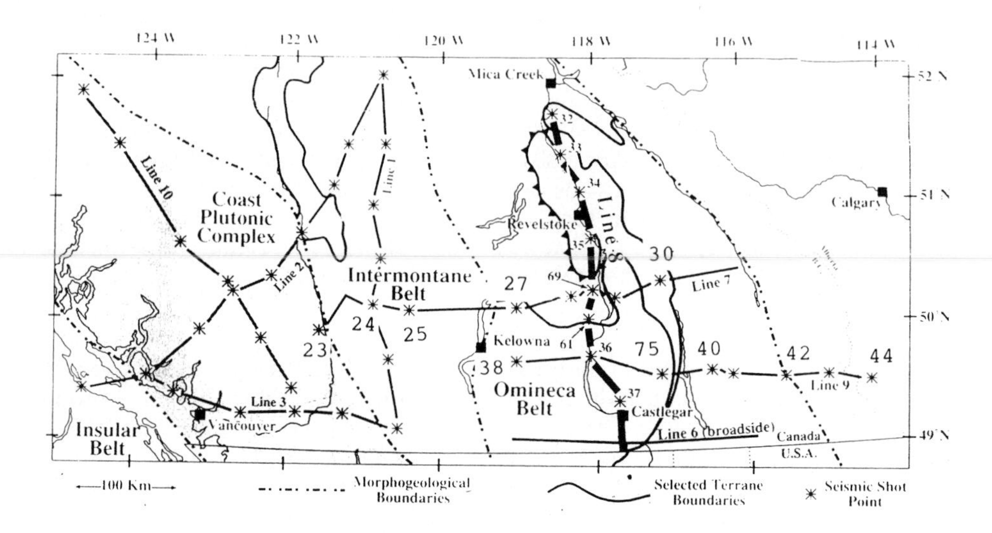

Figure 29. Location of regional Lithoprobe seismic lines in the Southern Canadian Cordilleran transect, in relation to the main belts of the Cordillera (simplified from Kanasewich et al., 1994). Highlighted is Line 8: from a detailed examination of seismic refraction data in it, Kanasewich et al. (1994) reported many discrepancies with the previous interpretation of seismic reflection data by Cook et al. (1992). These data are discussed in more detail in Chapter 6. The numbered "shot points" are referred to in the text as "stations", to distinguish them from shot points in the seismic reflection surveys. That the assumed terrane boundaries are shown in this Lithoprobe location map illustrates the *a priori* manner in which they are assumed in the conventional Lithoprobe interpretations.

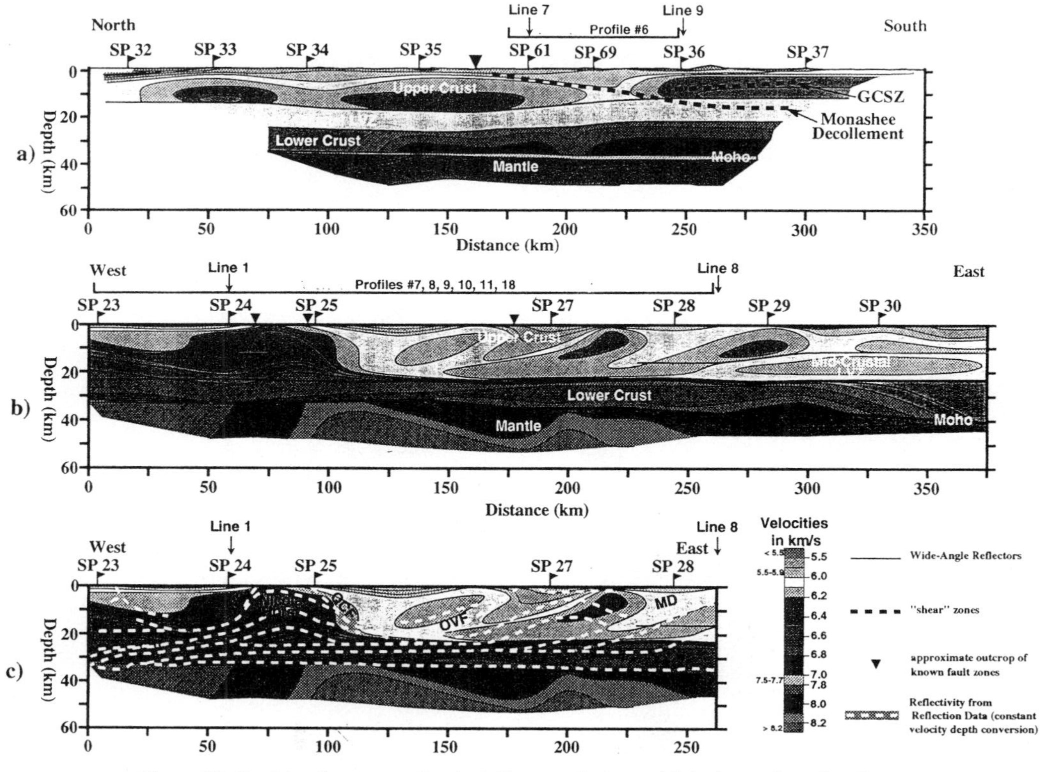

Figure 30. Crustal and upper-mantle seismic P-wave velocity models in the southern Canadian Cordilleran interior, obtained from refraction data in Lithoprobe lines 8 (**a**) and 7 (**b-c**). These results are largely inconsistent with the interpretation of seismic reflection data by Cook et al. (1992, white dashed lines in Fig. 30c). Modified from Burianyk and Kanasewich (1995). SP - seismic station; LVZ - low-velocity zone. "Profile" refers to Lithoprobe seismic reflection data along these lines.

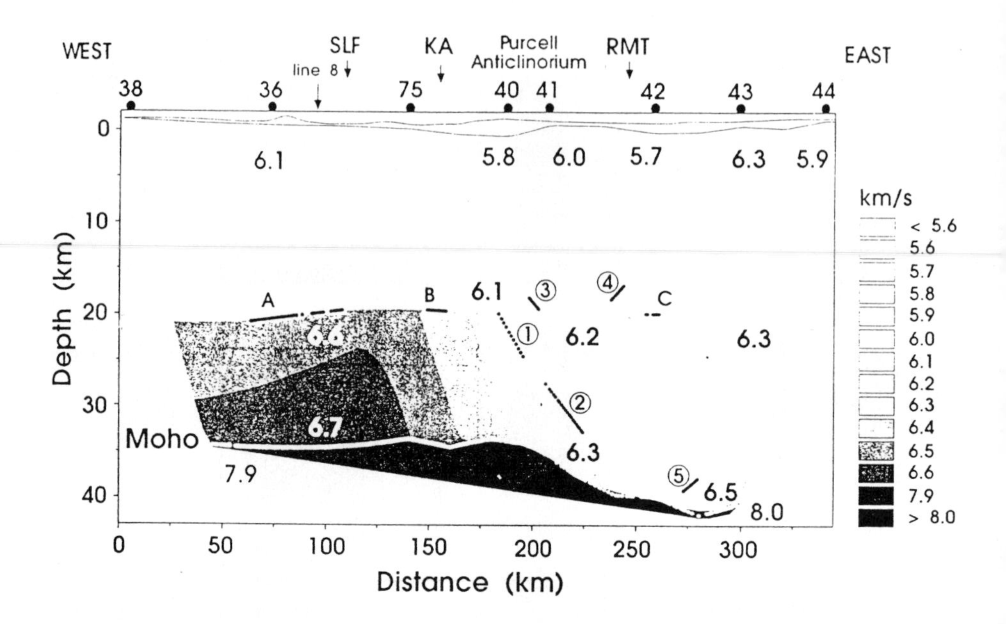

Figure 31a. Lower-crustal seismic P-wave velocity model for Lithoprobe Line 9 through the southern part of the Omineca miogeosynclinal orogenic belt and Rocky Mountain fold-and-thrust belt (modified from Zelt and White, 1995). Large lateral variations are present at all levels of the crust, including in the lower crust near the Kootenay Lake fracture zone. KA - Kootenay Arc; RMT - Rocky Mountain Trench fault; SLF - Slocan Lake fault. Dots numbered 36 to 75 at the top of the profile are seismic stations. Letters A, B, C mark mid-crustal reflecting points. Floating reflectors are numbered 1 to 5.

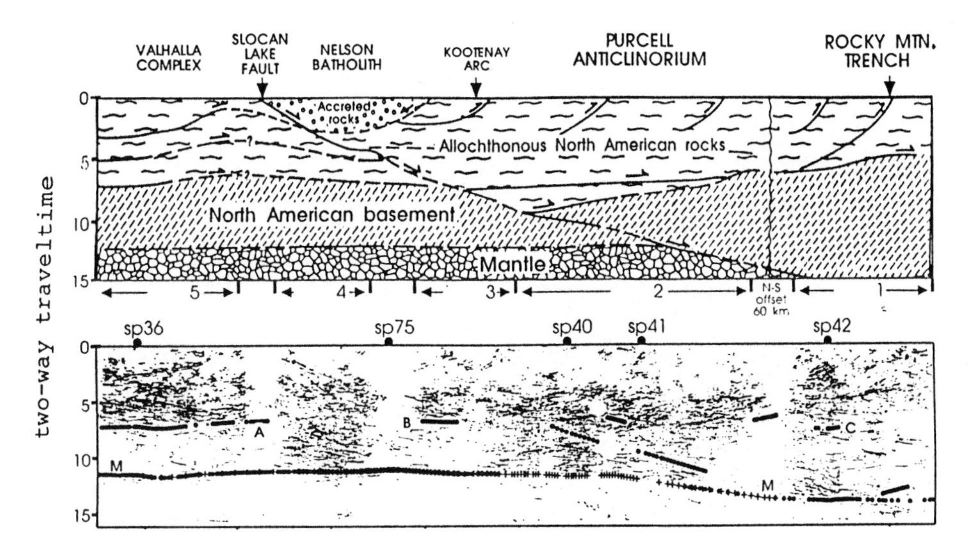

Figure 31b. Interpretation of seismic reflection data in Lithoprobe Line 9 (profiles 1-5) by Cook et al. (1992), as adapted by Zelt and White (1995). Migrated seismic reflection data are shown underneath. The Slocan Lake fault is assumed to be trans-crustal, the Southern Rocky Mountain Trench fault is shown to dip at a low angle, and the Nelson pluton is strangely labeled "accreted rocks". A-B-C - mid-crustal boundary; M - Moho; SP - seismic station.

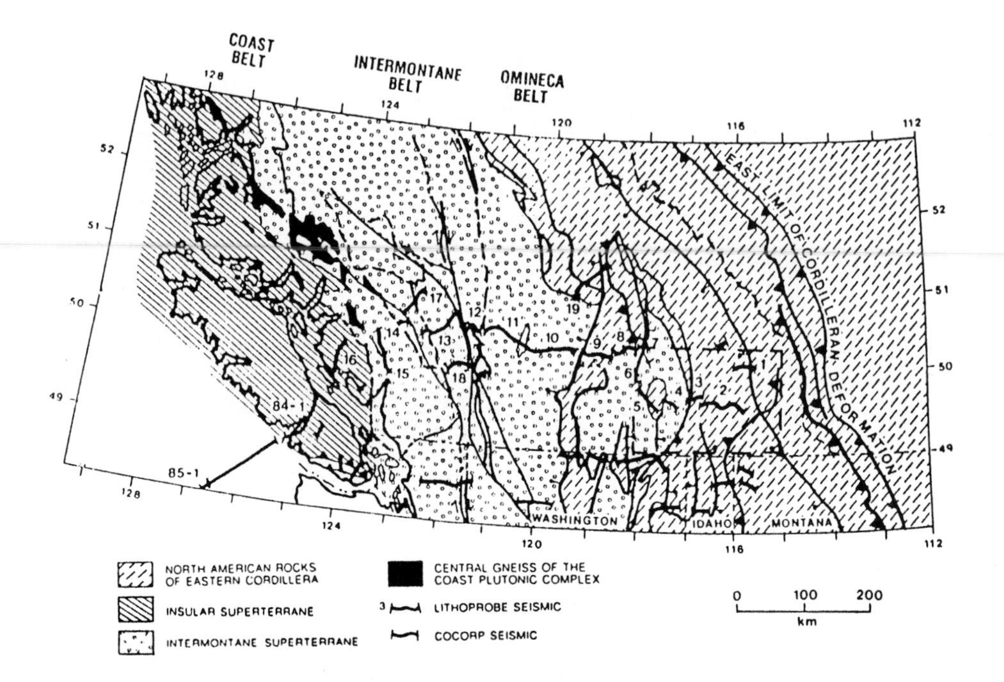

Figure 32. Location of Lithoprobe seismic reflection profiles in the southern Canadian Cordillera (modified from Cook, 1995b). The mistaken assumption that the Canadian Cordillera consists of two main accreted, exotic, composite "superterranes", incorporated into this Lithoprobe base map, has misled the Lithoprobe interpretations from the beginning. Location of the Lithoprobe seismic profiles in the Omineca miogeosyncline, with the details of surface geology of that area, is shown in Fig. 17.

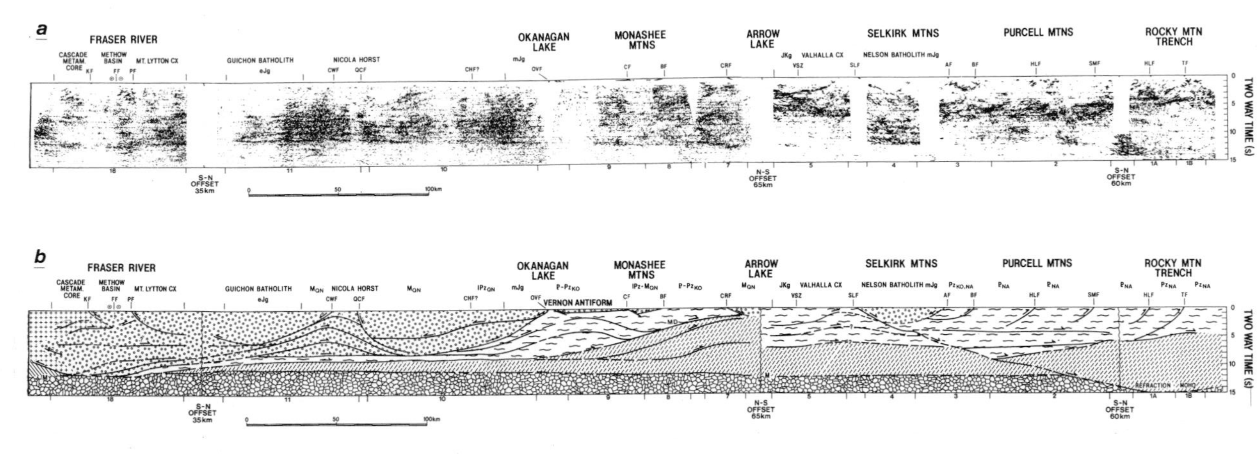

Figure 33. Lithoprobe seismic reflection data across the southern Canadian Cordillera (a) as processed and reproduced by Cook et al., 1992). Few details are shown in the uppermost crustal levels, complicating correlations of seismic reflection patterns with the mapped geologic features at the surface. The interpretation of these data (b) by Cook et al. (1992) assumes the Cordilleran crust is a stack of exotic-terrane trust slices, with an intact thin sliver of Proterozoic North American cratonic basement in the lower crust continuing as far as the eastern Coast Belt. The numerous known high-angle faults in the Cordillera are overlooked. Profile numbers shown at bottom; profile locations are given in Figs. 17 and 32. AF - Ainsworth fault; BF - Beavan fault (Monashee Mountains); BF - Benard fault (Purcell Mountains); CF - Cherryville fault; CHF - Chapperon fault; CRF - Columbia River fault; CWF - Coldwater fault; eJg - Early Jurassic granitic plutons; FF - Fraser fault; HLF - Hall Lake fault; JKg - Jurassic-Cretaceous granitic plutons; KF - Kwoiek fault; lPz-MQN - late Paleozoic-Mesozoic rocks of Quesnel terrane; MD - Monashee décollement; mJg - Middle Jurassic granitic plutons; MQN - Mesozoic rocks of Quesnel terrane; OVF - Okanagan Valley fault; OF - Pasayten fault; PNA - Proterozoic rocks of ancestral North America; PzKO - Paleozoic rocks of Kootenay terrane; PzNA - Paleozoic rocks of ancestral North America; QCF - Quilchena Creek fault; SLF - Slocan Lake fault; SMF - St. Mary fault; TF - Torrent fault; VSZ - Valkyr shear zone.

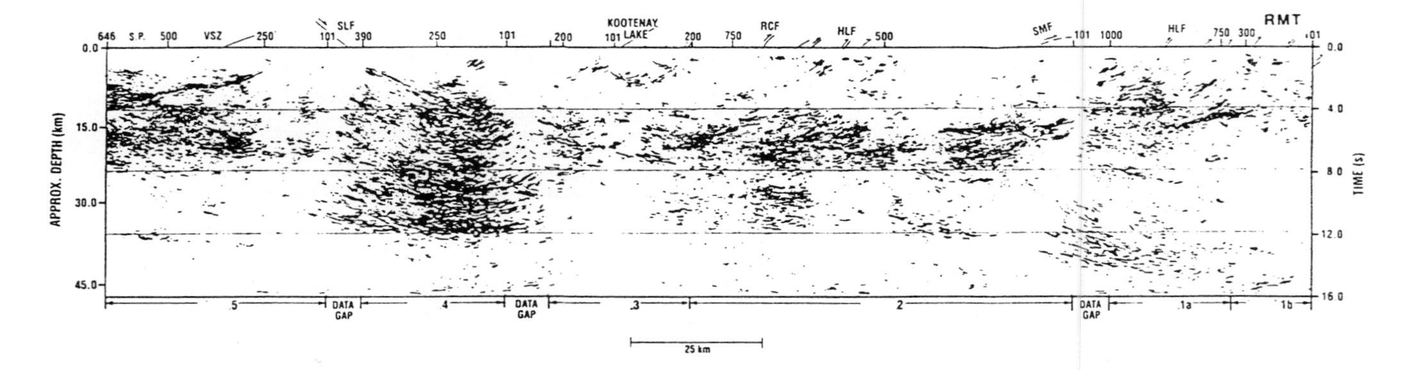

Figure 34. Details of Lithoprobe southern Cordilleran seismic reflection profiles 1 to 5 (as processed and presented by Cook, 1995b). Profile locations are given in Figs. 17 and 32. HLF - Hall Lake fault; RCF - Redding Creek fault; RMT - Rocky Mountain Trench; SLF - Slocan Lake fault; SMF - St. Mary fault; SP - seismic reflection shot point (different than seismic refraction stations); VSZ - Valkyr shear zone.

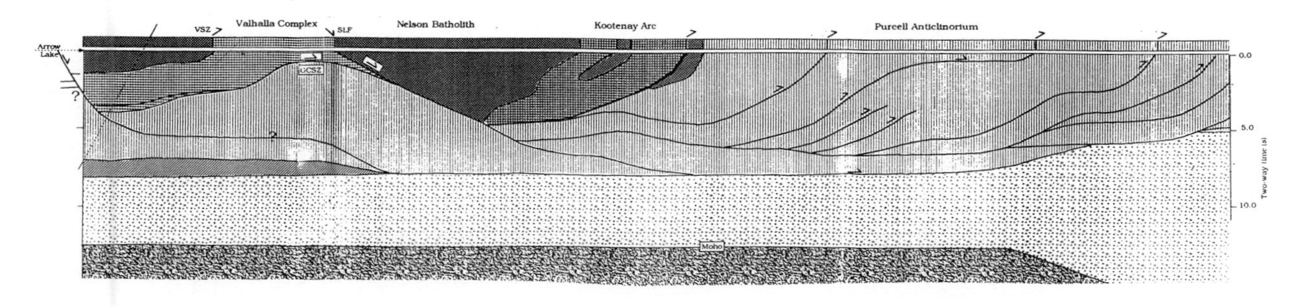

Figure 35. Interpretation of the Lithoprobe seismic reflection data in profiles 1-5 in the southeastern Canadian Cordillera (Fig. 34) by Cook (1995c). This interpretation assumes all major faults flatten out in the mid-crust and continue laterally for hundreds of kilometers before they merge into the assumed master detachment at the top of the lower-crustal sliver of cratonic crust. Faults such as Slocan Lake (SLF) are assumed to be much larger and more continuous than the geologic evidence shows them to be, while the very major high-angle faults and structural zones are overlooked. GCSZ - Gwillim Creek fault; VSZ - Valkyr shear zone.

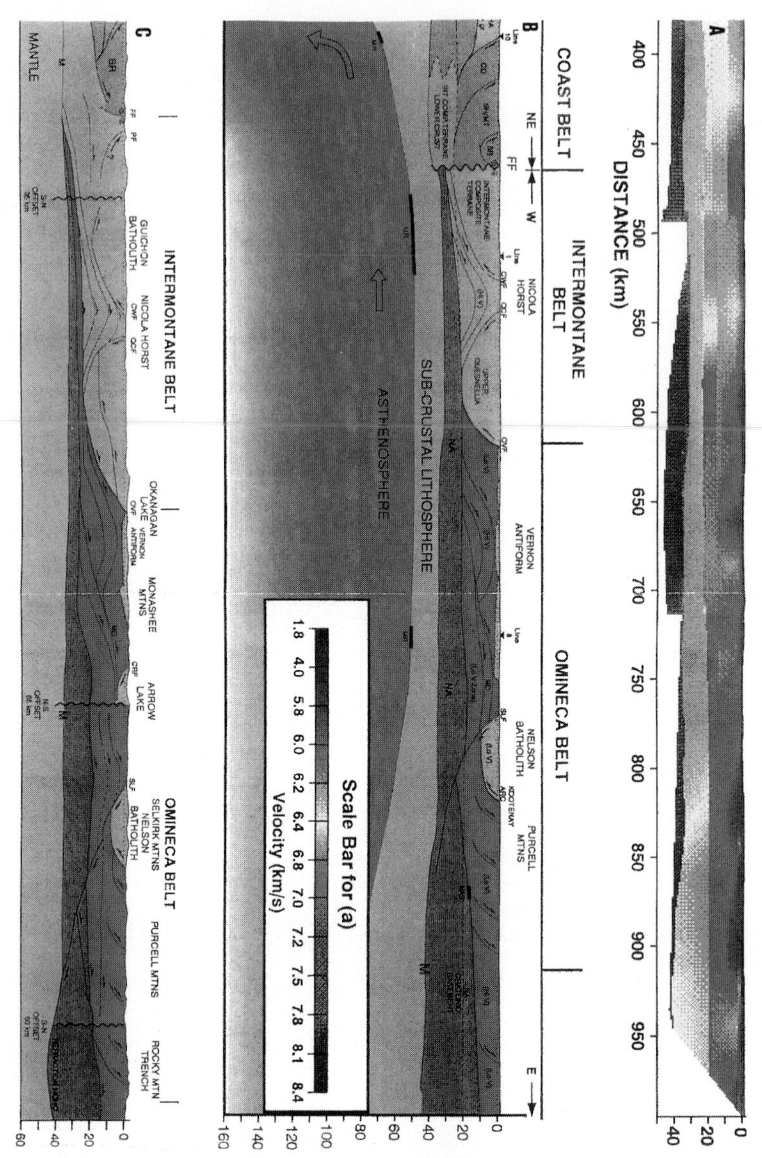

Figure 36. Crustal velocity structure of the southern Canadian Cordillera based on seismic data in the Lithoprobe transect (**a**), and two slightly different combined interpretations of refraction and reflection data (**b-c**), presented by Clowes et al. (1995): **b** - inferred lithospheric cross-section based predominantly on refraction data; **c** - crustal structure interpreted from reflection data. No vertical exaggeration; slight differences in the locations of some geologic features between Figs. 36b and 36c result from inexact coincidence of the refraction and reflection data along the Lithoprobe transect. Except for the hypothetical cratonic sliver in the lower crust, most of the Cordilleran crust is assumed to consist of exotic-terrane low-angle thrust slices. BR - Bridge River terrane; CD - Cadwallader terrane; CRF - Columbia River fault; CWF - Coldwater fault; FF - Fraser fault; HiV - high-velocity zone; LoV - low-velocity zone; M - Moho; MD - Monashee décollement; MT - Methow terrane; MR - mantle reflector; NA - North America; NAB - North American basement reflector; OVF - Okanagan Valley fault; QCF - Quilchena Creek fault; SH - Shuksan terrane; SLF - Slocan Lake fault.

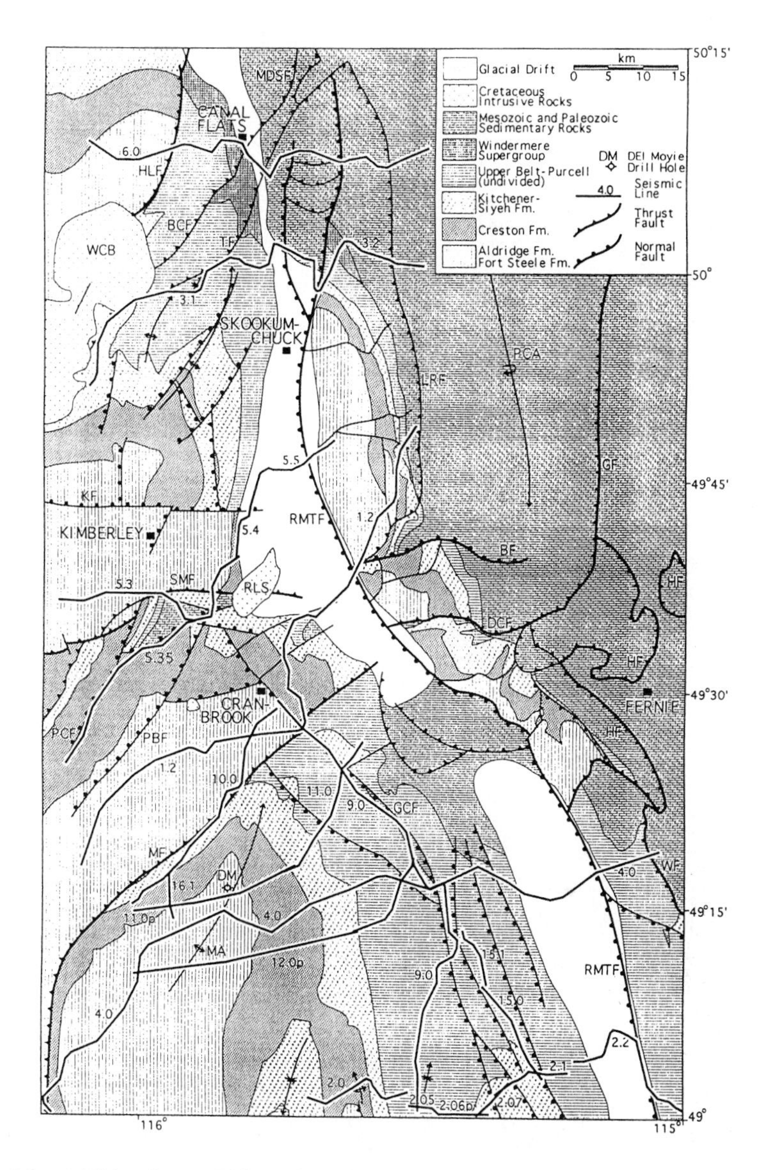

Figure 37. Additional, non-Lithoprobe seismic reflection sections in the southern Canadian Omineca miogeosyncline and Rocky Mountain fold-and-thrust belt (location map modified from van der Velden and Cook, 1996). As elsewhere in the Canadian Cordillera, many of the seismic sections run parallel to large high-angle faults. Sections 1, 2, 5 and 6, which cross the Southern Rocky Mountain Trench (RMTF), are presented in Figs. 38a-d. BCF - Buhl Creek fault; BF - Boulder fault; DCF - Dibble Creek fault; GCF - Gold Creek fault; HF - Hosmer fault; HLF - Hall Lake fault; KF - Kimberley fault; LRF - Lussier River fault; MA - Moyie anticline; MDSF - Mount DeSmet fault; MF - Moyie fault; PBF - Palmer Bar fault; PCA - Porcupine Creek anticlinorium; PCF - Perry Creek fault; RLS - Reade Lake stock; SMF - St. Mary fault; TF - Torrent fault; WCB - White Creek pluton; WF - Wigwam fault.

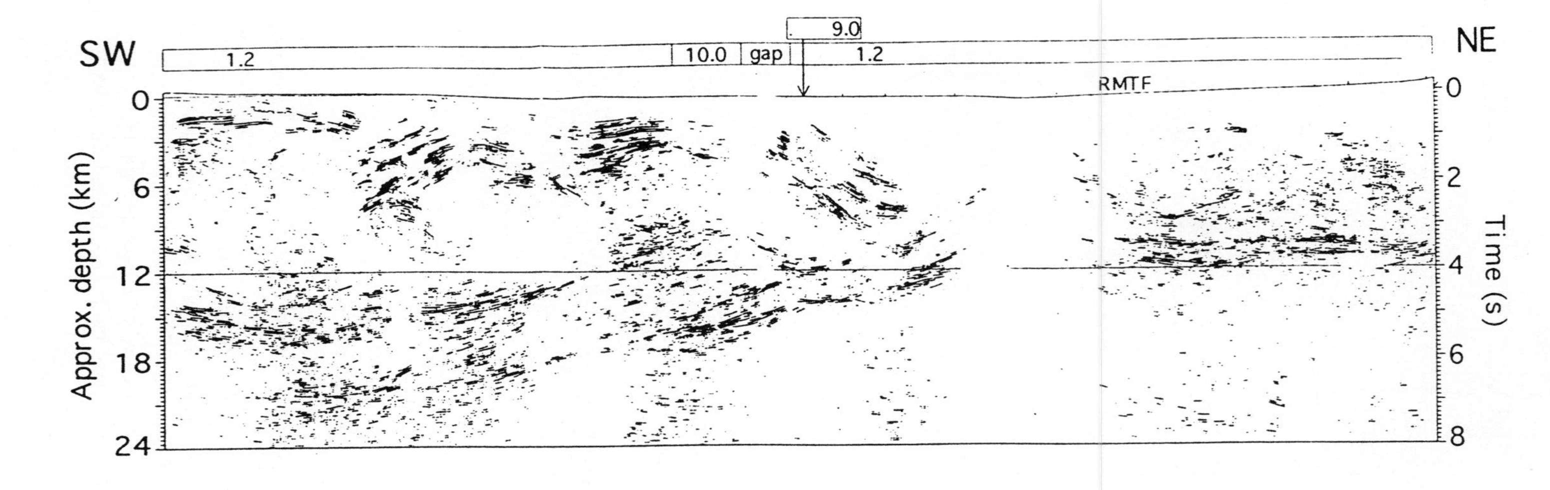

Figure 38a. Seismic reflection section 1, with its constituent numbered sub-sections, crossing the Southern Rocky Mountain Trench (RMTF) in the southern Canadian Cordillera (as processed and presented by van der Velden and Cook, 1996). Seismic-section location is given in Fig. 37.

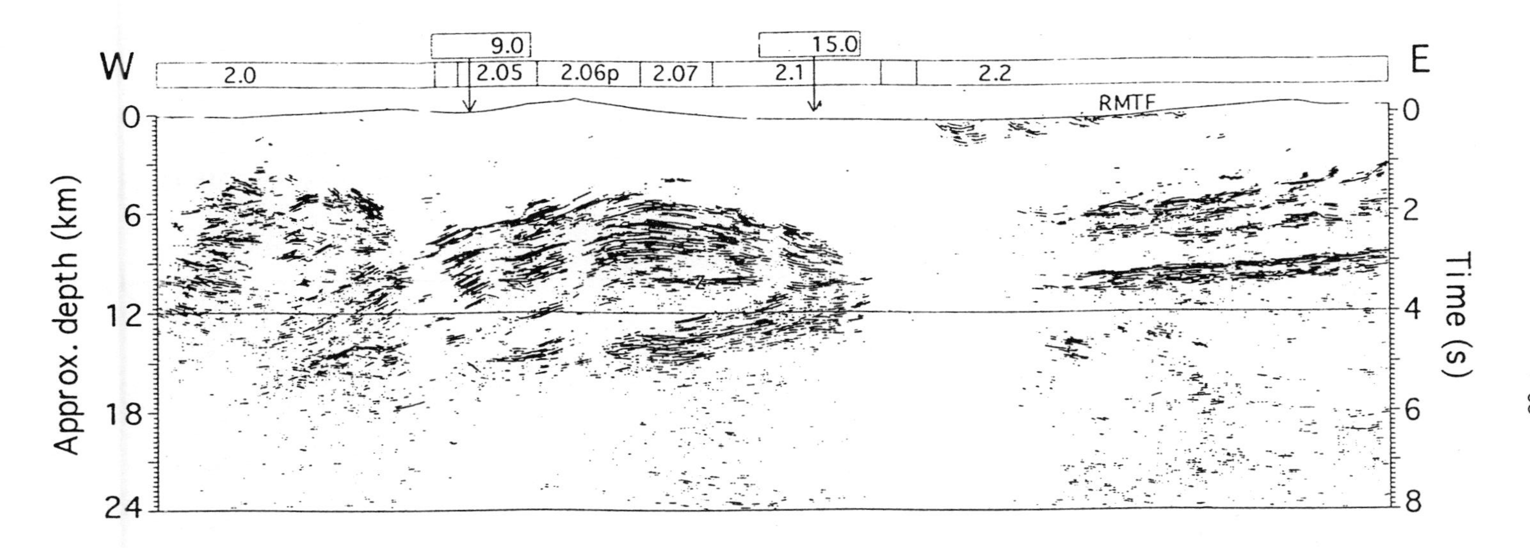

Figure 38b. Seismic reflection section 2, with its constituent numbered sub-sections, crossing the Southern Rocky Mountain Trench (RMTF) in the southern Canadian Cordillera (as processed and presented by van der Velden and Cook, 1996). Seismic-section location is given in Fig. 37.

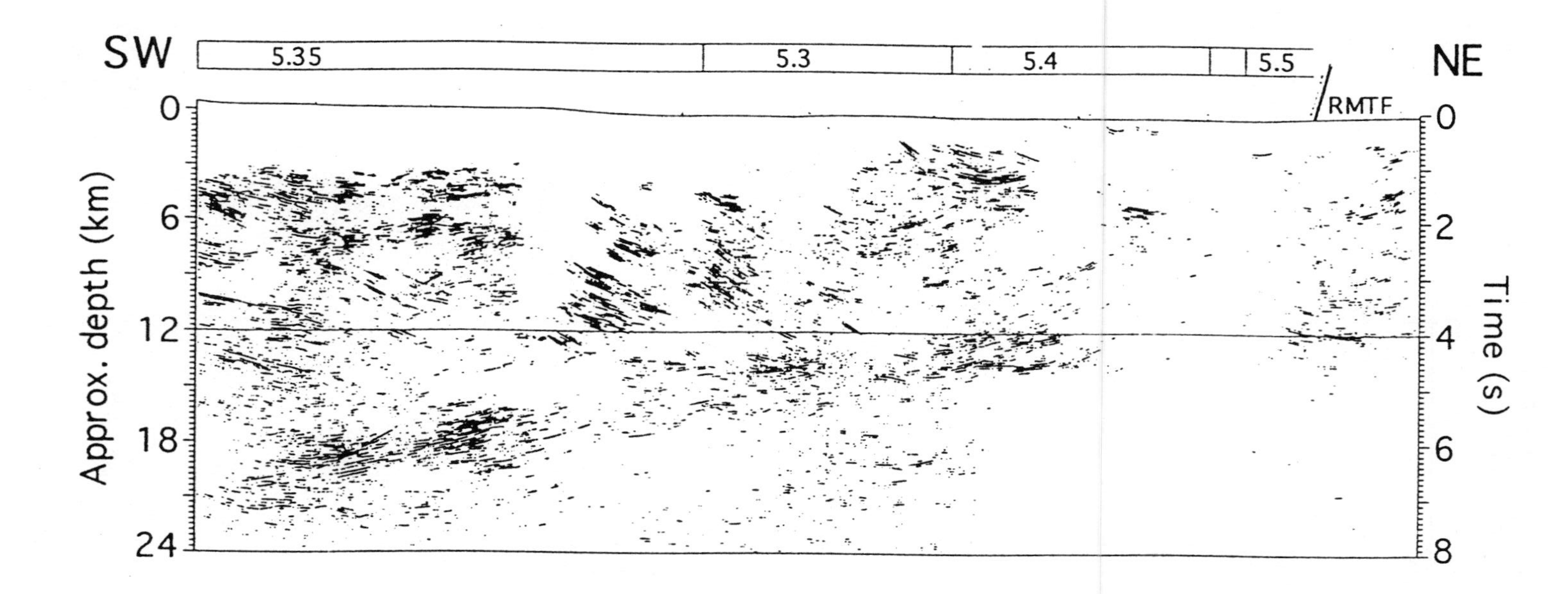

Figure 38c. Seismic reflection section 5, with its constituent numbered sub-sections, crossing the Southern Rocky Mountain Trench (RMTF) in the southern Canadian Cordillera (as processed and presented by van der Velden and Cook, 1996). Seismic-section location is given in Fig. 37.

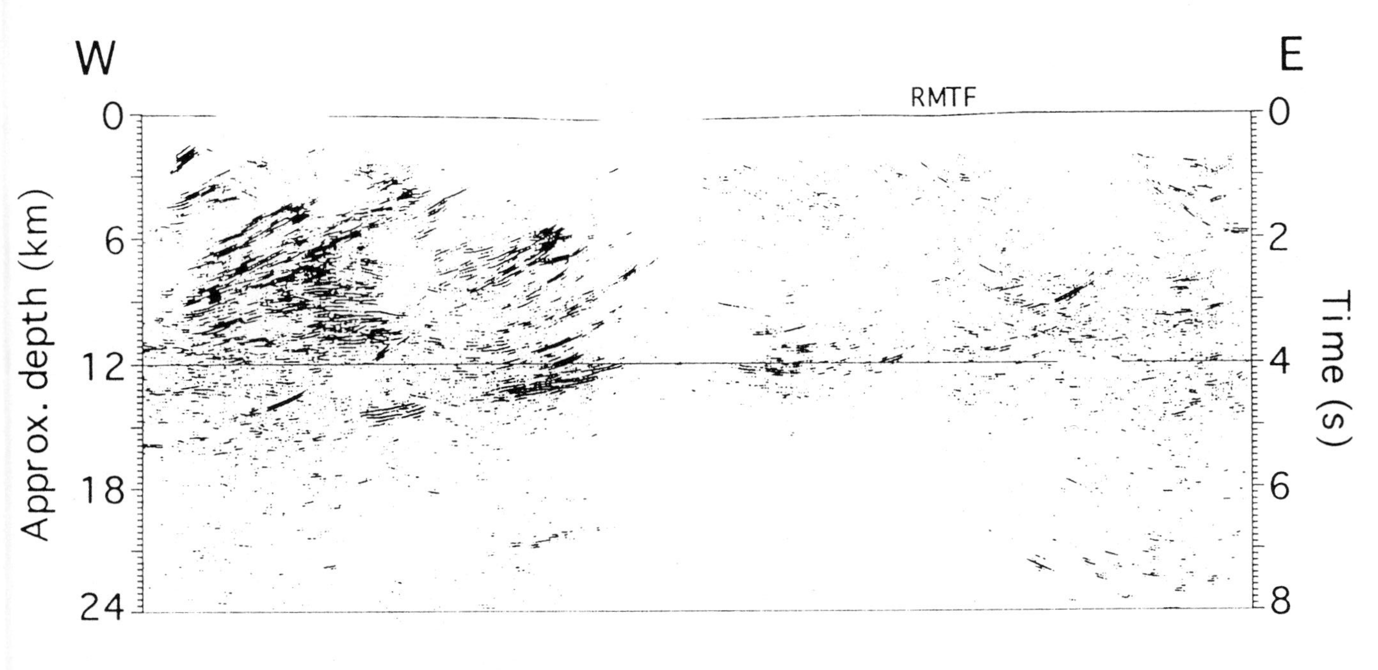

Figure 38d. Seismic reflection section 6, crossing the Southern Rocky Mountain Trench (RMTF) in the southern Canadian Cordillera (as processed and presented by van der Velden and Cook, 1996). Seismic-section location is given in Fig. 37.

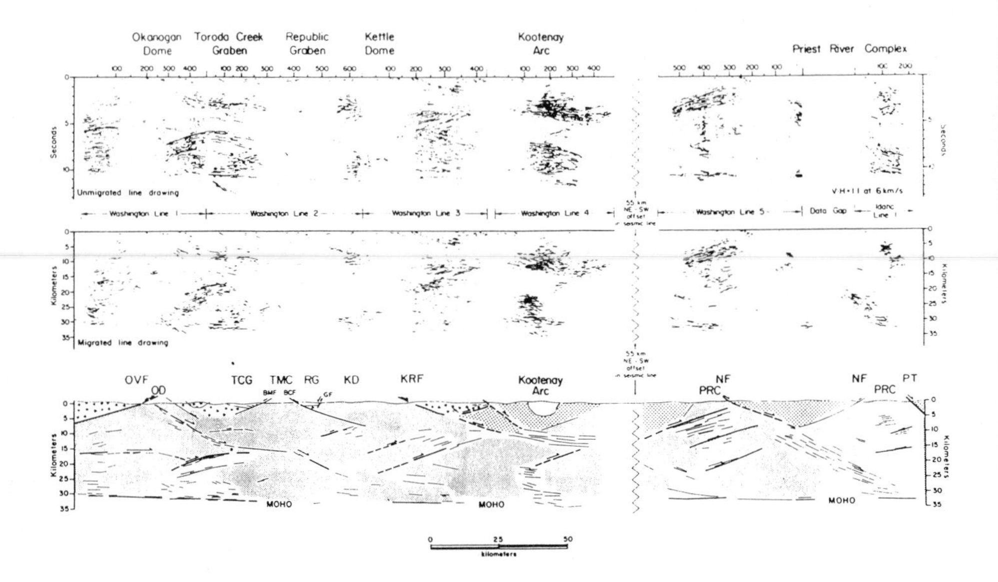

Figure 39. Interpretation of COCORP seismic reflection data across the Cordillera just south of the Canada-U.S. border, by Potter et al. (1986) and Monger et al. (1994). The flat and semi-continuous Moho was interpreted as Cenozoic, in contrast to the Lithoprobe interpretations in Canada just to the north (Cook et al., 1992; Cook, 1995a-c) which presumed the Cordilleran Moho to be Proterozoic. BCF - Bacon Creek fault; BMF - Bodie Mountain fault; GF - Granby fault; KD - Kettle dome; KRF - Kettle River fault; NF - Newport fault; OD - Okanagan dome (spelled Okanogan in the U.S.); OVF - Okanagan Valley fault; PRC - Priest River complex; PT - Purcell trench; RG - Republic Graben; TCG - Toroda Creek graben; TMC - Tenas Mary complex. Location of these seismic data, in relation to the main surface-mapped geologic features, is given in Fig. 24.

greements remain between the interpretations of these refraction and reflection data (cp. Kanasewich et al., 1994 and Cook et al., 1992).

Among several points of controversy, Kanasewich et al. (1994, p. 2661) noted that some of the reflections could be related to "impedance contrasts within fault zones that cut across the previously determined velocity structure". Also, "there is a marked discrepancy between the reflectivity trends presented by Cook et al. (1992) and present interpretation in the middle and lower crust" (op. cit., p. 2665). In particular, refraction data made it possible to detect a previously unnoticed low-velocity zone in the middle crust, with P-wave velocities of just 6.0-6.1 km/s. This low-velocity lens underlies the axis of the southern part of the miogeosynclinal Omineca orogenic belt and extends west into the Intermontane median-massif belt (also Burianyk and Kanasewich, 1995). The crust-mantle interface was found to lie in a transition zone several kilometers thick, and an acoustic discontinuity in the upper mantle was found with refraction data at a depth of 45-50 km.

Importantly, some discontinuous low-angle reflections, arbitrarily projected by Cook et al. (1992) and Cook (1995b-c) upward to some mapped steep and low-angle faults on the surface and inferred to continue into the lower crust, in fact have no links with geologically mapped faults and dissipate in the upper or middle crust (Kanasewich et al., 1994, their Figs. 7a-b). Kanasewich et al. (1994, p. 2665) noted that known shear zones, such as the Monashee décollement, can be correlated at depth with some "low-velocity trends", but they regarded this possibility with some caution: "if this is the case..." More strongly, they stated that "our 'Monashee Décollement' does not match well with that interpreted from the vertical incidence [reflection] data" by (Cook et al., 1992). "Cook et al. (1992), with geologic constraints, interpret the foot wall of the Monashee Décollement to be North American cratonic rocks", and "the velocities in the footwall of the Monashee fault zone tend to be lower than those in the hanging wall by 0.1 to 0.2 kms^{-1}", but "since this is the edge of the resolving power of the velocity measurements, one must be careful not to read too much into this difference" (op. cit.).

Seismic and other geophysical data, by themselves, cannot be interpreted uniquely (Pakiser and Mooney, eds., 1989). In the Interior Platform in Alberta and Saskatchewan, various authors (Chandra and Cumming, 1972; Kanasewich et al., 1987; Stephenson et al., 1989) have found the base of the cratonic crust to lie at depths of 42 to 47 km. Kanasewich et al. (1994) generalized the seismic velocity structure of the North American craton in western Canada as consisting of a 20-km-thick upper crust with P-wave velocities of 5.5-6.0 to 6.4 km/s, a 15-km-thick middle crust with 6.6 km/s, and a 10-km-thick lower crust with velocities of 7.1 km/s. Below lies the upper mantle, whose P-wave velocity is 8.2 km/s. The crustal seismic-velocity structure west of the Foothills, in the Rocky Mountain Belt, is considerably different: whereas the upper mantle still has velocities of 8.2 km/s, the Moho interface lies at depths of 50 km or more; the lower crust is twice as thick as in the Plains and has a slightly higher velocity of 7.15 km/s; the middle crust is ~30% thinner and slower, with 6.5 km/s; the upper crust is some 20 km thick, like in the craton beneath the Plains, but its velocity is less variable with depth and usually ~6.0 km/s.

Still farther west, in the miogeosynclinal Omineca and Intermontane belts, the average upper-lithosphere physical characteristics are different again. In the axial zone of the miogeosynclinal orogen, total thickness of the modern crust is just 35-37 km and seismic velocity of the uppermost mantle is only 7.9 km/s. Above the Moho, the lower crust is ~12 km thick and its P-wave velocities vary from 6.6-7.0 km/s at the bottom to 6.4-6.5 km/s at the top. The 10-km-thick middle crust has low velocities of 6.2-6.3 km/s at the bottom to 6.02-6.1 km/s at the top. In the 15-km-thick upper crust, the velocities are again higher, from 6.2 km/s at the bottom to 5.6 km/s near the surface (see also Burianyk and Kanasewich, 1995).

Zelt and White (1995) analyzed the crustal velocity structure in a 350-km-long, E-W, wide-angle refraction seismic profile across the southernmost Canadian Cordillera just north of the Canada-U.S. border. There, they found considerable variations in the upper-crust velocities: 5.6 to 6.3 km/s in the western Rockies and Purcell Anticlinorium, and 6.1 to 6.2 km/s in the part of the Omineca Belt west of the Kootenay Arc. The westward decrease of crustal thickness from 42 km to ~35 km over a distance of 80 km in the eastern Omineca Belt was confirmed (see also Berry et al., 1971; Cumming et al., 1979; Cook et al., 1992), and a westward increase of the velocity and density of the uppermost mantle was modeled.

Of particular interest is a fairly local upwarp in the crust-mantle interface between ST (stations) 40 and 41 in the refraction Line 9. This deep-seated upwarp coincides roughly with the crest of the Purcell Anticlinorium, which formed during the early stages of the Late Cretaceous Laramian orogeny due to inversion of the northern part of the Middle Proterozoic Belt-Purcell Basin. This anticlinorium plunges to the north and widens to the south, and south of the Moyie fault system it has a shape of an asymmetric box (McMechan, 1991, p. 628). The top of the middle crust, as modeled by Zelt and White (1995), lies at 20 km depth, similar to the estimated total thickness of the Belt-Purcell Basin (Peterson, ed., 1986; Constenius, 1996). The apparent mismatch, in the Purcell Anticlinorium's upper crust, between low velocities and high densities indicated by gravity data was attributed to the presence of this basin's thick metasedimentary succession intruded by diabase dikes. The small velocity gradient in the top ~20 km of the crust in this area is correlatable with high crustal temperatures (indicated by an anomalously high surface heat flow) and zones of high electrical conductivity in the crust (Gupta and Jones, 1995; Jones and Gough, 1995). Successive Belt-Purcell and Windermere rifts have been invoked to explain the geologically observable and geophysically inferred changes in crustal composition east and west of the

Kootenay Arc. The upper crust in the Purcell Anticlinorium was also assumed to be more mafic than in the miogeosynclinal zone west of the Kootenay Arc. The opposite, however, was modeled for the lower crust: more felsic under the Purcell Anticlinorium and more mafic under the miogeosynclinal orogen to the west (Clowes et al., 1995; Zelt and White, 1995).

From various seismic data, it has been established that crust with a typical cratonic thickness (>40 km) but a different velocity structure continues in the southern Canadian Cordillera as far west as the Rocky Mountain Trench. The crust thins westward to 35-37 km beneath the Purcell Anticlinorium and 30-32 km beneath the western Omineca Belt and the Intermontane Belt (Cook, 1995b; cp. Cumming et al., 1979). The upwarp of the crust-mantle boundary under the Purcell Anticlinorium has a broader counterpart that underlies the entire Intermontane median-massif belt (Cook, 1995b).

Reasons for the general lack of a sharp discontinuity at the base of the crust under the Rocky Mountains are unclear. Cook (1995b, p. 1521) simply dismissed its absence as unimportant: "There is no reason to believe that the lack of a distinct Moho reflection beneath the Rocky Mountains is related to the nature of the tectonic regime at the surface..." Summing up the evidence from the Lithoprobe Southern Cordilleran Transect, Clowes et al. (1995, p. 1505) noted, more cautiously, that "starting with the Foreland [i.e. the Rocky Mountain fold-and-thrust] Belt and eastern Omineca Belt of the eastern end of the [trans-Cordilleran] cross section, the clearest observation is the lack of definition of North American basement by the seismic velocity structure... In contrast, the top of the lower crust at a depth of about 20 km shows a prominent velocity change... What it represents lithologically and structurally remains uncertain."

What exactly is meant by "tectonic regime at the surface" (Cook, 1995b, p. 1521) is not explained. But some whole-crust peculiarities of the Rocky Mountain, Omineca and Intermontane belts, including seismic expression of their Moho interface, pre-megabelt basement and so on should be taken into account in the regional tectonic analysis. Tectonic processes may or may not have a clear surface expression, but they nonetheless operate at depth. The whole-crust belts of the Cordillera have been subjected to these processes during all their history, including recurrent crustal extension accompanied by regional delamination and rifting evident from geologic mapping. As is now proved, deep metamorphic and structural transformations of the crust in the Cordillera were partly localized in well-defined zones of Tertiary extension such as the Colorado Corridor in the U.S. (Beratan, ed., 1996; Friedman and Huffman, coords., 1998).

Lithoprobe data confirm the general westward thinning of the crust under the Omineca miogeosynclinal orogenic zone (inferred previously by Chandra and Cumming, 1972).

From the Lithoprobe seismic reflection data, Cook et al. (1992) located a sharp thinning of the continental crust west of the craton near the western Kootenay Arc. The same position was modeled by Zelt and White (1995). It is hard, however, to justify its postulated correlation with a fault zone mapped at the surface along the eastern side of the Valhalla metamorphic core complex, many tens of kilometers to the east.

In that interpretation, the rather local Slocan Lake fault was presumed to be a crustal-scale boundary between two assumed geologic (and tectonic) domains with fundamentally different characteristics (Parrish et al., 1988; Cook et al., 1992). Such an interpretation arose from the hypothesis of the Cordilleran crustal structure as a stack of crustal slices (Monger et al., 1994) displaced on a series of whole-crust (and "whole-orogen") listric thrust faults merging in the lower crust into a grandiose trans-Cordilleran detachment (Cook et al., 1992; Cook and Varsek, 1994; Cook, 1995a-c). In seismic images, like at the surface, the Slocan Lake fault dips to the west at a low angle (Carr, 1995), but only to a shallow depth. These seismic events lose definition in the middle crust, at a distance of only a few tens of kilometers from its mapped surface location. Between this fault and the mantle upwarp east of the Kootenay Arc lies the Valhalla metamorphic core, and several steep faults cut the entire crust. The most significant of these structural zones runs along the western boundary of the Kootenay Arc at longitude ~117°W as a series of closely spaced N-S-oriented steep fault strands, igneous bodies on strike with them (Late Triassic-Early Jurassic volcanic fields, several generations of Mesozoic batholithic granites), and the long and narrow Kootenay Lake in the modern topography. Vertical offsets of the Moho can be correlated with that prominent, whole-crust Kootenay Lake fault zone, which bounds the western blocks of the Omineca miogeosynclinal orogenic zone.

6 - OMINECA OROGENIC BELT AS TECTONOTYPE OF THE EASTERN CORDILLERAN MIOGEOSYNCLINE

Definition of *orogenic grain*

The compound term *orogenic grain*, though it is used often in descriptions of continental mobile megabelts, is loose and carries a great deal of uncertainty. Long before the deep-seated processes of crust orogenization (reworking) are manifested at the surface, they already operate at depth. Their surface expressions can be rifts, upwarps, downwarps, differential crustal block movements. Accumulation of initial post-basement sedimentary and volcanic rock successions (basins) creates a sharp contact between rocks of the pre-existing basement and the younger unmetamorphosed cover.

Further evolution of a mobile megabelt commonly involves dramatic subsidence, accompanied by accumulation of thick stratal packages and discordant intrusive rocks. The latter tend to be emplaced along subvertical fault conduits penetrating the entire crust and upper mantle. Presence of such huge fault zones also enables block displacements of large amplitude. Supracrustal rocks are subjected to high-grade burial metamorphism deep in the crust, transforming them into crystalline rocks similar to those of the megabelt's old basement. The distinction between the two main vertical tectonic units, basement and cover, becomes blurred or vanishes.

Orogenic processes in a mature mobile megabelt involve the entire lithosphere. This happens within the lateral boundaries of the megabelt, making it a first-order continental lateral tectonic unit distinct from the adjacent penecontemporaneous craton. These fundamental lateral units are commonly separated by great subvertical faults. Bounded by these lithospheric or crustal discontinuities, a mature mobile megabelt can no longer be treated as containing an old grain of the basement separate from a wholly post-cratonic grain of the cover, as both these étages and grains are fused together into a new entity of crustal or lithospheric proportions. The maximum reworking of the old grain and fusion of the basement and cover are reached in the tectonically most active parts of the megabelt - the orogenic zones, especially those lying far away from the continuous cratonic mass. In orogenic zones closer to the craton, as well as in median massifs, tectonic reworking and fusion are less profound.

In miogeosynclines, orogenized parts of a mobile megabelt lying close to the craton are usually less magmatized than hinterland orogens, but no less deformed. At late stages of megabelt evolution, deformation extends past the orogen's boundaries, giving rise to shallow (thin-skinned) fold-and-thrust belts overlapping the unreworked basement. The biggest such belts are formed over marginal parts of cratons, but smaller ones develop also on the flanks of median massifs in the megabelt interior.

In orogenic zones far in megabelt hinterlands, the basement's orogenic grain is indistinguishable from that of the cover, and only some isotopic signatures indicate the inheritance of reworked rocks from some older basement. In the outer, miogeosynclinal orogenic zones, by contrast, the basement and cover rocks are separable petrologically, and basement remnants are revealed in metamorphic core complexes, as well as less directly in anomalous zircon families and xenoliths. Median massifs, which lie between these dissimilar orogenic zones, contain rocks metamorphosed to only low grades, so the volcano-sedimentary cover in them is clearly separable from the older crystalline basement. The orogenic grain in these cases consists of two distinguishable components with different recorded memories of their dissimilar histories. The shallow fold-and-thrust belts, spectacular as they may look, are orogenic only in the sense that their forming forces were supplied by an orogen, but they are not indigenously orogenic and have no deep roots in the crust. Rocks composing the fold-and-thrust belts are commonly pieces of cratonic basement and (more usually) sedimentary cover; only in some cases do these belts also contain rocks displaced from the orogenic zones proper.

Need for correct distinction of regional tectonic étages and stages based on formalized criteria and proper procedures of examining the rock record

Recognition of structural-formational étages as fundamental vertical tectonic units in a regional geologic record was discussed recently by Lyatsky et al. (1999). An étage is a rock unit comprising all preserved rocks, stratal (sedimentary and volcanic) and non-stratal (igneous intrusive), which appeared during an interval of time - tectonic stage - when certain tectonic conditions prevailed. These paragenetic rock complexes have their characteristic compositional (formational) and structural (both internal and external) peculiarities. Tectonic time stages, with their corresponding rock-made étages, are separated by pulses of drastic regional restructuring. Each rock-made étage and corresponding tectonic time stage start at the bottom of the étage's lithostratigraphic succession, but the time span represented by an étage's rock assemblage is usually reduced by erosion of the upper levels during restructuring at the end of the tectonic stage. Intrusive rocks emplaced at deeper levels, which were safe from erosion, may help to fill the gap in the rock record, but magmatism is only episodic and intrusive rocks usually do not span the entire missing part of the stage. The period of restructuring-related erosion is assigned to the tectonic time stage which it closes, and a stage is considered to last from the bottom of its corresponding étage to the bottom of the next one (Fig. 40).

Structural-formational étages are defined in the rock record independently from preconceived prejudices. The biggest étages, mega-étages, are the basement and the post-

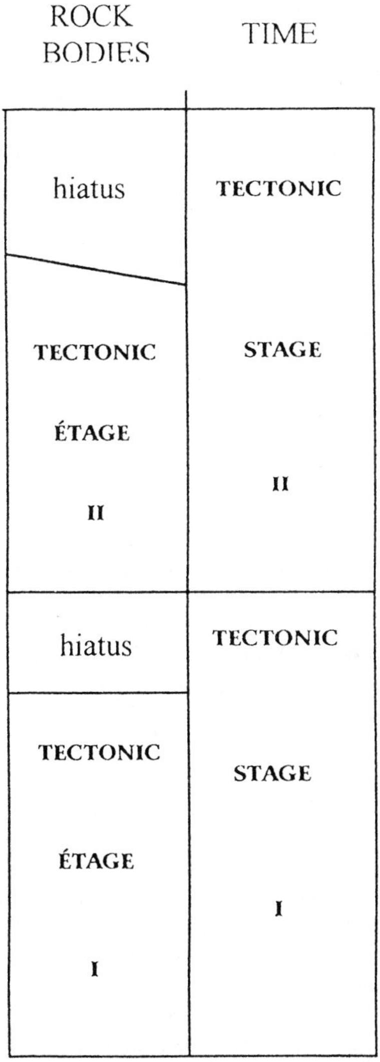

ROCK BODIES

TIME

hiatus

TECTONIC

TECTONIC ÉTAGE II

STAGE II

hiatus

TECTONIC

TECTONIC ÉTAGE I

STAGE I

Figure 40. Tectonic étages and stages, and their time spans. The span of a tectonic time stage includes the span of its corresponding rock-made structural-formational étage, plus the span of a hiatus at the étage's top (after Lyatsky et al., 1999). The beginning of the next tectonic stage is set at the base of the next étage.

basement cover. Both contain étages of a lower hierarchical order: e.g., the super-étages formed before, during and after a major super-stage such as inversion. In the 1970s, it was customary to distinguish in the Canadian Cordillera a "pre-tectonic crystalline basement", an "orogenic complex" and a "late to post-orogenic succession" (Wheeler and Gabrielse, coords., 1972). But misconceptions existed in this artificial division: all these étages are tectonic, and the last two are both orogenic.

For the purposes of regional tectonic analysis in orogenic provinces, the fundamental unit in the hierarchy of structural-formational étages is an étage bounded by markers that reflect breaks in the tectono-sedimentary, tectono-metamorphic, tectono-magmatic and tectono-deformational regional patterns all at once. Such breaks are conventionally, if loosely, called *orogenies*. In the Cordilleran mobile megabelt, each orogen - Omineca, Coast Belt - has its own characteristic set of orogenic episodes, or orogenies. But despite these differences from one belt to another, some events were common to the entire megabelt and are recorded in all its parts. These megabelt-scale tectonic events produced different manifestations in different lateral tectonic units (orogens and median massifs), but they were coeval.

But generally, the collection and composition of étages vary from one lateral tectonic unit to the next, owing to dissimilar tectonic stages and histories of vertical tectonic movements. Only some of the étages may be common to several different lateral tectonic units, if for a period of time these units experienced the same tectonic (mainly, tectono-sedimentary) regime. Magmatism and deformation in a lateral tectonic unit may be indigenous, induced by reworking at depth, or teleorogenic, as is common in median massifs. Some stratal rock assemblages may have internal local discontinuities, such as unconformities related to local and temporary uplifts - but if no significant restructuring occurred, these discontinuities do not serve as étage boundaries: they merely separate sub-étages formed in the same tectonic regime.

Regional metamorphism and magmatism are not coeval with the age span of a stratal-rock assemblage. Some magmatic suites may be syn-depositional, others post-depositional; some may contain comagmatic extrusive and intrusive rock complexes. But as long as they reflect the same tectonic regime, stratal and discordant intrusive rock complexes must be regarded as complementary parts of the same structural-formational étage. Metamorphism may affect deep crustal levels during the deposition of a particular étage or later, but as a tectonic phenomenon, it must be assigned to its corresponding tectonic stage, and must end before that stage may be considered finished. Deformation may be imposed on rocks of a particular tectonic étage and also on the étages below. It may be syn-depositional or post-depositional, involve predominantly folding or faulting, be deep-rooted or rootless, be induced by crustal warping or block movements. Like sedimentological, magmatic and metamorphic peculiarities, deformational styles are phenomenological expressions of a tectonic regime characteristic of a particular region. In that region, each pulse of deformation is assigned to a particular tectonic stage.

Definition of fundamental lateral and vertical tectonic units in the Canadian Cordillera from results of geologic mapping

A general outline of lateral tectonic units in the Canadian Cordillera was published in the first comprehensive synopsis of the geology of Canada (Douglas, ed., 1970). Two decades later, an outstanding map of vertical tectonic rock assemblages was produced under the auspices of DNAG on a scale of 1:2,000,000 (Wheeler and McFeely, comps., 1991). In it, vertical tectonic units (called tectonic assemblages) were arranged in time separately for each orogenic and median-massif zone - Omineca, Intermontane, Coast Belt, Insular. But unfortunately, these designated vertical tectonic units included only stratal rock packages; intrusive rocks were shown independently, even though they are also ascribed to specific time intervals. This separation diminishes the tectonic content of the defined units, as in reality stratal and non-stratal rock bodies together form a fossil record of tectonic regimes that existed during particular times in different regions. Confusion also occurs in the rock-unit nomenclature, as names used by Wheeler and McFeely (comps., 1991) for the assemblages sometimes overlap with names of previously recognized formations.

Quite unconvincing is the inclusion into these authors' assemblage definitions of subjective genetic attributions, such as Cap Mountain "rift-embayment sediments", Purcell-Wernecke "continental margin sediments", Muskwa "passive continental margin sediments" and so on. It is not clear on what rock criteria they distinguished "continental margin sediments" from "passive continental margin sediments", or "platformal continental margin sediments" from "clastic continental margin sediments" from "offshelf continental margin sediments". No criteria are presented to justify attributing the Middle Proterozoic Purcell (Belt) assemblage to a "continental margin", the Late Proterozoic Windermere assemblage to a "clastic continental margin", the Late Proterozoic Rapitan assemblage to a "rift", and the Lower Cambrian Gog assemblage to a "rifted and passive continental margin". In the Cordilleran hinterland, volcano-sedimentary assemblages were assigned variously to "oceanic arcs", "island arcs", "back-arc basins", "marginal basins", "oceanic volcano-sedimentary rock successions", "oceanic crust", "accretionary prism".

Nonetheless, in our description of the eastern Cordilleran miogeosynclinal zone and its Omineca orogenic belt, we make a wide use of Wheeler and McFeely's (comps., 1991) map, as it represents the most recent comprehensive summary of extensive regional geologic mapping in the last several decades. In using it, we make several reservations: no genetic characterizations are attached to stratal rock assemblages and intrusive rock suites; both these main rock-body types are combined together into structural-formational tectonic étages related to regional tectonic stages; descriptions of the stages include not just their sedimentary and magmatic aspects but metamorphic and deformational aspects as well; and much additional mapping-derived information is included from the DNAG volumes describing the Cordilleran mobile megabelt in Canada (Gabrielse and Yorath, eds., 1991) and adjacent regions, and from more recent publications.

Framework of main structural-formational étages making up the post-basement orogenic grain in the eastern Cordilleran miogeosyncline

Precambrian crystalline rocks in the Canadian Cordillera fall into three main U/Pb age groups: 2,100-1,850 Ma, 1,200-1,100 Ma and 800-700 Ma (Parrish, 1991a-b). The first cluster points to the ancient age of the Omineca orogenic zone's ex-cratonic basement. The second cluster, reported from the northern Omineca Belt, may reflect a Middle Proterozoic (pre-Windermere) orogenic event. At any rate, the Omineca orogenic zone is much older than just Mesozoic.

The oldest supra-basement structural-formational super-étage in the eastern Cordilleran miogeosyncline is a huge, ~20 km thick, succession of sedimentary and igneous stratal rocks of Mesoproterozoic age with a general name Belt-Purcell Basin. Great sills of mafic rocks, up to 6 km thick, are embedded in this succession. Granitic plutons are also recognized, and some deformation episodes have been reported. But owing to the complex Laramian thrusting and displacement, the composition of this enormous basin is hard to decipher even with the benefit of modern palinspastic restorations (e.g., Peterson, ed., 1986). It is apparent, though, that several structural-formational étages make up this basin's internal tectonic structure.

Chronostratigraphic correlations Belt-Purcell Basin rocks with rocks in the Mackenzie Mountains region are still uncertain (see in Stott and Aitken, eds., 1993). A consensus exists, on the other hand, that the Cordilleran mobile megabelt became a fully distinct tectonic province on a pre-existing cratonic basement at ~780 Ma (Late Proterozoic), when a rift 7-8 km deep was filled with the Windermere Supergroup clastic succession and its stratigraphic correlatives (e.g., Monger, 1989, 1993).

In Windermere time, a system of grabens was filled by a volcano-sedimentary assemblage dated at ~780 Ma and younger. Together, these grabens are commonly recognized as the Windermere rift, which separated the crust of the incipient Cordilleran mobile megabelt from the reduced craton. Reworked remnants of this separated ancient crust are commonly thought to underlie the Cordilleran interior. The earliest known possible precursor of the Cordilleran mobile megabelt actually lies to the east, in the crystalline basement of the Interior Platform in Alberta, where the NNW-SSE-trending Kimiwan zone of anomalous isotopic signatures in basement rocks indicates an episode of thermo-tectonic crustal reworking around 1,760 Ma (Burwash et al., 1995; Chacko et al., 1995). By Windermere time a billion years later, the locus of tectonism shifted several hundred kilometers to the west.

Lower and Middle Cambrian rocks at the base of the sedimentary cover in the Interior Platform of the North American craton lie, with a profound unconformity, on top of the Archean-Early Proterozoic crystalline rocks of the cratonic basement. In some localities in the Rocky Mountains on trend with the Athabasca-Peace River-Skeena transcurrent zone of weakness, though, Lower Cambrian sedimentary rocks overlie the Upper Proterozoic rocks conformably (McMechan, 1990). Thickness and age span of the Lower Cambrian Gog Formation, which is present only in the platform's western part, increase to the west; the horizons below have been found to contain Ediacaran fauna (Hofmann et al., 1985; Teitz and Mountjoy, 1989). Younger formations also tend to be thicker in the subsided troughs.

During the Paleozoic, the times and settings of sedimentation in the cratonic Interior Platform and the orogenic Omineca zone had much in common. Created in similar conditions, some formations are indistinguishable between these tectonic regions, as they record the same regional tectonic regime during specific intervals of time. The carbonate platforms recognized all over the Interior Platform and in many areas of the Rocky Mountain and Omineca belts are evidence of common paleogeography (Lyatsky et al., 1999). The early Paleozoic sedimentary cover did continue westward from the Interior Platform, and some of its étages blanketed the Omineca Belt region. Remnants of that old cover are now found on both sides of the large Omineca-Belt anticlinoria and of the Northern and Southern Rocky Mountain Trench faults. Composed of Late Proterozoic to Early Carboniferous sedimentary rocks, with thick carbonate packages resistant to erosion, these remnants have long been denoted as separate platforms - MacDonald, Cassiar, Cariboo - or discontinuous basins such as Selwyn. In the orogenic pulses that affected the Omineca Belt, these remnants were substantially deformed. Still, their distribution suggests original continuity.

Intermittent development of local troughs (elongated relative depressions) and carbonate platforms, recorded in the Ordovician to Devonian stratal rocks in the Omineca Belt, has been correlated with the appearances of local rift and platform settings (Cecile et al., 1997). Small manifestations of magmatism punctuate the rift-related intervals. Troughs were more persistent during the Carboniferous and Permian, when the Prophet and Ishbel troughs developed along the eastern Cordilleran miogeosyncline (Henderson, 1989; Richards, 1989). They were shallow, and thickening of stratal rocks in them was only in the low hundreds of meters. These troughs were two-sided, with provenance areas to the east and west. Both sides, facing the craton on the east and the median massif(s) on the west, received sediments of North American derivation, and there is no evidence of a continental margin along the eastern Cordilleran miogeosyncline (Aitken, 1993a-c).

To correlate miogeosynclines with continental margins of modern Atlantic type has been fashionable since the 1960s. Dietz and Holden (1967) compared them with the shelf-slope-rise system flanking modern continental landmasses. Bally (1989) equated the North American miogeosynclines in the Appalachians and the Cordillera, and the miogeosyncline-continental-margin association became a template for the descriptions of the Cordilleran mobile megabelt (Gabrielse and Yorath, eds., 1991; Burchfiel et al., 1992). Recently, though, Aitken (1993a-c) voiced big doubts about the correctness of this hypothesis.

The late Paleozoic Mississippian to Permian Cache Creek rock assemblage, as well as several younger assemblages in the Cordillera, have been described in terms of ancient oceanic floor, oceanic basins, accretionary prisms, mélange, terranes and so on (Fig. 41; e.g., Howell, ed., 1985; Gabrielse and Yorath, eds., 1991). Gabrielse and Yorath (1991, p. 6) recommended applying the term *Ancestral North America* "...to the craton, which throughout the Phanerozoic, remained relatively stable. The western edge of Ancestral North America, from mid-Proterozoic to early Mesozoic time, is thought to have approximately coincided with the western part of the Omineca Belt... Since the early Mesozoic, successive episodes of terrane accretion have led to the modern outline of the western part of the [North American] continent." These authors added: "...The miogeosyncline represents a broad, flat to gently west-sloping depositional surface upon which mainly shallow water carbonates and associated terrigenous clastics were deposited." Therefore, *miogeosyncline* "refers to the westward expanding then tapering wedge of supracrustal rocks that accumulated upon the westerly sloping Precambrian crystalline basement of Ancestral North America from mid-Proterozoic to Middle Jurassic time."

These assertions contradict the fundamental meaning of the tectonic terms *craton* and *mobile megabelt*, *miogeosyncline* and *eugeosyncline* (Stille, 1941; Spizharsky, 1973). Craton is not the same as continent; a miogeosyncline is not a "wedge of supracrustal rocks" and certainly not a "depositional surface" (see also Lyatsky et al., 1999). Rather, a miogeosyncline is a zone of deep crustal reworking, resulting in orogenization of both the basement and the post-basement cover into a single coherent orogenic zone within a mobile megabelt.

The assumed configuration of a Proterozoic or Phanerozoic ancestral North American continent, or the Mesozoic accretion of terranes, are not evident in the Canadian Cordil-

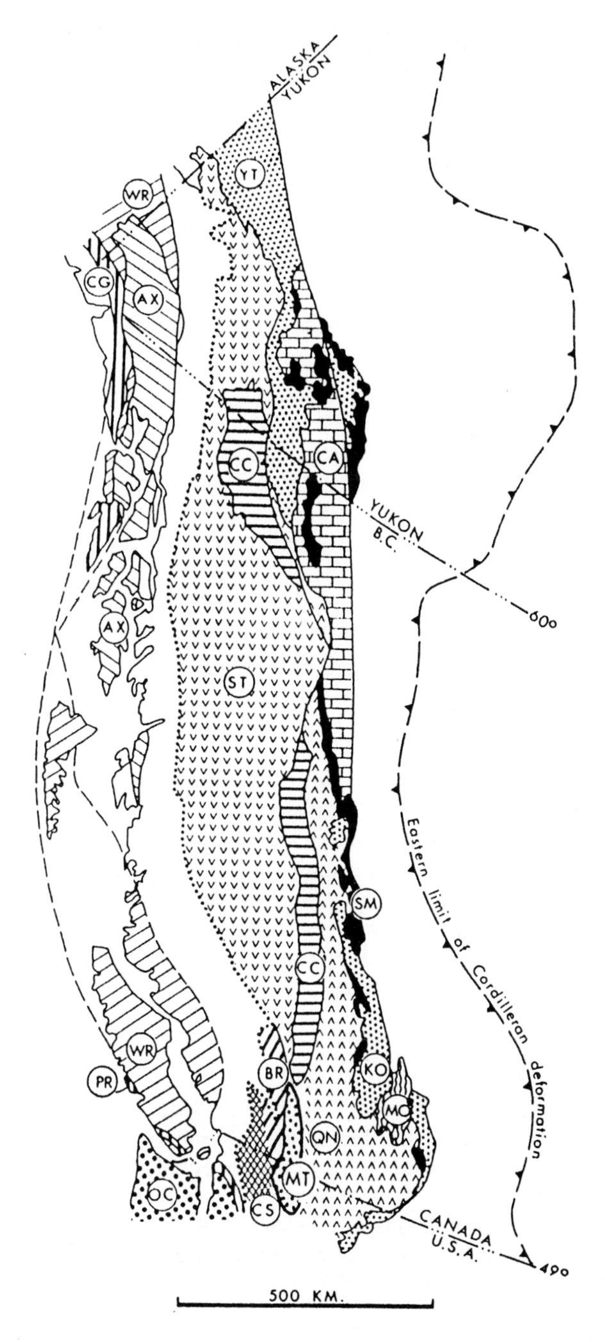

Figure 41. Map of assumed terranes in the Canadian Cordillera, as defined by Monger et al. (1994). Terranes: AX - Alexander; BR - Bridge River; CA - Cassiar; CC - Cache Creek; CG - Chugach; CS - Cascade; KO - Kootenay; MT - Methow-Tyaughton; OC - Olympic; PR - Pacific Rim; QN - Quesnel; SM - Slide Mountain; ST - Stikine; WR - Wrangell; YT - Yukon-Tanana.

lera (contrary to Monger et al., 1982, 1994). Instead, rock-based analysis of lateral and vertical tectonic units permits to reconstruct this megabelt's tectonic history from observable facts rather than from shaky assumptions. Regional analysis of rock assemblages, indigenous structural-formational étages and tectonic stages produces a framework where the evolution of one lithospheric tectonic region is inseparably linked to the evolution of its neighbors.

Mesozoic and Cenozoic tectonic evolution of the southern Omineca orogenic zone

Some new and non-traditional information

Geologic field mapping in the Canadian Cordillera is continually resulting in new information that contradicts the conventional terrane-collage hypothesis of Monger et al. (1991, 1994), Clowes et al. (1995) and Cook (1995a-c). Recently, for example, it was reported that, in the southern Canadian Cordillera, rocks of the supposedly-exotic Quesnel "terrane" are in fact in structurally unbroken continuity with rocks assemblages known to be indigenous to North America. This has led to new-old suggestions that most or all of the Intermontane Belt is native to North America (Erdmer et al., 1999; Thompson et al., 1999). Increasingly, an overwhelming amount of geologic evidence confirms that tectonic zones of the Cordillera involved *in situ*, side by side.

The main Mesozoic orogeny in the Canadian Cordillera has been called Columbian, and it is now considered to be part of great, variously manifested Nevadan orogeny recognized elsewhere in the North American Cordillera (King, 1969, 1977; Burchfiel et al., eds., 1992). The tectonotype of the Columbian orogeny in the Canadian Cordillera could be the Omineca tectonic zone. Having undergone several consecutive orogenies since the Late Proterozoic (Douglas, ed., 1970; Gabrielse and Yorath, eds., 1991), this zone was inverted into a pronounced mountain belt in the Middle-Late Jurassic, providing clastic material for the Bowser Basin in the west and the developing foredeep in the east.

Physiographically, the southern Canadian Cordillera west of the Southern Rocky Mountain Trench contains, from east to west, several distinct terrains trending N-S: Purcell Mountains, Selkirk Mountains, Monashee Mountains, Okanagan-Shuswap Highlands. The Southern Rocky Mountain Trench is oriented NNW-SSE, as are the Continental Ranges just to the east. They truncate these N-S-trending terrains at an oblique angle. The southern Omineca Belt thus looks like a right triangle whose hypotenuse is the Southern Rocky Mountains Trench (Holland, 1964; Mathews, 1986, 1991).

The Southern Rocky Mountain Trench is a valley many kilometers wide, controlled by a NNW-trending fault system of the same name. The Purcell Mountains correspond to the northern nose of the Belt-Purcell Basin, inverted into the Tertiary Purcell Anticlinorium. The Selkirk Mountains are determined by a regional fan structure whose axis is punctuated by a N-S band of metamorphic core complexes (Figs. 23, 42). Similar complexes in other N-S bands run along the Monashee Mountains and Okanagan-Shuswap Highlands.

By the end of the Early Jurassic, metamorphic and magmatic orogenic processes were probably under way at lower-crustal levels, but at the surface they were not yet strongly expressed. In the late Middle Jurassic, the single early Mesozoic shelfal sea in that region was intermittently divided into an extensive marine basin on the west and a segmented one on the east; the eastern basin was segmented due to uplift of central Omineca-Belt blocks (Poulton et al., 1993). The burial-history curve for southeastern British Columbia shows a long period of relative stability from the Pennsylvanian to Middle Jurassic, with major disturbances coming only in the Late Jurassic (op. cit., their Fig. 4H.4). That was the onset of tectonic inversion and mountain building.

In Parrish's (1995) analysis, the Monashee complex, comprising the Malton, Frenchman Cap and Thor-Odin gneiss bodies derived from Early Proterozoic protoliths, is different from paragneisses and orthogneisses of the Shuswap complex. The former is metamorphosed to granulite grade, the latter only to amphibolite (Parrish, 1995). The deepest exhumed crustal horizons are thought to be exposed in the Malton Gneiss in the northern Monashee Mountains (Simony, 1991), but the relationships between various pulses of metamorphism, magmatism and deformation in this area remain hard to disentangle (e.g., Digel et al., 1998). From a recent study of northern Selkirk fan structure (Colpron and Price, 1995), not only the previously metamorphosed Early Proterozoic rocks but also stratified Late Proterozoic and early Paleozoic rocks were buried by 173±5 Ma (Middle Jurassic) to depths of 20-25 km, corresponding to pressures of 6-7 kbar (600-700 MPa). Parrish (1995) noted that, in the Monashee complex, these rocks were strongly metamorphosed in Paleogene to Early Eocene time, around 60 to 55 Ma. Colpron and Price (1995) and Colpron et al. (1996) have found that some 10 km of uplift in the northern Selkirk structure occurred early, by the Late Jurassic. Thermobarometry of post-kinematic mid-Cretaceous plutons also indicates another uplift, prior to ~100 Ma. Woodsworth et al. (1991) proposed for that time "a Hercynotype environment" characterized by block tectonism. For the Eocene, Parrish (1995, p. 1019) proposed crustal-scale faulting which broke "...the southern Omineca belt into blocks that individually resemble metamorphic core complexes". High-angle faulting and movements of crustal blocks with amplitudes of several tens of kilometers can be inferred from this information.

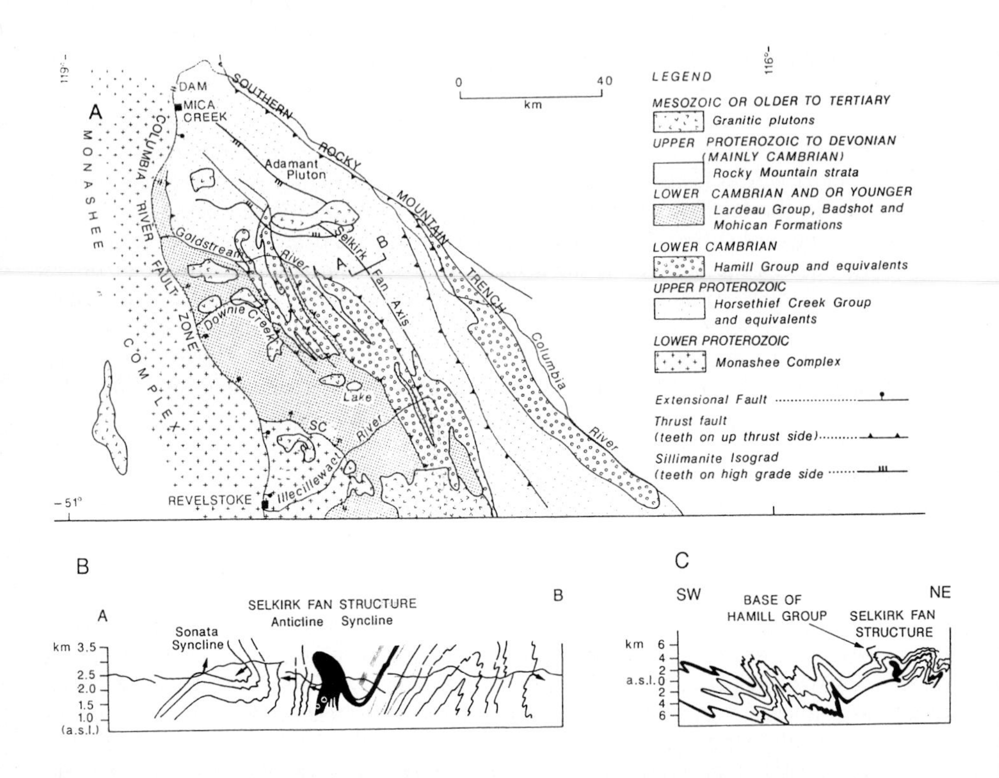

Figure 42. The Selkirk fan structure, in map and cross-section (modified from Brown et al., 1991; see also Fig. 23). SC - Standfast Creek fault. **(A)** - schematic map; **(B)** - cross-section A-B across the fan axis; **(C)** - generalized cross-section.

Ages of metasedimentary, metavolcanic and metaplutonic rocks in the Omineca miogeosynclinal orogenic zone have been established as Early to Middle and Late Proterozoic, Paleozoic, Mesozoic and, in some cases, Early Tertiary Eocene (Parrish, 1995). Rocks are metamorphosed to high grades in the Priest River complex just west of the Purcell Anticlinorium and Purcell Trench, and in the Kettle, Valhalla and other complexes along the axial zone of the Selkirk fan structure. On the western limb of the Omineca orogen, in the Okanagan and Shuswap complexes, the exposed regionally metamorphosed rocks are of lower grades. This indicates shallower burial and lower uplift amplitudes for the crustal blocks in the west.

Contractional structural deformation in the southern part of the Canadian Cordilleran miogeosyncline, including tight and recumbent folds and high- and low-angle reverse faults, is dated as mostly Middle Jurassic to Tertiary, but many such structures, especially in the lower horizons of metamorphic cores, are older (Fyles and Höy, 1991). Their origin is still not completely explained (Struik et al., 1991; Struik, 1991), but they were probably formed at ductile levels in the crust. Onset of an episode of contractional structuring in the Selkirk fan is now dated as Middle Jurassic, 173±5 Ma (Colpron and Price, 1995), when the affected rocks experienced the maximum burial. Late Cretaceous-Early Tertiary contractional structures in the Omineca orogen are correlated with the Laramian orogeny which is also responsible for the fold-and-thrust belt in the Rocky Mountains and Foothills.

The Rocky Mountain fold-and-thrust deformation front in the Foothills marks the latest position of the shifting boundary of the cratonic Interior Platform in Alberta. Between thin-skinned deformation in the Rockies and whole-crust mobilization in the Omineca orogen lies a zone of transition (McDonough and Simony, 1988, 1989; Simony, 1991) which predates the Tertiary Southern Rocky Mountain Trench fault system. In the Rocky Mountains of Montana, the big Lewis thrust was cut by steep Oligocene normal faults soon after thrusting had ended in the Eocene (Constenius, 1988, 1996).

The start of post-Paleozoic contractional deformation in the northern Selkirk Mountains is linked with the formation of SW-verging low-angle nappes in the Middle Jurassic. These initial structures lie east of the Monashee complex, and Colpron et al. (1996) associated it with the western limb of the Selkirk fan. East-dipping normal and thrust faults along the eastern margin of the Monashee complex run approximately N-S for a distance of some 150 km (Colpron et al., 1996, 1998; also Brown et al., 1992).

Brown et al. (1992) inferred two episodes of thrusting in this area, in the Middle Jurassic and in the Late Cretaceous-Paleocene, with the Monashee complex having been uplifted between these episodes. The Monashee complex (or "terrane") was though to be overlapped by the Selkirk allochthon in a 10-25-km-thick nappe of Upper Proterozoic to

Early Paleozoic rocks. Some fold-and-thrust structures in these stratified rocks were thought to be twice overturned. Yet, their vergence east and west of the poorly demarcated axial zone of the Selkirk fan structure was created at different times. East-verging isoclinal folds in the central and northern Selkirk Mountains are probably pre-Jurassic (e.g., Brown and Read, 1983). West-verging folds are predominant in the allochthonous sheet (Brown et al., 1991), but they were thought to predate the allochthon's northeastward displacement (Brown et al., 1986). In that scenario, east-vergent structures on the Selkirk fan's east side postdate the west-vergent structures on the west side. The Selkirk allochthon was thought to have been ramped onto the Monashee "terrane", which found itself in this thrust's footwall but "was not deeply buried until Late Jurassic and Cretaceous time" (Ranalli et al., 1989).

A very different tectonic history of the Selkirk fan area was inferred by Colpron et al. (1996, 1998) from extensive thermo-barometric studies and detailed geologic mapping. According to these authors, SW-verging Middle Jurassic thrusting in that area occurred at 173±5 Ma, when the corresponding crustal block was already buried to 20-25 km depth. Thus, its burial must have occurred before the allochthon's structural overburden was placed over it. Within a very short time of only about 1 m.y., rapid uplift of this block led to removal of ~10 km of rocks from its top. "...Most of the structures in the supracrustal rocks in the Selkirk fan structure had developed by the end of Middle Jurassic time" (Colpron et al., 1996, p. 1373). Rapid denudation of the Selkirk fan took place between 173 and 168 Ma. Importantly, syn-orogenic Middle Jurassic crustal extension, associated with rapid decompression of the rising rocks, affected this area. Unroofing on the western limb of the Selkirk fan occurred at a rapid rate exceeding 2 km/m.y., similar to the rate estimated for the Himalayas in the Miocene.

The Middle to Late Jurassic orogeny is one of the best-recognized in the Cordillera. Granitic plutons with Middle and Late Jurassic ages (180 to 160 Ma; Woodsworth et al., 1991) are abundant the Omineca miogeosynclinal orogenic belt as well as in the Intermontane median-massif belt. The Kuskanax and Nelson batholiths, along with their many satellite intrusive bodies, belong to this family. Coeval with them are volcanic rocks of the Rossland Group (also Powell and Ghent, 1996). The younger plutons of this complex magmatic suite, which are ~165 Ma and younger (Armstrong, 1988), could be post-kinematic. One of them, the Adamant pluton in the northern Selkirk Mountains, is strongly elongated E-W, apparently related to a fault transcurrent to the main NNW-SSE regional structural grain. Colpron et al. (1996) supposed this pluton, ~169 Ma in age, was emplaced while the crust in this area was still deforming rapidly. A similar pluton in the hanging wall of the Columbia River fault, the Galena Bay stock, yields ages of about 160 Ma (Parrish et al., 1988).

Jurassic magmatism concentrated mainly along the Selkirk Mountains and continued south into the U.S. Mid-Cretaceous plutonic magmatism, by contrast, was most abundant in the Purcell Mountains to the east (magmatism returned to the Selkirk fan axial zone later, in the Tertiary). The Standfast Creek fault is cut by the mid-Cretaceous Albert stock and related dikes. The mid-Cretaceous plutons, dated at 110 to 90 Ma, clearly postdate the already-created regional structural fabric in the Selkirk Mountains (Colpron et al., 1996, p. 1375).

Big differences in the tectonic history of the neighboring Selkirk and Monashee mountains have a regional significance. The Columbia River fault which separates them continued to be active after the Standfast Creek fault had died. Reed and Brown (1981) suggested the Late Jurassic Galena Bay stock intruded the Columbia River fault, and Parrish and Armstrong (1987) dated this stock and correlative intrusive bodies at 162 to 157 Ma. Later, however, Parrish et al. (1988) argued that the Columbia River fault was active in the Eocene.

The Columbia River fault is actually a segment of a bigger fault system. If the stratigraphic correlations in this area (Crowley and Brown, 1994) are correct to relate the Selkirk allochthon with the Early Paleozoic block, named Clachnacudainn, which lies between the Standfast and Columbia River faults, then the Columbia River fault cuts the Lower Paleozoic Lardeau Group near the town of Revelstoke. The Clachnacudainn block is commonly included in a huge east-verging system of folds and thrusts. This system continues from the Omineca Belt into the Rocky Mountain Belt of thin-skinned deformation (e.g., Simony, 1991). The relatively high-angle Purcell thrust fault on the eastern flank of the Selkirk Mountains, and the southern Omineca Belt, are cut by the Southern Rocky Mountain Trench fault obliquely, at an angle of ~15°. Both these faults dip to the west or SW, though the former is reverse and the latter normal. The Argonaut Mountain fault, on the west side of the Selkirk fan structure, is also steep, normal and west-dipping. This fault cuts the Downie Creek fault, which is a thrust like the Purcell fault but dips in the opposite direction, to the east. The Standfast Creek fault is another east-dipping thrust. Thus, thrust sheets of rocks assigned to the Upper Proterozoic Windermere Supergroup and the Lower Paleozoic Hamill and Lardeau groups underlie the northern Selkirk Mountains, with thrusts dipping inward, towards the axis of the Selkirk fan structure.

West of the Columbia River normal fault, metamorphic rocks are exposed in a big thrust sheet, which is thought to be antiformal, with the Monashee core complex dominating it. Along the western side of the Monashee complex, a major décollement has been proposed, subsequently projected through the entire Cordilleran crust (Brown et al., 1986, 1992). This presumed huge thrust, dipping WSW at low angles, is thought to be the bottom of the detached, 10-25-km-thick Selkirk allochthon. Its "integral part" is the Clachnacudainn "terrane" which partly overlaps the Monashee complex (Crowley and Brown, 1994). However, in map view, no such single great fault is evident. The Monashee fault separating the Monashee complex in its footwall and the

Selkirk allochthon in its hanging wall is not correlative with the Standfast Creek fault on the eastern side of the Monashee Mountains: the latter, though also a low-angle thrust, juxtaposes Lower Paleozoic stratified rocks over Devonian orthogneiss (Colpron et al., 1996). The Monashee and Standfast Creek faults both dip outward from the Monashee complex, but the rocks they juxtapose are different.

Ranalli et al. (1989, p. 1649) thought that "...the Monashee terrane was not deeply buried until Late Jurassic and Cretaceous time, when the thickened crust of the Selkirk allochthon was thrust northeast and ramped onto the terrane". This contradicts the evidence of Colpron et al. (1996). Later, Johnson and Brown (1996, p. 1597) revised the timing of events considerably: "Accretion of the Intermontane superterrane in the Early and Middle Jurassic involved large-scale northeast-directed thrusting along the suture, followed or accompanied by southwestward thrusting and backfolding along and inboard of this boundary in the Omineca belt". Parrish (1995, p. 1619) was skeptical: "Models of Jurassic and Cretaceous displacement on the Monashee décollement (Brown et al., 1986) contrast with documented Late Cretaceous to Paleocene evolution of the Valhalla complex and its contractional faults..."

Parrish (1995) described the Selkirk allochthon in terms of five zones (or levels), the zone of shearing ascribed to the Monashee décollement being zone 1 and the highest part of the allochthon being zone 5 (Fig. 43). Numerous igneous bodies in zone 1 are granitic, variously late kinematic or post-kinematic. Pegmatite dikes yielded Tertiary ages of 60 to 55 Ma (Late Paleocene to Early Eocene). Zircon crystallization during metamorphism and migmatization in the 1-km-thick Monashee shear zone was found to have occurred at 75 to 59 Ma (Late Cretaceous to Paleocene).

Zone 2 lies between 1 and 4 km above this base. Foliated pegmatite in it has crystallization ages of ~120-115 Ma (post-Neocomian Early Cretaceous). The metamorphism is dated at 89-86 Ma (early Late Cretaceous), matching a regional peak of metamorphism and strong deformation, and it may be correlated to the climax of the Columbian orogeny which affected vast parts of the Canadian Cordillera. Cooling ages of rocks have been found to be 74 to 53 Ma (Late Cretaceous Maastrichtian to Early Eocene).

Leucocratic granites, typical for extensional tectonic episodes, are concentrated in zone 3. They fall into three age groups: ~136 Ma (Neocomian), 105 to 96 Ma (mid-Cretaceous) and 67 to 57 Ma (Paleocene). A peak of palingenic transformation of rocks and granite formation occurred in the mid-Cretaceous, when some rocks were buried at ~700°C and 700 MPa (7 kbar, or about 20-25 km depth).

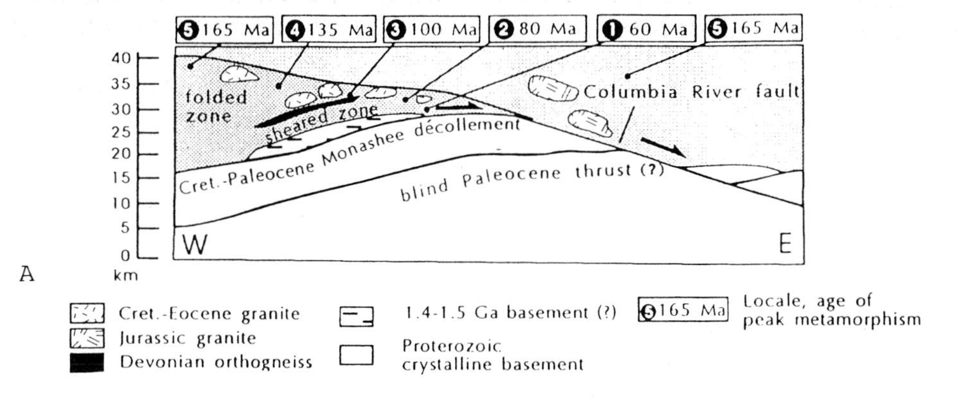

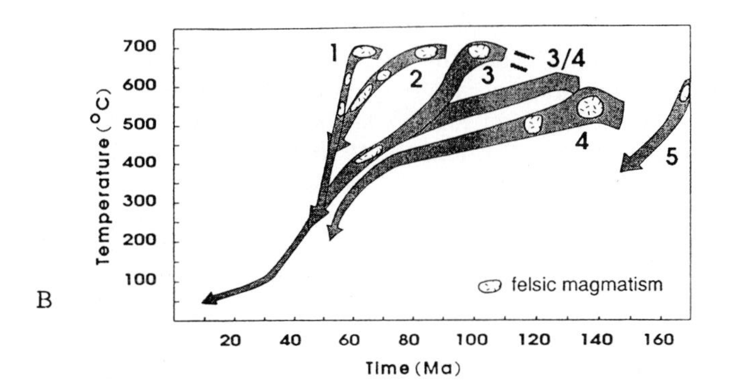

Figure 43. Metamorphic history and structure of the Monashee complex, as inferred by Parrish (1995): (**a**) ages of metamorphism at different structural levels; (**b**) reconstructed metamorphic time-temperature paths of rocks in these levels.

For rocks in zone 4, a major deformational and thermo-metamorphic episode has been dated at ~154 Ma (Late Jurassic), and a second thermal peak at ~135 Ma (Early Cretaceous Neocomian). These rocks did not cool below 400°C till after 80 Ma. However, elsewhere in the Monashee and Cariboo Mountains, a major thermo-metamorphic and deformational episode occurred at 175 to 165 Ma (late Middle Jurassic), coincident with the main episode of regional metamorphism all over eastern Cordillera in Canada (also Greenwood et al., 1991). That episode is well recorded in rocks in zone 5.

Restorations of temperature and pressure paths from studies of K/Ar systems and fission-track data (Fig. 43; Parrish, 1995) show the Selkirk allochthon and the Monashee metamorphic complex underwent several, partly overlapping, periods of uplift and cooling. The most distinctive of these orogenic tectonic events were three: in the Middle Jurassic, mid-Cretaceous and Early Tertiary. The first of these episodes was probably the most regional (also Colpron et al., 1996, their Fig. 12). The second episode (recorded also in zone 3 of the Selkirk allochthon) was the most protracted (cp. Greenwood et al., 1991). The third, in the Early Tertiary around 50 Ma, was the quickest and most abrupt. Each episode was characterized by extensive burial of country rocks some 15-20 km deeper than their position during the preceding uplift, by subsequent felsic magmatism, and by uplift. This suggests repeated vertical movements of separate crustal blocks, up and down.

The northernmost of the metamorphic core complexes, Malton, reveals local peculiarities in its thermal history and the deepest exhumation; the Valhalla complex is also found to be distinct (also Scammell, 1991, 1992; Spear and Parrish, 1996). The major crustal block south of the North Thompson fault appears to be rather coherent, although it may be composite, consisting of smaller sub-blocks. Even between the Malton Gneiss and the Mica Dam Antiform just 35 km away, variations in cooling curves demonstrate that block movements in these areas were not identical (Fig. 43, levels 3 and 3/4).

Parrish (1995) emphasized that the succession of the thermal peaks in that area demonstrates a downward younging through the Selkirk allochthon. This may suggest an intermittent progression in Mesozoic to Tertiary crustal heating. Also variable in time and space was the magnitude and intensity of vertical block movements. Conclusions such as these, however, remain poorly constrained (cp. Colpron et al., 1996).

In the Monashee Mountains area, in an E-W transect from the Columbia River fault to zone 2 in the Selkirk allochthon west of the Monashee shear zone's surface trace, Parrish (1995) also noted a progression in the age of metamorphism downsection, from 170-165 Ma at the high levels to 70-60 Ma at the lowest level. (His data for levels 4 and 5 were supplemented by analyses from the Cariboo Mountains northwest of the

North Thompson fault and 60-80 km from his E-W profile, but because the areas north of the North Thompson fault lie in another major crustal block with a dissimilar tectonic history, these analyses should not be regarded as relevant to the considered profile.)

Middle Jurassic metamorphism is well manifested all through the eastern Cordillera (Greenwood et al., 1991). It affected rocks in the Monashee, Cariboo and Selkirk Mountains (McDonough and Simony, 1988; Colpron et al., 1996) variably, in the brittle and ductile regimes. Mid-Cretaceous metamorphism, though it peaked around 100 Ma, was spread in time from 130 to 80 Ma. It was also widespread, all along the southern and central Omineca Belt (in the Selkirk, Monashee, Cariboo, Cassiar, Omineca Mountains and farther north; Greenwood et al., 1991).

In Parrish's (1995) zones 2 and 3, where regional thermal metamorphism culminated at 105 to 90 Ma, manifestations of the Early Tertiary regional metamorphism are subdued. However, this juvenile metamorphism is widespread in the eastern Cordillera (Greenwood et al., 1991). It is strong in zone 1, which is considered to be affected by pervasive shearing associated with the Monashee décollement. Carr (1992) supposed that tectonic activity on this fault zone in the Thor-Odin area ended at 58 Ma. Younger dates (55 Ma) of undeformed pegmatites that cross-cut and stitch the mylonite zone may indicate that deformation on the Monashee shear zone was somewhat diachronous (Parrish, 1995). New U-Pb determinations on monazites date the thermal metamorphism at about 52-50 Ma (Crowley, 1995).

Parrish (1995, p. 1631) regarded the Selkirk allochthon as a tectonic slice "progressively buried under a tectonic load in the hinterland of a foreland thrust-fold belt as it progrades toward the foreland", but this is not proved. He noted a periodicity of anatectic granitic magmatism there: at 170-160 (average 165) Ma, 140-130 (average 135) Ma, 120-115 (average 117.5) Ma, 100-92 (average 96) Ma, and 70-55 (average 62.5) Ma. Excluding the peak at 117.5 Ma (which is the least evident), these ages suggest a periodicity of ~35 m.y. in the recurrence of thermo-tectonic processes affecting the rocks at depths of 25-35 km. This periodicity is specific to the Omineca miogeosynclinal orogenic belt. No such phenomenon is recorded in the Intermontane median massif to the west or in the Rocky Mountain Belt to the east.

Parrish (1995) noted that, according to his observations, pressures at the peaks of regional thermal metamorphism were 5 to 8 kbar (500 to 800 MPa), suggesting depths of 15-25 km. Such burial could have been achieved by "near steady state evolution" (also Brown et al., 1991) as the allochthon was transported eastward. But the episodic recurrence of burial, heating, metamorphism and granitic magmatism suggests fluctuations in the regional tectonic regime and block movements.

The exposed Monashee complex contains the above-noted suites of anatectic granites, metamorphic rocks of various protoliths attributed to an Early Proterozoic basement, and Late Proterozoic(?) paragneisses overlying the oldest basement. Above the Monashee thrust fault lies an old unit dated as 1,500 to 1,400 Ma (Mesoproterozoic). In the Frenchman Cap culmination in the northern Monashee Mountains area, a migmatized gneissic package contains augen orthogneiss dated at 2,080 Ma. Metamorphic age of schists is 2,060 Ma. Now-deformed granitic dikes postdate them, having an age of ~1,910 Ma. Brown et al. (1986) proposed that all regional metamorphic and deformational events in this part of the Cordillera occurred in middle Mesozoic to Early Tertiary time. But the oldest recognized Phanerozoic metamorphic episodes in various areas fall outside this range. A pre-Late Devonian tectono-magmatic and tectono-deformational episode has been reported from northern parts of the southern Omineca Belt (Okulitch, 1984; Gordey et al., 1987; Gordey, 1991). A pre-Late Mississippian tectono-metamorphic, -magmatic and -deformational episode has been recognized in the Kootenay Arc (Klepacki and Wheeler, 1985).

New geologic facts at odds with the previous hypotheses about the southern Omineca Belt

Deformed Devonian orthogneisses crop out on the west and east sides of the Monashee Mountains uplift north of latitude 51°N, west of the Okanagan Lake fault and east of the Columbia River fault. In the east, these gneisses follow the curvature of the east-dipping Standfast Creek thrust fault. The Clachnacudainn block ("terrane"), bounded by the Standfast Creek fault and presumed to be an "integral part" of the Selkirk allochthon, lies over these Devonian orthogneisses east of the Tertiary Columbia River fault. There, Lower Paleozoic rocks are juxtaposed over these altered Devonian intrusives. Carbonatite intrusions dated at 360 Ma are also found within the Monashee complex (Parrish, 1995, his Fig. 8).

From the above, several important tectonic conclusions can be drawn. The main tectono-metamorphic episodes in the Cordilleran evolution, though severe, were not strong enough to completely reset the original U/Pb systems, and the narrow but big strain zones in the pre-Cordilleran-megabelt basement reveal a pre-existing or synkinematic pattern of crustal blocks (Crowley, 1995). Basement-related Proterozoic crustal blocks were restructured in the Devonian (orthogneiss bodies on the flanks of the Monashee complex have E-W elongation; Wheeler and McFeely, 1991) as well as during the subsequent tectonic episodes.

A common assumption about the southern Omineca Belt is that thermal alteration of the Monashee metamorphic rocks, whose protoliths are dated to be as old as 2,100 to 1,900 Ma (e.g., Parrish, 1991a; Crowley, 1995), occurred only in Middle Jurassic to Early Tertiary time. In fact, the eastern Cordilleran miogeosyncline between latitudes 50° and 52°N experienced episodes of orogenic reworking in the Mesoproterozoic, Neoproterozoic, Cambrian, Middle Silurian, Middle Devonian, pre-Late Mississippian, and later. Zircons and rocks metamorphosed during those times are found in the southern part of the Omineca Belt. In the Monashee Mountains area, Parrish (1995) noted that the Selkirk allochthonous body contains not only stratified but also intrusive rocks with Devonian and younger Paleozoic ages. Klepacki and Wheeler (1985) reported manifestations of strong mid-Paleozoic tectonism in the Goat Range area in the Kootenay Arc north of 50°N (also Fyles and Höy, 1991). Devonian to Early Mississippian granitic (orthogneiss) plutons are found at the latitude of the town of Revelstoke.

According to Read and Brown (1981) and Brown et al. (1986), rocks with Paleozoic metamorphic overprint were in the past located >80 km west of their present position, someplace in the Intermontane median-massif belt. This idea was compatible with the overall concept of terrane-collage construction of the Cordillera that prevailed in the 1970s and 1980s, but it has not been shown to be correct.

Read and Brown (1981) hypothesized that the huge Selkirk allochthon was transported eastward on the regional Monashee décollement as it was pushed by the docking of an accreted Intermontane Superterrane (also Monger et al., 1982). However, under the weight of rock evidence, the ideas of Mesozoic terrane accretion in the North American Cordillera are coming under increasing skepticism (e.g., Ernst, 1988; Rubin and Saleeby, 1992; Dickinson and Butler, 1998). Neither the large eastward displacement of the Selkirk rock mass nor the downward time progression of metamorphism in it are proved (see below). The underlying pre-megabelt Precambrian basement was not reworked uniformly, and dramatic strain disparities occur between differently reworked old crustal blocks (Crowley, 1995). Parrish (1995, p. 1634) observed that in the Monashee complex area "no structural discontinuity has been found at the basement-cover boundary"; and that boundary has instead been described as an unconformity at the top of basement orthogneiss (Scammell and Brown, 1990).

The Monashee zone of shearing lies some 2,500 m above the orthogneiss, in the metamorphosed cover of an ancient craton. This shear zone is impressive: about 1 km thick, mylonitized and containing slickensides that indicate some offsets. Tectonic shearing at this boundary occurred till the Middle Eocene. Abundant leucogranites and pegmatites bear traces of palingenic origin; but in the exposed upper 3 to 3.5 km, Parrish (1995, p. 1634) has found evidence of "prolonged Cordilleran high-temperature conditions like those on the overlying Selkirk allochthon". Pre-megabelt basement orthogneisses dated at 2,100 to 1,900 Ma, and the unconformably overlying cover rocks, were metamorphosed variously. Parrish (1995, p. 1634) regarded "the overlying allochthon as the heat source, with a presumed thermal inversion" (also Journeay, 1986). However, the heat more likely came from below, and episodic thermal events might have resulted due to release of heat trapped under cratonic crust (cp. Cooper, 1990).

The Selkirk tectonic entity (presumed allochthon) containing Late Proterozoic (Windermere) and early Paleozoic rocks may be less structurally offset than is sometimes assumed. To the north, van der Velden and Cook (1996) have supposed that a system of faults (including Purcell) which cuts the southern Omineca Belt or runs along it was formed in the Late Jurassic or Early Cretaceous. These faults sliced the crust, and in Laramide time individual crustal slices were displaced some 100-170 km southwest to their present position on the basal Rocky Mountain detachment. In an E-W profile across the Southern Rocky Mountain Trench some 100 km northwest of the Malton Gneiss, Brown et al. (1986) estimated a shortening of 125 km. McDonough and Simony (1988) estimated the displacements on the pre- to syn-metamorphic Bear Foot thrust to be at least 50 km, and on the post-metamorphic Purcell thrust at least 15 km. Read and Brown (1981) and Crowley and Brown (1994) suggested that the Selkirk allochthon moved over the Monashee complex for a distance of at least 80 km, and normal faults in that area postdate the Monashee thrusting. In an E-W profile slightly south of Revelstoke, Johnson and Brown (1996) have estimated an eastward displacement of 30 km on the Okanagan-Eagle River fault and 15 km on the Columbia River fault. Even the definition of these faults is in places unclear; still less certain are the estimates of assumed lateral displacements across them. The Standfast Creek fault, for instance, has been described variously: east-directed in the Jurassic, west-directed in the Late Jurassic to Early Cretaceous, compressional in the Mesozoic, extensional in the Early Tertiary, but all the same having a minor tectonic significance (cp. Read and Brown, 1981; Parrish et al., 1988; Crowley and Brown, 1994).

Rocks of the Selkirk package were at depths of 20-25 km just 55-60 m.y. ago, and an inverted lower-pressure metamorphic overprinting near the Monashee shear zone might have occurred at that time (Parrish, 1995). But decompression zones are not unusual at mid-crustal depths, and Journeay (1986) thought the inverted pressure curve in this area might reflect normal temperature conditions. Later, Gibson et al. (1998, p. 193) attributed this inversion "to synmetamorphic noncoaxial progressive deformation" rather than heating from the overlying allochthon. Parrish (1995), however, supposed that not only pressure but also temperature inversion is recorded in the minerals in the Selkirk zone 1 to 5, particularly in the felsic rocks that intruded them from Jurassic to Tertiary time.

Parrish (1995) regarded the Late Jurassic-Late Eocene paleotemperature diagrams for these five zones as evidence of thermal inversion due to extremely "rapid cooling" or "quenching" of the area. These laboratory data suggest basement temperature increased slightly between the Late Jurassic and Late Cretaceous, reached its maximum of ~550°C around 60 Ma (mid-Paleocene), and after that dropped to ~250°C in just 5-10 m.y. These changes were explained by deep burial first under a "foreland basin" (meaning, probably, a foredeep) and since the mid-Cretaceous under the assumed overriding thrust sheets. Subsequently, crustal extension and formation of a new generation of faults were followed by rapid uplift and tectonic exhumation of the high-grade Monashee metamorphic core. Journeay's (1986) scenario implies no dichotomy between the Selkirk and Monashee rock assemblages, whereas Parrish's (1995) presumes a metamorphic inversion that is difficult to explain and is now "refuted" (Gibson et al., 1998).

Factual geological information from this region provides no evidence for the idea of depositional burial followed by thrust loading. No thick Jurassic foreland or foredeep basin is known in the Omineca Belt or vicinity. The small Fernie Basin near the British Columbia-Alberta border was rudimentary, thin and local (Poulton et al., 1993). Total thickness of the Jurassic Fernie Formation is just ~400 m, and that of the Kootenay Group in that area only ~1,000 m. The Fernie Basin was local, associated with a transverse fault zone that cut the Rocky Mountain Belt and created the Crowsnest deflection in the Cordilleran grain. Away from it, the corresponding rock units are thinner (Stott et al., 1991; Poulton et al., 1993). Apart from volcanic rocks of the Rossland Group in the southernmost Omineca Belt, no other Jurassic or later Mesozoic stratified rocks are known in this region. The Jurassic system reaches its maximum total thickness of ~2 km only in the northeasternmost British Columbia, in the Interior Platform of the modern North American craton. During Jurassic time, a rather shallow shelfal sea existed along the eastern flank of the Omineca Belt, but geanticlinal rise occurred there by the end of the Jurassic. Middle Jurassic volcanism was minor and local. No deep "foreland basin" overlapped the older miogeosyncline. For the Late Jurassic, Poulton and Aitken (1989) suggested the axis of the embryonic foredeep lay just west of the Foothills-Rocky Mountains boundary, far to the east of where Parrish's (1995) foredeep would have been.

There is no consensus on the amount of Mesozoic and Cenozoic tectonic shortening in the eastern Canadian Cordillera, and large-scale translation of the Selkirk allochthon from the west is also problematic (though slickensides indicate some eastward displacements in the Monashee shear zone). McDonough and Simony (1988) found that in the western Rocky Mountains both the cover and the basement were involved in Laramian shortening. In an area straddling the Canada-U.S. border, the Kishenehn Formation of latest Eocene-earliest Oligocene age lies in a graben in the Flathead valley whose base is downdropped some 4 km. A west-dipping listric fault responsible for the appearance of this graben cuts the Laramide thrust stack (Price, 1994), as do similar middle and Late Tertiary faults in Montana (Constenius, 1988, 1996). An angular unconformity at the base of the Kishenehn Formation, dated at 36-35 Ma, marks the end of the Laramian fold-and-thrust deformation. On the Monashee shear zone, the last displacement occurred earlier, no later than 50 Ma.

In the Interior Platform around the city of Calgary, the top of the Precambrian crystalline basement lies at ~3,500 m depth below the surface; in the Foothills it lies at ~4,000 m. The top of the Upper Devonian in the Foothills west of Calgary is at ~3,100 m depth. In that region, Laramide fold-and-thrust shortening in the Devonian Palliser Formation is estimated to be around 65 km, and in the Cretaceous and Tertiary formations a few tens of kilometers. Cratonward translation of thrust sheets is well mapped in the Front and Main Ranges of the Rocky Mountains At roughly the latitude of Calgary, Price (1981) estimated the probable total shortening to be ~200 km. In the poorly accessible high mountains, the realism of these estimates is hard to assess. The entire Rocky and Purcell Mountains have a width of ~250 km near the Canada-U.S. border, but at 52°N they are only ~150 km wide. North of that latitude, Brown et al. (1986) estimated the shortening to be just 125 km.

In the easternmost Cordillera west of Calgary, Bally et al. (1966) showed that Laramian deformation did not involve the cratonic basement. After that, the same assumption was applied throughout the Rocky Mountains, but it may not be true in the Rocky Mountains' inner belts. Except where steep transverse faults are present, active basement influence could be hard to demonstrate. Near the Rocky Mountain-Omineca Belt boundary, McDonough and Simony (1988) estimated that on the Bear Foot and Purcell thrusts basement rocks were displaced a minimum of 65 km. This estimate is similar to the shortening proposed for the Upper Devonian Palliser Formation west of Calgary, where the palinspastic reconstructions are more reliable owing to a better biostratigraphic control.

Some 250 km northwest of Calgary, Late Devonian Frasnian barrier reefs in the cratonic platformal Alberta Basin (Fig. 44) are transected by prominent NE-SW-oriented paleogeographic features. Generally the Paleozoic platformal deposits in the Alberta Basin are laterally extensive and have a fairly uniform thickness, consistent with cratonic rather than continental-margin paleo-conditions (Lyatsky et al., 1999). Facies variations were related mostly to variations in local conditions, induced by steep basement faults. In the Late Devonian, to the northwest and southeast of one of the transverse NE-SW features (the so-called Cline Channel) massive stromatoporoid-dominated buildups, Cairn and Fairholme, are elongated along the Rocky Mountains trend. The NE-SW trend is discordant with the old and modern trends in the Foothills and Main Ranges. The NNW-SSE-trending Western Alberta Arch (Ridge), along whose trend the Cairn and Fairholme reef complexes are arranged, was a high-standing, elongated cratonic feature during the Silurian and most of the Devonian. Upper Devonian rocks in that area overlie the deeply eroded lower Paleozoic rocks, mostly Cambrian. The thick, NW-SE-trending Robson Basin was characterized by lateral changes in lithofacies which are hard to explain except by syndepositional faulting (Young, 1979). The Columbia Basin, west of the modern Southern Rocky Mountain Trench fault system, is roughly coeval with the Robson Basin, but contains the ~1,500-m-thick Hamill Group of quartzitic rocks whose lower strata are Late Proterozoic.

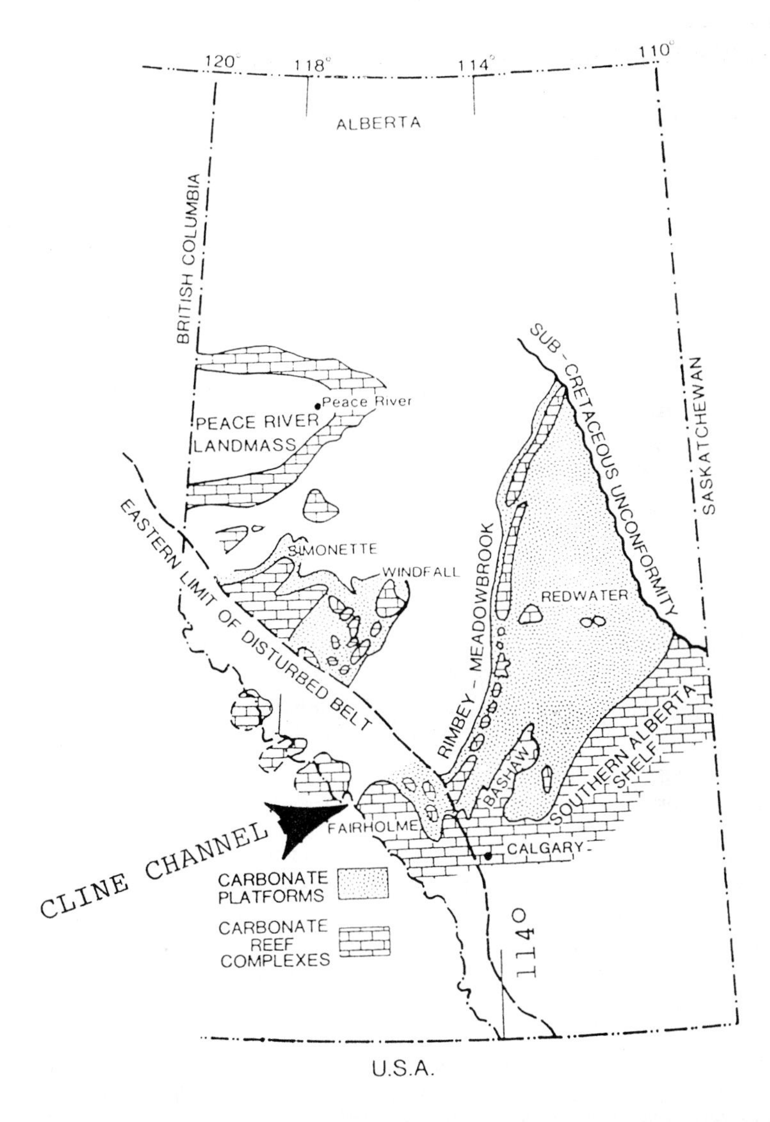

Figure 44. Paleogeographic map of the Upper Devonian Woodbend Group in the cratonic platformal Alberta Basin, showing the main reef trends (modified from Stoakes and Wendte, 1987; Osadetz, 1989). The NE-SW basement-controlled trend is strongly manifested in the distribution of reef chains, carbonate platforms and shale basins; the controlling steep basement faults are expressed in gravity and magnetic anomaly maps (Edwards et al., 1998; Lyatsky et al., 1999).

The Cline Channel succession between the Fairholme and Cairn reefal buildups is exposed in four major Laramian thrust sheets for a distance of some 60 km across the Rocky Mountains (Workum and Hedinger, 1987). Regional persistence of these lithostratigraphic relationships points to the great scale and influence of ancient, Late Archean-Early Proterozoic, NE-SW-oriented faults on the Late Devonian structural-formational patterns on the site of the Rocky Mountains (Mountjoy, 1980). The NW-SE trends were also active in that area in the Silurian and Ordovician (Norford, 1969) and Cambrian and late Precambrian (Fritz, 1991). After channels such as this were depositionally smoothed over, broad shelfal conditions spread all over the western Interior Platform in Alberta and the eastern Cordilleran miogeosyncline, and similar but shaly platformal conditions continued into the Early Carboniferous (Geldsetzer et al., eds., 1988). The old transcurrent NE-SW structural trends that continue from the craton and cut the Cordilleran grain were still active during the vigorous orogenic episodes that affected the Cordillera in the late Paleozoic, Mesozoic and Early Tertiary, and their *in situ* persistence is evidence against large lateral dislocations of crustal blocks.

At latitude 52°30'N, McDonough and Simony (1988) estimated the shortening to be 240 km from the Foothills-Front Ranges boundary to the eastern Cariboo Mountains. Of that number, only 10 km of shortening was found in Cretaceous rocks of the western Alberta Platform, and 65 km in the platformal Devonian Palliser Formation. Most of the shortening was thought to be in the Main Ranges and farther west. Translation of the basement was thought to be in excess of 200 km, which is >2.5 times the 80 km eastward translation proposed for the Selkirk allochthon by Read and Brown (1981). In a cross-section perpendicular to the trend of the Inner Rocky and Cariboo Mountains (which was presumed to include the accreted Slide Mountain and Quesnel terranes) just north of 52°N/120°W, Struik (1987) estimated the shortening since the Triassic to have been even larger, 526 km. He indicated some basement-level thrusting, but showed the basement as being less shortened than the cover (although he regarded as the Devonian orthogneisses in the Quesnel Lake area to be a disturbed pre-Triassic cratonic basement).

McDonough and Simony (1988, p. 1699) supposed that "presence of [the Proterozoic] basement that has been translated in excess of 200 km suggests that shortening in the Rocky Mountains must also be balanced at the level of basement southwest of the present position of the Monashee Complex". In Struik's (1987) palinspastically restored 276-km-long cross-section, the Malton Gneiss was treated as a horst ("Malton Arch"), and the shortening between it and the Southern Rocky Mountain Trench was estimated to be on the order of 100 km. In the western ~100 km (west of the "Malton Arch") of Struik's cross-section, the shortening was estimated to be 380 km. Later, though, Struik (1991) acknowledged that, in the area where the Slide Mountain "terrane" supposedly overthrusted the Omineca miogeosyncline, thrust faults are expressed "poorly" and "the sense of thrusting is unknown". In the Quesnel "terrane", although it was partly thrusted eastward, the folds are mostly upright, open to tight; they become more asymmetric to the east, but not drastically, and shear zones are roughly parallel to bedding.

Until a consensus is reached about the eastern Cordilleran tectonic shortening on the basis of objective geologic mapping, speculations about the distance of lateral translations of rock units should be treated with caution.

Need for caution in balancing cross-sections in orogenic zones

The now-standard worldwide practice of balancing geologic cross-sections primarily originated in the Canadian Rocky Mountains and Foothills in the 1950s and 1960s from the practical needs of the petroleum industry in Calgary (Bally et al., 1966; Dahlstrom, 1969, 1970). The principle of the method was explained by Woodward et al. (1989, p. 1) as follows: "The 'balance' occurs when bed-lengths, or cross-section areas, are equal in both the deformed and the undeformed state". Recognizing the limitations of this technique, they added a caveat for the unwary: "A balanced cross-section is not necessarily truth, it is simply a model which satisfies a larger number of reasonable constraints" (op. cit., emphasis theirs).

Balancing, employed to make palinspastic restorations and estimate structural shortening, relies on a set of geometrical rules and mechanical assumptions about rock deformation. It requires maximum advance knowledge of correlations of stratigraphic rock units. Knowledge of regional tectonic history, episodes of restructuring, tectonic displacements and structural configurations is also needed. Uncertainties arise also due to factors missed by the assumptions: primary lateral changes in unit thicknesses and lithofacies, as well as secondary phenomena such as dissolution of carbonates or salt and movements of rocks in and out of the plane of cross-section.

In orogenic interiors, to make such reconstructions is especially difficult. Unconfirmed assumptions are often made about near-flatness of detachment surfaces, uninvolvement of the crystalline basement in the deformation, and simplicity of fault geometries (e.g., simple planes rather than anastomosing faults). Presence of basement-derived igneous rocks, which are common in orogenic zones but not accounted for in conventional balancing, greatly alters the distribution of rocks both above and below the assumed detachment. Involvement of basement in the deformation may make the definitions of master detachment rely too heavily on subjective assumptions. Tilting and internal deformation of fault-bounded blocks and slices may also complicate the restorations considerably, and so can a lack of adequate knowledge about the history of displacements on each fault. As well, variations in the trend and strike of major structural features in the Foothills and outer Rocky Mountains, both local and regional, suggest vergence changes in space and time, permitting rocks to enter and leave the cross-section plane. Under these conditions, two-dimensional cross-sections may not be fully adequate for palinspastic reconstructions, especially in the complex inner parts of the Rocky Mountain fold-and-thrust belt. Continuity into the Cordilleran mobile megabelt of old, transcurrent, NE-SW-trending crustal weakness zones creates blocks with dissimilar sets of structural-formational étages, in which shortening was accommodated differently (cp. Sloss, 1988; Burwash et al., 1993; McMechan and Thompson, 1993; Cecile et al., 1997).

Systems of faults and folds that develop in severe compressional tectonic regimes are extremely complex and largely unpredictable. Even small variations in lithology and structure can cause faults to deviate from their theoretically estimated geometries, as model assumptions tend to average rock properties unrealistically. In fold-and-thrust belts, compressive deformations produce weak initial folds in stratal rocks, then recumbent folds with a predominant vergence, local thrust faults with splays, big low-angle thrust faults separating detached rock sheets of variable thickness, secondary folds superimposed on thrust sheets and faults, younger forward and back thrusts, and so on. A master low-angle thrust-fault plane, called décollement or detachment, underlies the frontal part of this complex stack; it deepens towards the mobile-megabelt hinterland and commonly loses definition there. Deep-seated, steep pre-existing zones of crustal weakness may localize the outward-vergent thrusts and folds (cp., e.g., Perry et al., 1983; Thompson, 1989; Rusmore and Woodsworth, 1991; McMechan and Thompson, 1993; Lawton et al., 1994; Lebel et al., 1996; Trop and Ridgway, 1997).

Low-angle faults are not always clearly apparent in mappable outcrops, especially if they are parallel to bedding. Master detachments usually remain hidden in the subsurface, unseen. So do some fault splays, the so-called blind thrusts. Commonly, only steep thrust-fault splays are clearly identifiable. The classic *charriers*, described in Europe's Alpine externides, are less obvious in the North American Rocky Mountains, and only deep drilling made it possible to confirm the idea of stacked sets of thrust sheets with an inverted age order (Bally et al., 1966). The absence of a single structural style along the deformation belt of the Rocky Mountains in the U.S. and Canada makes it difficult to construct balanced cross-sections based on a single set of assumptions and principles.

Cratonward progression of fold-and-thrust belts can make the outer margin of a mobile megabelt very spectacular, but in the megabelt interior such zones also occur on a smaller scale. Internal differences develop even within individual fold-and-thrust zones. The cratonward edge of the Rocky Mountain Belt differs structurally from this belt's inner parts; in particular, it contains triangle zones some 10 km wide (Gordy et al., 1977) which are bounded by east-dipping faults on the east and west-dipping faults on the west. The cratonward-vergent thrust stack in that area is dominated by faults dipping towards the Cordillera, but locally there are countless imbrications, backthrusts, duplexes, blind faults, etc. The imbrication style strongly depends on the lithology of disrupted rocks, and particularly on their mechanical competence and strength (Dahlstrom, 1970; Price and Fermor, 1985; Douglas and Lebel, 1993).

Several thrust-fault zones in the Canadian Rocky Mountain fold-and-thrust belt are very prominent and continuous: from east to west, Brazeau-Mill Creek, McConnell, Lewis (especially in the south), Bourgeau, Simpson Pass, Chatter Creek-Purcell. Some of them have surface traces continuing along the eastern Cordillera in Canada and the U.S. for hundreds of kilometers. Ramps are created mostly by mechanically competent Paleozoic carbonates, in which thrust splays become steep. On the other hand, shaly stratigraphic units, such as some Paleozoic intervals and the Jurassic Fernie Group, provide good low-angle glide horizons (Boyer, 1992; Cooper, 1992).

But even in the extensively mapped, drilled and seismically imaged Foothills and Front Ranges of the Rocky Mountains, the geometry and kinematics of Late Cretaceous to Tertiary Laramian thrusting remain difficult to unravel (cp., e.g., Bally et al., 1966; Price, 1981; Thompson, 1989; Mountjoy, 1992; Lawton et al., 1994; Lebel et al., 1996; Fermor, 1999; McMechan, 1999). Beyond this area, the controls available for cross-section balancing in other parts of the Rocky Mountains and elsewhere in the Cordillera are far more limited, and estimates of shortening and very imprecise.

Some of these estimates in orogenic areas rely too heavily on deep seismic images available from the COCORP and Lithoprobe programs. Seismic techniques are generally unable to detect steep discontinuities in the crust, including those related to ubiquitous subvertical faults. What these techniques reveal is low-angle discontinuities, of unknown geologic nature and age, across which the acoustic properties of rocks change. Reflections in seismic profiles may have very different origins: lithologic changes in a normal stratigraphic succession, old or modern metamorphic fronts and rheologic brittle-ductile transitions, igneous rock bodies, erosional unconformities, shear or fault zones, and so on. To discriminate between the multitude of genetic alternatives from seismic data alone is impossible. In addition to primary reflections, seismic images misleadingly contain noise of various sorts, artifacts of processing, and other geologically meaningless but coherent features that are easily misinterpreted as useful signal.

After many years of speculations about heavy thrust slicing of the crust, supposedly recorded in seismic reflection images (e.g., in the Appalachians; Ando et al., 1983; Cook et al., 1983), the fact remains that low-angle thrust faults are mainly localized in rootless fold-and-thrust belts rimming some orogenic zones and most mobile megabelts' cratonward peripheries. Beyond these specific zones, low-angle faults (reverse or normal) are rare (Dunne, 1996). In some areas well controlled by geologic mapping and drilling (for instance, in Pennsylvania), presumed low-angle thrust faults have actually turned out to be synclinal hinges in Paleozoic stratal rocks. Drillhole and surface geologic data have also negated models of low-angle normal crustal-scale faults in the Cordillera (Anders and Christie-Blick, 1994; Beratan, ed., 1996), although such models were previously popular (Wernicke, 1985). Estimation of shortening (and extension) in mobile megabelts is a very complicated task, which should rely not on assumptions and models but on real data from observed rocks.

Basing conclusions on observable geological facts as a sound alternative to fitting data to pre-conceived models

In a general physical model of oceanic-lithosphere evolution in the Pacific (e.g., Engebretson et al., 1985), during the last 180-185 Ma ~13,000 km of oceanic plates must have been subducted under the western margin of the North American continent and recycled into the mantle (e.g., Monger, 1993). No transcurrent fault in the Cordillera has an offset so much as approaching such an enormous amplitude, and no shortening is estimated with such an enormous value. Even in the most adventurous terrane-accretion scenarios, only a ~500-km-wide crustal belt west of the Omineca miogeosyncline is thought to consist of terranes whose origins are in Southeast Asia or South America (e.g., Davis et al., 1978; Monger et al., 1982). These "suspect" terranes, supposedly accreted in a kaleidoscopic, accidental fashion and made up of ocean-floor pieces, seamounts and island arcs, form the presumably-exotic accretionary part of the western Cordillera, which is defined as a collage (Davis et al., 1978; Coney et al., 1980). This so-called terrane-tectonic theory was popular in the 1980s, for two reasons: it was formally consistent with the presumed kinematics of Pacific-Ocean plates, and it greatly facilitated the then-popular form of structural (some called it tectonic) analysis of orogenic areas in the entire Circum-Pacific.

The Lithoprobe program in Canada still regards terranes "...as fundamental building blocks of the [North American] continent" (Clowes, 1996, p. 111). This once-fashionable concept fails to correspond to a growing volume of real geological facts (for details, see Lyatsky, 1996), with which it increasingly often finds itself in disagreement (e.g., Ernst, 1988; Woodsworth et al., 1991; van der Heyden, 1992; Dover, 1994; Tagami and Dumitru, 1996; Dickinson and Butler, 1998). Yet, many Lithoprobe publications still force-fit newly obtained data, especially geophysical and geochemical, to tectonic templates whose very foundations are now undergoing fundamental revision.

Monger et al. (1994) still believed the Lithoprobe and COCORP transects support the idea of general structure of the southern Canadian Cordillera as two parallel, NNW-SSE-elongated stacks of crustal slices separated by enormous low-angle thrust faults. In the west (including the Intermontane median-massif belt), these thrust sheets supposedly dip eastward due to assumed Mesozoic and ongoing subduction under the supposedly far-traveled accreted terranes. The terranes themselves were thought to have somehow come together into the Intermontane and Insular "superterranes" before their accretion to North America (also Monger et al., 1982). The east-dipping crustal-scale thrust sheets involved the assumed subducting oceanic crust of Triassic and younger ages. This simple picture overlooks much of the geologic reality in the Cordillera, including the bilateral geometry of the Coast Belt orogenic zone in western British Columbia (e.g., Rusmore and Woodsworth, 1991).

The eastern Cordilleran stack of west-dipping crustal slices, including the Omineca miogeosyncline and the Rocky Mountain Belt of imbricated thrust sheets above the crystalline basement, experienced an assumed shortening "in excess of 300 km" (Monger et al., 1994, p. 393). Formation of this stack "...apparently involved subduction of (probably) attenuated and normal continental crust in a back-arc position relative to subduction-related magmatic rocks farther west".

Monger et al. (1994, p. 371) argued that the entire "origins and evolution of Cordilleran crust" can be considered in terms of just passive-margin and active-margin tectonic regimes: "...the mid-Proterozoic to Middle Jurassic (1500 to 170 Ma) largely passive cratonic margin, dominated by crustal extension; the Middle Jurassic to Paleocene (170 to 60 Ma) active margin in which the dominant features of Cordilleran crust were established in a contractional and transpressive regime; and the Eocene (60 to 40 Ma) active margin, which features extension and transtension within a broad magmatic arc..." This assessment of long-know facts about the Cordillera is incorrect. Recurrent orogenic episodes, recorded due to their magmatic-intrusion, metamorphic and deformational manifestations, are dated as Late Proterozoic, middle and late Paleozoic, as well several events in the Mesozoic and Cenozoic (e.g., McMechan, 1991; Goble et al., 1995).

Monger et al. (1994, p. 371) stated that the "Middle Proterozoic to Middle Jurassic... cratonic margin" (sic - not continental margin) contained a crystalline basement of Archean-Early Proterozoic age which is a "southwestern extension of the Canadian Shield". This definition is flawed, because a shield is a specific lateral tectonic unit, characterized at the surface by a lack of volcano-sedimentary cover and limited by its deep-seated boundaries (Lyatsky et al., 1999). In this definition, shield rocks cannot be recognized in a mobile megabelt and specifically in the Omineca-Belt core complexes.

Arbitrarily, in the eastern Cordillera Monger et al. (1994, p. 371-372) lumped together into a single "package" rocks of the Belt-Purcell, Windermere and Cambrian through Jurassic successions. Another arbitrarily defined "package of rocks" includes Early Proterozoic and younger formations with an unspecified "strong Mesozoic structural fabric", whose "geological record differs" from that in the first rock package. The Monashee and Malton metamorphic core complexes and their surrounding Paleozoic rocks were considered to be constituents of the second package. Some other core complexes, however, have been included into a third package, farther west, particularly in the Intermontane Belt (e.g., Friedman and Armstrong, 1988, 1991), but because they are surrounded by Mesozoic rocks assigned to other terranes, their significance was dismissed as "tenuous".

In the interpretation of Monger et al. (1994), the beginning of the eastern Cordilleran miogeosyncline is linked to the initiation of the Belt-Purcell sedimentary basin, which originated in the Mesoproterozoic around 1,500 Ma (see also McMechan and Price, 1982). This is not correct, as the Belt-Purcell Basin has been found to be intracratonic (Harrison, 1972; Peterson, ed., 1986; Aitken, 1993a-c). Later episodes of metamorphism, granitic magmatism and folding there might have been induced from other parts of the Cordilleran mobile megabelt. The earliest possible precursor of this megabelt (the Kimiwan geochemical anomaly in the cratonic basement of the Alberta Platform; Burwash et al., 1995; Chacko et al., 1995) may be as old as ~1,750 Ma.

Without strong evidence, Monger et al. (1994) hypothesized that, as the entire Belt-Purcell Basin was dissected during the Windermere rifting in the Late Proterozoic, pieces of an ancient North American continent were transported to "another part of the globe" (see also Sears and Price, 1978; Ross, 1991). Monger et al. (1994, p. 373) speculated also that "...extension, separation, and possible removal of parts of the Late Proterozoic continental terrace wedge" occurred between the latest Proterozoic-early Paleozoic and Jurassic (see also Price, 1994). This idea relies on an unjustified model of passive-margin evolution of Bond and Kominz (1984) for the site of the Omineca Belt. Other studies have shown fundamental incorrectness of that model's analysis of the timing and rate of basin subsidence in the eastern Cordillera: no breakup unconformity or hinge zones are apparent parallel to the assumed Paleozoic shelf-slope-rise system, and these sedimentary basins were not west-facing but two-sided (Henderson, 1989; Richards, 1989; Morrow, 1991; Aitken, 1993a-c; Henderson et al., 1993; Richards et al., 1993).

The idea of an ancestral North America with an ancient continental slope serving as a buttress to randomly docking terranes requires several questionable assumptions: (a) permanent crustal ramp somewhere in the western Omineca Belt, where the ancient continental margin supposedly lay (Cook et al., 1991a-b, 1992; Cook, 1995b-c); (b) several sutures between squeezed remnants of oceanic crust (Monger et al., 1972; Tempelman-Kluit, 1979); (c) incompatible structural-formational characteristics between crustal blocks assigned to different "tectono-stratigraphic" terranes (also Monger et al., 1982). Three lines of evidence have been listed in support of the of terrane-collage model for the Canadian Cordillera (Price et al., 1981; Monger et al., 1982): (a) biostratigraphic, based on an erroneous interpretation of the Permian Cache Creek section in the Intermontane Belt (Monger and Ross, 1971); (b) paleomagnetic, based on dubious estimates from understudied rocks in complex orogenic areas (e.g., Irving and Yole, 1972; Irving and Monger, 1987); and (c) structural, based on exaggerated estimates of strike-slip motions on some major faults in the Cordillera (e.g., Wheeler and Gabrielse, coords., 1972). None of these lines of evidence has passed the test of time. Interpretation of the Cache Creek and other faunas has been revised (e.g., Nelson and Nelson, 1985; also Newton, 1988); the paleomagnetic data have been reinterpreted or found to be dubious (Butler et al., 1989; Monger and Price, 1996; Dickinson and Butler, 1998; Mahoney et al., 1999); geologically unconstrained estimates of strike-slip displacements have been shown to be in error (McDonough and Simony, 1988).

Tectonic interpretation of geochemical and isotopic data, constrained by regional geology

In the eastern Cordillera, the "assemblages of rocks" of Monger et al. (1994) and the "rock assemblages" of Wheeler and McFeely (comps., 1991) are different in their very essence. Monger et al. (1994) in effect confused them with lateral tectonic units of whole-crust extent, whereas Wheeler and McFeely (comps., 1991) defined big vertical tectonic units in the regional rock record. In the latter definition, they are approaching the notion of structural-formational tectonic étages, except that an étage should also include igneous rock suites of the same tectonic stage.

In the interpretation of Wheeler and McFeely (comps., 1991), some vertical tectonic units extend across the lateral-tectonic-unit zonation. This indicates that, at times, similar regional tectonic regimes existed in different crustal zones, despite the overall dissimilarity of these zones' history of sedimentation, magmatism and restructuring. In that system of rock units, the Late Archean-Early Proterozoic metamorphic rocks, orthogneisses and paragneisses, are regarded as pre-megabelt continental-crust remnants (basement) preserved despite the tectonic reworking. These remnants can be recognized petrologically (by determining protoliths), mineralogically (by analyzing mineral grains) or geochemically (by detecting alien elements in isotopic systems). Far less reliable is an assumption of basement presence in orogenic areas from regional geophysical characteristics, because these characteristics are not uniquely interpretable nor indicative of the rock age.

Late Archean ages have been found from some Sm-Nd crustal residence systems, but most Precambrian ages in the eastern Cordillera are Proterozoic. Metamorphic rocks in the core complexes were reworked at deep levels in the crust, but based on their protolith ages they are attributed to a craton that predated the Cordilleran megabelt. From the Interior Platform east of the Rocky Mountains, the top of the modern craton's basement can be traced fairly reliably in seismic profiles as far west as the Southern Rocky Mountain Trench and, in places, the Purcell Trench. Geochemically, basement-rock signatures are also detected much farther west (Reed et al., eds., 1993; Doughty et al., 1998).

Armstrong (1988), Armstrong and Ghosh (1990) and Armstrong et al. (1991) sought the limit of pre-megabelt continental crust based on $^{87}Sr/^{86}Sr$ ratios (by convention, this ratio is 0.706 or higher for continental crust). Monger et al. (1994) questioned the relevance of these results to the tectonic analysis of the Cordillera, on the grounds that crustal slices were presumably displaced a great distance laterally and are not *in situ*. But considerable displacements are not geologically proved in the Omineca miogeosynclinal zone nor in the Intermontane median-massif belt. On the other hand, the $^{87}Sr/^{86}Sr=0.705$ contour, which presumably outlines mantle-derived rocks, does not correspond to the eastern limit of supposedly accreted terranes. In the collage hypothesis, the terrane limit must lie much farther east, somewhere in the eastern Omineca Belt (Gabrielse and Yorath, eds., 1991; Monger et al., 1994).

To reduce the significance of geochemical and petrological evidence for ancient North American affinity of Cordilleran regions, Monger et al. (1994) suggested that the amount of Precambrian material in the Cordilleran metamorphic cores is unclear. This uncertainty exists, but it does not undermine the importance of the presence of basement-derived rocks and minerals.

Fitting tectono-metamorphic data to the model of Cordilleran evolution by long-lived eastward subduction and terrane accretion, Parrish (1995) interpreted the Selkirk allochthon and the underlying Monashee complex as two structural levels juxtaposed across the Monashee décollement shear zone during a 100-m.y. time span, from the Middle Jurassic to Paleocene. But the Selkirk allochthon experienced a polymetamorphic history, with three peaks - in the Middle Jurassic, mid-Cretaceous and Early Tertiary - corresponding to those in the entire Omineca miogeosynclinal orogenic zone. The Mesozoic-Cenozoic history of the underlying Monashee complex is similar, although prior to that these old rocks had been metamorphosed in the Early Proterozoic around 1,900 Ma, became part of a pre-Cordilleran craton, and reworked repeatedly, including in a Devonian orogenic episode.

Among the many regional orogenies in the Canadian Cordillera are tectonic episodes in the Late Triassic-Early Jurassic and mid-Cretaceous. The former is best studied in the Teslin zone (Fig. 45; e.g., Stevens and Erdmer, 1996), and the latter is represented widely across the region in granitic magmatism and block faulting (Struik, 1991; Woodsworth et al., 1991). The corresponding metamorphic overprints in the Monashee complex (Journeay, 1986; Parrish, 1995) may be related partly to decompression at the end of the corresponding orogenies. Journeay (1986) reported that paleopressure inversion in the upper levels of the Monashee complex was not associated with an inversion of temperature. Parrish (1995, p. 1635), on the other hand, argued for "a thermal inversion [to accompany the pressure inversion] because [this thermal episode's] short duration would be very difficult to explain otherwise". Several manifesta-

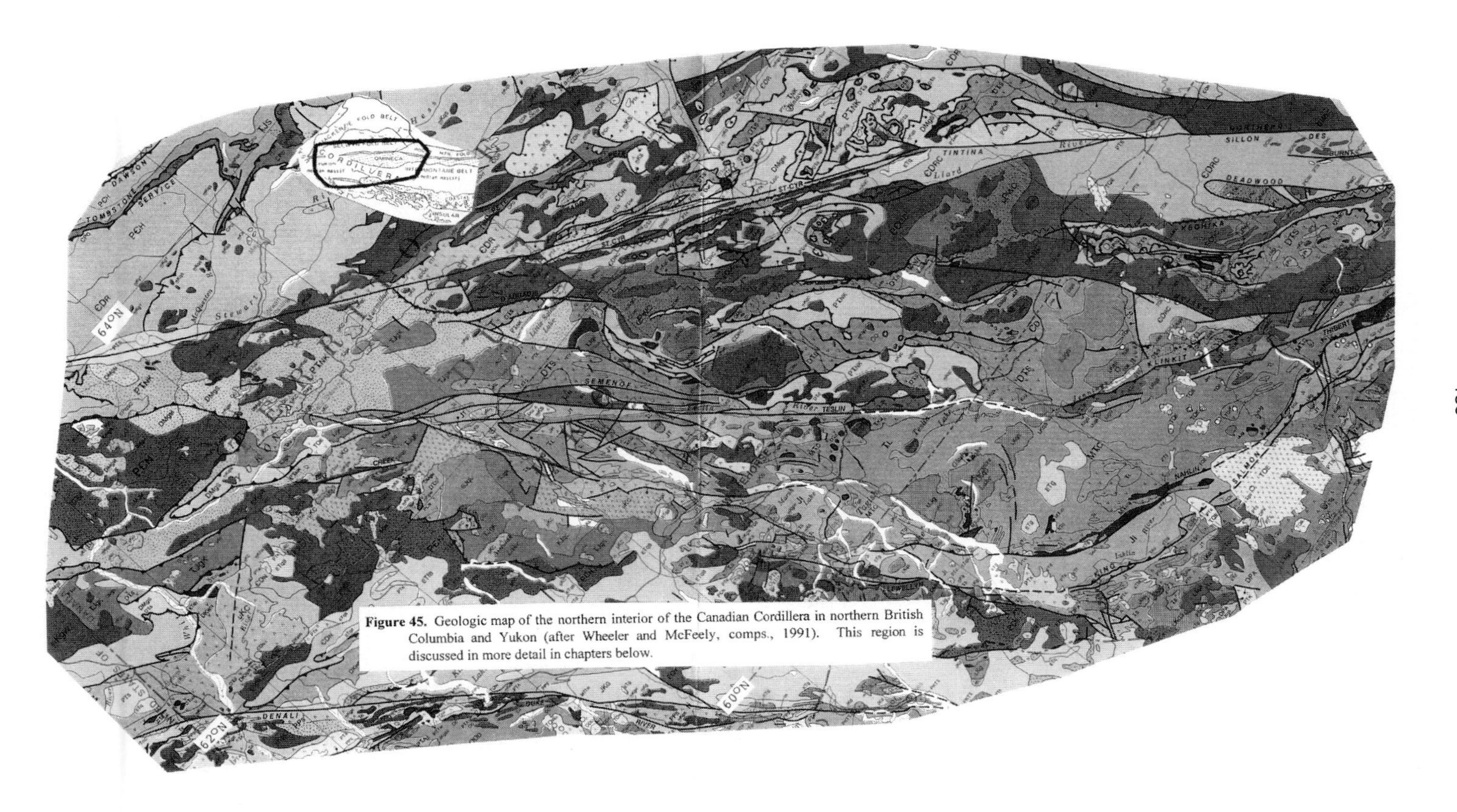

Figure 45. Geologic map of the northern interior of the Canadian Cordillera in northern British Columbia and Yukon (after Wheeler and McFeely, comps., 1991). This region is discussed in more detail in chapters below.

tions of anatectic granitic plutonism indicate that thermal regime indeed fluctuated in time, but this does not contradict episodic decompression.

Rocks with ages around 1,500 Ma have been found at the floor of the Belt-Purcell Basin (e.g., McMechan and Price, 1982; Evans and Zartman, 1990). Mafic Moyie intrusive rocks of 1,467±3 Ma bear isotopic signatures of the Archean Wyoming craton; they intrude the lower 1.5-2 km of the Belt-Purcell sediments and are thought to be penecontemporaneous with Belt-Purcell sedimentation (Höy, 1989; Anderson and Davis, 1995; Anderson and Goodfellow, 1995). Goble et al. (1995) reported geochemical differences in three distinct suites of sills and dikes in the Lewis thrust sheet, dated at 1,500 to 1,400 Ma, >963 Ma and ~800 Ma (using K/Ar and Ar/Ar methods). The inception of the Belt-Purcell Basin may have been related to Middle Proterozoic rifting, but sediment provenance from the southwest, from the so-called Western craton somewhere in Washington state (Harrison, 1972), implies this basin had an intracontinental and intracratonic nature.

Read and Brown (1981) proposed that the Selkirk rock assemblage (allochthon) used to lie >80 km southwest of its present position. This would put it on a possible northern continuation of the Western craton. In the U.S., south and north of the Lewis and Clark line, rocks with ages 1,576 to 1,370 Ma and younger are considered native and formed in situ; the same applies to the Canadian part of the Belt-Purcell Basin. In Parrish's (1995) schematic cross-section (Fig. 43), basement(?) rocks of 1,500-1,400 Ma age are shown above the assumed Late Cretaceous-Paleocene Monashee detachment, which puts them in the Selkirk allochthon. With the allochthon, these rocks were displaced from the southwest, from the site of the Intermontane median-massif belt that may also be part of the ancient Western craton.

From Parrish's (1995) data, the Monashee metamorphic complex shows no sign of Mesozoic metamorphism before Late Cretaceous. In the Valhalla complex, some 80 km southeast of the Thor-Odin dome in the southern Monashee complex, ages of orthogneisses cluster around 120 and 110 Ma (Early to late Early Cretaceous); but some samples yield Jurassic ages of 162, 195 and 200 Ma. Pelitic paragneiss there also contains detrital monazite grains recrystallized around 1,400 Ma (Spear and Parrish, 1996). All the rocks experienced thermal metamorphism, accompanied by emplacement of anatectic granites, around 70 Ma (Late Cretaceous Maastrichtian). For the Valhalla core complex, Spear and Parrish (1996) proposed a thermal history including rapid elevation and erosion of the roof rocks during only about 10 m.y. ending in the Paleocene, when both temperature and pressure dropped from deep-crustal values of ~850°C and 800 MPa (8 kbar) to mid-crustal ~500°-600°C and 500-600 MPa (5-6 kbar). Steadier unroofing during the next 20-30 m.y. brought these rocks to even shallower levels, with only 100°C temperature, by the mid-Oligocene. At that time, according to these authors, extensional conditions prevailed, in contrast to the compression of Mesozoic time.

Parrish (1995) tried to establish dissimilar tectonic histories of the rocks above and below the Monashee décollement shear zone, and speculated that burial metamorphism of the Monashee complex was greatly enhanced by "tectonic loading" to temperatures of ~350°C. He stated (op. cit., p. 1635) that "the rocks of the [Selkirk] allochthon and the Monashee complex... only share the latest Cretaceous-Eocene part of their thermal histories, indicating that pre-latest Cretaceous evolution of the allochthon occurred elsewhere, presumably considerably farther west into the [Cordilleran] hinterland". On current evidence, no such great displacements are indicated.

The unconformity that separates the (metamorphosed) supracrustal rocks of the Omineca miogeosyncline from the Early Proterozoic basement is quite evident. The tectono-magmatic record reveals an orogenic episode predating this unconformity (Scammell and Brown, 1990; Crowley, 1995), which seems to truncate leucogranites of 1,850 Ma age. The overlying succession is much younger (~740 Ma and less), coeval with the Windermere Supergroup whose oldest rocks have been dated at ~780 Ma. A structural discontinuity between the original crystalline basement and the Windermere Supergroup rift succession, whose age span lasted till the latest Neoproterozoic time, is quite obvious (e.g., Murphy et al., 1991). In and above the 1-km-thick Monashee shear zone, Parrish (op. cit.) showed up to 5 km of rocks denoted as basement(?) but having radiometric ages of only 1,500 to 1,400 Ma (Fig. 43), much younger than the Early Proterozoic basement beneath the pre-Windermere unconformity.

Ages of 1,850 to 2,100 Ma are reported for many of the gneisses in the N-S bands of the east Cordilleran metamorphic core complexes (Crowley, 1995; Spear and Parrish, 1996). Augen gneisses in the Priest River complex have been found to be 1,576 Ma (Evans and Fischer, 1986), and older rocks have been dated as ~2,650 Ma (Doughty et al., 1998). In the southern Belt-Purcell Basin, south of the transcurrent Lewis and Clark fault zone, in Idaho, deformation and metamorphism of granites and host rocks in a gneiss outlier have been dated at 1,370 Ma (Evans and Zartman, 1990). This mid-Meseproterozoic age has no equivalents in the Alberta Platform's cratonic basement. Thus, correlation of the basement east and west of the Canadian Rocky Mountains may only be partial, and with the current limited data base it is very uncertain and speculative.

The Late Cretaceous high-grade thermal metamorphism requires burial depths of 28-30 km. Some of the burial could have happened during the long period of deposition of the Belt-Purcell Basin sediments, whose total thickness is ~20 km in the U.S. and at least 10 km in Canada. Thick Late Proterozoic-early Paleozoic rock formations in the

Kootenay Arc area contain detrital zircons with ages clustered around 2,700-2,500 Ma, 2,000 Ma, 1,800 Ma and 700 Ma. Their sources are still unclear (Smith and Gehrels, 1991) but they could be blocks of the Archean Wyoming or more-remote Slave cratons, and the more-proximal Early Proterozoic Trans-Hudson Orogen (see Lyatsky et al., 1999 for references). In a crystalline-basement outlier in the southern Belt-Purcell Basin, detrital zircons have ages of ~1,850 Ma. Granites dated at ~1,370 Ma did not provide detritus for the Kootenay Arc sediments, but the granites of ~700 Ma did. The Lardeau Group has no evident clasts from the early Paleozoic sources to the west. It is overlain, with a pronounced structural unconformity, by Mississippian-Pennsylvanian stratified rocks whose clasts came from the eastern and western sides. None of these considerations require significant lateral tectonic dislocations, pointing instead to a polycyclic tectonic evolution *in situ*.

Klepacki and Wheeler (1985) suggested direct connections between the Kootenay-Arc successions with provenance areas in the presumed ancestral North America to the east. Smith and Gehrels (1991) also presented evidence that the supposedly accreted Quesnel terrane farther west is not allochthonous. Instead, they described it as "parautochthonous". They stated that in this area "Lower Paleozoic eugeosynclinal strata likely formed more or less at their present latitude and did not undergo significant post-depositional northward latitudinal displacement during Mesozoic time" (op. cit., p. 1282).

The Valhalla metamorphic complex and surrounding rocks developed *in situ*, without appreciable lateral translations. The geologically mapped low-angle Gwillim Creek and Valkyr shear zones reflect not large displacements of the upper rocks with respect to the lower ones, but *in-situ* ductile shearing at a deep crustal level. The general domal structure of the Valhalla complex could be a result of predominant vertical movement of rock material rising in a ductile fashion from lower-crustal levels. Some later movements on strands of the Valkyr and Gwillim Creek shear zones that envelop the metamorphic core, and on the steeper Slocan Lake fault, were probably results of local accommodation of displacements at the head of the rising metamorphic core which occurred in the Early Eocene (at 59-54 Ma). They began deep enough not to be fully brittle, and later were slightly reactivated in a brittle mode.

Spear and Parrish (1996) thought that the thermal anomaly of the estimated heating peak around 70 Ma, which produced the Late Cretaceous-Early Tertiary metamorphic rocks of the Valhalla complex, has "enigmatic" causes. But the ultimate thickness of the crust in the Early Tertiary was thought to have been as large as 50 km, and heating of rocks could have taken place in the lower crust and upper mantle. Surface geologic mapping shows that the N-S structural zones of metamorphic cores in the southern Canadian Omineca Belt lie 50 to 80 km apart. Domes in the N-S central band, from the Malton Gneiss to Frenchman Cap and Thor-Odin (in the Monashee complex) and Kettle culminations, are also spaced regularly, 80 to 120 km apart. Such regularity suggests a rather stationary tectonic regime in this region, with the crust dissected by faults in a regular network.

Further review of Mesozoic and Cenozoic tectonic manifestations in rocks in the southern Omineca miogeosyncline and adjacent areas

In the southern Omineca Belt, stratified Upper Triassic-Lower Jurassic and Middle Jurassic rocks cover older rocks, such as the late Precambrian-Early Cambrian Hamill and Lardeau assemblages, in the Valhalla complex area. The most common is the Middle Jurassic Rossland Group of argillite, siltstone and conglomerate intercalated with a very thick succession of basaltic flows and tuffs (e.g., Höy and Andrew, 1989). These rocks are cut by the Middle to Late Jurassic granitoids of the Nelson suite, commonly with broad contact-metamorphic aureoles (Powell and Ghent, 1996). Outside these aureoles, Mesozoic rocks are usually metamorphosed to no higher than prehnite-pumpellyite grade.

The oldest of the Jurassic batholiths is Kuskanax, situated between the Valhalla and Monashee metamorphic complexes. This batholith's age of ~173±5 Ma is coeval with one of the phases of folding in the Kootenay Arc. Slightly younger, 170-167 Ma, is the Adamant pluton and its satellites to the north. This pluton is elongated E-W, apparently along a block-bounding fault. Even younger, 159±0.6 Ma, are the Nelson Batholith and coeval plutons in the south (Woodsworth et al., 1991; Sevigny and Parrish, 1993).

The Kuskanax pluton was probably emplaced at mid-crustal levels with pressures around 5 kbar (500 MPa), whereas the emplacement depths of the Nelson and Bonnington plutons were shallower, with pressures of only ~2.5 kbar (220 to 280 MPa; Ghent et al., 1991). Rossland Group rocks were metamorphosed in hot contact with these plutons (Powell and Ghent, 1996), but regionally they were at shallow depths no more than 8 km and usually formed the plutons' roof.

Monger et al. (1994) placed the western edge of their ancestral North America (which they also called "unthinned North American craton", e.g., their Fig. 13) "approximately" near the axial zone of the Purcell Anticlinorium and, farther north, at the Purcell Trench. The Belt-Purcell assemblage and the much younger Windermere assemblage to the north lie west of this line. Next to them to the west, the Kootenay Arc (regarded as a suspect terrane) consists of highly metamorphosed and deformed Paleozoic rocks, with a "record of Paleozoic deformation, metamorphism, and intrusion that is not recognized in rocks to the east" (op. cit., p. 375). Different conclusions were drawn by Smith and Gehrels (1991) from their mapping: they treated the Paleozoic rocks of the Kootenay Arc as native (or parautochthonous), because its Late Proterozoic and Paleozoic rocks contain detrital zircons derived from the North American craton. A compromise position was taken by Wheeler and McFeely (comps., 1991), who called the Kootenay Arc a terrane but qualified it as "pericratonic".

The NE-SW-oriented Waneta fault bounds the Kootenay Arc on the west. This prominent, steep, reverse fault truncates another fault, Champion Lake, which is traced northward along the western margin of the Bonnington pluton. Displacements on the Champion Lake fault near the point of truncation, close to the Jurassic Trail pluton, are 1-2 km. This fault was probably active in the mid-Cretaceous, at the time of formation of granodioritic gneisses dated at 110-100 Ma in the southern Valhalla complex (Parrish, 1995), and later. Farther north, this fault lines up with the Slocan Lake fault that bounds the Nelson Batholith on the west but is terminated by the Kuskanax pluton (see also Powell and Ghent, 1996).

Detailed mapping indicates mid-Mesozoic activity along the Waneta fault involving deep crustal levels. Later the tectonic activity in this zone declined, though some shearing of Cretaceous igneous rocks suggests at least minor reactivation (Höy and Andrew, 1989; Powell and Ghent, 1996). Probably, this fault also defined the western boundary of a 200-km-wide, N-S-trending belt of pronounced extension in the Early Tertiary; this belt's eastern boundary coincides with the Flathead fault. Most of the extensional grabens in the eastern Cordillera (Figs. 46-47) lie in this belt, which may continue as far south as Utah (cp. Gabrielse, 1991a; Constenius, 1996). The total Eocene and Oligocene extension is this belt (also Janecke, 1994) may be up to 25-30 km (McMechan and Price, 1984). In the Purcell Anticlinorium, young steep faults cutting the older structures are Hope, Moyie and others.

A combination of old thrust faults and steep normal faults created a mosaic of large blocks, which defined the Tertiary structural background of the Purcell Anticlinorium (e.g., McMechan, 1991). The most prominent of these blocks are in the south, straddling the Canada-U.S. border, bounded by the Dibble-Moyie fault system, and in the north, bounded by the transcurrent Hall Lake fault. The latter was active at roughly the same time as the Waneta fault, as evidenced by a series of mid-Cretaceous plutons emplaced along it. In the north, the Hall Lake fault meets the Purcell Trench fault near the southern part of the long Kootenay Lake, which is controlled by an eponymous fault zone running N-S for about 70 km while being only a few kilometers wide. In the Goat Range area, between the Nelson and Kuskanax batholiths, folds in the late Precambrian-Early Cambrian Hamill and Lardeau groups within this zone become increasingly upright. The axial planes of these tight folds dip steeply to the SW, although Fyles and Höy (1991) noted significant variations in fold styles in outcrops along the Kootenay Lake.

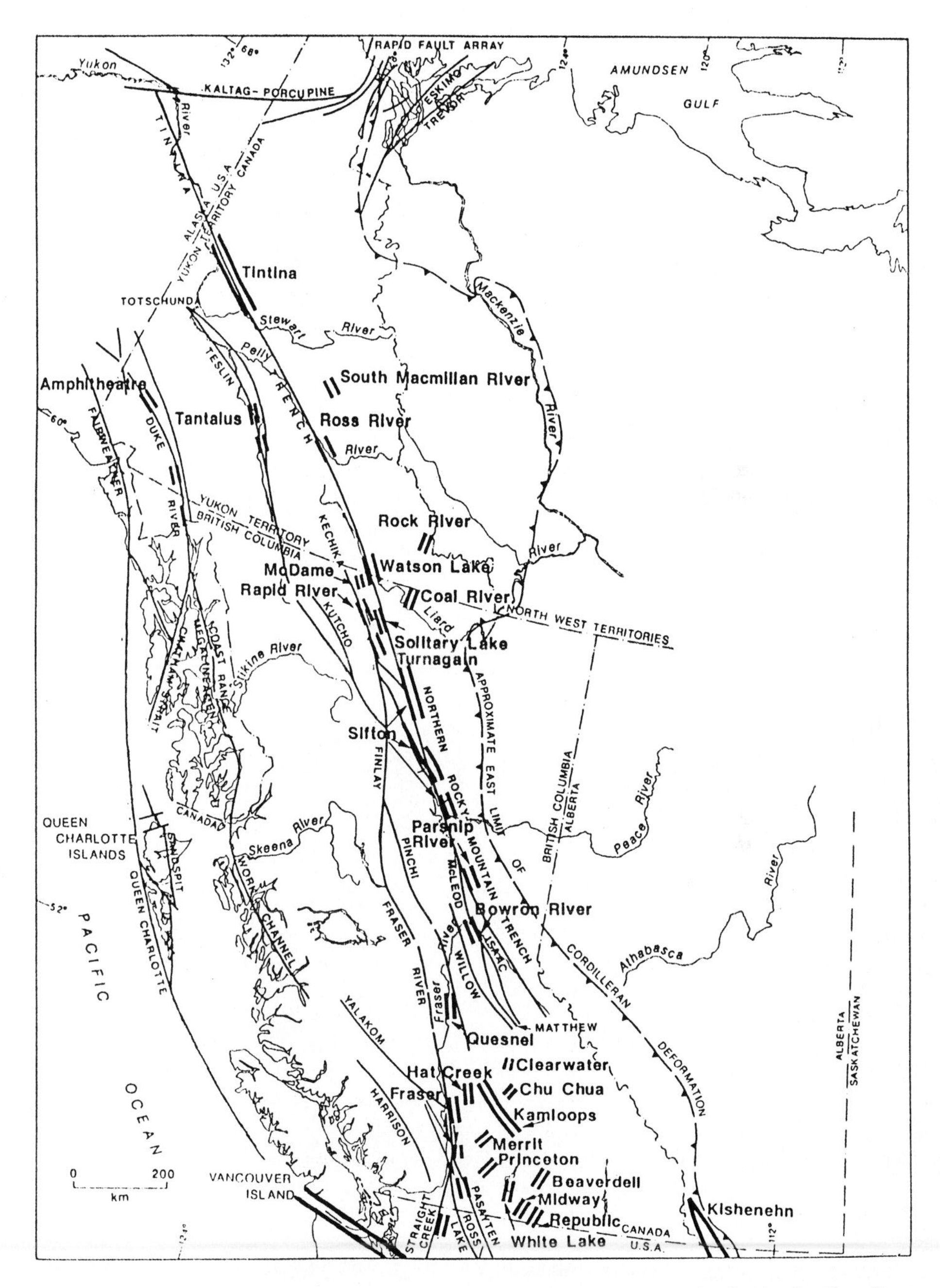

Figure 46. Major crustal faults and fault-controlled Tertiary sedimentary basins in the Canadian Cordillera (modified from Gabrielse, 1991a). Heavy double lines mark the grabens.

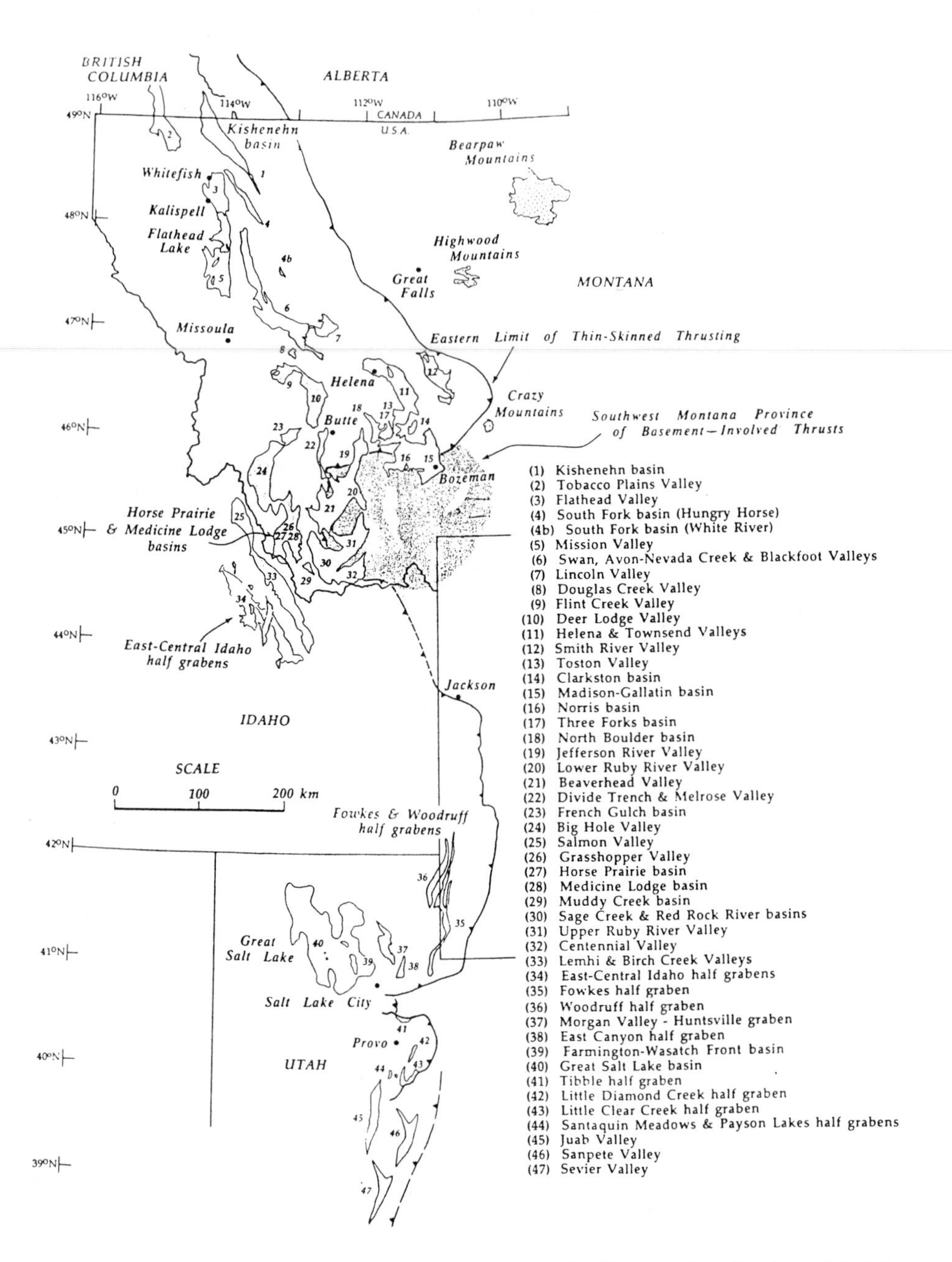

Figure 47. Tertiary sedimentary basins in the eastern Cordillera in the U.S. (simplified from Constenius, 1996).

The overlying Mississippian-Pennsylvanian stratified rocks, separated by a strong unconformity, are also intensely folded (Klepacki and Wheeler, 1985). Fyles and Höy (1991) noted that some folds of different phases along the lake are coaxial.

The latest phase of regional folding in the Kootenay Arc affected the stratified Triassic rocks. This deformation episode was associated with Jurassic magmatism, which was almost coeval with a regional metamorphic event. Klepacki and Wheeler (1985) believed that the apparent curvature in the regional structure also developed at that time.

Younger folds are also recognized. Their age is undetermined, but their vergence is to the west (Fyles and Höy, 1991). Later faults, many of them transcurrent, represent the mid-Cretaceous tectonic event which was associated with abundant magmatism in the Kootenay Arc as well as in the Purcell Anticlinorium.

North of the Goat Range and Kuskanax batholith area, the Columbia River fault defines the western side of the northern Kootenay Arc. This big fault terminates the Waneta-Kootenay Lake fault system. It dissects the easternmost part of the Selkirk tectonic entity (the Clachnacudainn terrane of some authors) and cuts the Standfast Creek fault. The Columbia River fault is young (Tertiary) and low-angle normal (Brown et al., 1992; Crowley and Brown, 1994).

The lower Paleozoic Lardeau Group is not found east of the Standfast Creek fault, but it has been mapped on both sides of the Columbia River fault. Correlated across both these faults are regional west-verging folds of the Selkirk fan structure. The Standfast Creek fault is cut by a mid-Cretaceous pluton.

In the Cariboo Mountains, in the headwaters of the North Thompson river, the large west-vergent Scrip Nappe is well seen in the Windermere Supergroup rocks (Simony, 1991). In the northern Selkirk Mountains, near Mica Dam, large phase-1 folds of this nappe are overprinted by later folds. Folds and thrusts, including the Purcell thrust fault, related to the phase-2 deformation, were roughly coeval with the climax of Middle Jurassic metamorphism. The transition from ductile to brittle deformation is gradual (Pell and Simony, 1982). The band of phase-2 folds has been recognized in a large area from the northern Monashee Mountains to the southeast through the Selkirk Mountains into the Purcell Mountains (Simony, 1991).

Significant changes in the structural and metamorphic styles have been reported across the NNE-SSW-trending North Thompson fault, on which the southeast-side-up vertical offset may be as much as 12 km (Fig. 48). Metamorphic grades change from green-

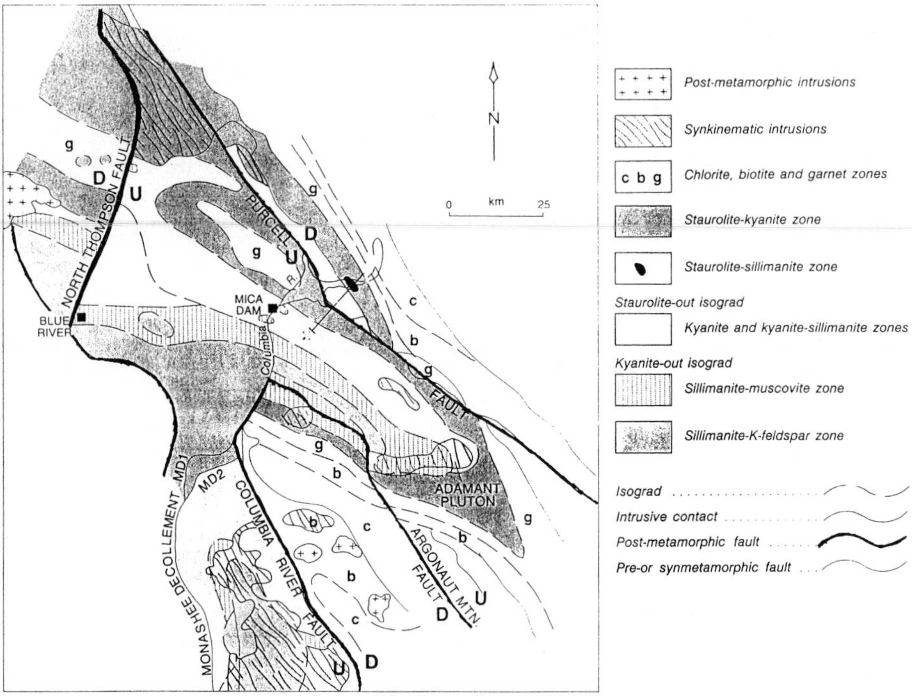

Figure 48. Abrupt changes in the metamorphic grade of surface rocks across the North Thompson fault (simplified from Greenwood et al., 1991). D - downthrown side; U - upthrown side.

schist in the Cariboo block northwest of this fault to amphibolite in the Selkirk block to the southeast (Greenwood et al., 1991). The maximum depth of burial of metamorphosed rocks changes from ~10 km northwest of the North Thompson fault to ~25 km and possibly >30 km to the southeast (also Parrish, 1995). Simony (1991) noted also that regional deformation of phase 2 occurred in the southeastern block first.

Another big crustal block in this area, with greenschist grade of metamorphism at the surface, is bounded on the west by the Purcell fault. This fault is probably steep and crustal-scale, though it has low-angle splays in the upper crust. From existing descriptions, the same seems to be the case with the North Thompson fault. These two big faults accommodated differential burial and uplift of these three crustal blocks - Cariboo, northern Selkirk and Purcell. The deepest burial of now-exposed rocks was reached as late as in the Early Tertiary (Parrish, 1995). The most dramatic subsequent (Eocene-Oligocene) uplift occurred in the northern Selkirk block.

Folds and faults after phase 2 (Simony, 1991) are recorded best in the Monashee Mountains and northern Selkirk Mountains. The large NE-verging open fold called Mica Dam Antiform is a typical example of these structures. Parasitic secondary folds mask its overall anticlinal structure only partly. Overthrusting, verging ENE, is the most obvious east of this antiform, on both sides of the Southern Rocky Mountain Trench fault system. But neither the Purcell nor the Southern Rocky Mountain Trench fault marks a clear structural boundary between the Omineca miogeosynclinal orogenic zone and the Rocky Mountain fold-and-thrust belt (Simony, 1991).

Simony (1991, 1995) emphasized the "complete gradation" of ductile to brittle deformation observable in this region, based on mapping of exposures of various levels of the crust. In the Omineca Belt, the top 15-20 km of the crust are exposed for direct observation. Vertical and lateral links between tectono-metamorphic and tectono-deformational manifestations have been stressed for the eastern Canadian Cordillera (Simony et al., 1980; Simony, 1991). These links include coeval magmatism (Woodsworth et al., 1991).

Besides the North Thompson and Purcell faults, big faults like Argonaut Mountain, Columbia River and Southern Rocky Mountain Trench also mark major block boundaries in the southern Omineca Belt. The Argonaut Mountain fault runs SSE from the North Thompson fault, after a gap. Possibly, it is offset by the E-W fault that controls the elongation and position of the mid-Cretaceous Adamant pluton. The southern continuation of the Argonaut Mountain fault is probably the Tangier River fault in the area of the Selkirk fan structure; there is an interruption there due to the mid-Cretaceous Battle Range pluton.

Locally, in the Slocan Lake area, the Champion Lake-Slocan Lake fault system has low dips to the east. This system juxtaposes the Tertiary Valhalla metamorphic dome and the Jurassic Nelson pluton. Farther west, on the western side of the Valhalla dome, the Valkyr shear zone runs towards the prominent Columbia River fault in the north and the Kettle fault in the south. A series of Middle Eocene syenite plutons punctuate this structural trend. In the topography, this trend is accentuated by the elongated Upper and Lower Arrow lakes. A local deviation from the linear N-S trend of the Kettle-Columbia River fault system (including the Valkyr shear zone) is the westward curve highlighted by the Lower Arrow Lake, around the west side of the Valhalla dome. Between the Lower Arrow Lake and the surface trace of the Valkyr shear zone, there is a swarm of leucocratic dikes of Paleocene to Early Eocene age. The time from 59 to 54 Ma saw the last major deformation episode in this region (Spear and Parrish, 1996), prior to subsequent block movements.

All along its trace, the Kettle-Columbia River fault system is usually considered to be low-angle, normal. But no consensus about it exists: Ghosh (1995, his Fig. 3) depicted it as a thrust; Johnson and Brown (1996) artificially curved and connected it with the Champion Lake-Slocan fault system to the east; Wheeler and McFeely (comps., 1991) and Colpron et al. (1996) showed its segments as dipping variously east and west. Such diverse surface expressions along a single trend suggest a steep, deep-seated zone of crustal weakness, with various shallow splays.

The steep, west-side-down shear zones on the west side of the Valhalla dome are thought to be extensional, and they cut all older structures (Spear and Parrish, 1996, their Fig. 2). Following a peak of metamorphism at 72-67 Ma, when the metamorphic rocks were at 28-30 km depth in the crust, these rocks were steadily raised and unroofed during 20-30 m.y. (op. cit., their Fig. 16). Parrish (1995) invoked hypothetical "underlying thrust faults" that supposedly "elevated the complex and induced erosion". On the other hand, post-orogenic extension and uplift in this region in the Early Tertiary, Cretaceous and Jurassic could have occurred without such hypothetical mechanisms, but rather by movements of blocks bounded by steep faults.

Amplitudes and rates of downdropping and uplift were not the same for different crustal blocks in the southern Omineca Belt, whether in the Mesozoic or Tertiary. The northern Selkirk block, east of the Columbia River fault, was in the Late Jurassic at a depth of 20-25 km. It was later elevated and exhumed (Colpron et al., 1996). The Battle Range pluton and its satellites, emplaced at shallow depths of 7-10 km, were uplifted slowly between ~100 and 85 Ma, just prior to the Late Cretaceous Campanian (also Crowley and Brown, 1994), as well as later (Colpron et al., 1996, their Fig. 12).

From the analysis of metamorphic isograds in the northern Selkirk Mountains, Greenwood et al. (1991) confirmed that metamorphic patterns vary significantly across the Selkirk fan structure. Southeast of Mica Dam, the most dramatic variations occur across the Purcell and Argonaut Mountain faults. Chlorite to garnet facies are found between Mica Dam and the Adamant pluton, but the dominant metamorphic facies in the central zone are staurolite, kyanite and sillimanite (Ghent et al., 1982). This metamorphic pattern persists for ~150 km southeast of the North Thompson fault, to about the Trans-Canada Highway.

Timing and estimates of post-metamorphic displacements on North Thompson, Argonaut Mountain and Purcell faults vary (Greenwood et al., 1991). Tertiary movements across the North Thompson fault were 3-4 km, southeast side up, whereas on the Purcell fault displacements in the Tertiary were minor. Yet, in the Mesozoic the block west of the Purcell fault had risen by about 7 km. On the Argonaut Mountain fault, this block exhibits a similar upthrow.

Struik (1991) pointed out that the uplift of the Omineca Belt in mid-Cretaceous time was "strong". It was accompanied by emplacement of extensive granitic plutons between 130 and 85 Ma (late Neocomian to earliest Campanian). The mid-Cretaceous igneous-rock suites are among the major ones in the eastern Canadian Cordillera. The Cassiar suite in central British Columbia and the Bayonne suite in the south are similar in composition but markedly different in style. The Cassiar suite bears clear connection with the regional tectonic kinematics, including NNW-SSE elongation of stocks and along-strike deformation (Woodsworth et al., 1991). Plutons of the southern, Bayonne suite are mostly isometric or elongated E-W. In the Shuswap Lake area, the E-W-elongated mid-Cretaceous plutons are clearly different in their shape from the fairly isometric plutons east of the Columbia River fault. Mid-Cretaceous plutons in the Selkirk fan and the Kootenay Arc were emplaced at shallower crustal levels (Archibald et al., 1983, 1984) than their counterparts in the Purcell Anticlinorium.

In mobile megabelts such as the Cordillera, inter-regional phenomena like these commonly exist despite the differences between individual blocks, belts and zones. Regional tectonic regimes interacted complexly with smaller-scale lateral tectonic units in whole-crust orogenic zones, whose individual patterns existed against an inter-regional background of broader similarities in tectonic setting. During some tectonic episodes, especially during thermo-tectonic relaxation accompanied by regional crustal extension and differential uplift and subsidence, block tectonics frequently predominated in large orogenic crustal zones.

Many faults and blocks in the southern Omineca Belt were restructured during the Late Cretaceous-Early Tertiary Laramian orogeny, and many pre-existing faults were reactivated to accommodate Tertiary crustal movements. Among the rejuvenated major faults were North Thompson, Southern Rocky Mountain Trench and Columbia River. These faults are composite and contain segments with variable ages and senses of movement. The Purcell thrust fault accommodated major west-side-down displacement in Tertiary time, and the North Thompson fault had northwest-side-down Tertiary movements. The Tertiary Columbia River fault is described as low-angle and east-dipping; estimates of displacement on it range widely, from 1 to 40 km.

Tertiary displacements on the Southern Rocky Mountain Trench fault system are difficult to estimate because the trench is wide and correlations of rocks across it are uncertain. Displacements have been estimated, variously, to exceed 1-2 km (according to McDonough and Simony, 1988, their Fig. 12) and to reach 10 km (van der Velden and Cook, 1996). In the topography, the trench is 600 km long, from 48°-49° to ~55°N, running as a broad and straight valley. Its discordance with the surrounding terrains is highlighted by its deep incision, up to 1 or 2 km. In places, the trench valley is tens of kilometers wide, U-shaped or graben-like, but elsewhere it is relatively narrow and V-shaped, 5 to 12 km across at the bottom. This topographic gap separates distinct high mountain ranges. No significant strike-slip offsets are found on the Southern Rocky Mountain Trench fault system (McDonough and Simony, 1988). Its dip-slip movements are thought to be all Tertiary, but its conformity with the older structural trends has been noticed for decades (Henderson, 1959; Leech, 1967).

Van der Velden and Cook (1996) related the Southern Rocky Mountain Trench to a hypothetical west-facing ramp in the pre-megabelt basement associated with an ancestral North American continent as well as the North American craton (also Cook et al., 1991a-b, 1992; Cook, 1995b-c). In that interpretation, seismic data mark such a ramp as a 10-km-high step. This ramp was assumed to have been preserved within the megabelt's crust since mid-Proterozoic time, and to have influenced the shape of an orogen-scale basement detachment (also Cook and Varsek, 1994) which developed during the Mesozoic contractional deformations. In that scenario, crustal-scale slicing and east-vergent juxtaposition of the slices were assumed to be a defining process in the Mesozoic tectonic evolution of the Cordillera (in contrast to its Paleozoic passive-margin stage). After the subduction-induced crustal stacking process had reached its peak in the Early Tertiary, contractional thrusting ceased and crustal extension ensued. The Southern Rocky Mountain Trench fault, in this model, does not cut the basal detachment: rather, this fault is listric and flattens out to the west, where it merges with the master thrust fault underlying the whole Cordillera as far as the Pacific.

This speculative scenario is inconsistent with evidence from geologic mapping. Geologic evidence indicates repeated orogenies with large-amplitude vertical movements. In the Mesozoic and Cenozoic at least, each orogenic pulse was followed by an episode of extension. Preservation of Proterozoic crustal "ramps" is implausible, in the face of evidence for deep-rooted crustal reworking. An alternative interpretation of the Lithoprobe data is called for, more in line with regional geology.

Geologically grounded interpretation of Lithoprobe seismic data

In the mountains of the Cordillera, access for heavy seismic-acquisition equipment is possible mostly along roads, many of which were built along river valleys and mountain passes that tend to follow large NNW-SSE and transcurrent crustal faults. Seismic sections shot along steep faults suffer from a degraded signal-to-noise ratio, and they should be analyzed with great caution. Even with little noise, seismic data must not be regarded as fully reliable images of deep crustal structure, and extra noise compounds the unreliability. This is particularly true of segments of Lithoprobe Line 9 which run subparallel to the large St. Mary, Moyie and other faults.

Despite considerable difficulties in profile 1, Eaton and Cook (1988) depicted a cratonic basement east of the Southern Rocky Mountain Trench to be dipping towards the mobile-megabelt interior at an angle of only 2°. The corresponding reflection was correlated with another one in seismic profiles, on the western side of the Southern Rocky Mountain Trench, dipping to the west at about 13° (Figs. 29-39; Cook et al., 1988; Van der Velden and Cook, 1996).

In the south, the Southern Rocky Mountain Trench is in a complex relationship with the Moyie-Dibble Creek fault system. In the Fernie area east of the trench, the Dibble Creek fault dips to the north. The Southern Rocky Mountain Trench fault system truncates the Dibble Creek fault, and the Moyie fault begins on the trench's western side. It strikes NE-SW in the Cranbrook area, where its steep dips are consistently to the NW; this fault is treated as reverse. Across the Canada-U.S. border, this fault seems to turn N-S (Wheeler and McFeely, comps., 1991; Constenius, 1996, his Fig. 6). To the west of this fault, in northern Montana and Idaho the supposed basal thrust detachment seems to rise from ~20 km depth to <10 km (Fillipone and Yin, 1994; Constenius, 1996). The high-angle Moyie fault is missed in the Lithoprobe near-vertical-incidence seismic data, but some noise in Line 9 may be diffractions from this fault system.

The Moyie-Dibble Creek fault system bounds a crustal block some 50 km wide, whose eastern boundary is the Southern Rocky Mountain Trench fault. A series of steep reverse faults and low-angle thrusts related to eastward overturning of stratal rocks from the nearby Kootenay Arc are prominent east of the big N-S Kootenay Lake fault system. These faults, as the entire late-stage structure of the Purcell Anticlinorium, were given their final form by the Laramian orogeny during the Tertiary (Archibald et al., 1984). Carr (1995, p. 1723) stated that in the Purcell Anticlinorium "the crustal geometry determined from mapping and construction of cross sections is consistent with geometries observed in the seismic reflection data". But she also noted some uncertainties: "west-dipping packages of reflectors that project to the surface are interpreted to be generated by thrust faults, or stratigraphic horizons or mafic sills", even though she assumed all of them to be "thrust parallel within thrust sheets". With the seismic data across the Purcell Anticlinorium, these postulations are hard to prove. Many of these seismic reflection profiles are kinked or discontinuous; profile 6 was shot N-S along the Kettle-Columbia River fault system, and the E-W profile 2 partly runs along the big St. Mary fault.

In the top 20 km in profile 2, variously dipping seismic events suggest several antiforms and synforms. About 40 km to the south lies one of the few wells in this region, the DEI Moyie, which is 3,477 m deep. It penetrated the lowest formations of the Middle Proterozoic Belt-Purcell Basin but did not reach the probable Archean Wyoming-craton basement (Anderson and Davis, 1995; Anderson and Goodfellow, 1995). These lower units were intruded by mafic sills before they were buried deeply or even solidified. The sills, whose total thickness is up to 6 km, distorted the original depositional stratigraphic pattern of the basin. Sills such as these could easily account for many observed seismic reflections in Line 9, weakening the need for structural explanations. Sills could also cause multiples at longer traveltimes, contaminating the seismic images and complicating interpretation.

Clowes et al. (1995, p. 1499) believed that "many, but certainly not all, velocity structure variations in the upper and middle crust across the Cordillera can be related to known geotectonic features". Such an assumption-driven approach opens the door for easy speculations about the unsampled deep crust, where no direct geologic observations are possible. The priority should always be to incorporate seismic interpretations into a framework derived from observable, mappable geological facts. Without this, the inherent non-uniqueness and uncertainty of deep-crustal geophysical interpretations (cp. Pakiser and Mooney, eds., 1989) may allow the interpreter's fancy to go astray. Vertical seismic resolution at deep crustal levels is only about 2 km, and lateral resolution is poor as well (also Kanasewich et al., 1994). Considerable lateral geologic variations complicate the velocity structure of the Cordilleran upper and middle crust. Generally, the degree of tectonic reworking and restructuring of the Cordilleran crust increases towards the interior of the megabelt, and with that increases the crust's complexity. The main pitfall, though, is that geophysical data do not reveal the geologic nature and age of anomaly sources.

A short "floating" reflector in wide-angle data from Line 9 was interpreted by Clowes et al. (1995, p. 1499) as "the top of North American basement" because it coincides with a reflection in the near-vertical-incidence data that has been interpreted this way (Cook et al., 1992). This is circular reasoning, and Clowes et al. (1995) acknowledged that this assumed crustal boundary "...does not have a distinguishable velocity signature".

Uncertainties in the interpretation of such "floating" reflections were discussed by Kanasewich et al. (1994).

Based on the Lithoprobe seismic data, Cook et al. (1992) and Clowes et al. (1995) extended the Slocan Lake fault from its outcrop trace downward and eastward, for tens of kilometers, even though the seismic reflection profile 4, which they projected to the surface fault trace, loses definition in the mid-crust. In fact, this fault can be traced in the seismic data from the surface down to only ~10 km depth under the Nelson batholith. The interpretation of its assumed continuity farther down is based on inconsistent correlations of discontinuous seismic events across a data gap, under the Kootenay Arc, in the area where several proved steep faults merge into the Waneta-Kootenay Lake fault system. Importantly, Clowes et al. (1995, p. 1505) cautioned that "the schematic thrust faults and Nelson Batholith are not identified explicitly by R-WAR [refraction and wide-angle reflection] data". The Slocan Lake fault is actually a moderate-scale fracture, not unusual in such a heavily deformed area.

Many other faults, reverse and normal, have been mapped to the east, in a 40-km-wide heavily fractured belt between the Nelson pluton and the Hall Lake fault. This belt falls is crossed by profile 2 in Lithoprobe Line 9 near ST (station) 40. The two west-dipping, NE-SW- to N-S-oriented faults running from ~50°N, Hall Lake and Moyie, are old. The Moyie fault is stitched by a mid-Cretaceous pluton that was emplaced at a depth of ~15 km (Archibald et al., 1984; Woodsworth et al., 1991); the Hall Lake fault was rejuvenated in the U.S. later, around 71-69 Ma (Fillipone and Yin, 1994).

The age span of faults to the west is even wider: from Permo-Carboniferous (Fyles and Höy, 1991) to Late Paleocene-Early Eocene (65-55 Ma, coeval with the formation of the Purcell Anticlinorium; Archibald et al., 1984). In the fracture zone along the Kootenay Lake, different faults and their strands dip variously east and west, but they are steep, with normal and reverse senses of displacement. Local complications are associated with the pair of normal faults, Purcell Trench and Newport, bounding the Priest River metamorphic core complex: they dip outward away from the core, and the crustal slice they bound wedges out a short distance north of the Canada-U.S. border. The east-dipping Purcell Trench fault controls the position of the Kootenai River and of the southern arm of the Kootenay Lake. This fault is related to a deep-seated N-S structural trend in the southern Omineca Belt. Several generations of other faults of different origins, senses of displacement and original dips are also concentrated in the 40-km-wide fractured crustal zone east of the Jurassic Nelson pluton.

In the crustal block between the Southern Rocky Mountain Trench and Dibble-Moyie faults, there are lateral changes in rock composition, and structural patterns of many generations. Several steep faults are indicated by apparent discontinuities of seismic reflection pattern. Seismic sections across the Southern Rocky Mountain Trench contain big blank segments up to several tens of kilometers long, coinciding with the trench. No reflections, especially from the upper crust, are present in these blank segments, across which the crustal reflection character changes dramatically. Van der Velden and Cook (1996) interpreted this fault as normal and listric, flattening out to the west near the top of the preserved cratonic basement there. Near the Canada-U.S. border at 49°N, the Southern Rocky Mountain Trench fault was interpreted by these authors to have an offset of ~10-12 km both vertically and horizontally, but even at 50°15'N these values decrease rapidly. In contrast, Zelt and White (1995) used the Lithoprobe Line 9 to interpret the Southern Rocky Mountain Trench fault as an east-vergent thrust, and a similar interpretation was favored by Carr (1995, her Fig. 2). Yet, the blank zone and reflection-character change across it suggest this fault remains steep through most of the crust (cp. Hajnal et al., 1996). At any rate, correlations of the supposed crystalline basement across such gaps (Eaton and Cook, 1988; Cook et al., 1992; Cook, 1995b; van der Velden and Cook, 1996) are tenuous.

To the west, a broad, open anticlinal fold characterizes the upper 10 km of the crust. Given the presence of Middle Proterozoic mafic sills in the Belt-Purcell Basin, strong upper-crust reflectivity in this part of the Cordillera is not surprising. A pronounced west-dipping band of reflections can be traced from ~4 s to 6-6.5 s traveltime, i.e. 18-19 km depth, under the DEI Moyie well location; farther west this band rises to 14-15 km. A similar westward rise is recognized also south of the Canada-U.S. border (Yoos et al., 1991; Harrison et al., 1992; Fillipone and Yin, 1994). This seismically imaged rise is found west of the Moyie fault.

In their regional cross-section (Fig. 36), Clowes et al. (1995) showed all major faults between the Nelson pluton and the Southern Rocky Mountain Trench as thrusts. Inconsistently, though, in these authors' middle cross-section the "North American cratonic basement" was suggested to be dipping gradually to the Kootenay Arc with a depth increase from ~10 to 22 km, whereas in the bottom cross-section this surface was shown with two ramps and an irregular deepening from ~15 to 22 km in several steps. The thrusts in that area were shown in the bottom cross-section as flattening out at just over half the depth of their flattening in the middle cross-section, and lying above a smooth, slightly west-dipping crustal discontinuity 7 to 10 km above the presumed top of the cratonic basement.

Van der Velden and Cook (1996) tied these supposed movements on the Southern Rocky Mountain Trench fault with the ill-justified interpretation of the Slocan Lake fault as a trans-crustal low-angle feature: these authors transferred the assumed displacement from one fault system to the other via a hypothetical orogen-scale detachment (Cook and Varsek, 1994) under the Purcell Anticlinorium and Kootenay Arc. Carr (1995, p. 1734) also believed that, in order to understand the thrust faulting all across the southern Omineca Belt, "...one must correlate structures across the Slocan Lake fault". In fact, this local fault has a limited significance: it represents activity only in the area of the Valhalla metamorphic core and the Nelson pluton. The major change in crustal reflection character occurs at the long, 40-km-wide zone of faults along the Kootenay Lake, where a change in velocity structure has also been modeled in the upper and lower crust (Zelt and White, 1995).

The 40-km-wide Kootenay Lake fracture belt contains many faults and fault strands of different ages, nature and sense of displacement. Common for them is their steepness. For example, the Whitewater fault is Carboniferous to Permian; it is overlain by Permo-Triassic conglomerates which are in turn overlain by Upper Triassic sedimentary rocks. The Whitewater fault is distorted by Middle Jurassic deformation and cut by a Middle Jurassic pluton. The Schroeder fault was active after the Early Triassic but is also cut by a Middle Jurassic pluton (Klepacki and Wheeler, 1985). These structures were later complicated further by two phases of folding (Fyles and Höy, 1991). Large, continuous, upright phase-2 folds, slightly east-vergent, have been mapped southeast of the Kuskanax pluton in association with steeply dipping faults. These folds tighten to the south along northern Kootenay Lake, where reverse faults shear the overturned folds. Along southern Kootenay Lake, by contrast, the folds have vertical to east and SE-dipping axial planes. In the north, these folds generally plunge to the north, whereas in the south they plunge to the south. This configuration suggests broad squeezing during the Middle Jurassic orogeny (cp. Archibald et al., 1983; Fyles and Höy, 1991).

Folds of later phases along the Kootenay Lake are generally coaxial with the older ones, but more variable in style. Along the central part of the lake, the latest (phase-4) folds are open to fairly tight. Along the eastern side of the Nelson pluton, several steep, asymmetric to slightly overturned anticlines and synclines have been mapped (Fyles and Höy, 1991, their Fig. 17.48c), but the general structure of this area is dominated by a continuous big syncline.

The E-W Lithoprobe seismic profile 3 and the western part of profile 2 (in Line 9) cross this complex boundary zone. The Nelson pluton is generally acoustically transparent on both sides of ST 75, and only the synformal base of this huge granitic body seems to be faintly marked in the seismic images at about 10 km depth (depending on how exactly the seismic images were processed and displayed). In Lithoprobe interpretations, the western part of the pluton's base is correlated with the Slocan Lake fault (Cook, 1995b; Zelt and White, 1995), but the seismically imaged and geologically mapped (Fyles and Höy, 1991) syncline east of the Nelson batholith has been missed in these interpretations (Cook et al., 1992; Cook, 1995b, his Fig. 3).

Between the Kootenay Lake and Hall Lake faults, Cook and van der Velden (1995) included in their interpretation variously five (their Fig. 6I) or three (their Fig. 5B) faults, all shown as low-angle, listric and dipping to the west. But in the geological E-W cross-section of Gabrielse (1991b, Fig. 19), these faults dip eastward. Because the Lithoprobe seismic sections in this area, as reproduced by Cook (1995b) and Cook and van der Velden (1995), contain few reflections to a depth of ~10 km, it is not clear how fault dips and trajectories were interpreted from these data.

Conspicuously, the N-S Lithoprobe Line 8, which ties the E-W lines 9 and 7 some 70-90 km west of the Kootenay Lake, has been reinterpreted by Kanasewich et al. (1994) in a way that differs considerably from its earlier interpretation by Cook et al. (1992). Of all the points raised, two are the most important for this discussion. The upper crust, as modeled by Kanasewich et al. (1994) from refraction and wide-angle refraction seismic data, has a distinct block structure. In the north (south of Mica Dam) a block 70 km wide lies between ST 32 and 34. Farther south there is a block ~120 km wide, and south of ST 69 another block is apparent. The central upper-crustal block is thicker than others: up to ~18 km, as opposed to ~15 km in the north and ~13 km in the south. The middle crust in this whole region has an abnormally low P-wave velocity of ~6.1 km/s, lower than the velocity of the overlying crustal levels. This velocity inversion was probably caused by anomalously high heating from below. The Moho is shallow (35-37 km) and rather smooth. Thickness of the lower crust varies from 20 to 25 km due to relief at the lower-middle crust boundary. The lowest middle-crust velocities (6.02-6.03 km/s) occur at the crustal-block boundaries, indicating these blocks are rooted in the mid-crust or deeper.

Several discontinuous, "floating" reflections in Line 8 were recorded in both near-vertical-incidence (Cook et al., 1992) and wide-angle (Kanasewich et al., 1994) seismic surveys. Two of them, under and south of ST 61, were interpreted by Cook et al. (1992) as the Monashee décollement, and the ones under and north of ST 37 as the Gwillim Creek shear zone. These two floating events in Line 8 dip towards each other but do not meet: under ST 36, a gap between them is ~55 km in width and ~5 km vertically. Kanasewich et al. (1994, p. 2661) were cautious in interpreting these floating events, which have a "limited lateral extent" and "little constraint on velocity parameters".

The N-S Line 8 is ~350 km long, and it depicts three crustal blocks. The upper crust in it thickens towards the middle: from ~13 km in the southern block and ~15 km in the northern block, to ~18 km in the central block. The northern block lies south of Mica Dam, where the upper 15-20 km of the regional stratigraphic record are exposed for direct observation at the surface. Two more blocks are evident to the north of Line 8. Near Mica Creek, the amphibolite grade of rock metamorphism requires former burial of the northern Selkirk block some 12 km deeper than that of greenschist-grade rocks typical for the Cariboo Mountains block across the North Thompson fault (Greenwood et al., 1991). Seismic velocity structure in the northern part of Line 8, some 30 km south of Mica Dam, suggest that upper-crust rocks between ST 32 and 34 form another distinct crustal block; the steep fault controlling the NE-SW orientation of the Columbia River in this area may form the boundary between these blocks.

Kanasewich et al. (1994, p. 1661) argued that "the absence of evidence, from Pg [shallow] waves, of juxtaposed geologic blocks with differing seismic properties is consistent with discrete zones being the source of the P_{R1} and P_{R2} reflections. That is, these reflections can be interpreted to be from impedance contrasts within [low-angle] fault zones that cut across the previously determined velocity structure". Thus, if the "floating" reflections on both sides of the vertical block-bounding discontinuity at ST 69 represent the same geologic surface, the pre-existing steep crustal boundary must have been cut by the recorded low-angle discontinuous seismic events whose nature remains undefined. Neither geological not geophysical data in the Omineca Belt suggest a thrust-stack structure of the crust.

Kanasewich et al. (1994) reported significant disagreements in the interpretations of crustal geometry in Line 8 between ST 61 and 37, as imaged with the wide-angle and near-vertical-incidence seismic methods. In the refraction and reflection images of the Omineca-Belt crust in the E-W Line 7, some crustal boundaries inferred from refraction and reflection surveys dip in opposite directions (Burianyk and Kanasewich, 1995). The refraction data in Line 8 showed the middle crust, with its velocity inversion, to deepen to the north rather gently, and the Moho to rise in the same direction even more gently. Also rising to the north are the "Monashee" floating event in the crust and an upper-mantle event at depths of 45-50 km. The discontinuous reflections of "limited lateral extent" (Kanasewich et al., 1994) in that area had previously been interpreted as an orogen-scale Monashee décollement (Cook et al., 1992). In this regard, Kanasewich et al. (1994, p. 2665) commented: "Our 'Monashee Décollement' does not match well with that interpreted from the vertical incidence data". To explain the "marked discrepancies" between the interpretations of reflection and refraction data, Kanasewich et al. (1994, p. 2665) supposed that these seismic techniques "are imaging different aspects of the deformational system". But in none of the images is a coherent "orogen-scale detachment" traceable convincingly.

The N-S Lithoprobe Line 8 intersects two E-W lines, 7 (at ST 69) and 9 (at ST 36). Both these E-W lines cross the Kettle-Columbia River and Waneta-Kootenay Lake fault systems as well as the Central Selkirk uplift between them. In Line 7, the Kettle-Columbia River fault system lies west of ST 29, and the Waneta-Kootenay Lake fault system east of ST 30. Refraction (Burianyk and Kanasewich, 1995) and reflection (Cook et al., 1992) data have been interpreted along Line 7. On the whole, the signal-to-noise ration in Line 7 is estimated to be lower than that in Line 8. This produces ambiguities in interpretation (Burianyk and Kanasewich, 1995), especially in the segment covering the Kootenay Lake zone of fracturing.

Best constrained in Line 7, from the arrivals of shallow waves (Pg), is the velocity structure of the upper crust to the east of the intersection with Line 8. From the tie point, the bottom of the crust in Line 7 rises to the east from 16 km to 12 km around ST 29. Thickness of the middle crust increases from ~9 km at the tie point to ~12 km in the east. Between the two big N-S fault systems, the middle crust is more uniform than the upper crust; it contains a low-velocity zone with the above-noted velocity inversion. East of the Kootenay Lake zone of fracturing, the structure of the upper and middle crust remains fairly monotonous as far as Radium Hot Springs in the Southern Rocky Mountain Trench fault system. Velocity structure of the lower crust is more heterogenous, as P-wave velocity drops eastward from ~6.8 km/s under ST 29 to only 6.3 km/s on the eastern end of Line 7, with the change occurring near the Kootenay Lake fracture zone. The crust maintains its total thickness of ~40 km along the entire eastern part of Line 7. In Line 9, about 120 km to the south, crustal thickness has been modeled to reach 42 km near the Southern Rocky Mountain Trench (Zelt and White, 1995).

Cook et al. (1992) selectively highlighted some of the deep reflections and speculatively linked them with faults mapped on the surface (see also Cook and Varsek, 1994; Cook and van der Velden, 1995; van der Velden and Cook, 1996). These interpretations of crustal structure from seismic data are full of great low-angle faults slicing the entire Cordilleran crust. A simple tectonic explanation has been provided for this speculative interpretation (Cook, 1995c, p. 1822): due to persistent "late Paleozoic to Present" plate-convergence regime, "crustal blocks of the eastern accreted terranes were detached and perhaps rotated"; later, "once allochthonous crustal rocks became welded to the craton by the end of the Paleocene, convergence ceased in the Foreland and Omineca belts, and regional transtension ensued". As discussed herein, this scenario is not compatible with geological facts.

Tectonics of the eastern Cordillera from combined analysis of geological and geophysical data

Regional distribution of seismic properties at several levels of the crust in the southern Canadian Cordillera was computed recently by Clowes et al. (1995) and Burianyk et al. (1997) and presented in map form (Figs. 49-50). The data are limited, but the Lithoprobe corridor from ~52°N to the Canada-U.S. border is covered better than many other areas. Particularly useful may be the maps of Moho depth and average P-wave velocity of the upper and middle crust.

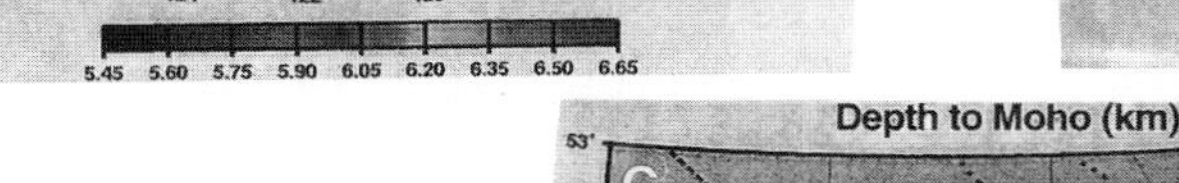

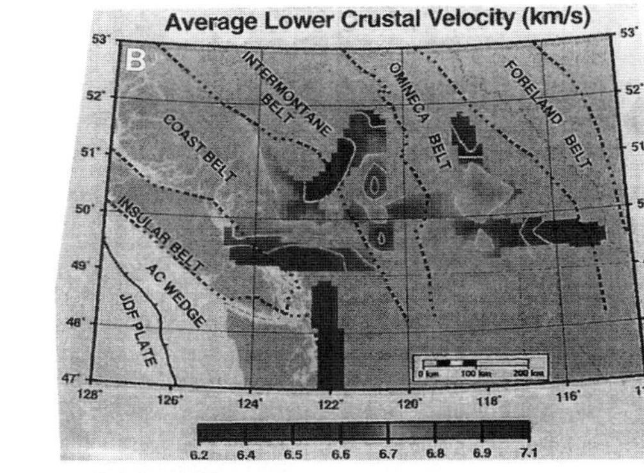

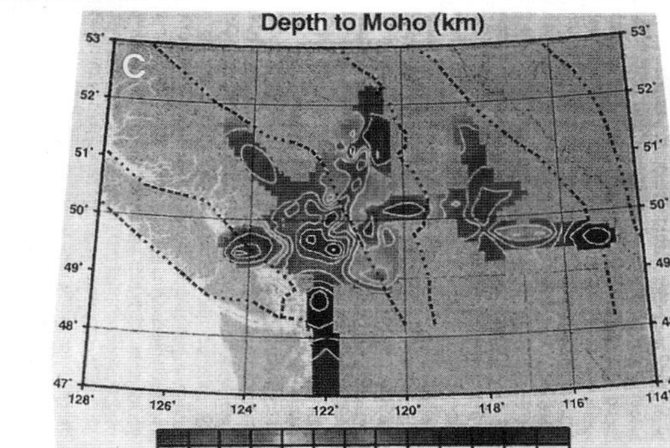

Figure 49. Crustal thickness and velocity structure in the southern Canadian Cordillera, interpreted from Lithoprobe seismic data by Clowes et al. (1995). Illustrating a common misunderstanding, the Rocky Mountain fold-and-thrust belt is incorrectly designated "Foreland Belt". AC - accreted; JdF - Juan de Fuca (these notations reveal incorrect assumptions about the structure of the continental-margin zone: in fact, no ongoing subduction is occurring off Vancouver Island; Lyatsky, 1996).

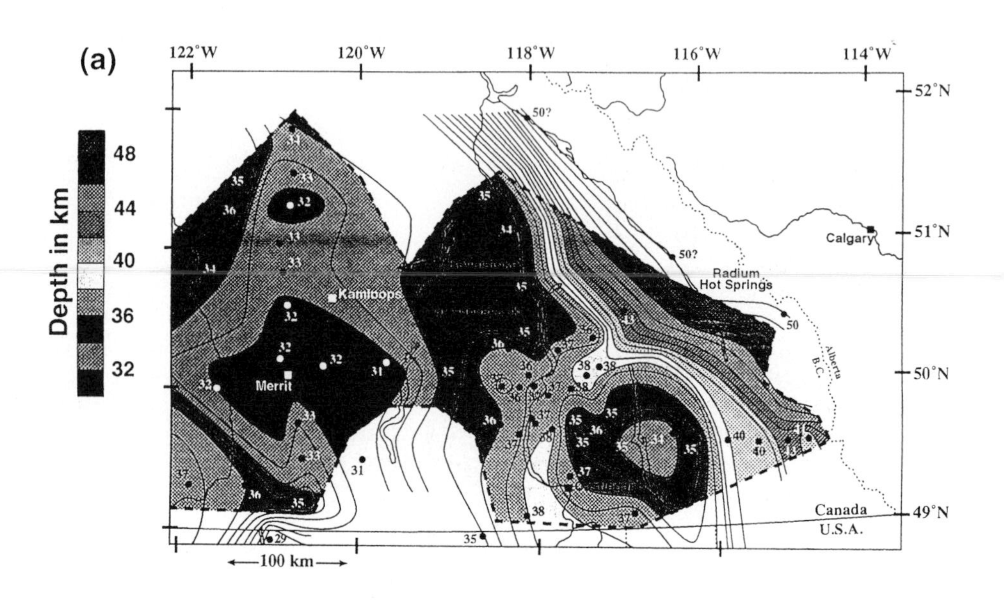

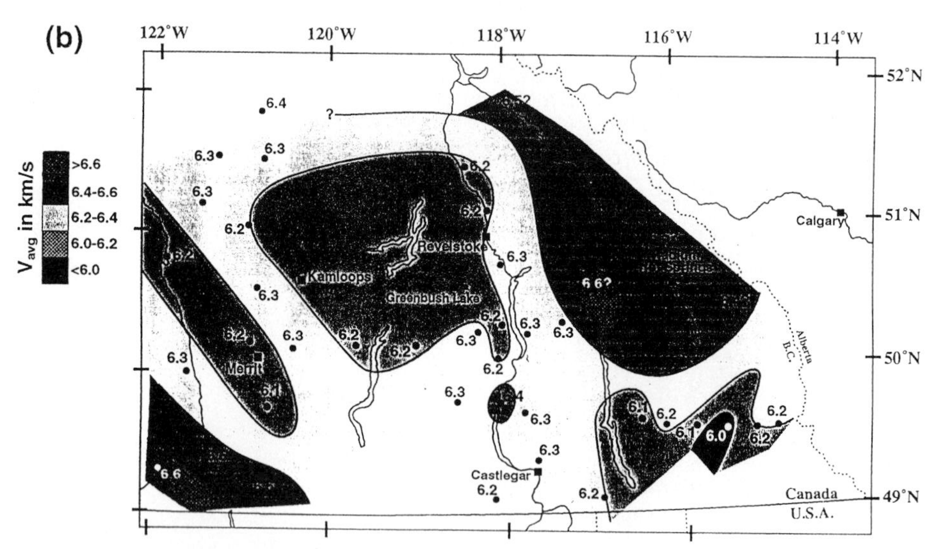

Figure 50. Crustal thickness (**a**) and average P-wave velocity (**b**) in the southern Omineca miogeosyncline and Intermontane median-massif belt, interpreted from Lithoprobe seismic data by Burianyk et al. (1997). Note the considerable local Moho-depth complexity in the Kootenay Arc-Purcell Anticlinorium area, as well as the slight shallowing of the Moho in the southern Intermontane median-massif belt.

In the Moho-depth map, north of ~50°N, E-W and NE-SW structural trends are apparent in the southern Cordillera in the Omineca orogenic zone and the Intermontane median-massif belt; the latter also has some occasional N-S trends. At the levels of the upper/middle (undifferentiated) and lower crust, there are stark differences in seismic properties between the eastern and western parts of the Omineca miogeosyncline: the lowest average velocities in the lower crust are recorded in the eastern part of the Omineca Belt, where the crust is thick and the Moho deep. That area contains remnants of the ex-cratonic basement, revealed in the core complexes.

In Lithoprobe publications, the apparent westward change in crustal properties is linked to the assumed major whole-crust Slocan Lake fault. To justify this fault's known easterly dip in a region where presumed crustal-scale thrusts are supposed to be dipping west, Cook (1995c, his Fig. 7b) offered a complicated interpretation of Line 9. In it, the Hall Lake fault flattens out and curves sharply under the western Purcell Anticlinorium and Kootenay Arc, undulating at ~5 s two-way traveltime (depth ~14-15 km). Underneath, other undulating and subhorizontal thrusts are shown at and above the flat top of the presumed "North American basement" at depths of ~21 to 23 km, serving in part as the basal décollement. The presumably major Slocan Lake normal fault is shown as cutting all the supra-basement thrusts and merging with the deep master detachment. The data in Line 9 are now available with different processing, and this geometry in it is not apparent.

From its surface trace on the east side of the Valhalla metamorphic core complex, the Slocan Lake fault may be coplanar with a seismic reflection dipping some 30° to the east in the upper crust. This event loses definition at the bottom of the Nelson pluton. Cook (1995c) and Cook et al. (1992) interpreted this fault to become very low-angle under this batholith. For that, they correlated it with a short (4-5) km "floating" reflector at ~15 km depth, which has no eastward continuity. This lack of continuity makes such a bold correlation far from obvious.

Presence of such a major low-angle fault under the Kootenay Arc is not evident. In seismic Line 9, the Kootenay Arc is composed of several anticlinal and synclinal forms, seemingly sheared at their axes. A big syncline is found in profile 3 in the top 2.5 s of the data, but there is no coherent east-dipping event in the middle and lower crust that could represent the Slocan Lake fault.

Instead, several steep seismic discontinuities exist in Lithoprobe profiles 3 and 2 beneath surface traces of shear zones west of the Kootenay Arc. Below a blank upper part of the seismic sections, these discontinuities disrupt reflections between 4 and >8 s traveltime. At least six such disruptions are apparent between the area west of the Kootenay fracture belt and the Hall Lake fault. Industry seismic sections crossing this fracture zone near the southern end of the Kootenay Lake also show a change in crustal characteristics across the Hall Lake fault (van der Velden and Cook, 1996). Section 3, shot along ~50°N, contains a band of events 0.8 s thick at about 4 s traveltime. They have been attributed to a "near-basement reflector". Two normal, steep, est-side-down offsets 1.5-2 km each have been suggested at the top of the cratonic basement, one of them under the Southern Rocky Mountain Trench. The western part of industry section 3 was shot N-S, parallel to an elongated anticline and fault on the east side of the mid-Cretaceous White Creek batholith, and a broad band of seismic events there may be off-line arrival, multiples or diffractions. Multiples may be caused by the reflective bottom of the batholith. (The cup-like shape of the reflection probably caused by this pluton's bottom is similar to that from the bottom of the Jurassic Nelson Batholith.) Diffractions at depth are consistent with the presence of steep faults nearby. The White Creek pluton was emplaced at a depth of ~15 km (Woodsworth et al., 1991), and lies along the Hall Lake fault which could have been a conduit for magma. Mineralogical zoning of the pluton indicates it was crystallized under stable conditions, but the overprinting foliation and evidence from its metasedimentary country rocks also indicate a later tectonic activity around 110 Ma.

About 7 km north of the White Creek pluton, industry seismic section 6 crosses the steep Hall Lake fault, which can be traced to a depth of ~10 km. Sideswipe from the pluton, beginning at 2-2.5 km depth, distorts the images in this section too. To the west, another fault seems to run from the pluton's bottom; it also dips to the west and is parallel to the Hall Lake fault. East of the Kootenay fracture belt, a "near-basement reflector", similar to that in section 3, lies at ~4 s traveltime (~12 km depth). If so, the cratonic basement correlated with that in the Interior Platform may continue at least to the middle of the Purcell Anticlinorium as was suggested earlier by Price (1981). The basement top seems to have steep offsets (van der Velden and Cook, 1996, their Plate 4d-e).

Individual strands in the Southern Rocky Mountain Trench fault system have only relatively minor vertical offsets and no significant strike-slip ones (McDonough and Simony, 1988). The appearance of the Tertiary topographic trench along this fault system may be related to intermittent activity of the McBride basement arch (Campbell et al., 1973). Along this arch, Upper Proterozoic and lower Paleozoic strata become thinner, though it happens without substantial facies changes (Struik, 1987). No west-facing basement ramp favored by Cook et al. (1991a-b, 1992) is suggested by geologic data. Basins in this region in the Middle and Late Proterozoic, Early Cambrian, Carboniferous and Permian (e.g., Stott and Aitken, eds., 1993; Figs. 51-52) were two-sided, and the corresponding NNW-trending rifts were parallel to this arch.

Cook (1995b-c) and van der Velden and Cook (1996) constructed a speculative picture where the Southern Rocky Mountain Trench fault is listric to a depth of ~10 km, with

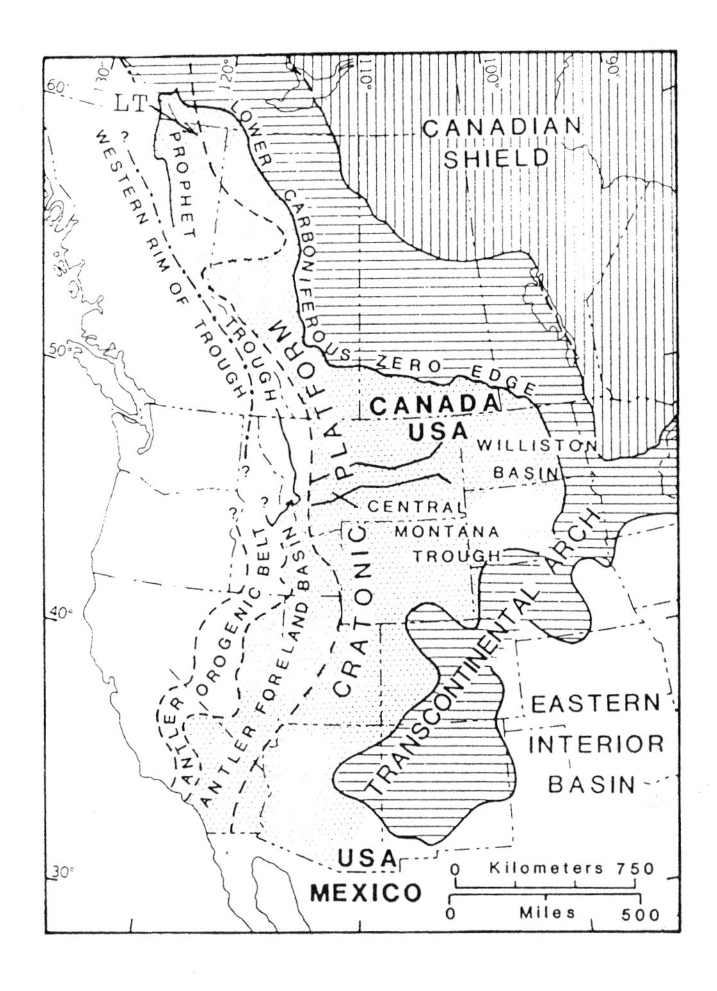

Figure 51. Position of the Antler orogenic belt, as well as the cratonic foreland basin and its Carboniferous Prophet Trough in Canada, in relation to the Antler orogen (modified from Richards, 1989). LT - Liard Trough.

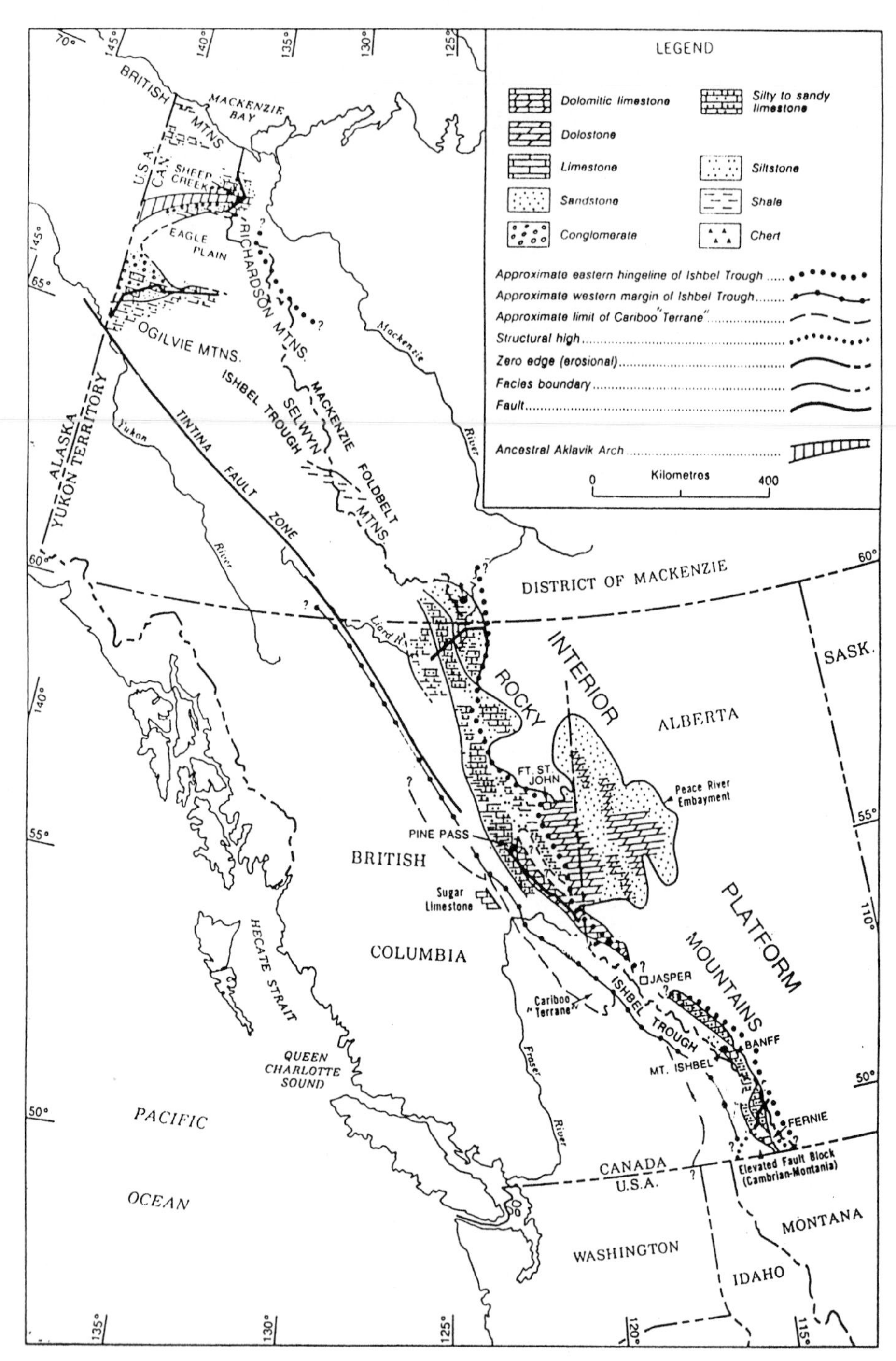

Figure 52. Position of the Early Permian Ishbel Trough in western Canada (modified from Henderson, 1989). Location of rocks in the Rocky Mountains is not palinspastically restored.

scissor-like vertical movements from zero at ~50°30'N to 10-12 km at the Canada-U.S. border. At depth, the westward-flattened Southern Rocky Mountain Trench fault system is linked with the eastward-flattened Slocan Lake fault, so that in Eocene to Miocene time displacement on one resulted in complementary displacement on the other. For this hypothesis, no geologic justification exists.

Across the Southern Rocky Mountain Trench line, the Moho deepens to the west (cp. Kanasewich et al., 1987, 1994; Clowes et al., 1995; Figs. 29-39). Various authors have estimated the Moho beneath the Rocky Mountains to lie at 45-46 km depth. Farther west, the position of the Moho is less well constrained, but many authors agree that under the Purcell Anticlinorium it arches upward (e.g., Zelt and White, 1995). In Lithoprobe Line 9, the Moho culmination seems to lie near ST 40, west of the anticlinorium's axis. Clowes et al. (1995) continued the Slocan Lake fault from its surface trace at the Valhalla dome-Nelson batholith boundary up to 110 km to the east, across these structures and the Kootenay fracture belt, to an assumed eastward drop in the Moho under eastern Purcell Anticlinorium (Fig. 36, bottom cross-section). In another published interpretation (Fig. 36, middle cross-section), the Slocan Lake fault is continued for only 85 km, without reaching the Moho and dissipating in the crust around ST 40. Due to low lateral seismic resolution, to correlate the Slocan Lake fault dissipation with the Purcell Anticlinorium Moho culmination, across the Kootenay Lake fracture zone (also Zelt and White, 1995), is unjustified. As was discussed above, geological and seismic-reflection considerations suggest the real Slocan Lake fault is more local and shallow, and it does not continue in the subsurface east of the Nelson pluton. Floating reflections under ST 40 and 41, east of the Kootenay Lake fracture belt, have no relation to this fault.

The top of the cratonic basement under the Purcell Anticlinorium was shown by Clowes et al. (1995, their Fig. 10c) as rather flat, with a ~2-km offset under the Kootenay fracture belt. Cook et al. (1992), Cook (1995b-c) and van der Velden and Cook (1996) showed the basement displacement under the Kootenay fracture belt as being larger, ~3 km. Whatever its exact amplitude, offset at the level labeled as the top of cratonic basement coincides with the ~40-km-wide zone of faults between the Kootenay Arc and the Purcell Anticlinorium. Changes in Moho depth are also coincident with this fundamental, steep, whole-crust fracture zone. Just south of the Canada-U.S. border, Potter et al. (1986) regarded the Moho's relative flatness under such a tectonically reworked region as evidence that the Cordilleran Moho was formed in the Cenozoic.

Under the Valhalla metamorphic dome, the Moho has been imaged at a traveltime of 11.5-12 s, or 34-36 km depth (Eaton and Cook, 1990). This is considerably shallower than under the Alberta Platform, but still 3 km deeper than under the Purcell Anticlinorium: the anticlinorium thus corresponds to a Moho rollover, which was modeled clearly and fairly reliably between ST 40 and 41 in Line 9, and less clearly between ST 29 and 30 in Line 7. Usually the Cordilleran Moho is shown in the Lithoprobe publications as continuous and smooth, but in detail it is broken by steep faults. It may seem "structurally uncomplicated" (Cook, 1995b, p. 1524) only in the first approximation. The Moho upwarp around ST 40 mimics the middle-upper-crustal antiform, suggesting that the Tertiary Purcell Anticlinorium (cp. Archibald et al., 1983, 1984) is genetically linked to that Moho upwarp. In the southern Omineca Belt, the Moho is disrupted by a "series of step-like offsets" (Cook, 1995b, p. 1524), and vertical offsets more than 2 km in amplitude are common. They can be correlated with offsets in the restructured top of the ex-cratonic basement under western Purcell Anticlinorium (Cook, 1995b, his Fig. 4) and in the Kootenay fracture zone.

Bouguer gravity anomaly maps of the Canadian eastern Cordillera indicate that beneath the Purcell Anticlinorium lies a crustal block bounded by NNW-SSE faults paired with NE-SW ones. The northern corner of this block corresponds to the anticlinorium's northward extension. A relative gravity maximum at ~49°N/116°W (~-125 mGal in Bouguer maps, compared to <-160 mGal in adjoining areas) is correlative with the Moho culmination at ST 40 in seismic Line 9. Two subparallel NNW-SSE magnetic anomaly trends also indicate these faults. The northern NE-SW block-bounding fault is seen best, but the southern NE-SW fault is less apparent in magnetic maps. The Kootenay Lake fracture belt, punctuated along its trend by several mid-Cretaceous plutons, Bayonne and others, is highlighted by a series of Bouguer and isostatic gravity lows east of and along 116°W. On the same trend, some magnetic lows are present as well (Figs. 53-54).

Zelt and White (1995) labeled the granites of the Jurassic Nelson batholith "accreted rocks". This is incorrect: even in the very mobilistic interpretation of Monger et al. (1985), rocks west of the Omineca Belt (including rocks of the Kootenay Arc and the Intermontane Belt) were ascribed to "parautochthonous terranes".

The N-S-elongated Nelson pluton lies west of and parallel to the Kootenay fracture belt. Mid-Cretaceous Bayonne-suite plutons lie along this belt from 48°N to 51°N. The Early Tertiary magmatism again shifted to the west, roughly to the position of the N-S Kettle-Columbia River fault system. The pattern of N-S and E-W structural trends in this area persisted through the entire Mesozoic and Cenozoic time. It is superimposed on the Late Archean-Early Proterozoic NE-SW pattern predating the Cordilleran mobile megabelt, and on the NNW-SSE patterns of the megabelt itself. The early Tertiary N-S rifting came after deep crustal reworking in the Nevadan (Late Triassic-Jurassic), Columbian (mid-Cretaceous) and Laramian (Late Cretaceous-Early Tertiary) orogenies. These orogenies reworked the crystalline basement east of the Kootenay fracture zone only partly, but west of it the transformation of the original cratonic basement was more profound.

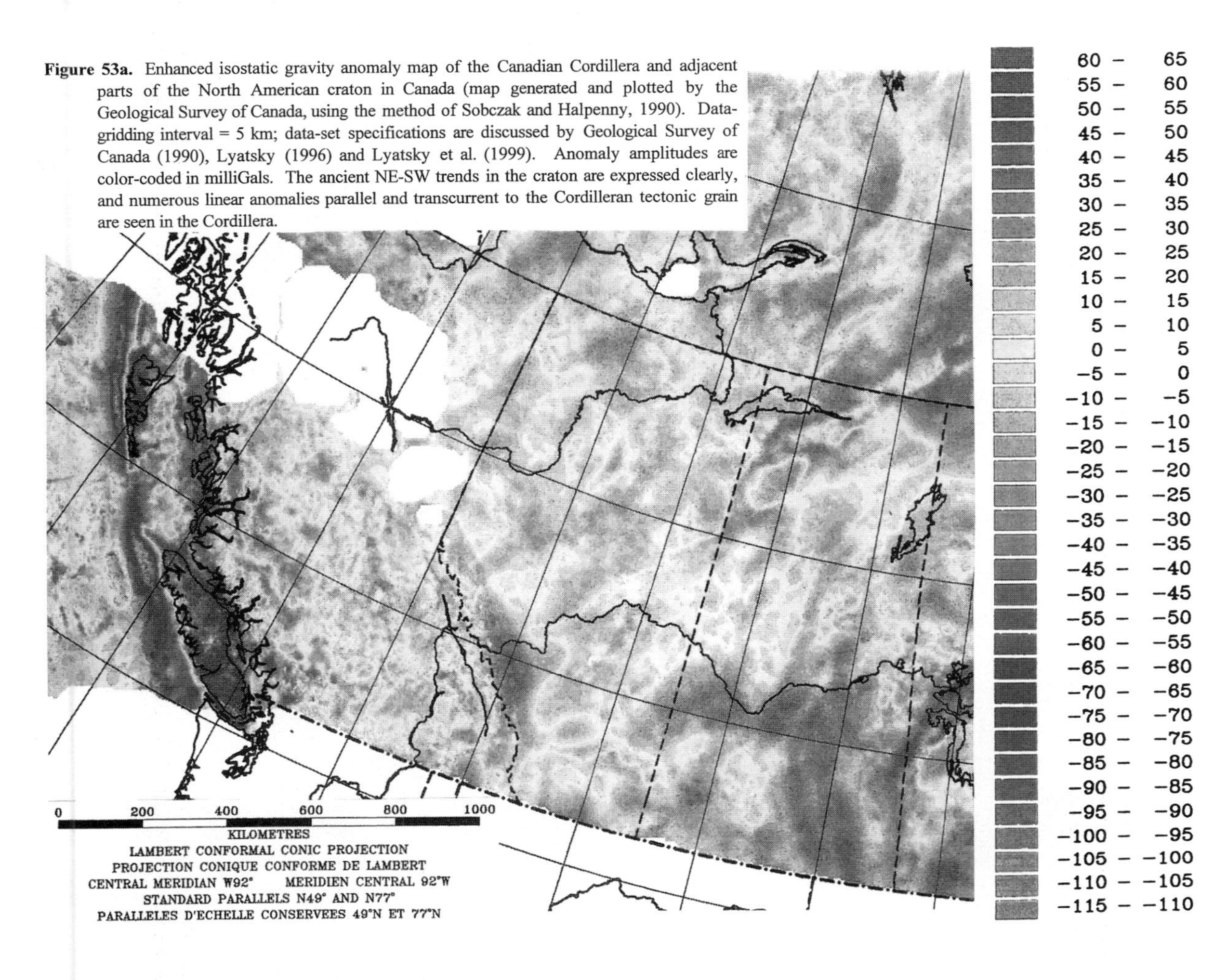

Figure 53a. Enhanced isostatic gravity anomaly map of the Canadian Cordillera and adjacent parts of the North American craton in Canada (map generated and plotted by the Geological Survey of Canada, using the method of Sobczak and Halpenny, 1990). Data-gridding interval = 5 km; data-set specifications are discussed by Geological Survey of Canada (1990), Lyatsky (1996) and Lyatsky et al. (1999). Anomaly amplitudes are color-coded in milliGals. The ancient NE-SW trends in the craton are expressed clearly, and numerous linear anomalies parallel and transcurrent to the Cordilleran tectonic grain are seen in the Cordillera.

B

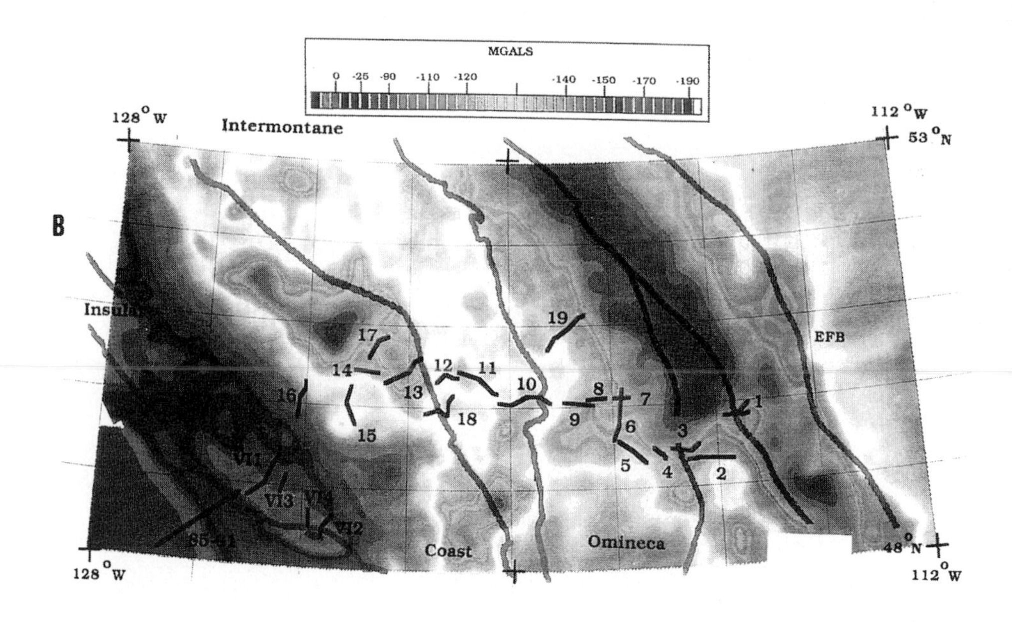

C

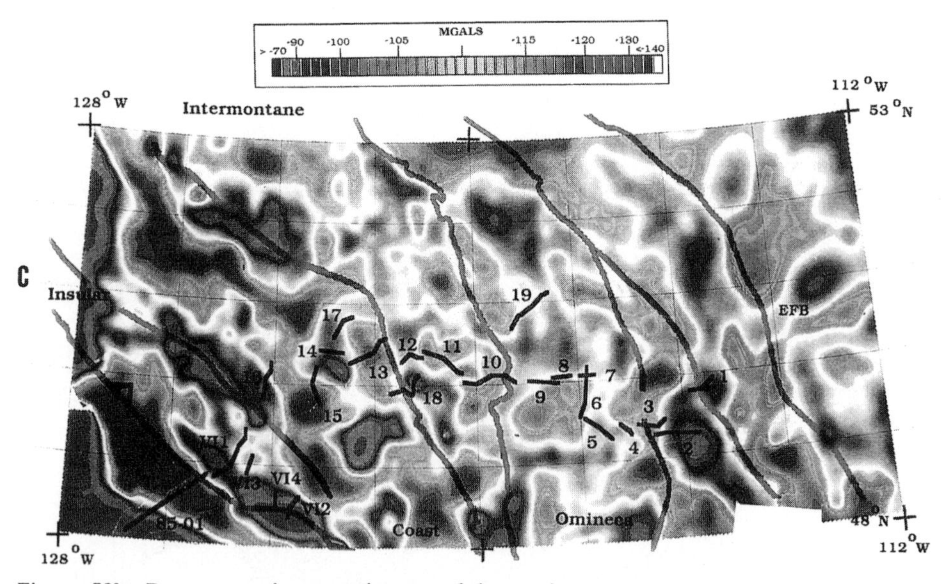

Figure 53b. Bouguer gravity anomaly map of the southern Canadian and northernmost U.S. Cordillera (modified from Cook et al., 1995). Nomenclature as in Fig. 53c.

Figure 53c. Bouguer gravity anomaly map of the southern Canadian and northernmost U.S. Cordillera, wavelength-filtered at 12.5/25-133/200 km (modified from Cook et al., 1995), with an overlay of the main Cordilleran geologic belts. Numbered heavy lines mark the Lithoprobe reflection seismic profiles. Map plotted in the Lambert conformal projection, with a data-gridding interval of 5 km. EFB - eastern boundary of Laramian deformation.

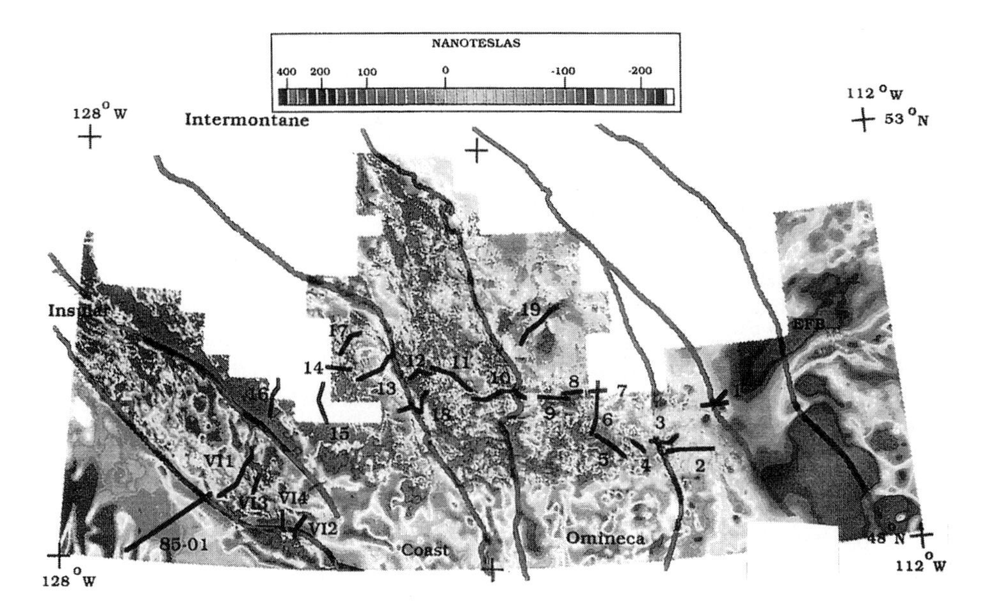

Figure 54. Total-field aeromagnetic anomaly map of the southern Canadian and northernmost U.S. Cordillera (modified from Cook et al., 1995), with an overlay of the main Cordilleran geologic belts. Numbered heavy lines mark the Lithoprobe reflection seismic profiles. Map plotted in the Lambert conformal projection, with a data-gridding interval of 812.8 m. The data have been corrected for the International Geomagnetic Reference Field. EFB - eastern boundary of Laramian deformation.

In the modern North American craton, the crystalline basement lies at a depth of ~3.5 km beneath Calgary and dips gently to the west (Mossop and Shetsen, comps., 1994). The same basement is traced westward, where it lies considerably deeper under the Rocky Mountain Belt of east-vergent thrust sheets. Still the same basement may also be suggested to continue under the eastern Cordilleran miogeosyncline, but there it is more broken-up and mechanically separated from the basement to the east. The crustal block around ST 40 underlying the Purcell Anticlinorium also belongs to the miogeosyncline. But whether this basement is the same as that in the modern North American craton, and whether both are remnants of a single Early Proterozoic craton or products of cratonic masses that were distinct in the time of the Mesoproterozoic Belt-Purcell Basin, is difficult to say.

Gravity and magnetic lineaments

Deep-seated processes affecting the depressed cratonic basement in the Rocky Mountain fold-and-thrust belt varied across and along this belt. The Whitewater Creek gravity gradient runs NE-SW from the location of the Bayonne granitic pluton to roughly the Devonian Fairholme reef complex in Alberta. This gradient highlights a structural trend that determined the position of the Hall Lake fault in the Omineca miogeosyncline and the Whitewater barrier reef in the Paleozoic Alberta Basin. A round Bouguer gravity low west of 51°N/114°W may represent a large granitic pluton in the crystalline basement which in this area lies under a stack of Laramian thrust sheets. At 50°N/116°W, a NE-SW structural trend marks the location of another big mid-Cretaceous batholith (Battle Range), which lies at an intersection of two gravity lineaments, NE-SW and NW-SE.

The old NW-SE trend, oblique to the general NNW-SSE grain of the Cordillera, is expressed in the position of a series of mid-Cretaceous plutons aligned from Hall River in the southeast to the Goldstream pluton in northern Monashee Mountains. Most of these plutons are stocks discordant with the general structural fabric of the region. The Battle Range-Goldstream structural trend cuts the regional Selkirk fan structure. These plutons were emplaced at various depths, but commonly with significant metamorphic aureoles overprinting the country rocks, in the northern Purcell Anticlinorium and along the western side of the Selkirk fan structure from the Battle Range pluton to the North Thompson fault. The mid-Cretaceous fault pattern, highlighted by the distribution of aligned plutons in southern Omineca miogeosyncline, was substantially different than the Jurassic fault pattern.

Expressed in the gravity and magnetic maps are primarily the mid-Cretaceous patterns. The NE-SW Whitewater Creek structural trend is punctuated by several elongated, narrow positive magnetic anomalies that run across the Rocky Mountain Belt and western Alberta Platform (cp. Edwards et al., 1998; Lyatsky et al., 1999). Several NW-SE-trending lineaments are apparent in the gravity and magnetic anomaly maps covering the Omineca and Intermontane belts. Some of these lineaments coincide with the Omineca-Rocky Mountain belt boundary, others with the Omineca-Intermontane belt boundary north of 52°N. Two NW-SE lineaments transect the Omineca-Intermontane boundary north of 49°N and 50°N, where this boundary is conventionally drawn along the N-S trend of the Tertiary rifting.

Even better than the Bouguer gravity anomaly maps, the isostatic gravity anomaly maps show the NE-SW and NNW-SSW trends in the Alberta Platform and Omineca miogeosyncline. The Jurassic NNW-SSE structural trend which controls the Omineca-Intermontane belt boundary between 50°N and 52°N lacks a strong gravity expression: there, a set of small, aligned positive anomalies bounds a large low that lies partly in the Omineca and partly in the Intermontane Belt. Another NNW-SSE-oriented set of small highs is located in the middle of the Intermontane Belt between 50°30'N and 51°30'N. Both these isostatic gravity lineaments are related to Jurassic faults along the main structural grain of the Cordillera.

A pronounced NE-SW isostatic gravity lineament, running from 49°30'N on Vancouver Island to ~54°N in the Alberta cratonic platform cuts the older structures. Another such trend, from ~49°30'N in the southern Omineca Belt to ~52°N in the Alberta Platform, corresponds locally to the mid-Cretaceous Battle Range batholith and its associated stocks and dikes. They (e.g., the Albert stock; Colpron et al., 1996) cut, in particular, the Standfast Creek fault as well as the Clachnacudainn thrust sheet which is often treated as an integral part of the Selkirk allochthon (Johnson and Brown, 1996). Importantly, though, continuity of NE-SW trends in the Cordilleran mobile megabelt with Late Archean-Early Proterozoic trends in the Alberta Platform points to ancient inheritance of some fundamental Cordilleran structures.

In the early Middle Jurassic (173 to 168 Ma), rocks composing the axial zone of the Selkirk fan were still at large depths with pressures of 6-8 kbar (600-800 MPa). Uplift of the central Selkirk horst, bounded by the Purcell fault on the east and the Argonaut Mountain fault on the west, from the corresponding depth of ~25 km (Ghent et al., 1979) might have occurred later (Colpron et al., 1996), during late the Columbian (mid-Cretaceous) episode of extension, magmatism and upwarping. Decompression episodes have been established from thermo-barometric studies for later stages of both Nevadan and Columbian orogenies (cp. Ghent et al., 1991; Colpron et al., 1996). In the Albert stock area, decompression probably occurred soon after the stock was crystallized around 104 Ma (U-Pb zircon dating; Crowley and Brown, 1994). From $^{40}Ar/^{39}Ar$ dates, uplift took place around 85 Ma (Colpron et al., 1996).

A change in the regional stress regime reflected in the distribution of active faults could have happened around that time, too. It predates the Late Cretaceous, Laramian orogeny. The Argonaut Mountain fault, dipping steeply to the SW, separates two crustal blocks with contrasting P-T-t paths. East-side-up vertical displacements across this fault are thought to be large, on the order of 6-8 km (Greenwood et al., 1991; Scammell, 1993). Tertiary high-angle faults, such as Kettle, Hope, Moyie, Hall Lake, Purcell, Southern Rocky Mountain Trench, developed in the conditions of late Laramian regional extension. Most faults of that generation have a normal sense of offset, cutting formations of the Omineca miogeosyncline and the Intermontane median-massif belt as well as thrust sheets of the Rocky Mountain Belt. Rapid extension and differential warping passed gradually into low-amplitude extension that continued in places till the Late Oligocene-Early Miocene (Constenius, 1996). The most tectonically active in this area was the miogeosynclinal orogenic zone. The 180-200-km-wide belt of considerable crustal extension in the eastern Cordillera runs N-S and crosses the Canada-U.S. border, overprinting all the older structures. The maximum crustal restructuring in this N-S zone occurred in the rejuvenated Kootenay Lake fracture belt.

Further improvements in the knowledge of regional tectonics by integrating geological, geophysical and geochemical data

The Moho makes a strong seismic reflection under the Interior Platform, but beneath the Rocky Mountains the reflection Moho is not well defined. The top of the crystalline basement is traced under the outer and inner Rocky Mountain fold-and-thrust belt more reliably, and some Paleozoic structural-formational units continue from the Alberta sedimentary basin across the Rocky Mountains into the Omineca miogeosyncline. Before the Cordilleran mobile megabelt developed, an older craton had continued far to the west. Subsequent orogenic processes affected it from below, and the original crystalline basement west of the Purcell Anticlinorium was substantially reworked. The basement was probably greatly restructured in the Omineca orogenic zone, especially in and west of the Kootenay Lake fracture zone, though its remobilized remnants are still recognized in the metamorphic core complexes. The entire lithosphere, and more certainly crustal rocks exposed at the surface in the eastern Cordillera, in the Omineca miogeosyncline and Kootenay Arc, evolved during the Phanerozoic *in situ* (e.g., Smith and Gehrels, 1991).

According to Monger et al. (1994), rocks west of the Purcell Anticlinorium are considered “western-facies equivalents” of those in the east only “traditionally” (see, e.g., Wheeler and McFeely, comps., 1991). This western package is “isolated”, has a different geologic record and bears “a strong Mesozoic structural fabric” (Monger et al., 1994, p. 371-372). But that would not justify denying an autochthonous nature of these rocks. No proof exists that the deeper crust is exotic or wedged onto a hypothetical ancestral North America. The three orogenies in the Mesozoic and Cenozoic produced deep reworking of the western part of the North American continent, with an apparent recurrence of tectonic regimes. Reworking began in the Proterozoic and was well under way in middle to late Paleozoic (Antler orogeny) time. Great tectono-metamorphic and tectono-magmatic maturity was reached in the Cordillera in the mid-

Mesozoic Nevadan orogeny. All four types of tectonic manifestation were also strong in the late Mesozoic, after the Late Jurassic. The mid-Cretaceous magmatic style and rock composition reveal genetic connections with continental crust, whose deep sources are indicated for both Bayonne and Cassiar suites (Woodsworth et al., 1991). In the eastern Cordilleran miogeosyncline, large reworking and restructuring of the ex-cratonic basement took place in the Early Tertiary. On some faults, deformation continued from Jurassic to early Cenozoic time (Gabrielse, 1985); other faults became inactive, while new ones appeared. Syn-plutonic faulting was common for many mid-Cretaceous batholiths (Woodsworth et al., 1991).

Detailed field studies (Corbett and Simony, 1984) show the Champion Lake fault continues north to connect with the Slocan Lake fault. It served as a conduit for plutons of the Nelson suite, some of which it also sheared. A pre-Tertiary predecessor of the Slocan Lake fault could have also been a conduit for the Kuskanax pluton, but there is no evidence the that Tertiary Slocan Lake fault cuts that pluton's rocks. On the east side of the Valhalla metamorphic core complex, the Slocan Lake fault bears signs of activity in the Eocene. The Valkyr shear zone on the opposite, western side of this metamorphic core complex is steep and dipping to the west. The rocks it displaces include the Ladybird pluton granites 59-55 Ma (Middle Eocene) in age. The Slocan Lake fault, low-angle (~30°) and east-dipping, has recorded displacements dated at 59-58 and 44-40 Ma (Late Eocene; Beaudoin et al., 1991a-c; Carr, 1995). Johnson and Brown (1996) treated the "Columbia River-Slocan Lake-Champion Lake system" essentially as a single fault. However, the Columbia River fault is much better aligned with the Kettle fault, which passes some 60-80 km west of the Champion Lake-Slocan Lake system.

The Kettle-Columbia River and Waneta-Kootenay Lake-Purcell fault systems contain segments that were active in the Jurassic, Cretaceous and Tertiary. The Columbia River fault is mapped as low-angle, on the east side of the Monashee complex. The Valkyr fault runs parallel to the Upper Arrow Lake, between the Valhalla and Monashee complexes. Variations in the direction and angle of dip are seen also in the Waneta-Kootenay Lake-Purcell fault system. The Purcell fault was active after Middle Jurassic but before Tertiary time, as a thrust dipping at a moderate angle to the west (McDonough and Simony, 1988). Steeper westward dips are recorded on faults in the northern Kootenay Lake area. Farther south, the Kootenay fracture belt contains many faults that could be Mesozoic or Eocene, alternating from one area to the next: normal, west-dipping or east-dipping reverse (see, e.g., Gabrielse, 1991b).

From seismic and drillhole data, Simpson and Johnson (1989) described a typical fault pattern over a big core complex in Arizona. Faults with different dips interact there complexly, in an anastomosing manner. In the well-studied Valhalla and Monashee core complexes in British Columbia (Parrish, 1984; Carr et al., 1987; Parrish et al., 1988; Scammell, 1993; Schaubs and Carr, 1998), rocks with a huge age span from Proterozoic to Mesozoic were found to have been metamorphosed at depths of almost 30 km. Some of them were still at these depths even at the end of the Cretaceous, and Early Tertiary unroofing by erosion due crustal uplift brought them to the surface (Spear and Parrish, 1996). Events like these recurred in the eastern Cordillera more than once.

Exhumed pieces of Mesozoic and Early Tertiary lower crust show that their protoliths were supracrustal rocks, buried to great depths with metamorphic conditions reaching granulite grade. Rock alteration was accompanied by general creep and selective zonal flow, due to changes in physical conditions and rheological properties of the reworked rocks. Zones of metamorphism were superimposed across the original layering, and shears cut both stratigraphic bedding and metamorphic zonation. Partial melting gave rise to palingenic granites that intruded the upper-crustal levels; a large volume of plutons and volcanic flows requires that a substantial part of the buried rock mass was melted. As the metamorphosed, deep-seated rocks rose, restructuring occurred. During the rise, rock behavior changed from ductile to brittle. The ductile fabrics were largely low-angle, and they developed in and over the head of a rising metamorphic dome. The newer, brittle fabrics were steeper, commonly related to dome-bounding faults. The position of the three parallel N-S-trending zones of metamorphic core complexes (from east to west, Priest River, Kettle and Okanagan) in the southern Omineca miogeosyncline was controlled by the deep N-S-oriented faults. These deep-seated faults are now masked by low-angle shears and fractures formed in and over the rising domes. The Kettle-Columbia River fault system marks the axial zone of Tertiary extension imposed on the Omineca miogeosynclinal grain, so both rifts and core complexes are found in the southern Canadian Cordillera.

Monger et al. (1994) speculated that the "domain of extreme Eocene extension" in the Omineca miogeosyncline may be paired with a "domain of Eocene transtension" in the Intermontane Belt (which they regarded as a "superterrane" of oceanic-crust affinity). In fact, the continental-crust basement of the Intermontane median-massif belt is Proterozoic, and metamorphic rocks from it are found in the Nicola dome, Takla Lake outlier, and Cascade metamorphic cores (Misch, 1966; Armstrong, 1982; Friedman and Armstrong, 1988; Monger et al., 1994). From the distribution of $^{87}SR/^{86}Sr$ ratios, Armstrong et al. (1991) put the western extent of continental lithosphere as far as the middle of the Intermontane median-massif belt. They based this conclusion on laboratory studies of igneous-rock samples, including those from abundant teleorogenic plutonic relatives of the Coast Belt orogen to the west (see, e.g., Woodsworth et al., 1991). The ex-cratonic Proterozoic basement more likely continues to the Takla Lake metamorphic core and the metamorphic cores of the North Cascade Mountain.

Numerous extensional Late Cretaceous and Early Tertiary graben basins in the Omineca and Intermontane belts (Gabrielse, 1991a) indicate that Tertiary extension affected both the median massifs and the miogeosynclinal orogenic zone, although its manifestations were not the same. Friedman and Armstrong (1988) pointed out some shortcomings in the rather geometrical model used by Price and Carmichael (1986) to explain the Tertiary extension in the Intermontane Belt by an interplay of big strike-slip faults. The explanation of Friedman and Armstrong (1988) is also imperfect, as thermal weakening of the crust which they considered to be the cause of extension was regarded by them "as a result of magmatism". In reality, deep crustal thermal reworking was responsible for weakening of the crust, magmatism and metamorphic-core formation (cp., e.g., Beratan, ed., 1996). Metamorphic rocks in core complexes elsewhere in the Cordillera bear distinct patterns of ductile deformation predating the upward passage of these rocks through the brittle-ductile transition level. Brittle deformation occurred later, and brittle patterns were superimposed on previous ones as the metamorphic rocks reached the upper part of the crust. In the Intermontane Belt, this happened after ~47 Ma; in the Coast Belt orogen and particularly in the North Cascades, it happened near 35 Ma (see Lyatsky, 1996 for review).

Friedman and Armstrong (1988) invoked external causes, presumably related to plate interactions in the Pacific, to explain the Tertiary extensional tectonic regime in the Cordillera. But similar regimes had also existed in the Cordillera before, each time following a peak of orogenic activity in a particular zone, as crustal reworking induced inversion, extension and regional uplift. In the Canadian Cordillera, such episodes happened in the Jurassic Nevadan, mid-Cretaceous Columbian and Late Cretaceous-Early Tertiary Laramian orogenies.

From magnetotelluric data, Jones and Gough (1995) concluded that the Kootenay Arc lies at the eastern limit of the regional Canadian Cordillera lower-crustal electric conductor (also Gough, 1986). The modern lower crust under the Purcell Anticlinorium is modeled with a high resistivity of 100 to 300 ohm*m, and under the Valhalla dome with a low resistivity of only 5 to 30 ohm*m (Jones and Gough, 1995). This electrical-properties break matches with the crustal-scale Kootenay Lake zone of steep faults. In contrast, shallow low-angle faults flanking the Monashee, Valhalla and Okanagan core complexes have no significant conductivity anomalies at depth (see also Marquis et al., 1995). Jones and Gough (1995) concluded that neither seismic nor magnetotelluric data imply "a simple geometric extension of unmodified North American basement" across the central Cordillera to the Fraser River fault or beyond (as suggested by Cook et al., 1992; Cook, 1995b-c). From seismic characteristics, Burianyk and Kanasewich (1995) supposed that crust with cratonic parameters may continue to the Okanagan Lake. In Lithoprobe Line 9, Zelt and White (1995) showed a variable P-wave velocity in the upper crust under the Purcell Anticlinorium, ~5.7 to 6.3 km/s, vs. a more uniform velocity to the west. Mid-crustal velocity increases by some 0.4-0.5 km/s in the southern Omineca Belt relative to that under the Rocky Mountains. In the lower crust, the velocity drops from 6.4 km/s in the eastern unreworked basement to 6.3 km/s in the crystalline basement under the Purcell Anticlinorium, but increases to 6.7 km/s west of the Kootenay Lake fracture belt.

Middle Jurassic and Eocene tectono-magmatic activity strongly affected most of the southern Canadian Cordillera as far east as the Kootenay Arc. Mid-Cretaceous activity spread farther east, involving also the Purcell Anticlinorium. In the corresponding orogenies, rocks of the Omineca miogeosyncline were buried deep enough to reach amphibolite metamorphic grades, whereas rocks in the Rocky Mountain and Intermontane belts were metamorphosed far less, sometime negligibly.

The patterns of migration of Cordilleran metamorphism and magmatism are complex and fit no simple model of continuous, relatively steady plate convergence at the tectonically inert continental margin, with terrane accretion and collisions (cp. also Woodsworth et al., 1991). Polyphase deformation in the miogeosyncline varied is style, location and timing (e.g., compare Kootenay Arc and Purcell Anticlinorium; Fyles and Höy, 1991), which is also hard to match to a template based on assumed one-sided (west-sided) evolution of this region. The Cordilleran continental crust and lithosphere were nothing like inert, and they developed mostly due to their own, internal causes, since the Proterozoic onward. The orogenies that punctuated its tectonic history were caused for the most part indigenously.

In recent Lithoprobe publications, cratonic crust in the Canadian Cordillera thins westward to zero near the Fraser fault at the Intermontane-Coast belt boundary (Cook et al., 1992; Cook, 1995b-c). Jones and Gough (1995), however, were skeptical that a thin cratonic-crust "tongue" can continue from the Interior Platform through much of the Cordillera. Indeed, presence of several tectonic zones with very contrasting tectonic histories and great vertical movements to achieve burial and exhumation precludes such simple continuity.

Lithoprobe's tectonic scenario for the Cordillera supposed long-lived (since late Paleozoic or Mesozoic) regional compression caused by subduction from the west, followed by Tertiary extension. This simple scenario is contradicted by many observations suggesting indigenous complexity of the Cordilleran evolution.

7 - BROAD LOOK AT GEODYNAMICAL MECHANISMS OF CRUSTAL RESTRUCTURING IN THE CORDILLERAN OROGENS AND MEDIAN MASSIFS

Low-angle normal and thrust structures in mobile-megabelt interiors

Impressive tectonic structures in the U.S. Cordilleran interior have been discussed for decades (see Oldow et al., 1989 for overview). On a big low-angle thrust plane, the huge middle-late Paleozoic Roberts Mountains allochthon, composed of rocks generally considered to have eugeosynclinal affinities, rests on top of a miogeosynclinal rock succession (Figs. 55a-b). This thrust fault and eastward-transported rocks package are mapped in the southern U.S. Cordillera, up to the Cenozoic Columbia plateau-basalt province in Oregon, Washington and southern Idaho. Parallel to and slightly west of the Roberts Mountains thrust, the less impressive earliest Mesozoic Golconda allochthon is another testament to the huge tectonic stresses applied to this region laterally, from the west, from at least the region around Nevada or eastern California and Oregon. It seems attractive to assume the cause of these stresses is subduction from the Pacific, but other possibilities are worth examining as well.

After all the discussions about the assumed far-traveled terranes, including the Franciscan complex in the western California Coast Ranges, Tagami and Dumitru (1996) concluded from local evidence that sedimentary rocks of this complex, accumulated since the Late Jurassic, were derived from the North American continent, with sources in the Sierra Nevada and the granitic Idaho batholith. These authors were firm that, contrary to conventional assumptions, their observations provide "...no suggestion that these parts are highly allochthonous" (Tagami and Dumitru, 1996, p. 11,353). How exactly parts of the Franciscan complex were buried to depths of 10 to 30 km, reached crustal levels with high-pressure but low-temperature metamorphic conditions, and then were returned back to the surface remains a matter for discussion. Elsewhere in coastal and Baja California, paleomagnetic evidence for some of the presumed terrane displacements has also been found to be false (Dickinson and Butler, 1998).

The Roberts Mountains and Golconda thrusts have curved surface traces, sharply changing their orientation from predominant N-S to almost E-W, quite discordant with the NNW-SSE orientation of the Franciscan formation to the west. The huge Sierra Nevada batholith between them and the Franciscan region is also oriented NNW-SSE. Much younger is the Sevier Desert fracture. It is mapped farther east, in southern Utah. Sense of motion on it is unclear: some workers regard it as a thrust, others as a low-

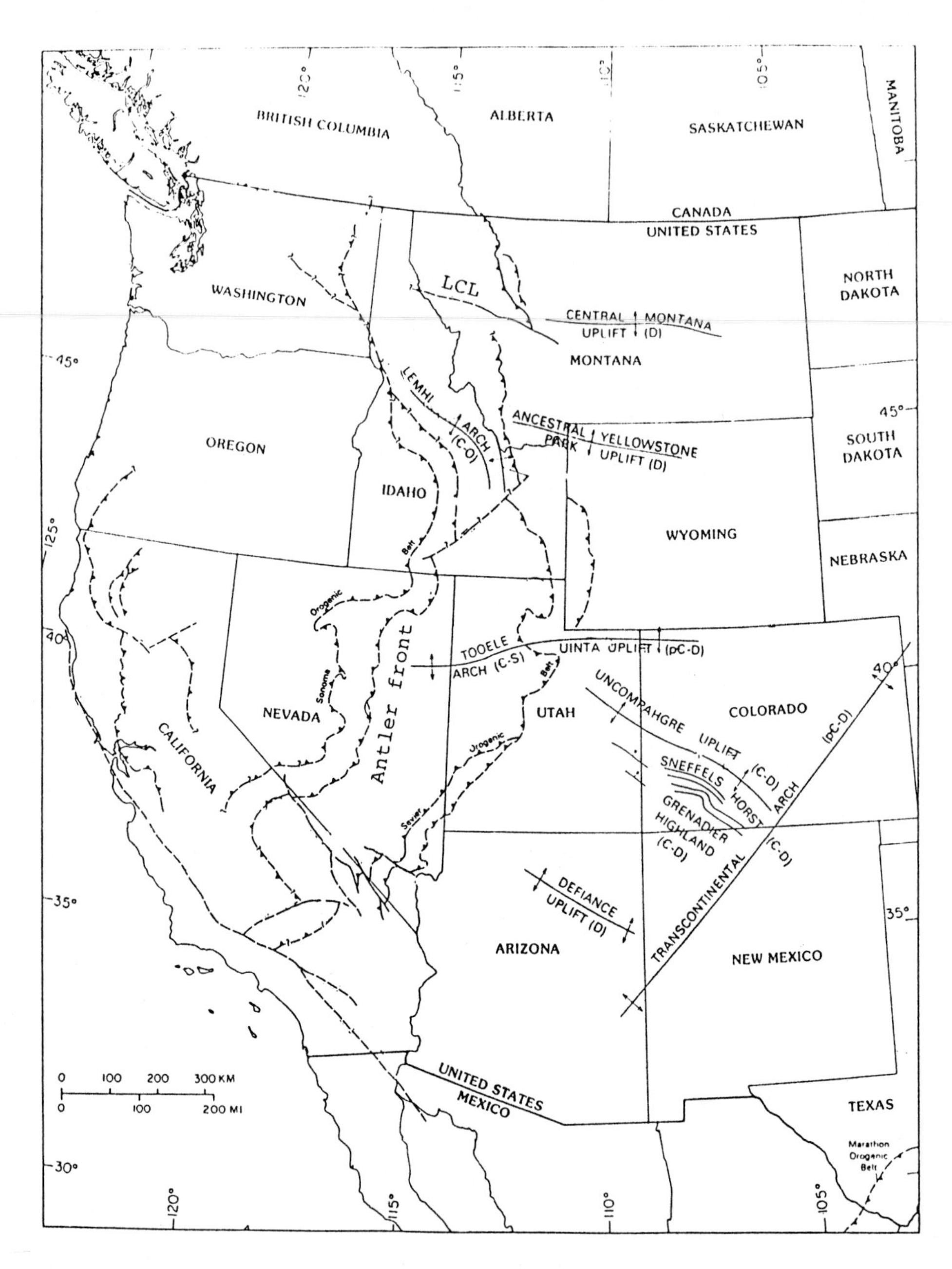

Figure 55a. Index map of major Phanerozoic thrust zones and uplifts in the U.S. Cordillera (modified from Poole et al., 1992). LCL - Lewis & Clark lineament.

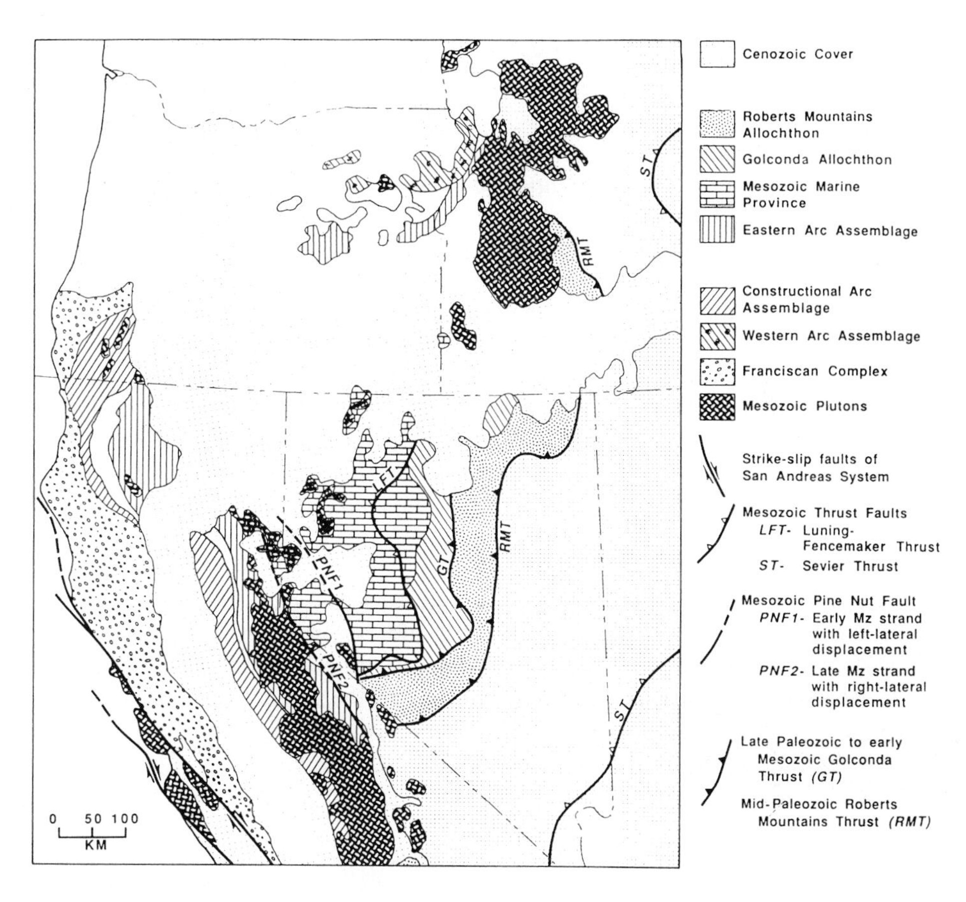

Figure 55b. Location of the Roberts Mountains and Golconda thrusts and the Franciscan rock assemblage in the U.S. Cordillera (simplified from Oldow et al., 1989).

angle normal fault, still other suggest it may not be a major detachment at all (see Allmendinger et al., 1986; Allmendinger, 1992; Anders and Christie-Blick, 1994; Allmendinger and Royse, 1995; Anders et al., 1995). An important point acknowledged and underlined in this debate is the inherent non-uniqueness of geological interpretations of geophysical data. This non-uniqueness calls for great caution in reading seismic images to derive a reliable picture of the structure of the crust (see also Pakiser and Mooney, eds., 1989; Beratan, ed., 1996).

To the north, the Golconda thrust disappears beneath the Tertiary Columbia basalt province. The Roberts Mountains thrust continues a little farther north, but only as far as the Idaho batholith. None of the tectonic features related to the Antler orogeny in the U.S. - the Roberts Mountains allochthon, the inverted Antler mountainous high, the foreland and foredeep basins - can be traced directly into Canada (Oldow et al., 1989). However, coarse clastic sediments of the Late Devonian Earn Group seem to be associated with an Antler-age rift zone in Canada, which contains several mapped grabens (Gordey, 1991). Older Devonian orthogneisses, exposed in the area of the Selkirk fan structure in the Omineca miogeosyncline, also indicate deep orogenic reworking at that time (Wheeler and McFeely, comps., 1991). The younger two-sided Prophet/Ishbel sedimentary basin (Henderson, 1989; Richards, 1989) is a foredeep basin probably responding to Late Carboniferous-Permian inversion in that region (Lyatsky et al., 1999).

Contrary to the numerical model of Bond and Kominz (1984; see its critique in Aitken, 1993a-c), orogenies have affected the Omineca miogeosyncline since Mesoproterozoic time (McMechan, 1991), contributing all forms of tectonic manifestation - metamorphic, magmatic, sedimentary and deformational. In the Mesozoic-Cenozoic geologic record in the Omineca orogenic zone, recurrent orogenies are well recognized in cycles of basin formation, deep subsidence, inversion and late-orogenic uplift.

Repeated downward and upward crustal block movements in mobile-megabelt interiors cannot be explained in terms of accretion of "suspect terranes" of unknown origin. Illustrating the uncertainties of their definition and even existence, these terranes have been ascribed variously to fringing-arc systems (Burchfiel and Davis, 1972) or more exotic areas of origin (Monger et al., 1982), and regarded as either a collage of completely separate entities (Davis et al., 1978) or pre-collisional aggregations called "superterranes" (Monger et al., 1982, 1994).

Middle Devonian to Early Mississippian formations whose origin is assigned to fringing-island-arc settings have been described in the western Sierra Nevada foothills and in the Klamath Mountains (Irwin, 1981; Merguerian and Schweikert, 1987). Coeval rocks of the Kootenay Arc may have a similar origin, although, according to Smith and Gehrels (1991), they are native to North America, though they were later thrusted eastward over the assemblages of the western Omineca Belt (Wheeler and Gabrielse, coords., 1972). A big unconformity lies at the base of the Late Mississippian-Pennsylvanian Milford Group, which is also strongly deformed (Klepacki and Wheeler, 1985).

Monger et al. (1994) supposed that a "fossil" continental margin of an ancestral North America lies beneath the eastern Selkirk Mountains, at longitude 117°30'W. This is hardly possible, as deep crustal reworking in the Omineca orogenic belt by the Antler and younger orogenies makes preservation of older crustal structures unlikely. Still, Monger et al. (1994, p. 380) speculated that "...the Kootenay Arc is a crustal scale monoclinal flexure of about 20 km amplitude that can be interpreted as a fault-bend fold in which the entire parautochthonous supracrustal sequence of this part of the Cordillera is draped over a footwall ramp marking the 'fossil' edge of the North American Craton (Price, 1981)".

The 10-20-km-thick Middle Proterozoic Belt-Purcell Basin succession is thought to have formed in an intracratonic (Winston, 1986; cp. also Oldow et al., 1989) or at least intracontinental setting, so no continental margin existed in the eastern Canadian Cordillera. Devlin and Bond (1988) and Gabrielse and Campbell (1986) reported the age of volcanics assigned to lower Windermere Supergroup strata to be 770 to 750 Ma. If this is also the age of Windermere rifting, imposed on various pre-existing tectonic features including the Belt-Purcell Basin, the previous Middle and Late Proterozoic magmatic and deformational tectonic manifestations (McMechan and Price, 1982; McMechan, 1991) were succeeded by these younger ones around that time. Additional differential block movements there during Cambrian time (Monger et al., 1994) complicated this region's structural history further.

Steep crustal-scale faults oriented WNW-ESE to E-W and Montana and Idaho and NE-SW in Alberta and British Columbia are very old, inherited from a pre-Cordilleran-megabelt craton of Archean-Early Proterozoic age (Winston, 1986; Lyatsky et al., 1999). This structural network is well recognized in the modern North American craton as well as in the Cordilleran mobile megabelt. These ancient faults, transcurrent to the megabelt's general NNW-SSE structural trend which evolved strongly since at least Windermere time, are control anomalous structural lineaments and zones in the Intermontane median-massif belt (e.g., in the Skeena and Stikine arches). In the Cordillera east of the Intermontane Belt, these anomalous lineaments are exhibited even more clearly all along the Omineca and Rocky Mountain belts (Peterson, ed., 1986; Stott and Aitken, eds., 1993). Differential block movements across these fundamental faults are described in many parts of the Cordillera in Canada and the U.S. (see review in Lyatsky et al., 1999). Along-strike variations in the geologic record in the eastern Canadian Cordillera (McMechan and Thompson, 1993; Cecile et al., 1997) are best known for Cambrian and Devonian time, but they are also recognized in younger tectonic étages (Lyatsky et al., 1999). Their connections with distinct whole-crust blocks separated by NE-SW-trending faults are now acknowledged commonly (Cecile et al., 1997).

Despite tectonic reworking in the mobile megabelt, these ancient basement-related zones are still recognizable in discontinuous steep faults, elongated plutonic bodies and discordant anomaly trends in gravity and magnetic maps. Several big crustal blocks recognized in the Omineca miogeosyncline from the analysis of seismic data, metamorphic facies and thermo-barometric information also indicate the persistence of inherited block patterns.

In the region of the Selkirk fan structure, SW-verging ductile folding and imposed thrusting on the western limb likely began in the Middle Jurassic, when the affected rocks were still at 20-30 km depth (Colpron et al., 1996). Rapid uplift by ~10 km is estimated for the Late Jurassic, and another episode of vertical movements is suggested by metamorphic-grade considerations for the mid-Cretaceous (Greenwood et al., 1991; Colpron et al., 1996). These movements took place across steep crustal-scale faults (Struik, 1991). To the west, a younger episode of thrusting, also ductile, affected the Monashee shear zone in the Late Cretaceous (see review in Parrish, 1995). It continued till ~80 Ma (Campanian). Since ~74 Ma, the entire Monashee complex and overlying rocks (but not blocks of the Selkirk fan structure) underwent rapid uplift that continued till at least 53 Ma (Early Eocene). Yet, displacements on the Monashee shear zone had ceased earlier, around 58 Ma (Late Paleocene; Carr, 1992).

From these observations, some important conclusions can be drawn: (a) timing of vertical tectonic movements in the Selkirk and Monashee Mountains differed; (b) ductile and brittle deformation happened repeatedly; and (c) broad regional uplift could have involved these blocks after shearing on the Monashee zone had ended. Apatite fission-track studies of the upper Monashee complex indicate that, after migmatization in the shear zone around 59 Ma, common for the Monashee and Selkirk blocks was cooling from 35 to 25 Ma (Oligocene). Transitions between ductile and brittle tectonic conditions are usually not clear-cut or abrupt, and during upwarping and downwarping many rocks passed this transition repeatedly. Parrish (1995, p. 1631) noted that, during the last shearing episode, some shearing affected "the entire Selkirk allochthon".

Crustal extension followed each of the last three major Cordilleran orogenies, Nevadan, Columbian and Laramian. The largest extension episode is thought to have been the one in the Early Tertiary: Brown and Carr (1990) estimated it to be ~80% in the southern Omineca Belt, whereas Parrish et al. (1988) offered a more conservative estimate of ~30%. From formal thermal reasoning, several events have been postulated there for Laramian and post-Laramian time: a great increase in crustal thickness to 50-60 km, accommodated by huge whole-crust listric faulting, followed by collapse of the overthickened crust which probably also involved these faults (e.g., Ranalli et al., 1989). But neither geologic mapping nor geophysical data in the Canadian Cordillera give evidence for great delamination on listric faults or spectacular whole-crust collapse.

The idea of extensional collapses of slice-stacked crust in orogenic areas, as developed by Dewey (1988), has come under criticism from some regional geologists (e.g., in the Norwegian Caledonides; Fossen, 1993). Such collapse, driven by vertical gravity forces acting on the crust, would be a logical consequence of crustal thickening in areas where the lithospheric mantle is detached or completely removed by subduction. Without roots, the orogenic area loses its support from beneath, and collapse follows (Stille, 1941 used a German word *Umbruch*, but he applied it in different situations; see, e.g., Lyatsky, 1967). Intraplate extensional regimes do not necessarily require influences from far-away plate margins. Post-orogenic and non-orogenic extensional shearing is recorded in many Precambrian cratonic terrains where middle- and lower-crust levels are exposed for observation (e.g., Langenberg, 1983).

Extensional deformation of the crust can apparently occur on steely or gently dipping brittle normal faults or by ductile flow. Zones of extensional shearing in and beneath the brittle-ductile transition tend to re-use some of the many pre-existing zones of weakness - intervals with reduced rock strength, boundaries of plutonic bodies, mechanically weak fault planes, etc. Propagating extensional shear zones may also cross and overprint pre-existing structural fabrics.

Dangers of fashionable prejudices

The Late Cretaceous-Early Tertiary Laramian orogeny is the latest of several in the history of the Cordilleran megabelt. Its manifestations are the most obvious, as they are the least overprinted by subsequent tectonism. Spectacular fold-and-thrust imbrication has been obvious to investigators for decades. Ironically, in the Rocky Mountains these grand manifestations of orogenic power have turned out to be rather shallow and rootless. The orogenic processes operating deep in the crust, though more hidden, include major aspects of crustal reworking - metamorphic, magmatic, deformational - and produce whole-crust warping, ductile displacements, block movements and so on.

The Devonian to Mississippian Antler orogeny was recognized in the western U.S. only in the mid-20th century (Roberts, 1949, 1964). Its extent in the Canadian Cordillera is still unclear (Gordey et al., 1987; Oldow et al., 1989). Later orogenies obscured its manifestations but did not fully eliminate them in the Omineca miogeosynclinal orogenic zone, Kootenay Arc, or Intermontane median-massif belt (Armstrong, 1988; Wheeler and McFeely, comps., 1991).

The Permian to Triassic Sonoma orogeny is recognized to have been responsible for dramatic reworking of the western Cordillera in the southwestern U.S. (Silberling, 1973; Walker, 1988). Its significance is often discussed in terms of reorganization of a western North American continental margin as well as the formation of the Golconda allochthon, although the significance of this tectonic episode remains underappreciated: it was one of the fundamental orogenies in some of the zones of the Cordilleran megabelt. The Havallah Basin is assumed to have lain west of the Golconda allochthon and to have been the source of this allochthonous sheet. Most probably, it was a fairly local, inboard marginal basin in a fringing system (Burchfiel and Davis, 1975). Sediments were supplied to this basin from the Antler High in abundance (Miller et al., 1989), indicating the basin's intimate links with the North American continent (see also Speed, 1977, 1979; Speed and Sleep, 1982).

In the Canadian Cordillera, huge translations have been suggested (Monger et al., 1972, 1982) but not proved (Nelson and Nelson, 1985) for the supposedly oceanic-crust rocks, such as the Permian through Triassic-Jurassic Cache Creek which is regarded as both an assemblage and a terrane (e.g., Monger, 1993). Speed (1977, 1979) also favored a similar independent microplate or terrane Sonomia in the U.S. Cordillera, but obvious faunal correlations in Permian rocks in Nevada and Oregon (Wardlaw et al., 1982) are evidence against this supposition.

Oldow et al. (1989) concluded the age of the Sonoma orogeny in the U.S. is the most reasonably bracketed between the Late Permian and Early Triassic. In the Kootenay Arc area in British Columbia, the Permian Kaslo Group lies over the erosionally truncated Milford rock assemblage (Klepacki and Wheeler, 1985). This structural-formational étage is folded (comparatively mildly) and intruded by dioritic dikes. Early Late Triassic conglomerates above contain Carnian fossils and belong to a younger étage.

Farther north, the assemblage usually called Slide Mountain terrane contains rocks correlatable with groups, including Kaslo, dated as Carboniferous and Permian. Similarities in Precambrian zircon ages (the ~1,700 Ma cluster is the most distinct) suggest genetic links of Slide Mountain and Kootenay rocks, despite their later disruption by thrust faulting. Silurian granitic clasts are also found in Carboniferous Slide Mountain rocks (Roback et al., 1994), and small granitic bodies of that age have been noted in the Omineca miogeosyncline (Cecile and Norford, 1993). Roback et al. (1994) favored an affiliation of the Permian Kaslo Group with a marginal basin (also Klepacki and Wheeler, 1985; Miller et al., 1992), in sharp contrast with the idea of a far-traveled terrane (e.g., Harms, 1986). From detailed mapping, Thompson (1998) found that the Slide Mountain and Quesnel terranes and/or rock assemblages (assigned to the Intermontane Superterrane; Monger et al., 1982, 1994) are not as "exotic" as was supposed before, and their supposedly thrust boundaries with the Omineca miogeosyncline are, at least in places, stratigraphic.

From regional isotopic data, Ernst (1988) was skeptical of the ideas about accretion of "preexisting ancient continental fragments" in the western Cordillera in Proterozoic and Phanerozoic. In the western U.S. Cordillera, he noted three big blocks of continental-crust nature: Salinian (west of San Andreas fault), Nevada-Utah (east of the Sierra Nevada) and North Cascade (straddling the Washington-British Columbia border).

The oldest rocks in the North Cascades are associated with a cratonic basement (Misch, 1966; Harrison, 1972). Its position matches that of the so-called Western craton, inferred from studies of provenance areas of the Belt-Purcell Basin (Harrison, 1972; Peterson, ed., 1986). Later ideas about the genesis of crustal structure there included contraction accompanied by development of abundant anatectic granitic bodies (Zen, 1988) or thrust-fault thickening (McGroder, 1991). Magmatic loading combined with diapiric behavior of granitic bodies was viewed as another possibility (Brown and Walker, 1993). All these mechanisms produced a variety of manifestations, including various folding styles, metamorphic-facies patterns, pluton petrology, low-angle and steep faults (Lyatsky, 1996). Attempts have been made to reconcile the observed regional metamorphic patterns with two-sided thrusting east and west of the axial zone of the North Cascades, and with its diachronous uplift (McGroder, 1991; see also Woodsworth et al., 1991).

The North Cascade area had long been a part of a much larger Coast Belt orogen, which continued through western British Columbia into Alaska for a distance of ~2,000 km. Only in the Tertiary did dextral strike-slip displacements on the N-S-trending Straight Creek-Fraser fault distort this original configuration. By the mid-Cretaceous, the Coast

Belt orogen (Rusmore and Woodsworth, 1991) was a huge zone of deep burial (fossiliferous Pennsylvanian rocks were metamorphosed to amphibolite grade before being returned to the surface). It was impregnated with voluminous mid-Cretaceous plutons. The orogen's axial zone was rimmed by outward-verging fold-and-thrust belts, facing the Insular Belt on the west and the Intermontane median-massif belt on the east. These fold-and-thrust belts in the Cordilleran interior are shallow, and they mask the position of the Coast-Intermontane and Coast-Insular belt crustal boundaries (e.g., Wheeler and Gabrielse, coords., 1972).

Monger et al. (1982) referred to the eugeosynclinal Coast Belt and miogeosynclinal Omineca orogens as "tectonic welts". This notion, which arose in the context of the terrane-tectonic hypothesis, has never been adequately defined. Still, these so-called "welts" do correspond to orogenic zones, highly metamorphosed and magmatized, deeply buried and inverted, flanked fan-like by fold-and-thrust structures. These two orogenic zones lie between less mobilized rock assemblages in the Rocky Mountain, Intermontane and Insular belts.

Two-sided crustal structure in the Coast Belt and Omineca orogenic zones apparently arose due to the presence of more-rigid crustal blocks on these orogens' flanks. Slightly metamorphosed or unmetamorphosed cover over a stiff crystalline basement suggests these rigid blocks may be remnants of craton(s). The distribution of Precambrian rocks in the metamorphic core complexes suggests the western limit of old, formerly cratonic continental crust in the Cordillera lies someplace west of the Intermontane Belt (cp. Armstrong, 1988; Ernst, 1988; Oldow et al., 1989; Monger et al., 1994). Quite possibly, parts of this old ex-cratonic basement belonged to the Western craton.

The *in situ* median massifs with Precambrian basement of old cratonic affinity, which lie between the elongated orogenic zones that developed in the Late Proterozoic and Phanerozoic, explain Cordilleran geology in ways the terrane-collage concept cannot. Interpretation of seismic data should not ignore the absence of geologic evidence to suggest that steep faults at the Omineca-Intermontane belt boundary flatten downward.

Because seismic reflection sections are full of continuous and discontinuous events, both useful signal and undesirable noise, of different amplitudes, dips and traveltimes, many various artificial ways may be proposed to connect these events, producing various geometries to fit preferred tectonic models. But no clear images of steep faults usually appear in seismic data, although high-angle faults can sometimes be indicated by lateral changes in seismic reflection patterns, discontinuous reflections, or certain types of noise such as diffraction trains (cp. Hajnal et al., 1996). Usually, though, picturing crustal structure from seismic reflection images gives a false impression that low-angle discontinuities are predominant or even the only ones present.

To explain the supposed subhorizontal-slice structure of the Cordilleran crust, Monger et al. (1994, p. 361) speculated about "the apparent dominance" in the "structural fabric of the Cordillera" of the products of "crustal convergence". This picture and explanation overlook the steep faults, which are a fundamental property of continental crust in cratons and mobile megabelts. Some big high-angle faults are inherited from premegabelt time: in the Cordillera, many old weakness zones trend WNW-ESE in the U.S. and NE-SW in Canada. Other steep faults formed during the megabelt's own evolution, and the most persistent and common lateral zonation in the Canadian Cordillera is NNW-SSE.

Monger et al. (1994, p. 361) stated that "...as interpreted from surface structural relationships, much of the [Cordilleran] crust appears to consist of enormous, downward-younging stacks of thrust sheets that are soled by basal thrusts or décollements". It is not clear which "surface structural relationships" these authors refer to. If this crustal geometry is inferred from deep seismic data, this inference suffers from the common misinterpretation of low-angle reflections as thrusts and from the basic inability to image steep faults. Seismic methods detect low-angle discontinuities in certain physical properties of rocks. The geologic nature of these discontinuities may vary: they can be lithologic, rheological, structural, each category containing many subclasses of its own. No easy distinction between these very different causes can be made from seismic data, which reflect the presence of a discontinuity but not its origin or age.

The relationships between rock assemblages mapped on the surface in the field (e.g., as shown in the map of Wheeler and McFeely, comps., 1991) suggest that only marginal belts of particular orogenic zones contain fold-and-thrust stacks - and those stacks are usually shallow-rooted. Such thrust belts are observed on both sides of the Omineca (the Selkirk fan structure, for example) and Coast Belt orogens (Fig. 56; Rusmore and Woodsworth, 1991; Evenchick, 1991, 1992). In the other lateral tectonic units of the megabelt, compressional and extensional, burial and uplift structures are found in complex relationships to one another. They are formed in numerous recurrent generations, as is apparent in the Omineca orogenic zone, the Intermontane median-massif belt (and the Yukon-Tanana massif in Yukon and Alaska), and the Coast Belt (Coast-Cascade) eugeosynclinal orogenic zone. From recent geologic mapping, Thompson (1998, p. 201) found evidence that in the Intermontane Belt "eugeosynclinal [in his designation] rocks stratigraphically overlie miogeoclinal strata", that "they were deposited into one

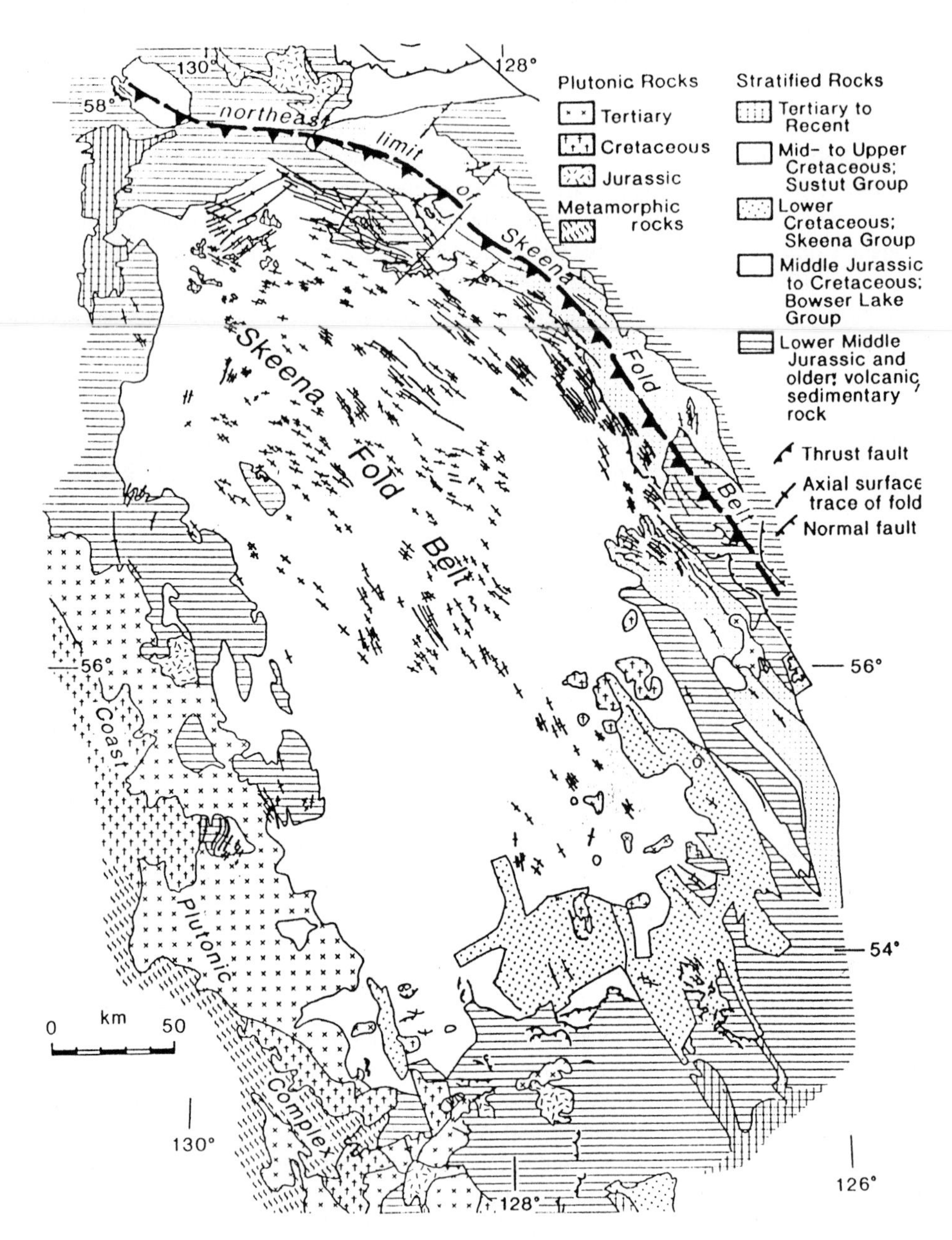

Figure 56. East-vergent fold-and-thrust belt on the east side of the Coast Belt orogen, expressed as the so-called "Skeena foldbelt" in the Bowser Basin in the Intermontane median-massif belt (modified from Evenchick, 1991). Position of the Bowser Basin between the transverse Skeena and Stikine arches in the Intermontane Belt is shown in Fig. 60.

or more marginal basins superimposed on what had been a relatively stable Eocambrian shelf", that a subsequent episode included " 'the bottom dropping out' from beneath the outer portion of the craton margin early in Cambrian time", and these stratal rock bodies include those "previously interpreted as exotic".

Median massifs: semi-reworked remnants of ancient craton(s) within the Cordilleran mobile megabelt, with emphasis on the northern Cordillera

The median massifs' Precambrian basement

Whether all Mesozoic and Cenozoic geologic bodies west of the Cordilleran miogeosyncline were accreted due to plate convergence or developed *in situ* remains a subject of discussion (Yorath and Gabrielse, eds., 1991; Monger et al., 1994; Plafker and Berg, eds., 1994). For many of them, cratonic or near-cratonic origin seems probable, due to continuity of ancient transverse structural trends, geochemical signatures or presence of indigenous detrital zircons with Canadian-Shield or Western-craton ages. In many parts of the Cordillera, and particularly in the Yukon-Tanana median massif in the Yukon and Alaska, common are Early Proterozoic zircon ages of 2,300-2,100 Ma. Assignment of the Stikine and Yukon-Tanana continental-crust blocks, with all their geologic similarities, to a single belt of median massifs within the Cordillera is a considerable departure from the terrane-collage hypothesis.

Late Proterozoic ages have been reported for rocks in northern, west-central and east-central Alaska (Foster et al., 1994; Moore et al., 1994; Patton et al., 1994). These rocks are often assigned to the Cordilleran miogeosyncline (Plafker and Berg, 1994b), implying the possibility of the craton's considerable former continuity into the region now occupied by the northern Cordillera. Plafker and Berg (1994b, their Figs. 5a-c) suggested that an old North American craton continued during Paleozoic-early Mesozoic time at least as far west as the Porcupine Plateau, which lies south of the eastern Brooks Range and north of the Yukon-Tanana Upland (Fig. 57). Yet, they distinguished a separate slice of the craton, ~2,500 km long and only up to 200-250 km wide, labeled Yukon-Tanana-Stikine composite terrane, to allow room for the supposed Cache Creek sea which they showed as having been 350 to 800 km wide during a ~130-m.y. interval from the Mississippian to Middle(?) Jurassic (they indicated that distances of presumed separation are unknown; their Fig. 5b). Plafker and Berg (1994b) suggested that, to open the Cache Creek sea, the northwestern continental and cratonic margin of ancestral North America experienced "extension and subsidence". They noted also that the late Paleozoic supposed island-arc magmatism recognized in British Columbia is not known in the Yukon-Tanana massif.

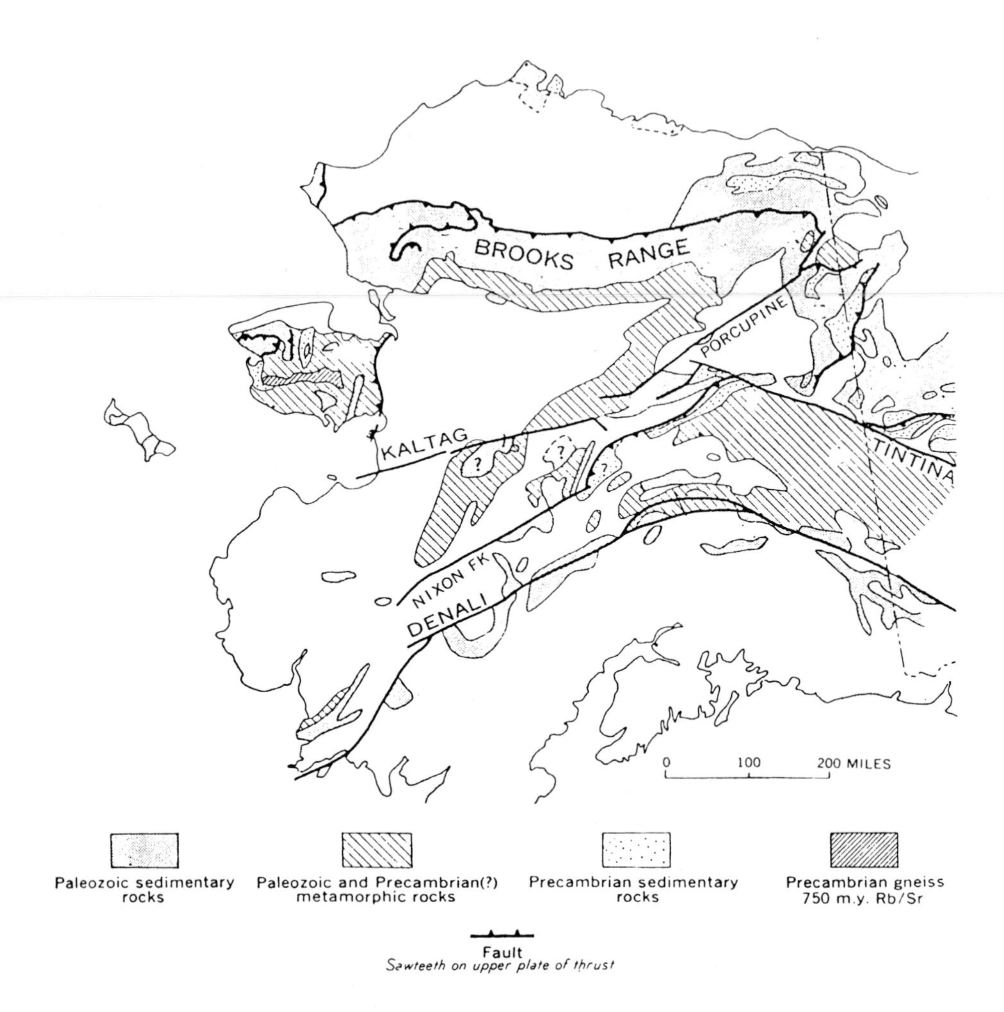

Figure 57a. Major faults and outcrop areas of Precambrian and Paleozoic rocks in Alaska, as identified by geologic mapping (simplified from Brosgé and Dutro, 1973 and King, 1969). Paleozoic rocks on the Arctic coast are from drillhole data.

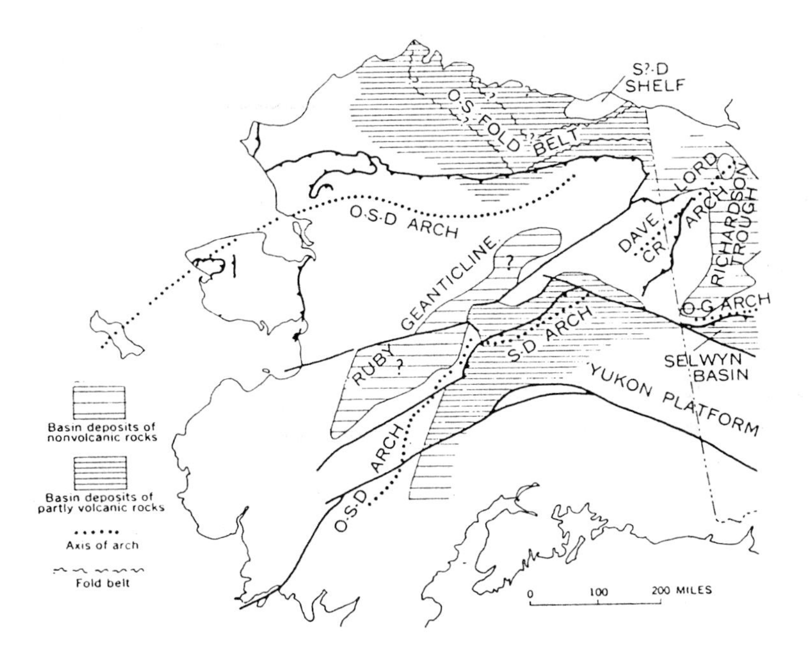

Figure 57b. Major early Paleozoic tectonic elements in Alaska, as identified by geologic mapping (simplified from Brosgé and Dutro, 1973). D - Devonian; O - Ordovician; O-G - Ogilvie; S - Silurian.

The Yukon-Tanana median massif (sometimes regarded as a terrane) is distinctly block-like. Uplifted to topographic elevations of up to 2 km, it is hilly, with hundreds of meters in local relief. Abundant Mesozoic and lesser Cenozoic igneous formations are mapped across the upland areas (also Hansen and Dusel-Bacon, 1998), but the core and basement of this massif are composed of Proterozoic metamorphic rocks. The basement is overlain by a volcano-sedimentary cover, partly metamorphosed but unhomogenized. It is dominated by limestones, dolomite and quartzites of the 6-km-thick Tindir Group of Late Proterozoic-Cambrian age, overlain by a variable, 5-km-thick, Paleozoic to Cretaceous volcano-sedimentary succession. Structural unconformities have been noted at the base of and within the cover: at the bottom of the upper (transitional to Cambrian) Tindir Group, at the bottom of the Late Devonian arenitic succession, and before the Permian, Jurassic and Cretaceous packages (Dover, 1994). Mesozoic thrusts, mapped around the massif's perimeter, usually verge towards the interior of the massif, as is common for rigid crustal blocks surrounded by orogenic zones within mobile megabelts.

Following the fashion to postulate supposedly exotic terranes, the Yukon-Tanana massif has been subdivided into a number of presumably independent tectonic entities (Silberling and Jones, eds., 1984; Silberling et al., 1994). Local variations in the stratigraphy in highly disrupted mobile-megabelt areas, if emphasized over the regional similarities, can indeed lead to the conclusion that contrasting terranes are present. But structural-formational (stratigraphic, magmatic, metamorphic, deformational) similarities in continuous belts unite many "terranes" into coherent tectonic zones of common whole-crust evolution. Small "terranes", if discriminated properly, commonly indicate the presence of not accreted crustal slices but blocks of different crustal depth. Larger tectonic belts, of whole-crust or whole-lithosphere depth extent, define the fundamental historical characteristics of smaller blocks which make up those belts.

From a mostly stratigraphic framework, Dover (1994, p. 186) observed that "the similarities between many terranes are at least as impressive as the differences, and these similarities are of critical significance in evaluating relations among terranes". Also skeptically, Woodsworth et al. (1991, p. 494) stated: "...Most plutonic rocks are associated with metamorphic belts in which the accretionary history of the terranes (and even the identity of the terranes themselves) is uncertain". But whereas terrane assignments of rock packages in the Cordillera are confused and vary from one author to the next, evidence from rocks shows the Intermontane median-massif belt is largely underlain by semi-reworked ancient continental crust.

Current geological concepts tend to reduce mobile-megabelt zoning to just a few stock choices: e.g., fore-arc, island-arc, back-arc, inter-arc, continental volcano-plutonic arc and other big subduction-related zones. In reality, however, this generalized and oversimplified scheme can be misleading. In Alaska, a segmented belt of cratonic origin from the Seward to Yukon-Tanana massifs is well recognized from petrological and geochemical studies (Dusel-Bacon and Aleinikoff, 1985; Aleinikoff et al., 1986; Smith and Rubin, 1987; Dover, 1994). Although these massifs have been interpreted as ancestral continental-arc or Andean-type-arc or island-arc entities, they could also be peripheral parts of an ancient proto-North American craton separated from the main cratonic mass during the early stages of mobile-megabelt evolution in Late Proterozoic or Paleozoic time.

An Early Proterozoic cratonic basement is thought to exist all across the Yukon-Tanana massif, but its exposures are scattered (Aleinikoff et al., 1986). By contrast, the volcano-sedimentary cover of Late Proterozoic-Cambrian and younger ages is widespread (Dover, 1994; Hansen and Dusel-Bacon, 1998). Structural-formational étages of the cover across the Yukon-Tanana massif have typical cratonic platformal lithologies: late Precambrian shale and stromatolitic limestone; Late Proterozoic-Cambrian to Early Devonian carbonate and clastic rocks; Late Devonian to Pennsylvanian terrigenous successions; Permian to Triassic, facially diverse sequences of conglomerate, limestone and oil shale; Jurassic to Early Cretaceous pelitic to arenitic rocks with abundant quartz; local Cretaceous to Early Tertiary conglomerates.

Some of these tectonic étages are coeval with the melting of continental crust, indicated by in granitic suites in the Cordillera in Alaska and British Columbia (e.g., Beikman, 1980). In the central Alaskan belt of crystalline terranes, the best-recorded episodes of granitoid plutonism are Middle Devonian, Late Devonian to Early Mississippian, Late Triassic to Early/Middle Jurassic, Cretaceous to Early Tertiary (Aleinikoff et al., 1986; Foster et al., 1994). Several widespread metamorphic events also took place in the Late Proterozoic and Phanerozoic; in Paleozoic time a number of them coincided with the plutonic pulses (Dusel-Bacon and Aleinikoff, 1985; Aleinikoff et al., 1986; Dusel-Bacon, 1994, her Plates 4a-b).

Eclogitic rocks have been mapped locally in an overthrusted slice parallel to the Yukon-Tanana massif's boundary zone with the Kuskokwim orogen. Radiometric K/Ar dating on amphiboles from the glaucophane-bearing schists indicate a possible metamorphic age of 470±35 Ma (Foster et al., 1994). Also metamorphosed are all Paleozoic plutonic rocks, and even Late Devonian-Early Mississippian granitoids were transformed into augen orthogneiss. Younger igneous rocks across the Yukon-Tanana massif are regionally unfoliated.

Mesozoic and Cenozoic granitic plutons are divided into three age groups (Foster et al., 1994): 215 to 188 Ma (Late Triassic to Early Jurassic), 95 to 90 Ma (Cretaceous) and 70 to 50 Ma (latest Cretaceous to early Tertiary). Regional metamorphism of Mesozoic and Cenozoic rocks is usually burial-type and low-grade, except in local areas of dynamic and contact metamorphism that accompanied the motion on deep fault zones and intrusion of granitic bodies. Cretaceous metamorphism, as recorded by metamorphic dating, is the most widespread (e.g., Aleinikoff et al., 1986; Dusel-Bacon, 1994), although the transformation of Late Devonian-Early Mississippian plutons into augen gneiss probably preceded the Cretaceous plutonism (Dusel-Bacon and Aleinikoff, 1985). Basement-derived zircons in the augen gneisses have yielded Early Proterozoic ages (2,300 to 2,100 Ma).

Plutonism in Mississippian, Cretaceous and Early Tertiary times was accompanied by abundant volcanism. Comagmatic volcanic and plutonic rocks, coupled with coeval or slightly older burial metamorphism, indicate several pulses of subsidence and uplift, usually involving the entire Yukon-Tanana massif. Regional episodes of cooling are reported for the late Early and Middle Jurassic and Late Cretaceous-Early Tertiary (Hansen et al., 1991; Foster et al., 1994). The Late Cretaceous and Early Tertiary nonmarine sedimentary rocks also indicate the massif was uplifted at that time, and the high modern topographic elevations show it is uplifted today.

In the terrane-tectonic interpretations, special attention is paid to ultramafic rocks. All too easily, ultramafic bodies are presumed to be ophiolites (though even real ophiolites are actually rare; e.g., Malpas et al., eds., 1990). In Alaska, Patton et al. (1994, their Fig. 1) showed the distribution of undifferentiated mafic-ultramafic complexes, but classified them separately as having variously structural or intrusive boundaries. The intrusive complexes are particularly common in southeastern Alaska, from the Chugach area to Dixon Entrance north of the Canadian Queen Charlotte Islands. North of the southern Alaska Ranges, three most coherent belts of such igneous complexes are mapped parallel to the Kuskokwim orogen - in the Ruby Anticline, on the northwestern boundary of the Yukon-Tanana massif, and in the Salcha-Seventymile transect across this massif. All these belts have a roughly similar NE-SW trend, suggesting their connection to the Kuskokwim orogenic event of Jurassic time.

Patton et al. (1994, p. 671) stressed that only in northern and western Alaska can the known mafic-ultramafic complexes "clearly... be labeled ophiolites", whereas others "are not ophiolites" or "have uncertain affinities". The complexes in the Ruby Anticline are imbricated in thrust sheets, but former deep-sea geologic settings there are indicated by radiolarian chert. The Salcha-Seventymile belt is apparently imposed on the Yukon-Tanana schist formations and strongly dismembered along a fault zone. The age of these mafic and ultramafic rocks is still not definite but is almost certainly pre-Late Devonian (serpentinite clasts are found in the basal conglomerate of the late Paleozoic succession of the Yukon-Tanana sedimentary cover).

Even less clear is the tectonic position of mafic-ultramafic and felsic igneous bodies in more-reworked and less-exhumed parts of the Cordilleran median-massif belt farther south, where the Yukon-Tanana massif is conventionally correlated with the Stikine block of the Intermontane Belt. Several ultramafic, peridotitic bodies have been mapped along the Tintina fault. Foster et al. (1994, p. 231) regarded them as Alpine-type igneous complexes of complicated genesis: "originally derived from the mantle" and "emplaced into orogenic crust". Patton et al. (1994, p. 682), on the other hand, cautioned that these mafic-ultramafic complexes (Salcha-Seventymile and others) are "too complex or too poorly understood to speculate on their tectonic setting".

In the heyday of the terrane-tectonic ideas, Tempelman-Kluit (1979) described similar mafic-ultramafic and metamorphic rocks in the southwest of Canada's Yukon Territory, and it was thought that the transcurrent Teslin fault separates the Yukon-Tanana and Stikine terranes of island-arc origin. The striking similarity in these terranes' geology - abundant Paleozoic schists, gneisses and quartzite, porphyritic peraluminous granites (augen gneisses) - make it possible to extend the conclusion about the Yukon-Tanana massif's continental-crust origin into the Yukon Territory and probably to the Stikine massif (based on metasedimentary crustal protoliths for the augen orthogneisses, inherited Early Proterozoic zircons, high initial $^{87}Sr/^{86}Sr$ ratios, and other indicators of continental crust; Aleinikoff et al., 1986; Miller, 1994).

Not only Paleozoic but also Mesozoic sialic magmatism in that region has been linked to continental-crust settings (Mortensen, 1988). The same probably holds true for the most voluminous, Cretaceous magmatism in the Yukon-Tanana massif. Miller (1994) inferred a thickened continental crust across the massif in the Early and mid-Cretaceous, and perhaps till the end of active Alaskan plutonism in the Late Cretaceous-Early Tertiary (for the latter period, though, links to Alaskan-Aleutian subduction might be more plausible). A thinned low-velocity continental crust, indicated by its seismic properties (Beaudoin et al., 1992), is a result of these multiple processes and events.

Genetic links between assumed terranes

In Tempelman-Kluit's (1976, 1979) descriptions, the Stikine terrane was considered a fragment of an ancestral North American continent. An island arc, in this model, was formed on a margin of this continental fragment, which later, in the Jurassic, was re-joined back with the continent. Alternatively, a continental arc of Andean type has been proposed for these continental-crust blocks (Burchfiel et al., 1987; Dover, 1994).

Plafker and Berg (1994a-b) and Silberling et al. (1994) have interpreted the contact between the Yukon-Tanana and Stikine massifs to be complex (Figs. 58-61). Paleozoic rocks, usually metamorphosed, are commonly exposed in uplifted blocks; subsided blocks contain exposures of Mesozoic rocks, metamorphosed only slightly or not at all. In southwestern Yukon Territory, in the upper Yukon River area, a boundary of the miogeosynclinal Omineca Belt with the presumed accreted terranes is shown near the confluence with the Stewart River (Gabrielse et al., 1991a, their Fig. 2.4). This area is strongly disrupted by several NW-SE-trending faults and imbricated in a series of nappes (Tempelman-Kluit, 1979). However, Gabrielse and Campbell (1991, their Figs. 6.1-6.2) suggested a correlation of the miogeosynclinal Windermere Group with a Precambrian part of the upper Tindir Group in the Yukon-Tanana stratigraphic section and with some sedimentary successions southwest of the northern portion of the Stikine terrane. The transport of clasts into the Tindir sedimentary basin in the Yukon-Tatonduk area was from the west (Young, 1982), suggesting the western Cordillera at that time contained highlands.

In the Nasina Basin, west of the Selwyn Basin in the Cordilleran miogeosyncline (Tempelman-Kluit, 1979), the Cambrian through Silurian succession comprises mainly limestone and shale, with the clastics becoming coarser upward (also Fritz et al., 1991, p. 218). Whether correlative rocks are present in the volcano-sedimentary cover of the Stikine massif is a matter of speculation, but the Devonian-Mississippian tectonism, accompanied by emplacement of abundant granitic plutons in northern Yukon, was expressed in faulting and predominant uplift in western miogeosynclinal zones. An influx of west-derived chert-bearing clasts, preserved in coarse sedimentary rocks of the Devonian Earn Formation in some narrow grabens along the eastern Cordillera, indicates uplifts in the inner parts of the Cordilleran mobile megabelt (Gordey et al., 1991, their Fig. 8.11). The westward extent of these provenance areas is unclear.

The terrane-collage hypothesis imposes near-mandatory interpretations of ultramafic complexes as remnants of oceanic crust squeezed between accreted terranes, and of blueschist minerals in rocks as evidence for paleo-subduction zones. Revision of these

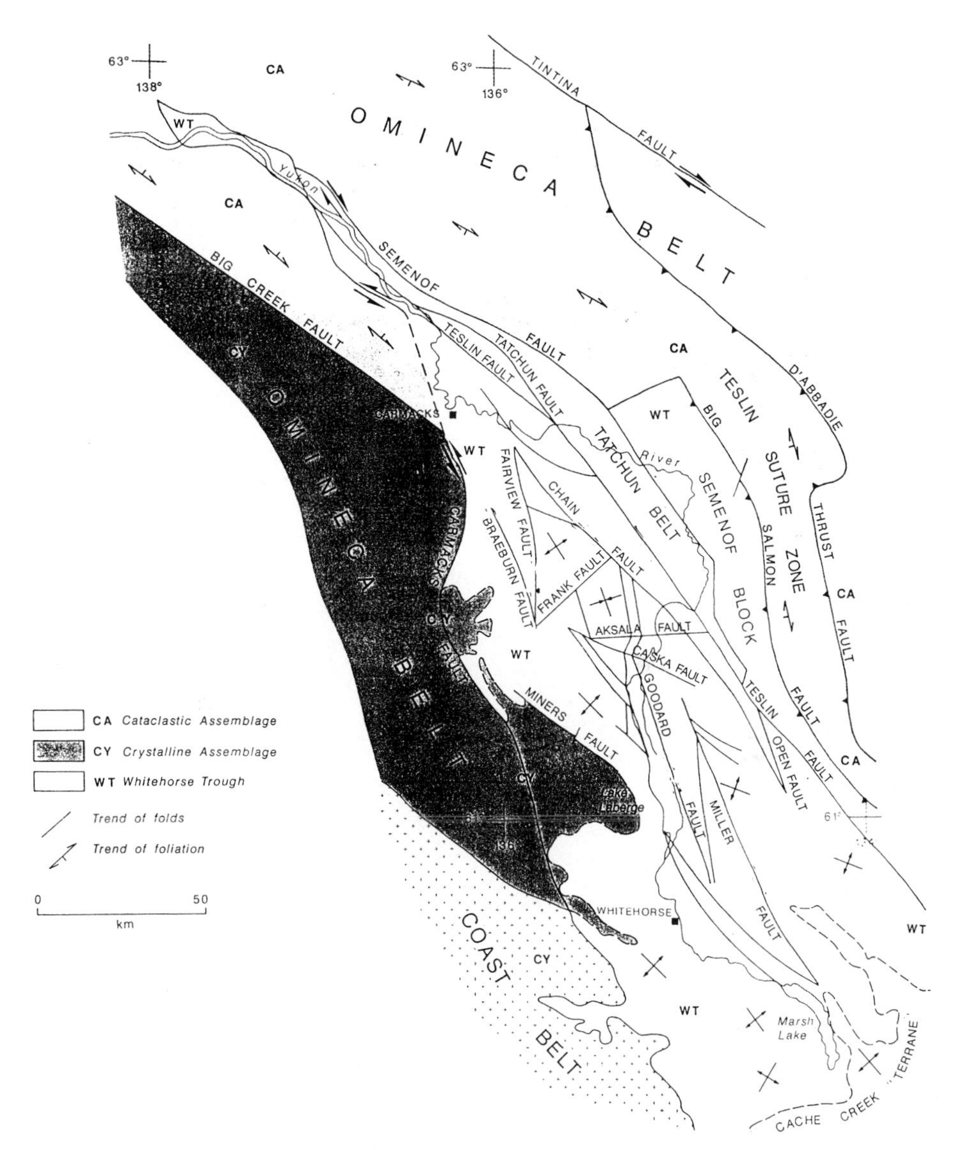

Figure 58. Large faults and folds in the northern Intermontane median-massif belt, as identified by geologic mapping (simplified from Gabrielse et al., 1991b).

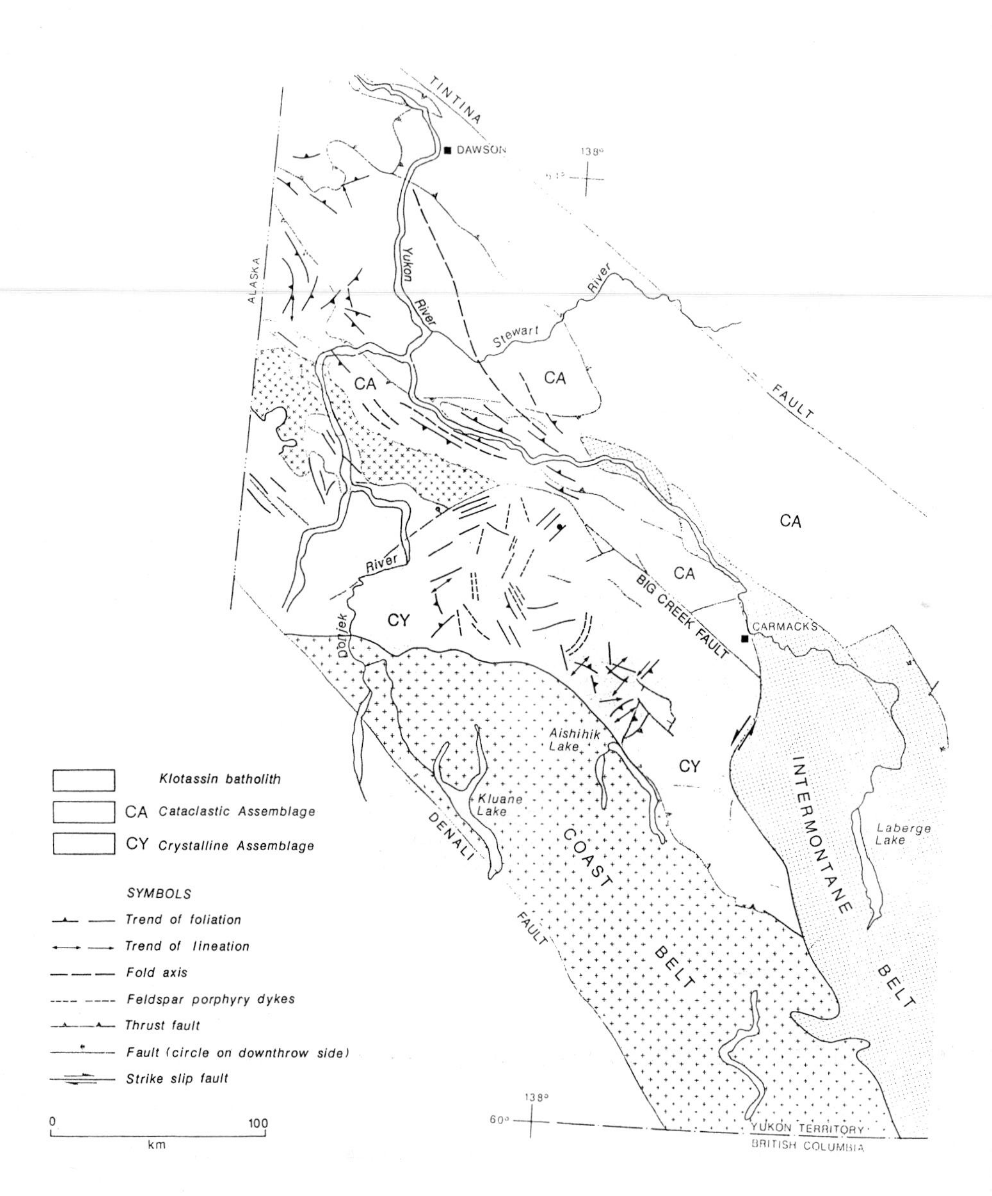

Figure 59. Large faults and folds in the northern miogeosynclinal Omineca orogenic belt, as identified by geologic mapping (modified from Tempelman-Kluit et al., 1991).

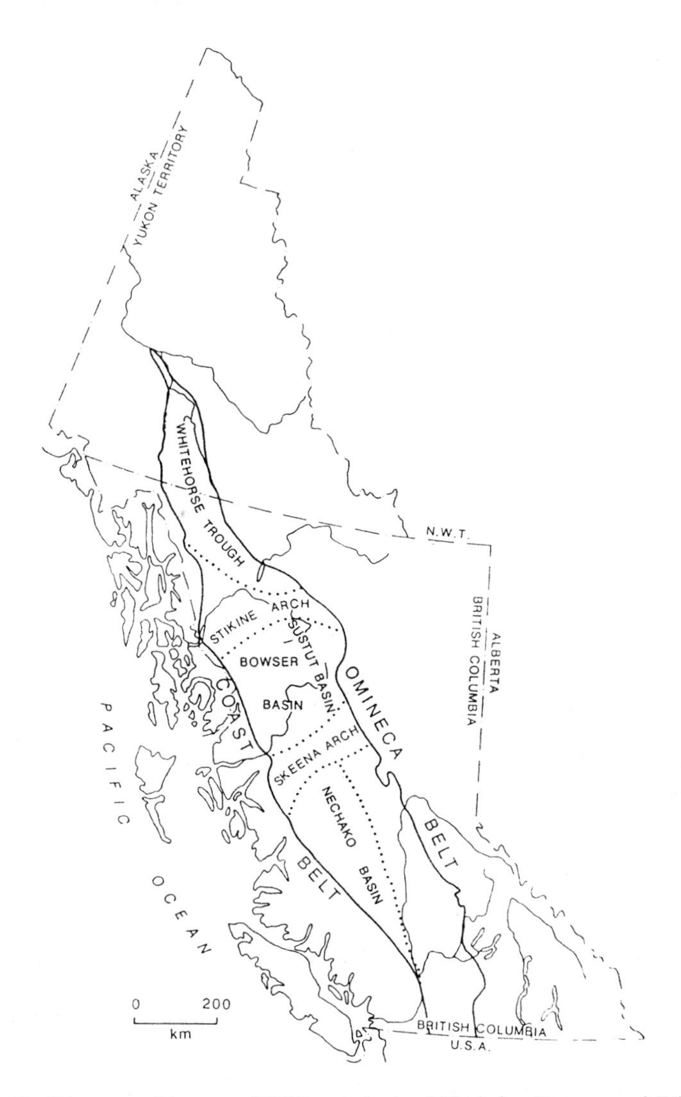

Figure 60. Transverse Skeena and Stikine arches and Nechako, Bowser and Whitehorse Trough basins between them, in the Intermontane median-massif belt (simplified from Gabrielse et al., 1991b). A simplified geologic map of the Bowser Basin is shown in Fig. 56.

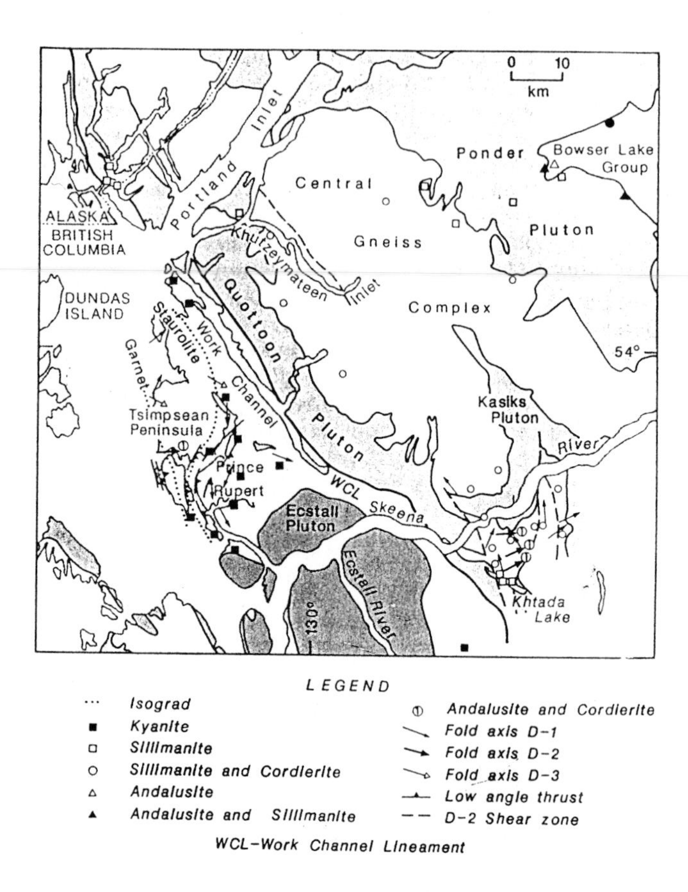

Figure 61. Structures and plutons along the NE-SW-trending, transverse Skeena River lineament in the eugeosynclinal Coast Belt orogen, as identified by geologic mapping (simplified from Greenwood et al., 1991). This lineament lies on trend with the southern flank of the transverse Skeena Arch in the Intermontane median-massif belt (Fig. 60).

and other mandatory postulates is possible in a broad context of rock and rock-body properties observed directly in outcrop and inferred indirectly from geophysical images.

No geological (mineralogical, petrological, stratigraphic, etc.) fact can be uniquely attributed to a specific original setting. Determination of rock and mineral genesis is inevitably uncertain, because different settings can produce similar observed characteristics and similar settings can produce different ones. Glaucophane-bearing blueschists require high-temperature, low-pressure physical conditions, but subduction zones are only one sort of setting where such conditions may occur. Ultramafic rocks require involvement of the mantle, but ultramafic lithology is not proof of ophiolite nature and oceanic-crust origin. Not all ultramafic complexes are ophiolites (which are actually quite rare), and not all ophiolitic formations are necessarily pieces of crust typical beneath the Earth's modern oceans. From geophysical characteristics, transitional (semi-oceanic) crust is modeled often in marginal and mediterranean seas. Even the stratotypical Troodos ophiolite on Cyprus seems to have an origin linked with multiple eruption phases and local subduction (see, e.g., Yang and Hall, 1996). The temptation to explain all events in Cordilleran evolution by tectonic activity in the Pacific Ocean is unjustified (Lyatsky, 1996).

The Intermontane Superterrane is presumed to combine several smaller terranes, including the moderately deformed and slightly metamorphosed Stikine and Quesnel terranes; the highly deformed, disrupted and metamorphosed Cache Creek terrane; and the allochthonous, overthrusted sheets of the Slide Mountain terrane (e.g., Monger et al., 1982, 1991, 1994). The Cache Creek and Slide Mountain terranes in this hypothesis are treated as oceanic-crust entities, and the Stikine and Quesnel terranes as island-arc ones. But the nappes combined into the Slide Mountain terrane are made up of Late Devonian-Mississippian to Permian and in places Triassic clastic rocks and fairly thick volcanic sequences with non-ophiolitic ultramafic intrusions (Alpine- and Alaska-type), deformed prior to and intruded by Late Permian or Triassic diorites (see descriptions in Klepacki and Wheeler, 1985; Monger et al., 1991, 1994; Silberling et al., 1994; and references therein). At least some clastic rocks included into the Slide Mountain terrane were formed in marine depositional environments near land.

Deeper-water Paleozoic depositional environments are evident for rocks assigned to the Cache Creek terrane. This rock assemblage contains shallow-water, fusilinid-bearing carbonates intercalated with deep-water radiolarian cherts of Carboniferous through Triassic-Jurassic age; clastic rocks of greywacke composition; volcanic successions and ultramafic complexes. Extensive outcrops of this assemblages are found in western Yukon and north-central British Columbia (Figs. 45, 62, 63). The best and largest

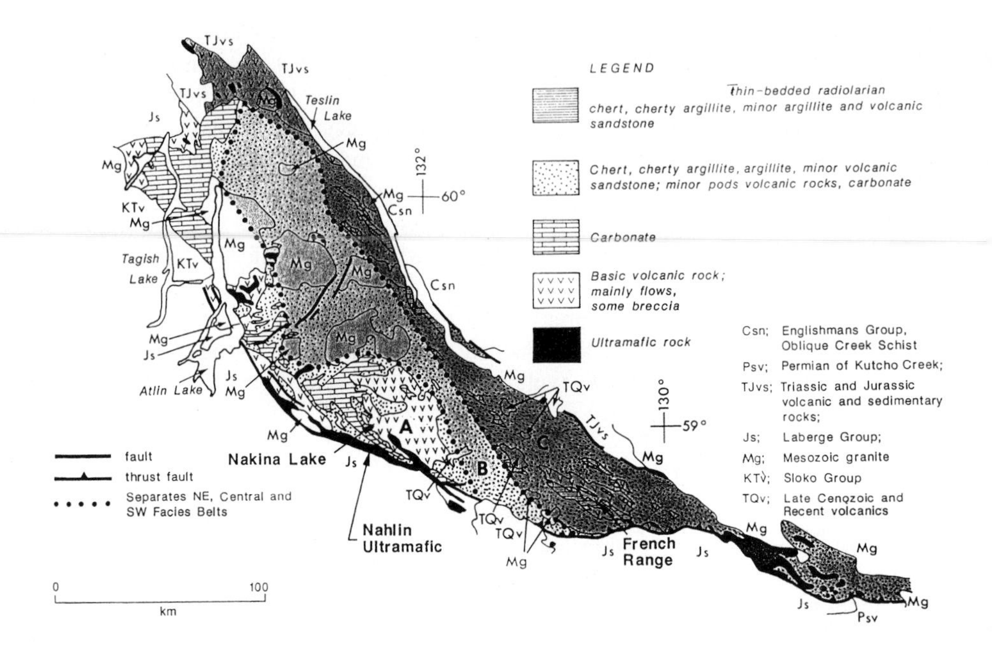

Figure 62. Simplified geologic map of the northern inlier of the Cache Creek rock assemblage (modified from Monger et al., 1991).

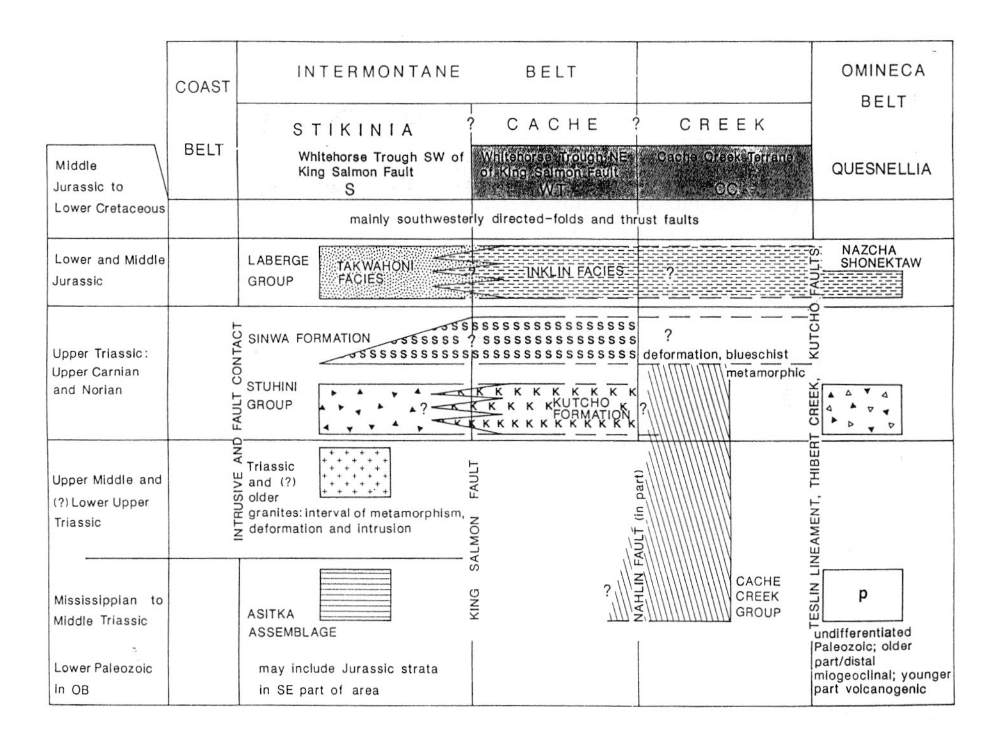

Figure 63. Facies equivalence of Mesozoic rocks of the Cache Creek assemblage with other rocks of the Intermontane median-massif belt (simplified from Monger et al., 1991).

exposure of ultramafic body is described in the western part of this area (the Nahlin body in the Nakina block; Terry, 1977). It is dominated by foliated peridotite and has occasional diabase dikes, but no typical ophiolitic sheeted-dike layer. Other outcrop areas and rock bodies included into the Cache Creek terrane and much smaller in size and very disrupted by major fault systems. North of Teslin Lake in the Yukon, some 400 km north of where these faults merge, this block is apparently terminated by a NE-SW fault. Another such transverse fault cuts this block northeast of southern Atlin Lake, towards and across Teslin Lake.

The NE-SW-oriented faults disrupting the Intermontane-Belt terranes continue into the Coast Belt on one side and the Omineca Belt on the other (e.g., Douglas, 1986). The fault zone controlling the Stikine Arch also has this trend, running across various assumed terranes but parallel to segments of the Dease and Stikine rivers. A sliver between two parallel thrust faults, Nahlin and King Salmon, has been assigned to the Cache Creek terrane by Gabrielse et al. (1991b), in contrast to other interpretations (Silberling et al., 1994). The Teslin fault, accentuated by the Teslin Lake, separates the Cache Creek terrane from Omineca Belt and the Quesnel terrane according to Gabrielse et al. (1991b), and from the southern continuation of the Yukon-Tanana terrane according to Silberling et al. (1994). East of Dease Lake, all rocks are disrupted into a tectonic mélange between two adjacent thrust faults just 15-20 km apart. Within this thrust zone and local thrust sheets parallel to it, stratal rocks are overturned south-southwestward onto the Stikine block (Gabrielse et al., 1991b, their Fig. 17.20; also Thorstad and Gabrielse, 1986). To the north, the King Salmon and Teslin faults are steep, with possible strike-slip components.

To the north, a block conventionally included into the Stikine terrane is mapped as an extensive area of Mesozoic volcano-sedimentary rocks. Steeply dipping faults, including the northern Nahlin fault, separate these rocks from the Cache Creek assemblage. The NW-SE-trending Whitehorse Trough is both bounded and internally disturbed by faults. West of Marsh Lake the Whitehorse Trough is in contact with the Coast Belt orogen (Gabrielse et al., 1991b; see also Rusmore and Woodsworth, 1991). Farther north, it is rimmed on the east and west by crystalline and sedimentary Omineca-Belt rocks. Silberling et al. (1994), on the other hand, proposed that next to the Whitehorse Trough lies the Yukon-Tanana massif.

Plafker and Berg (1994b, p. 995) offered yet another interpretation: “the Stikine terrane was originally depositional on the crystalline complex of the Yukon-Tanana terrane”. An interfingering contact of the Stikine-terrane rocks with rocks of the Yukon-Tanana massif (or, in another interpretation, with the Omineca Belt) is recognized in a semi-circular rim on both sides of the northern continuation of the Stikinian rock assemblages (see also Wheeler and McFeely, comps., 1991). Other investigations suggest that contact is not depositional but structural, along a NW-SE-trending strike-slip fault (e.g., Samson et al., 1991). Faults are indeed mapped in that area in abundance. They are predominantly steep, with local overturned splays and small (a few kilometers) strike-slip offsets. Earlier movements on these faults are usually obscured due to Middle Jurassic and younger movements related to the King Salmon and Nahlin fault activity.

The confluence area of the Yukon, Stewart and Donjek rivers contains the northwesternmost metamorphic core, flanked by a cataclastic assemblage of the Omineca Belt. The steep Big Creek fault separates it from a metasedimentary assemblage of quartzite, schist and marble lenses, which could be part of the Omineca Belt or the Yukon-Tanana massif (Tempelman-Kluit, 1976, 1991). To the south, the Intermontane median-massif belt is flanked by the Coast Belt orogen, and Stikine-type rocks are found in grabens bounded by steep faults. These grabens developed in the uplifted block-like Yukon-Tanana massif and in the anticlinorium-shaped Cordilleran miogeosyncline. Collinear arrangement of the small grabens in the northwestern wedge-out area of the Stikinian (Whitehorse-Trough) assemblage and the oval-shaped Nisling metamorphic culmination suggests an extensional regime at the time of their formation.

As interpreted by Gabrielse (1985), faults in the Whitehorse Trough are dextral strike-slip, but this supposition remains unproved by detailed structural analysis. Even for the well-studied Tintina fault area to the east, the type and time of motion are not entirely clear: pull-apart sedimentary basins and intrusive bodies correlatable with fault activity give a rather reliable time of movement as Triassic-Jurassic to Late Cretaceous-Early Tertiary (Gabrielse et al., 1991b), but extensive Mesozoic tectonism might have obscured evidence of older movements.

The entire northern block of the Intermontane median-massif belt, from the Yukon River to Marsh Lake in northern British Columbia, is separated from its southern counterparts by a block of typical orogenic rocks assigned to the Cache Creek terrane. Regional tilting during the deposition of this ~80-km-wide assemblage (from Atlin Lake in the west to Teslin Lake in the east) is suggested by (a) presence of the oldest, Early Mississippian, stratal rocks in the west, whereas only Late Pennsylvanian through Permian rocks are found in the east; and (b) predominance of carbonate sedimentation in the west, in contrast to deeper-water sedimentation in the east. Monger et al. (1991) interpreted this assemblage in terms of oceanic reefal paleoenvironments, analogous to the modern Bahamas. The Cache Creek ultramafic bodies in this area are elongated along the western bounding faults that cut the depositional zones obliquely. All this does not require a pelagic oceanic environment or subsequent subduction. Greywacke intercalation only suggests only an intra-orogenic setting, and Late Triassic glaucophane-bearing schists in this terrane (233-213 Ma by K/Ar dating) are associated with greenschist-grade metamorphic sequences (description in Gabrielse et al., 1991b).

The subparallel Triassic to Middle Jurassic Nahlin and King Salmon faults, which flank the northern British Columbia Cache Creek inlier on the west, probably belong to a single fault zone punctuated by several ultramafic bodies and a belt of Upper Triassic to Middle Jurassic rocks. The latter are juxtaposed obliquely against the western, carbonate upper Paleozoic succession of the Cache Creek terrane proper (cp. Gabrielse et al., 1991b; Monger et al., 1991). The Mesozoic sequence is usually treated as western marginal formations of the Whitehorse Trough and included in the Cache Creek terrane - but these rocks are also known to be interfingering facies equivalents of the Upper Triassic to Lower-Middle Jurassic volcano-sedimentary rocks southwest of the King Salmon fault, which are assigned to the Stikine terrane (Monger et al., 1991). If this fault coincides with a narrow but persistent zone of facies changes as suggested by these authors, both these rock successions should be regarded as equivalent, providing a link between terranes. In addition, the King Salmon-Nahlin fault system is coplanar with the western fault boundary of the crystallin block in the Donjek area which is probably a southern piece of the Yukon-Tanana massif. If so, the continental-crust Yukon-Tanana median massif may partially underlie the Whitehorse Trough, as well as the Stikine Mesozoic assemblage in a structural triangle between the trough, the northern British Columbia Cache Creek inlier and the Coast Belt orogen.

This triangular block lies in a remote area and is poorly studied. Coarse clastic rocks of greywacke composition are predominant there; other rocks are volcanic flows and tuffs (Takwahoni Formation of the Upper-Middle Jurassic Laberge Group). Mesozoic plutons in that area have high initial $^{87}Sr/^{86}Sr$ of >0.704 to 0.707 and are enriched with heavy lithophile elements. This indicates contamination of magmas with continental-crust material probably from a Precambrian crystalline basement (Day and Matysek, 1989) of the underlying Yukon-Tanana massif.

This triangular crustal block is partly correlative with the recently recognized Windy-McKinley terrane (Monger et al., 1991) as well as with parts of the Nisling terrane. It contains Precambrian, Paleozoic and Mesozoic rocks. The Stikine terrane farther south, according to Plafker and Berg (1994a-b) and Silberling et al. (1994), was rimmed by Paleozoic continental-margin blocks labeled Tracy Arm and Taku terranes; the Quesnel terrane was supposedly part of the same continental margin. The northern part of the

Intermontane median-massif belt, by contrast, seems less fragmented and more elevated, being cored by the rigid Yukon-Tanana massif. Regional uplift of the Intermontane Belt's western side from Nisling to southern Tracy Arm block has exposed lower levels of the volcano-sedimentary cover, revealing its contamination by continental-crust material (Plafker and Berg, 1994b, their Figs. 5f-g).

Predominance in the Stikine terrane of Paleozoic rocks in the north and Mesozoic rocks in the south may indicate a late-stage southward tilt, with uplift towards the NE-SW-oriented Stikine Arch. The known Precambrian through Paleozoic rocks in the Yukon-Tanana blocks and in the Nissling assemblages are found in the north, where the Cache Creek inlier also lies. South of the Stikine Arch lies the large, almost isometric Bowser Basin containing thick Jurassic sedimentary rocks. The southern boundary of this basin is with the NE-SW-trending Skeena Arch.

In this description, the northern Cache Creek inlier looks like a local structural complication in the huge Intermontane Belt of median massifs. This inlier is tilted to the east and south, so the oldest rocks in it, Mississippian volcanics and carbonates, are recognized only in the west and north (Monger, 1977). This tilt direction replicates that of the entire Intermontane Belt. At least two generations of folds and faults have been noted in the inlier: the common NW-SE-trending structures are Jurassic, and the N-S ones are older (Gabrielse et al., 1991b, p. 593-594). The Nahlin fault has generally steep dips, with some thrust-splay complications, in contrast to the Middle Jurassic King Salmon thrust fault. Uplift of the Cache Creek inlier along the Nahlin fault juxtaposed the inlier's Paleozoic rocks against the Mesozoic assemblages which cover much of the rest of the Intermontane median-massif belt. The inlier, including its mafic-ultramafic complex, provided abundant clasts for the Bowser sedimentary basin.

Traditionally, the Cache Creek rock assemblage has been described as typically oceanic. This conclusion was based on an interpretation of its Paleozoic radiolarian cherts as pelagic-oceanic, of the carbonates as reefal, and of the mafic-ultramafic igneous complex as ophiolitic (Monger, 1977). At present, it is recognized that radiolarian-rich rocks can form in marginal and mediterranean seas (e.g., Friedman and Sanders, 1991). The Paleozoic Cache Creek carbonates are not simply reefal but represent an extensive shallow-water carbonate platform; they are well-bedded, laterally persistent and long-lived. The Cache Creek mafic-ultramafic complexes are not typically ophiolitic but rather Alaskan-type. Like the non-ophiolitic Salcha-Seventymile complex in Alaska (Patton et al., 1994), basalts and ultramafics of the Cache Creek inlier lack typical ophiolitic rock associations and petrologic cross-sections. Similar rocks are known also in areas west and south of the Cache Creek inlier (Brown and Gunning, 1989; Nixon and

Hammack, 1991). From detailed mapping, Bloodgood et al. (1989) have concluded that in the Cache Creek inlier, a common lithologic composition is interbedded chert, limestone and basalt, with gradational contacts between them; and that the Cache Creek Group was deposited in a shallow marine basin, perhaps in a back-arc tectonic setting. The "stratigraphic continuity" of these rocks contradicts their interpretation as a tectonic mélange.

The eastern boundary of the Cache Creek inlier corresponds to the Teslin fault, which juxtaposes it with rocks assigned mostly to the Yukon-Tanana, Slide Mountain, Dorsey and Quesnel terranes. The extent and identity of many fault slivers are unclear. Their terrane assignments are dubious, as are the definitions of many terranes themselves. For example, rocks of the Quesnel "terrane" in the southern Canadian Cordillera have been shown by field mapping to be integrally continuous with rocks unquestionably indigenous to North America (Erdmer et al., 1999; Thompson et al., 1999). The pre-Permian Dorsey terrane may not exist at all as a separate entity (Stevens, 1996); certain marker beds mapped across the Stikine Ranges may be correlatable with the lower horizons of the Quesnel terrane (Stevens and Harms, 1995; Harms and Stevens, 1996).

Detrital zircons in sedimentary rocks of the Dorsey terrane have been traced to their sources in the interior of the North American craton, "thus confirming a North American affinity for these sedimentary rocks and their correlation with the Cassiar Terrane..." (Ross and Harms, 1998, p. 107). The Slide Mountain assemblage (terrane), on the eastern side of the Intermontane Belt, was previously described as the easternmost rock assemblage of a late Paleozoic to Triassic Cordilleran eugeosyncline (Monger, 1977). To the northeast, the Slide Mountain rocks have been fully correlated (Tempelman-Kluit, 1979; Gabrielse et al., 1991b) with the Anvil allochthon on top of the Yukon-Tanana crystalline rocks. Late Devonian to Early Carboniferous slivers are found (in the Sylvester allochthon) in fault contact with other Slide Mountain stratigraphic units (e.g., Harms, 1986; Monger et al., 1991, their Fig. 8.64). In the north, the Anvil allochthon is cut by gabbroic rocks bracketed between 360 and 340 Ma (early Mississippian; U-Pb dating on zircons; Grant et al., 1996). In the western Omineca Mountains, the same upper Paleozoic assemblage has been correlated variously with the Cache Group or with the Harper Range piece of the Quesnel terrane (Monger, 1977; Gabrielse et al., 1991b). The latter authors have reported uncertainties in the assignment of late Paleozoic rocks in many localities either to the Slide Mountain terrane or to parts of the Quesnel terrane (see also Wheeler and McFeely, comps., 1991).

The name Slide Mountain Group is derived from an area in the Cariboo Mountains where this group contains just one formation. In more complete sections, in the Cassiar Mountains to the north, this group is predominantly Mississippian through Permian in age, with variable lithologies and thickness: limestones, cherts, argillites and volcanics, with many lateral facies changes. The volcanics are mostly mafic; basalts locally have pillow structure and are sometimes intercalated with volcaniclastic rocks. A variety of ultramafic rocks (dunite, harzburgite, peridotite etc.) are also found. Slide Mountain Group rocks are metamorphosed mostly to greenschist grade, and ultramafic rocks are commonly serpentinized. Jade, blueschist minerals and eclogite are found elsewhere in the Slide Mountain assemblage (e.g., in the Sylvester allochthon) and in some areas in the Yukon-Tanana assemblage (also Plint and Gordon, 1996), in association with major steep faults - Teslin, Tintina, Finlayson Lake (Erdmer et al., 1996). The entire Slide Mountain terrane has been interpreted to be of oceanic-crust origin (Gabrielse et al., 1991b).

Tectonic interpretation in this region is made more difficult if large strike-slip displacements (cumulative up to 5,000 km) occurred on these faults between the mid-Cretaceous and Tertiary (Gabrielse, 1985). However, in Alaska the Yukon-Tanana massif was rather stable through time, and remained near the North American continent (as is evident from analysis of paleocurrents and sediment-provenance areas). Only limited rotation and east-west lateral translation have been proposed for this continental-crust massif since the Late Proterozoic (Plafker and Berg, eds., 1994). If blueschists and eclogites are assumed to indicate former subduction zones, rocks of the Slide Mountain terrane indicate a late Paleozoic oceanic-crust basin between the Yukon-Tanana massif and the North American craton; this basin has been named Cache Creek Ocean (Plafker and Berg, 1994b). But no other evidence exists to support this idea; the basin could have been no more than marginal, with only a small separation between the continental-crust masses (Nelson, 1993). Besides, blueschist minerals and eclogites indicate only the physical and chemical conditions of rock metamorphism (high pressure, low temperature, rapid cooling) rather than any specific tectonic setting.

High-pressure/low-temperature metamorphic rocks commonly bear lenses of eclogite. These lenses suggest emplacement of mantle-derived material into metamorphosed country rocks at lower-crustal levels with temperatures >400°C and pressures of 12-14 kbar (1,200-1,400 MPa). Such physical conditions exist under continents near the crust-mantle boundary, in areas where deep faults caused rapid thermal relaxation. In subduction zones, similar physical conditions may be created at the boundary between the downgoing oceanic-crust slab and the overriding continental-crust plate. In the Canadian Cordilleran hinterland, the presence of eclogites in both greenschist-grade and blueschist-grade rocks suggests inclusion of eclogite into the crust along faults. At any rate, to regard high-pressure rocks as "direct physical evidence of the subduction of former oceanic crust" (Erdmer et al., 1996, p. 61) could easily be misleading.

Near the British Columbia-Yukon border, eclogites have been reliably mapped along the Teslin and Tintina fault zones and in areas of auxiliary faults. The eclogites' country rocks are assigned variously to the Slide Mountain and Quesnel assemblages and, importantly, the Yukon-Tanana massif. In all localities, the eclogite bodies are schistose and mylonitized, bearing signs of retrograde metamorphism of greenschist grade. But to link metamorphism specifically to plate or terrane collisions and subduction is speculative, even adventurous. The time of eclogites' origin is uncertain; it seems to have been around 260-255 Ma (Late Permian; Wanless et al., 1978; Erdmer et al., 1996, 1998). The available blueschist ages are close, around 272-269 Ma (Creaser et al., 1996). In the Slide Mountain terrane, Mortensen (1992) found the common ages to be 370-360 Ma (Late Devonian; U-Pb dates on zircons). According to Hansen et al. (1991), however, metamorphosed supracrustal rocks from the Teslin fault zone yielded ages of 195-190 Ma (Early Jurassic; Ar-Ar on white mica); the oldest post-kinematic granitic plutons in that area are thought to be 188 Ma in age (Harvey et al., 1996) but may be 212 Ma (Late Triassic; Mortensen, 1992). If the late-stage regional retrograde greenschist metamorphism immediately predates the recorded uplift and cooling and occurred in the Late Triassic and Early Jurassic, the processes of eclogite formation as well as blueschist and widespread greenschist metamorphism might have lasted from ~270 Ma to 212-188 Ma. This would be in agreement with the geology of the entire northern Canadian Cordillera.

Mediterranean seas and local subduction

Sometime in the Late Triassic, probably diachronously along the Teslin orogenic zone, the orogenic activity that had lasted since the late Paleozoic (Mississippian?), came to an end. The timing of that activity fits the Antler orogeny in the U.S., suggesting that Antler restructuring, which was strong in the U.S. Cordillera, was significant in Canada as well. If so, the Intermontane Belt, which has existed since perhaps the Proterozoic, was fragmented by the Antler-Teslin sequence of orogenies in the late Paleozoic. However, at the end of this strong orogenic episode, the massif, split into the Stikine and Quesnel blocks, was fused back together, having also acquired orogenic rock formations recognized as the Cache Creek and other fragments.

A sharp break in the history of the northern Cordillera was proposed by Gabrielse (1996) to have occurred in the earliest Jurassic: pre-Jurassic rocks ceased to be foliated; regional tectonic uplift produced a widespread unconformity; block tectonics was accentuated by localized volcanic centers and plutons, in contrast to elongated plutons and volcanic chains of previous time; NE-SW-oriented faults and arches (Stikine, Skeena) bounded crustal blocks with different stratigraphic records and basins (e.g., Bowser Basin). Gabrielse (1996) ascribed this change in magmatic and tectonic activity to a change in subduction regime: north-directed subduction under the Quesnel terrane supposedly ceased in the Early Jurassic, whereas a SW-directed subduction zone under the northern Stikine terrane existed till the Middle Jurassic.

But, according to Erdmer et al. (1996), the metamorphic history of the region (Early Permian eclogite metamorphic peak around 270 Ma, followed by regional cooling to ~350°C around 260 Ma, and possible Early Jurassic metamorphic overprinting) casts doubt on the existence of subduction zones there in Late Triassic and Early Jurassic time. Erdmer et al. (1998, p. 615) noted high-pressure metamorphism in the Canadian Cordillera was "multiepisodic and diachronous". Indeed, if eclogites and blueschists from the Teslin and Tintina fault zones, which lie ~150 km apart, indicate subduction, then subduction zones must have existed in the Permian and earliest Mesozoic on both sides of the Yukon-Tanana massif. However, presence of the Cassiar carbonate platform between these faults suggests close connections of the massif with the North American craton. If the Yukon-Tanana massif is part of the miogeosyncline (Wheeler et al., 1991; Silberling et al., 1994), then eclogites and blueschists hosted by the Yukon-Tanana rocks within and near the Tintina fault zone would be interpreted, implausibly, as indicators of subduction within the miogeosyncline. Alternatively, subduction in the Permian and Triassic occurred only along the Teslin zone, which was a suture whose influence affected the pericratonic zone to the east and the eugeosynclinal Cache Creek and Stikine terranes to the west (Monger, 1977). That influence, in a zone 250-300 km wide, caused the appearance of eclogites and glaucophane schists to the east, and abundant mafic volcanism and ultramafic plutonism to the west.

Rocks in the Teslin zone are disturbed by numerous fault strands, but not enough to make a Franciscan-type mélange: steep fault and bedding dips and NW strikes are common, and deformation is not chaotic. Like the volcano-sedimentary formations, Devonian-Mississippian and Permian-Early Triassic plutons have been affected by tectonic deformation (Mortensen and Jilson, 1985). By contrast, Late Triassic-Jurassic and younger plutons are undeformed (also Mortensen, 1992).

Country rocks in the Teslin zone, and the nearby Late Devonian-Mississippian to Permian-Early Triassic volcano-sedimentary sequences assigned to the Slide Mountain (in the east) and Cache Creek (in the west) terranes (Gabrielse and Yorath, eds., 1991) formed at the same time but in dissimilar paleoenvironments. Monger (1977) ascribed these environments to Paleozoic pelagic oceanic settings - atolls, reefal islands, volcanic plateaus. Another explanation is that no big ocean existed, as shallow-marine carbonates, greywackes, quartz-bearing sandstones and conglomerates indicate shallow sea floor and proximity to sediment sources. Presence of cherts, some of them radiolarian, and of tholeiitic volcanics does not change this interpretation: these rocks are found also in the Stikine terrane which is not regarded as a product of pelagic ocean (also in Gabrielse and Yorath, eds., 1991; Plafker and Berg, eds., 1994). Sharp variations in sediment type and faunal assemblages are common in marginal and mediterranean seas, and mobile orogenic settings are conducive to the accumulation of very variable biocenoses and lithofacies in laterally varying paleoenvironments.

Nonetheless, rock assemblages in the Teslin zone and the adjacent Slide Mountain and Cache Creek entities have for years been described as oceanic or oceanic-crustal (Tempelman-Kluit, 1979; Gabrielse et al., 1991b; Harvey et al., 1996). Eclogites and blueschists have been treated as indicators of former subduction zones between oceanic crust to the west and continental crust to the east. Accretion of terranes, detached from the downgoing lithospheric slabs and attached to the continent in an accidental order, supposedly began in early Mesozoic time, producing a kaleidoscopic collage of genetically contrasting blocks. As part of this process, the former Teslin (or Cache Creek) ocean was squeezed out of existence as terranes on its flanks came together (Monger et al., 1982; Plafker and Berg, 1994b).

This hypothetical scenario is now being reconsidered. Continuity of lithostratigraphic units negates the ideas about a mélange or "cataclastic" fabrics (e.g., Erdmer and Helmstaedt, 1983; Hansen, 1992). Moderate (greenschist-grade) regional metamorphism in supracrustal rocks requires no dramatic tectonism. Two-sided geometry of the fan structure is not consistent with accretionary-prism origin. Eastward subduction of continuous Paleozoic Cache Creek oceanic crust on the west side of the Yukon-Tanana massif in Alaska and in the Teslin zone in Canada fails to explain many geological observations. As alternatives, Oliver et al. (1996) favored southwestward subduction in the Teslin zone, whereas Stevens and Erdmer (1996) and Stevens et al. (1996) preferred a model with transpressional (rather than head-on) dynamics at that subduction zone. Two-sided local subduction, on both sides of the mediterranean-type Teslin sea, could more easily account for the NNW-SSE-trending fan structure (which has upright, tight folds in the axial zone and outward vergence of thrusts and recumbent folds on the limbs; Gabrielse et al., 1991b; de Keijzer and Williams, 1996). Continental-crust blocks lie to the east and west; blocks on the west now belong to the Yukon-Tanana and Intermontane median-massif belt. In northern Canada, the median massif lies between two orogens on its flanks: the early Mesozoic Teslin orogen on the east, and the Cretaceous Coast Belt orogen on the west. This scenario draws from Stevens (1994) and Brown et al. (1995), but is closer to the former.

The Teslin zone is ~20 km wide, clearly much narrower than the original basin. Composite thickness of volcano-sedimentary successions that accumulated in the Teslin sea apparently exceeds 10 km. Fine clastic sediments accumulated in the axial zone of subsidence; coarser clastics and carbonates on the shallower-water flanks. For example, a 500-m-thick Upper Mississippian limestone unit (Sylvester Group) developed along the eastern side of the sea, and up to 2,000 m of Upper Mississippian through Permian limestone (Cache Creek Group) appeared along the western side. These rocks were uplifted in the Triassic as a result of deep heating, similar to that in the Omineca and Coast Belt orogens. (In the Cache Creek inlier in the southern Canadian Cordillera, presence of Early or Middle Jurassic radiolarian fossils [Cordey et al., 1987] suggests the marine basin there was not fully inverted till the Middle Jurassic). Although the causes of such heating are unclear (Spear and Parrish, 1996), it commonly produces mountain ranges along the axes of heated belts. In the Teslin mediterranean sea, inversion took place in the Triassic, and in places in the Early Jurassic. The Sylvester Group carbonates (Nizi Formation) are proved to be sliced and displaced a few tens of kilometers eastward, whereas the Cache Creek carbonates might have been shifted to the west only slightly on the thrust planes of the Nakina and King Salmon faults. These shallow-water carbonates indicate the coasts of the Teslin mediterranean sea were uneven and their distance from the axis varied along strike. This picture differs from that of Harms (1986), who suggested the Sylvester allochthon's rocks could have come from a vast region of the paleo-Pacific.

The Teslin orogen, like Coast Belt and Omineca, has outward-verging fold-and-thrust zones on its flanks. However, none of these bilateral orogenic belts is symmetric: the most prominent fold-and-thrust zones are found on the western side of the Coast Belt orogen and on the eastern side of the Omineca orogen; in the Teslin orogen, the biggest fold-and-thrust zone developed along its boundary with the Cordilleran miogeosyncline. Overlap of miogeosynclinal Paleozoic formations and cratonic carbonate platforms by rocks attributed to the Teslin orogen is an established fact (other bodies of overlapping rocks are presumed to belong to the Slide Mountain, Cache Creek and Quesnel terranes; e.g., Gabrielse and Yorath, eds., 1991; Silberling et al., 1994). Westward thrusting along the western side of the Teslin orogen overlapped the rocks of the Intermontane median-massif belt (specifically, the Stikine terrane). Late Middle-Late Jurassic granitic plutons crosscut these thrust sheets. In contrast, the Coast Belt orogen's thrust sheets are cut by mid-Cretaceous plutons, and sheets of the Omineca orogen by Late Cretaceous-Early Tertiary intrusions.

Other important differences between these orogens are noted as well. The Coast Belt orogen developed along several NNW-SSE-oriented lineaments, which follow deep faults that provided an enormous volume of magma of granodiorite composition; these rocks make up 80% of the orogen's entire volume. A pre-uplift extensional thermo-tectonic event there was followed by the intrusion of light Jurassic and younger plutons, which prevented a large topographic depression from forming. A similar situation developed in the Omineca orogen, where heating from ?subcrustal depths also helped prevent basin subsidence. In both belts, protoliths were lowered to crustal depths of 15-20 km and metamorphosed. Later, regional uplift in these orogens, which has continued throughout the Cenozoic, produced the Coast and Omineca mountains.

The Teslin orogen did not evolve quite according to this blueprint. Pre-uplift subsidence was not offset by a deep-sourced heating or intrusion of plutons, and produced a mediterranean sea. Gabrielse et al. (1991b) considered it likely that in the Slide Mountain terrane rocks were derived from near-continental and distal paleo-Pacific oceanic depositional areas (also Harms, 1986), but such overly mobilistic movements seem unjustified and unnecessary.

The Teslin orogen was gradually inverted since the Late Permian and Triassic time. Island-arc plutono-volcanic magmatism over the western local Teslin subduction zone was superimposed on the slightly subsided Intermontane Belt. The Triassic Stikine volcanic belt was oblique to the general modern N-S trend of Intermontane-Belt volcanic centers. Feldspathic xenoliths of granitic and metamorphic (schists, gneisses) rocks from the crystalline basement of the Intermontane Belt are common. Ultramafic xenoliths are more typical in the Yukon-Tanana massif, but they are present in the Intermontane-Belt regions to the south as well. Although in the bulk isotopic signatures of the Intermontane Belt's igneous rocks continental-crust contamination is rare, continental-crust basement does exist at least in the northern and middle parts of the Intermontane median-massif belt (also Plafker and Berg, 1994b). Deep faults apparently controlled the position of Triassic magmatic centers, and their parallelism and temporal coincidence with faults in the Teslin orogen suggest that teleorogenic influences were imposed widely on the Intermontane Belt. Similar teleorogenic impositions also occurred later, from the Omineca and Coast Belt orogens.

Tectonic position of the Bowser Basin

Regional elevation of the Intermontane Belt since the Early Jurassic was accompanied and followed by reactivation of basement faults inherited from the Precambrian. The Teslin and other Phanerozoic orogenic and teleorogenic influences had never obliterated those ancient fault patterns completely. In the absence of such external influences, these Precambrian faults revealed themselves in modest differential uplift, subsidence and tilting of blocks. Early Jurassic plutons disrupted the simple coherence of fault patterns, but volcanic centers of that age were commonly sited at fault intersections (Gabrielse,

1996). Lack of foliation in these rocks points to predominance of extensional tectonic regimes during that and subsequent times. Farther south, the Early-Middle Jurassic plutonic suite was an early manifestation of the transverse NE-SW-oriented Skeena Arch (Tipper and Richards, 1976).

Though Jurassic and Cretaceous block tectonics was moderate in the Intermontane Belt, it was enough to create large block-to-block variations in the Middle-Late Jurassic and Cretaceous stratigraphy. Major NE-SW-trending basement faults facilitated the development of the Stikine and Skeena arches and of the Bowser sedimentary basin between them (Figs. 56, 60). Raised blocks in the median massif(s) and especially in the flanking orogens served as provenance areas of clastic sediments for Mesozoic basins in the miogeosyncline and pericratonic regions to the east.

The end of local Teslin subduction was followed by a change of the magmatic regime in the median massif(s). Thick rhyolitic successions of the Early-Middle Jurassic Hazelton Group (below the Bowser Lake Group) contain breccias, greywackes and conglomerates, whereas the overlying sedimentary strata are continuous and facially more uniform (e.g., Eisbacher, 1974; Gabrielse and Yorath, eds., 1991; Bassett and Kleinspehn, 1997). The so-called "Skeena fold belt" in the northern Bowser Basin (Evenchick, 1991, 1992) is a later, Cretaceous-Early Tertiary superimposition of teleorogenic origin, accompanied by inversion and induced by the Coast Belt orogen.

The Bowser Lake Group contains abundant clastic material shed from the inverted Teslin orogen, including fragments of Cache Creek rocks in conglomeratic sequences. Tertiary plutons, dikes and numerous veins reflect influences of the Coast Belt and Omineca orogens; and Tertiary sedimentary basins there, usually coal-bearing, contains clasts from these uplifted belts. Much of the material eroded from the Omineca-Belt mountains was transported by streams to the east, providing sediments for the basins of the cratonic Interior Platform, but some of it was transported westward and laid down in the Intermontane Belt. Rocks of the sedimentary cover of the Intermontane median massif(s), from Upper Devonian to Tertiary, are metamorphosed to just low grades if at all. Teleorogenic (from the west) thrusts and folds of the "Skeena fold belt" are not a consequence of native, indigenous crustal processes within the massif. Rather, they are shallow and rootless, involving neither the basement nor even the lowest horizons of the volcano-sedimentary cover. The prominent Skeena and Stikine arches, which bound the Bowser basin on the north and south, are continuations of Late Archean-Early Proterozoic crustal weakness zones in the North American craton to the east (Lyatsky et al., 1999).

All the above suggests the continental-crust basement of the Intermontane massif(s) is only semi-reworked and retains much of its original structure.

Causes of Cenozoic tectonism

Although manifestations of tectonic processes recur, directionality of regional geologic evolution of the Cordilleran mobile megabelt, from initial rifting to orogenic maturity, is best revealed by magmatic activity. Diachronous eastward migration of magmatism has been noted across the Insular, Coast and western Intermontane belts from Jurassic to Eocene time (Hutchison, 1982; Armstrong, 1988; Woodsworth et al., 1991; van der Heyden, 1992). This suggests joint evolution of these three tectonic zones. Also directional is the increase in felsic content of magmatic rocks through time, apparent in the Coast and Omineca orogenic belts alike despite the differences between the timing and styles of magmatism in these two zones. Mid-Cretaceous plutonic magmatism occurred in these orogenic zones simultaneously (Woodsworth et al., 1991), even though they are separated by a stable median massif. The Intermontane Belt received coeval teleorogenic granites. The timing of Eocene magmatism in these three zones was also similar (Parrish et al., 1988; Souther, 1991; Woodsworth et al., 1991).

If voluminous granitization had anatectic sources in the middle and lower crust (cp. Chadwick et al., 1989), simultaneous intrusion of plutons in several tectonic zones in the same mobile megabelt may be due to common in-situ causes at depth. These tectono-magmatic episodes in the Cordillera coincided mainly with extensional tectonic episodes, suggesting common evolutions of the tectonic zones.

Extensional episodes normally follow orogenic compression. They are not entirely post-orogenic (as is commonly assumed) but partly late-orogenic, postdating most of the metamorphism and shortening. Also late-orogenic to post-orogenic may be delamination of material in the ductile crust and emplacement of new material from the mantle. Dissipation of orogenic, relatively felsic lower-crustal roots by lateral flowage or delamination is probably part of the orogenic cycle, although the exact nature of these processes is not clear (Rudnick, 1995; Rudnick and Fountain, 1995). Regional tectonic extension and uplift at the end of the Columbian and Laramian orogenies followed the orogenic overheating of the crust and development of crustal roots.

The Coquihalla volcanic province in southern British Columbia (Berman and Armstrong, 1980) contains granitic batholiths and stocks of Neogene age. Several such plutons lie along the Fraser-Straight Creek fault system; the large Chilliwack batholith of about 29 to 16 Ma is found in that area. The Garibaldi volcanic belt of three discontinuous, en-echelon lines of Pliocene-Quaternary volcanic centers trends NNW-SSE. Volcanic centers of this age elsewhere in the western Canadian Cordillera are generally common. The Neogene Anahim belt has a transcurrent E-W to ENE-WSW orientation, transecting the Coast and Intermontane belts (Bevier et al., 1979; Hickson, 1991; Woodsworth et al., 1991). These volcano-plutonic chains and volcanic plateaus are related to a grid of paired N-S/E-W and NNW-SSE/ENE-WSW faults (Lyatsky, 1991, 1993). Central and shield-like volcanoes of similar composition and age over the northern Bowser Basin, Stikine Arch, western Whitehorse Trough and southern Yukon-Tanana massif are related to major NW-SE faults, in an extensional modern tectonic regime (Souther, 1991).

The prominent elongated Quottoon granitic pluton in the Coast Belt's axial zone is diachronous, from 80 Ma (Campanian) to 67-64 and even 59 Ma (Paleocene). Whatever its deep origin, this pluton was probably emplaced along crustal lineaments. The system of faults responsible for these lineaments, accentuated by presence of sheared rocks, provided a convenient thermal conduit. Hollister et al. (1996) suggested a thermal gradient zone between the Paleogene intrusive rocks on the west and previously cooled rocks of the mid-Cretaceous orogen on the east. Younger Tertiary plutons in the Coast Belt orogenic zone have been related to an extensional crustal regime beginning in Late Eocene time (Woodsworth et al., 1991). But farther north, in Alaska, the superimposed Tertiary magmatism bears characteristics thought to be consistent with subduction-related origin (Samson et al., 1991; Moll-Stalcup, 1994).

The well-studied Eocene extensional episode is recognized across the Canadian Cordilleran interior. Differential block movements of many kilometers across steep normal faults are recognized in the orogenic zones and even in the median massifs. Commonly, uplifted fault-bounded blocks are cored by light granitic plutons. Some metamorphic-core culminations contain deep crustal material brought up from depths of 20-30 km. In the Coast Belt, metamorphic complexes are common between local pendants and large granitic plutons. Metamorphic complexes in the southern Omineca Belt are arranged in three N-S-oriented bands. There, two strong pulses of uplift, in the mid-Cretaceous and Early Tertiary, made sources of sediments for two thick clastic wedges in the foredeep to the east (the Blairmore and Belly River/Paskapoo assemblages; Stott et al., 1993a-b).

Contractional tectonics can thicken the continental crust greatly, to 50 and even >70 km as in the Himalayas and Pamir. In many Paleozoic orogenic provinces, the average crustal thickness today is 38 km, with the lower crust being 16-18 km thick (except in

Europe, where these thicknesses are smaller; Rudnick and Fountain, 1995; their Table I and Fig. 2). In Mesozoic and Cenozoic orogenic zones that have undergone subsequent extension, the modern crustal thickness is only 32-33 km, with the lower crust only 10-11 km thick (Durrheim and Mooney, 1991, 1994). Thus, the crustal processes associated with extensional tectonic regimes lead primarily to thinning of the ductile lower crust (Holbrook et al., 1992).

Dissolution, outflow or delamination of crustal roots probably begins as soon as the roots appear, and these processes compete with compressional thickening. When thickening ends, thinning becomes predominant. Isostasy requires that high mountains be underlain by abnormally light or, more likely, thicker crust. Yet, the Rocky Mountain fold-and-thrust belt was built not by deep-seated processes that thickened the lower crust but by rootless, shallow deformation induced by stresses from the Cordilleran interior, and these mountain ranges apparently still lack isostatic balance (Garland and Bower, 1957; Geological Survey of Canada, 1990). Elasticity of the lithosphere is one common way to support loads without deep local isostatic compensation (e.g., Wu, 1991).

In the Coast and Omineca orogenic belts, tectonism involved not rootless thrusting but whole-crust restructuring. Morphologically, at shallow crustal levels these zones are represented by bilateral fan-like structures, with a broad, folded axial zone and outward-vergent fold-and-thrust zones on the flanks. At the surface, topographic rise competes with exogenic terrain-lowering processes such as erosional denudation. The higher and more rugged the mountains, the faster the erosion.

Upper crust in a mobile megabelt is conventionally assumed to be a physical layer with P-wave seismic velocities ranging from 5.7-5.9 to 6.2-6.3 km/s. In Paleozoic orogenic provinces, the upper crust may be only half as thick as in Mesozoic and Cenozoic contractional regions. Block movements matter as well, and in the extended Basin and Range Province the upper-crust thickness varies from >8 to ~15 km. Crustal extension and uplift at the end of an orogenic cycle are accompanied by cooling, helped by erosional removal of the insulating upper-crust blanket. Existing methods of geothermometry are unable to distinguish between the specific causes and mechanisms of cooling, but they can tell at what depths the rocks had been. Granulite-grade rocks, which are common in metamorphic core complexes, had been at temperatures of 820±30°C and pressures of 8±1 kbar (800±100 MPa), corresponding to depths of 25-30 km. Amphibolite-grade rocks had been at 20-25 km.

In the Omineca orogenic zone, rocks of amphibolite and granulite metamorphic grades are exposed in the core complexes. In the Valhalla complex, Spear and Parrish (1996) have identified the youngest metamorphic peak around 72-67 Ma. After that, continuous cooling took place during Tertiary erosion of the uplifted terrain. Some Valhalla-complex metamorphic rocks are estimated to have reached the depths of 28-30 km before rapid cooling in the late Paleocene and Eocene. That cooling was not disrupted by appreciable tectono-magmatic processes (Spear and Parrish, 1996). In western Alberta, clastic rocks of the Paskapoo Formation record the maximum clastic influx from the Omineca Belt around the beginning of the Eocene. Later unroofing and cooling were steadier, and they continue to the present.

Whereas in the southern Coast Belt seismic data reveal a 3-4-km crustal root, no root is recognized under the Omineca Belt (Cook et al., 1992; Varsek et al., 1993; Burianyk and Kanasewich, 1995, 1997). If such roots were formed during orogenic compression, they may be a cause of subsequent uplift as the isostatic equilibrium is restored. As a mobile megabelt dies and its crust becomes cratonized, heat flow decreases to usual cratonic values and the crust acquires the normal cratonic seismic-velocity layering (Pakiser and Mooney, eds., 1989).

Insights into deep crustal structure of the Intermontane median-massif belt from potential-field and seismic data - a critical look

Systematic collection of regional geophysical data began in the Canadian Cordillera in the 1950s, and first summaries were produced in the early 1970s (Figs. 53-54). At that time, the investigators' attraction to modeling, driven by plate-tectonic ideas and assisted by modern computers, was not yet wholehearted. Geophysicists still focused on gathering factual material. They still retained a measure of skepticism regarding the fit of their conclusions to speculative models, geosynclinal or plate-tectonic. The early gravity maps (Garland and Tanner, 1957; Walcott, 1967; for modern maps see Geological Survey of Canada, 1990) highlighted the main anomalies related to the isostatic state and some internal variations of crustal properties. Generally negative gravity-anomaly values along the highest mountain ranges (Coast and Rocky Mountains) were clearly apparent. A gravity gradient zone was recognized between the Rocky Mountain and Omineca belts (see also Stacey, 1973), and the fairly uniform anomaly values over the Intermontane Belt correspond to the character of that region's geology and topography. Stacey (1973) suggested that most Cordilleran provinces are probably at or near the isostatic equilibrium; only in the Insular Belt did the lithospheric density structure seem complicated.

Limited aeromagnetic data were analyzed comprehensively in the 1970s (Berry et al., 1971; Haines and Hannaford, 1972, 1976; Hannaford and Haines, 1974; Coles et al., 1976; Coles and Currie, 1977). Two main sets of anomalies were recognized: those with a NNW-SSE trend, corresponding to the Cordilleran grain; and those trending NE-SW, continuing from the Canadian Shield and Interior Platform in the North American craton (also Garland and Bower, 1959; Walcott, 1968; Geological Survey of Canada, 1990; Edwards et al., 1998; Lyatsky et al., 1999). Prominent NE-SW magnetic (and gravity) anomalies continue into the Rocky Mountains, where they are less coherent. Farther west, NE-SW anomalies can be traced, with disruptions, across many Cordilleran tectonic zones, belts and major faults. In the Intermontane Belt, a complex pattern of magnetic anomalies differs from the high-intensity Coast-Belt pattern and the less-intense Omineca-Belt pattern.

Seismic data, being confined to transects, do not offer such regional coverage. In the Intermontane median-massif belt, some of the Mesozoic sediments accumulated in thick basins, and occasionally volcanic rocks formed thick pads around volcanic centers. In some areas, extensive stratigraphic horizons, 1-2 km thick, of Permian carbonates and early Mesozoic volcanics are remarkably uniform. Such horizons, and regional unconformities, should make strong seismic reflectors all across the Intermontane Belt, but few supracrustal-level events are included in the published Lithoprobe seismic data displays (e.g., Cook et al., 1992; Varsek et al., 1993; Clowes et al., 1995). On the other hand, where Intermontane Belt's lower Mesozoic horizons are available for direct observation, they show strong signs of considerable vertical tectonic movements. Jurassic conglomeratic sequences 400-700 m thick are found along elevated flanks of all large sedimentary basins - Bowser, Sustut, Tuyaughton. High-amplitude block-tectonic movements took place in the Mesozoic in the transcurrent Skeena and Stikine arches. But the resolution of Lithoprobe seismic images of geologic boundaries in the Cordillera is low, especially in the lower crust and upper mantle (with lateral uncertainties up to 25-50 km and vertical uncertainties of 1-2 km; Clowes et el., 1995).

In gravity and magnetic maps, elongated zones of gradients, contour offsets and breaks in anomaly patterns have long been recognized as commonly related to faults at the crystalline-basement level (Coles et al., 1976). Some NE-SW-trending faults and anomalies, with confirmed origin dating back to the Early Proterozoic, run from the craton across the belts of the Cordillera, variously with or without offsets and disruptions. Inevitably, interpretations of potential-field and other geophysical data suffer from non-uniqueness: for example, it is often hard to tell without the benefit of additional information whether a gravity low represents a thick sedimentary package or a granitic pluton. Information from other geophysical surveys, and above all geological control, are necessary to constrain interpretations.

Decades ago, seismic data were used in southwestern Alberta to confirm that the cratonic crystalline basement continues under the Foothills and Rocky Mountains (Bally et al., 1966). It was also revealed that this basement under the thrust stack is not involved in deformation, and the thrusts in that structural belt disrupt only the cover of the Interior Platform. But seismic data provide subsurface images only in narrow corridors along the profiles, and lateral extrapolation is usually based on gravity and magnetic anomalies. Still, seismic data constrain the choices of subsurface rock-body geometries for potential-field modeling and interpretation, and deep seismic and other geophysical techniques have been used widely since the 1950s for crustal studies in the Canadian plain and mountain regions (Garland and Tanner, 1957; Cumming et al., 1962; Cumming and Kanasewich, 1966). The first summaries, in the 1970s, provided initial insights into the structure of the crust of the craton in southern Alberta and across the Cordilleran megabelt (Berry et al., 1971; Forsyth et al., 1974; Berry and Forsyth, 1975).

An unexpected result of deep seismic imaging in the hinterland of the North American Cordillera has been the discovery that the Moho, depressed under the high mountain ranges in the Rocky Mountains and along the Pacific coast, is flat under the Basin and Range province in the south and the Intermontane Belt in the north (Pakiser and Zietz, 1965; Berry et al., 1971). Where the Moho is deep, as in the Rocky Mountain Belt in Canada, it lies at ~45 km, but its depth decreases near the Rocky Mountain Trench to 33-35 km under the Omineca miogeosynclinal belt. Near the continental slope, the continental crust is attenuated to only 18-20 km thick or less. But whereas the mid-crustal Conrad discontinuity is common in much of Eurasia, in North America its definition is less clear. The change in reflection pattern, from a mostly-transparent upper crust to a reflective lower crust, may correspond to the brittle-ductile transition. Velocity and density structure of the crust varies from region to region, in the craton and the Cordillera alike (e.g., Riddihough, 1979).

Whereas early deep seismic surveys had relied upon refraction methods, the 1970s also saw the use of reflection techniques in deep crustal seismic studies (Oliver et al., 1976). The possibilities for detection of low-angle discontinuities in the crust and upper mantle thus increased, and new computers permitted new methods of data processing and display. The first deep reflection surveys, arising from oil-industry practices, were made in the U.S., and in a few years deep seismic studies were implemented as national programs by many rich countries (the U.S. COCORP, Canadian Lithoprobe, German DEKORP, French ECORS, British BIRPS). This work led to the appearance of new crustal structural models, and low-angle reflections were readily interpreted as crustal-scale thrust slices (e.g., Brewer et al., 1981; Ando et al., 1983; Cook et al., 1983).

The concept of "thin-skinned" tectonics became popular in the Appalachian and Cordilleran regions.

Then came the recognition of deep seismic techniques' limitations. One major problem is the usual inability of reflection methods to detect steep faults. This happens because the seismic signal travels down and up, parallel to such faults. But major steep faults in the lithosphere define the blocks and weakness zones whose activity determined the regional geology. Large-scale thrusts should be regarded separately from this lateral zonation along steep boundaries: rootless thrust sheets overlap these deep, long-lived zone and block boundaries at very shallow levels. Lack of sharp images of subvertical features has resulted in disregard by geophysicists and tectonists of these old crustal mosaics.

Misconceptions arising from one-sided interpretation of seismic reflection data can be offset by analysis of potential-field data, primarily gravity and magnetic, which permit regional coverage and respond to lateral rather than vertical variations in rock properties. Thus, potential-field data are more suited for detection of steep discontinuities. Any geophysical data, seismic and potential-field alike, require geologic calibration in their interpretation. But interpretations tend to be influenced by the prevailing general tectonic concepts. In the Cordillera today, the prevalence of terrane-tectonic ideas has artificially narrowed the conceptual framework in which geological and geophysical data are interpreted tectonically.

Predominance of low-angle reflections in seismic images, coupled with ideas about terrane collages and orogenesis due to subduction (Monger et al., 1982, 1994), produced the interpretations of the Cordilleran crust as a stack of thrust sheets (Fig. 64), with their bounding thrust faults merging into a detachment in the middle and/or lower crust (e.g., Cook and Varsek, 1994; Varsek, 1996). Accreted terranes came to be regarded as tablets just 10-20 km thick, whose stacking on top of an attenuated basement with North American affinity was supposedly responsible for the construction of the Cordilleran mobile megabelt. In an accidental collage, these terranes were joined to (Monger et al., 1982) or pushed over (Cook, 1995a-c) the leading edge of the North American continental plate, as it advanced over and overrode the oceanic-lithosphere plates in the Pacific Ocean (Hyndman, 1995). Only in the latest Cretaceous did the terranes finally come together in something like their present-day arrangement, forming a tectonic entity regarded previously as the Canadian Cordilleran eugeosyncline (Douglas, ed., 1970) and now thought to be the terrane-made part of the Cordillera west of the miogeosyncline (Gabrielse and Yorath, eds., 1991). For practical purposes, this implies that prior to accretion, metallogeny of each terrane must have been quite differ-

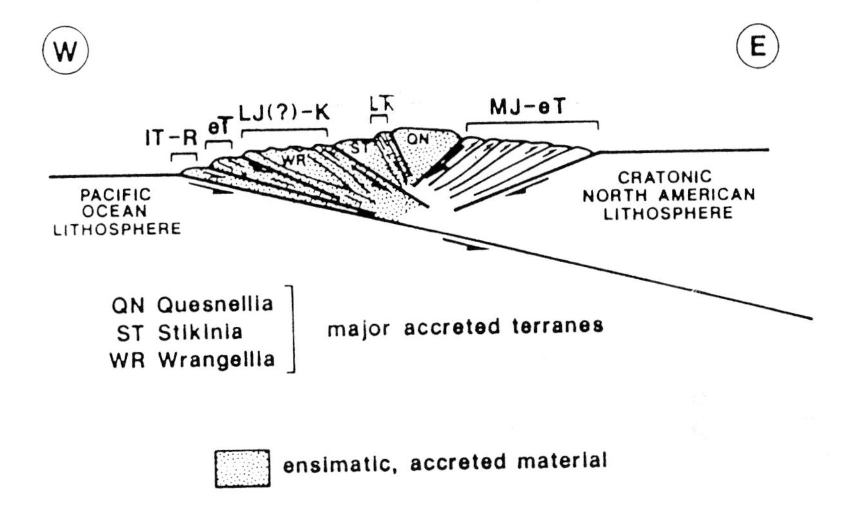

Figure 64. Hypothesis of the Cordillera as a thrust stack of accreted terranes, illustrated in an assumed schematic cross-section of the Canadian Cordilleran crust (modified from Monger et al., 1994). eT - Early Tertiary; L - Cretaceous; LJ - Late Jurassic; lT - Late Tertiary; LTr - Late Triassic; MJ - Middle Jurassic; R - Recent.

ent from and unconnected with that of the rest of the Cordillera (cp. Dawson et al., 1991; McMillan, 1991a).

The idea that the Cordillera formed randomly (by terrane assembly) opened the door for loose speculations. One such speculation has been that an ancient continental margin of an ancestral North America was preserved for a huge period of time as a ramp in the crust in the Omineca Belt, and another that ancient crustal-scale structural detachments have also been preserved in the Cordilleran crust for hundreds of millions of years (Cook et al., 1991a-b, 1992; Cook and Varsek, 1994). Both these ideas are rooted in the suppositions that the Cordillera and mobile megabelts in general form by accretion of terranes and crustal slices at a continental margin, and that the continental crust is tectonically inert.

The Bouguer gravity gradient zone linked to the presumed Omineca intracrustal ramp probably does represent some sort of deep lithospheric boundary, but not between predominantly continental and predominantly oceanic lithospheric domains. Probably, it separates a less-reworked lithosphere to the east from a more-reworked lithosphere to the west. The idea about long-lived Cordilleran crustal detachments assumes that terranes were accreted as thrust slices, but in such a manner that a "normal" continental-crust thickness was somehow maintained in the terrane thrust stack across much of the Cordillera. No breaks in the basal detachment were postulated, and faults mapped at the surface, even steep ones, were assumed to flatten out at depth. The correlations of mapped steep faults at the surface with low-angle seismic events deep in the crust are often tenuous, carried across discontinuities in reflections and gaps in the data. Besides, these correlations ignore much of the known tectonics in the Cordillera, notably the burial of supracrustal protoliths tens of kilometers into the crust and the subsequent rise of their metamorphosed products back to the surface. To suppose that old crustal structures, ramps, thrusts or any other, were preserved intact through such dramatic younger episodes is implausible.

Somehow, crustal-scale thrusts separating presumably distinct terrane slices were assumed to cross the main Cordillera tectonic zones, even though these zones had very dissimilar tectonic histories. From all these assumptions, an ungeological conclusion was reached that major topographic zones in the Cordillera are in poor correspondence with tectonic features, because the most fundamental tectonic features were assumed to be crustal-scale thrusts and detachments (Cook, 1995a-c).

Besides, large variations in the gravity anomaly values from region to region suggest the extent and mechanism of compensation may vary laterally. The gravity gradient zone along the Omineca Belt corresponds to a change in Moho depth from ~45 km under the Rocky Mountains to the east, to 35-36 km or less in areas to the west. These variations have not been properly reconciled with either presumed old crustal-scale detachments or supposed large-scale extension of the Cordilleran crust in the Tertiary.

From seismic reflection data along the Canada-U.S. border, Potter et al. (1986) concluded the Moho reflections are little disrupted and probably Cenozoic, this plausibly suggesting considerable tectonic reworking in the lower crust. In the southern Canadian Cordillera, in the southwestern Omineca Belt and along the Intermontane Belt, Moho reflections indicate a long, N-S-oriented arch with a vertical amplitude of 3-5 km at its apex (Cook et al., 1992). At upper, brittle crustal levels, block-tectonic fabrics are strong. The lower crust is, in contrast, ductile. This dissimilarity does not imply that a detachment exists in the mid-crust, but merely that rheological behavior of rocks above and below the brittle-ductile transition is not the same. Confirmed large (tens of kilometers) vertical movements of Cordilleran crustal blocks and zones in the geologic past could not have left older features undisturbed. The idea of mid-crustal or lower-crustal detachments extending across the Cordillera (Cook and Varsek, 1994; Marquis et al., 1995) ignores the geologic evidence of extensive tectonic reworking and is thus unjustified by regional tectonic analysis.

Also unjustified is the postulated (Cook, 1995a-c; Clowes et al., 1995) lack of correlation between topography and deep crustal structural features. Such a correlation should not be expected to be perfect, because the tectonically created relief is modified rapidly by exogenic erosion and by new generations of thermo-tectonic movements. Even in flat cratonic platforms, the tectonic topography is distorted by exogenic influences such as erosion and glaciation. In high mountains, the topography is rugged and shaped mostly by erosion. In the Cordillera, to simply equate geomorphologic zones of complex exogenic and endogenic origin with tectonic zones, and geomorphologic surfaces with tectonic ones, would be an error. Yet, since topography is partly a product of tectonism, a rough correlation is normal.

Another wrong idea is that the Moho in continental mobile megabelts is long-lived. Cook (1995a-c) postulated the Moho in the southeastern Canadian Cordillera is as old as Proterozoic. However, a change in crustal thickness from ~45 km beneath the Rocky Mountain fold-and-thrust belt to ~35 km in Cordilleran belts to the west (Chandra and Cumming, 1972; Cumming et al., 1979) corresponds to the transition from a zone of thin-skinned deformation to zones where the whole crust was reworked; vertical tectonic movements had amplitudes of 20-25 km as late as in the Early Tertiary. An Early Tertiary episode of crustal extension in a broad area of the miogeosynclinal Omineca Belt and the Intermontane median-massif belt is indicated by a variety of geologic evidence, including radiometric dating of abundant Cenozoic igneous rocks from plutons, dikes and veins (e.g., Spear and Parrish, 1996). Extensive thermo-tectonic processes are required to produce these phenomena and to cause the ongoing uplift of high-standing mountain belts of the Cordillera. South of the Canada-U.S. border, the reflection Moho beneath the western part of the miogeosyncline and the Intermontane Belt has been interpreted to be young (Potter et al., 1986). Young age of the Moho has also been much discussed farther south, in the U.S. Cordilleran Basin and Range Province in Nevada and Utah (e.g., Allmendinger et al., 1987; Hauser et al., 1986, 1987). Regeneration of the Moho is probably a result of tectonic reworking, and the ability of the lower continental crust to flow ductilely has been discussed repeatedly in the literature (e.g., Dohr, 1989; Lyatsky, 1993). The idea of the lower crust's long-term structural stationarity is groundless.

Seismic data offer no more than indirect images of the subsurface. It is impossible to determine the geologic nature of a seismic reflection without independent constraints - and such constraints are usually unavailable deep in the crust. Reflections may be caused by metamorphic fronts, old unconformities, structural discontinuities of various kinds, and many other features. Interpretation and correlation of seismic reflections requires a great deal of caution and must be founded on, above all, regional *geological* analysis.

Sweeney et al. (1991, p. 50) noted that indeed there are geophysical belts running parallel to the principal structural grain of the Cordillera. The eastern belt, west of the Rocky Mountains, is characterized by a relatively shallow Moho, elevated Bouguer gravity anomaly values, relatively smooth magnetic anomalies, low electrical resistivity of the crust, and high heat flow. The western geophysical belt, lying west of the Coast Mountains, is distinctly different, and the region between the Rocky Mountain Trench and the eastern Coast Mountains falls into a single, broad geophysical domain.

These geophysical belts do not match the bipartite division of the Canadian Cordillera into the miogeosynclinal and eugeosynclinal domains, whose boundary is conventionally placed at the western edge of the Omineca Belt. Neither do these geophysical boundaries simply fit the major physiographic belts, but they do match the division of the thin-skinned Rocky Mountain fold-and-thrust belt from the deep-rooted rest of the Cordillera. But the distribution of geologic and physiographic belts in the Cordillera has varied in geologic time. Interpretation of the modern regional distribution of physical properties of the crust, inferred from geophysical data, requires an understanding of the history of Cordilleran tectogenesis, now and in the past.

Sweeney et al. (1991, p. 53) argued that the cratonic crystalline basement is well imaged seismically beneath a detachment zone under the Rocky Mountain (also Bally et al., 1966) and eastern Omineca belts. In many areas, the basement surface is conventionally thought to be the surface of Laramian detachment, leaving the crystalline basement undisturbed. Kanasewich (1993, p. 66) noted that the crust thickens from the Interior Platform westward "...to over 65 km under the Rocky Mountains and the Rocky Mountain Trench". Cook (1995b, p. 1521), however, suggested that "beneath the Rocky Mountains... the Moho does not generally produce a distinct reflection". Under the Purcell Anticlinorium the Moho is nonetheless is depicted as deepening to the east, and Cook (1995b) linked this crustal-thickness change to a presumed low-angle continuation of the steep Slocan Lake fault mapped on the eastern limb of the Valhalla metamorphic core culmination. Step-like offsets of the Moho beneath the Omineca Belt have also been noted. Gravity gradient zones can generally be reconciled with seismic refraction (Cumming et al., 1979) and reflection (Cook, 1995a-c) results. Bouguer anomaly values of -200 mGal and less, typical for the high ranges of the Rocky Mountains, rise westward to -150 and even -100 mGal in the Omineca and Intermontane belts (Stacey, 1973). Whether the very low gravity anomaly values in the Rocky Mountains result from just deep Moho or also from anomalous densities of the upper mantle is not clear. Importantly, many prominent NE-SW-trending gravity and magnetic anomaly zones continuing from the Canadian Shield across the Alberta Platform and the Rocky Mountain Belt are disrupted (though not always completely) at the Rocky Mountain Trench.

Cook et al. (1995) underestimated the significance of the NE-SW-trending anomalies in the southern Canadian Cordillera west of the Rocky Mountain Trench. In the craton, these faults belong to an ancient structural network dating back to Late Archean or Early Proterozoic time. Cook et al. (1995) did emphasize the more-apparent correspondence of other anomalies with faults trending N-S or NNW-SSE (Fraser, Yalakom, Pasayten). But underestimation of NE-SW-trending anomalies and structures has led to a conclusion that the westward extent of a cratonic basement in the Cordillera substantially ends at the miogeosyncline, and farther west at most only a thin wedge of cratonic crust lies between the upper mantle (or subducting oceanic lithosphere) below and obducted exotic terranes above (also Clowes et al., 1995; Hyndman, 1995).

The lack of evidence for ongoing subduction at Canada's Pacific continental margin was discussed by Lyatsky (1996). Presence of thick continental crust in various parts of the

Cordilleran interior is now discussed widely (e.g., Plafker and Berg, eds., 1994). Persistent NE-SW-trending faults and potential-field anomalies transecting the NNW-SSE grain of the Cordillera are evidence that much of this crust retains some properties dating back to ancient Precambrian times.

Interruptions of NE-SW-oriented potential-field anomaly trends continuing into the Cordillera from the craton occur at the Rocky Mountain Trench and in the eastern Omineca Belt, where the crust changes from cratonic to reworked. Still, even west of this transition zone the NE-SW trends do not disappear completely, and they continue as far as the Insular Belt and the continental margin (Lyatsky, 1993, 1996). Westward continuation of these trends is expressed in mapped faults, potential-field anomaly shapes and contour offsets and topographic lineaments (river valleys, fiords). The Skeena and Stikine arches are products of rejuvenation of ancient basement structural trends in the Cordilleran interior. Similar inter-regional transcurrent features are known also in the conterminous U.S. and Alaska.

In median massifs of the North American Cordillera, the usual thickness of the continental crust is ~30-35 km (e.g., Pakiser and Mooney, eds., 1989; Beaudoin et al., 1992; Clowes et al., 1995). Cook and Varsek (1994) and Cook (1995a-c), however, modeled ex-cratonic crustal rocks in the Intermontane Belt to exist only in the modern lower crust, beneath the brittle-ductile transition, and presumed the overlying crustal horizons comprise exotic-terrane thrust slices. But whereas they placed the ex-cratonic basement at a depth of 20-25 km, Marquis et al. (1995) put it at only 10-12 km. The presence in the Intermontane median-massif belt of NE-SW-oriented potential-field anomalies, which in the craton are associated with Late Archean and Early Proterozoic crustal zonation, suggests that incompletely reworked basement rocks lie, and have lain for a long time, above the Curie isotherm. Similarity of ancient fault patterns in the craton and the Cordilleran median massifs contradicts the picture of the Cordillera as a random collage of exotic terranes.

From geologic observations (e.g., Gabrielse and Yorath, eds., 1991), the oldest exposed rocks of the Intermontane Belt's sedimentary cover are Devonian, and the late Paleozoic stratal rock package (including the Harper Ranch and Cache Creek assemblages) exceeds 4,500-5,500 m in composite thickness. A pronounced angular unconformity underlies these upper-crustal horizons. Although such a package should be seismically reflective, deep Lithoprobe seismic data are usually shown without details of their short-traveltime part. Besides, seismic reflection methods do not usually reveal steep crustal discontinuities. Reflection and refraction data permit coarse discrimination of crustal layers and zones with various acoustic characteristics, but available refraction models in the Canadian Cordillera are usually too smooth to assist significantly in the detection of high-angle faults.

By the late Middle Jurassic, in the northern Canadian Cordillera rocks of the Cache Creek assemblage were thrusted southwestward towards the inner parts of the massifs. In Middle-Late Jurassic time, in the southern Canadian Cordillera rocks assigned to the Cache Creek assemblage were thrusted eastward. These two dramatic events were connected with tectonic activity of the Teslin orogen. The Bowser Basin, by contrast, underwent long-time subsidence with amplitudes of 6 km and more, while differential uplift of other blocks produced the transcurrent arches in the Intermontane Belt. But, apart from the Nicola horst (in seismic Line 7), no signs of these movements are seen in the Lithoprobe seismic data.

The conventional Lithoprobe interpretation holds the Nicola horst to be a 25-km-thick stack of three or four crustal thrust sheets, lying on top of a regional detachment in the lower crust. A thin sliver of cratonic basement is supposed to lie underneath. Faults bounding the Nicola horst are supposed to be listric, flattening out rapidly with depth (Cook et al., 1992; Clowes et al., 1995, 1998). These interpretations are not unequivocal: for example, Marquis et al. (1995) discussed alternatives for deep crust under the Nicola horst and placed the top of the old North American basement at much shallower depth of only 10-12 km. Geologic estimates that the volcano-sedimentary cover in the Intermontane Belt has a total thickness of ~15 km suggest a basement depth similar to that modeled by Marquis et al. (1995).

New details are revealed in seismic data by reprocessing, and some reflection bands are interpreted in geologic terms differently by different authors. Disputes exist, for example, about whether certain events are intracrustal or subcrustal (cp. Varsek et al., 1993; Clowes et al., 1995), and “the lack of definition of North American basement by the seismic velocity structure” (Clowes et al., 1995, p. 1505) is acknowledged. Postulations of direct continuation of the North American basement across much of the Canadian Cordillera (as far as the Fraser fault, ~350 km west of the central Omineca Belt) in a slice with a total thickness of just 10-15 km overlook geologic evidence for large vertical movements and tectonic reworking of big crustal blocks and zones in the Phanerozoic. An ex-cratonic basement does exist in the Intermontane median-massif belt, but it is separated from the modern North American craton and semi-reworked. Tectonic events that could have separated the median massifs from the craton were, among others, the Antler orogeny in Devonian-Carboniferous time and the Teslin orogenic pulse in Mississippian through early Mesozoic time. A median massif is a distinct lithospheric lateral tectonic unit within a mobile megabelt, and its crustal structure differs from that of neighboring orogens. Geological considerations such as these should be the guide to interpreting the structure of the crust from geophysical data, because otherwise such interpretations risk being arbitrary, subjective and non-unique.

Compared to the surrounding orogens, the Intermontane median-massif belt was tectonically remarkably stable. Its subsidence was no more than ~15 km, and the basement and the younger volcano-sedimentary cover were not fused by strong metamorphism. The basement top may be marked by some seismic reflections at a depth of 10-12 to 15 km (4-5 s) in south-central British Columbia, particularly in the Nicola horst area. According to the data presented by Clowes et al. (1995), Moho depth across the Omineca, Intermontane and eastern Coast belts varies little in the southern Canadian Cordillera, between just 32 and 36 km. By contrast, depth to the lower crust (having average P-wave velocity of 6.6 to 6.9 km/s) is much more variable, from 12 to 27 km. Perhaps these intracrustal seismic-property variations are related to variations of rheological properties of rocks in the mid-crust. Ductilely creeping lower-crustal material could account for the fairly flat (and young) Moho, but the semi-brittle transitional zone in the mid-crust could be more complex. Vertical tectonic movements, which strongly affected the structure of the brittle upper crust, had less long-term effect on the lower crust where lateral variations were more easily smoothed out by ductile creep. It is the upper and middle crust that retain the most memory of tectonic movements.

The Moho discontinuity is usually defined well in seismic profiles, owing to a change in bulk velocity, density and internal reflectivity. The continental upper-mantle P-wave velocities beneath the Moho are usually around 8.1 km/s. Anomalously, in the Canadian Cordilleran interior at the base of the crust lies a transition zone several kilometers thick, with P-wave velocities from 7.3-7.4 km/s to 7.7-8.2 km/s. Clowes et al. (1991, p. 1501) pointed out that the E-W Lithoprobe lines 3 and 7, which cross the Omineca zone of mountain-building, Teslin orogen, Intermontane median-massif belt and eastern Coast Belt, reveal "...the greatest variation in upper mantle velocities" (7.5-7.65 to 8.2 km/s; see also Zelt et al., 1992, 1993; Burianyk and Kanasewich, 1995).

The inferred physical properties of the Cordilleran crust, including seismic velocities, vary from one set of authors to another (cp. Varsek et al., 1993; Marquis et al., 1995). Seismic interpretations sometimes differ even over short distances (e.g., across the Canada-U.S. border: cp. Potter et al., 1986; Clowes et al., 1995; Miller et al., 1997). In a N-S seismic profile in western Washington state (Miller et al., 1997), no continuous reflections seem to separate the upper and middle crust; the lower crust was modeled as a thick layer with velocities of >6.6 km/s and containing reflective bands with 7.3 km/s velocities; the uppermost mantle was assigned a low velocity of only 7.6 km/s; the reflection Moho was modeled with a rugged relief of ~8 km; the crustal structure was modeled as generally non-uniform; and topographic domains correspond to changes in crustal thickness. Underlining the probable subjectivity, these interpretations differ considerably from the Lithoprobe ones on the Canadian side of the border nearby.

But some robust similarities are nonetheless seen in seismic interpretations all over the North American Cordillera. Under median massifs, anomalous upper mantle has been detected from the Basin and Range province (e.g., Pakiser and Mooney, eds., 1989) through the Canadian Cordilleran interior (Clowes et al., 1995) to east-central Alaska (Beaudoin et al., 1992). Low average crustal velocities, 6.0 to 6.3 km/s, are common in median massifs. Similarly low upper-middle-crustal velocities in the Canadian and Alaskan Coast Mountains and the Washington Cascade Mountains correspond to the massive, deep-rooted granitoid batholiths forming the core of the Coast Belt orogen. In the Omineca Belt, crust with average velocities of ~6.2 km/s seems to be correlatable with metamorphic rocks exposed in a few core-complex culminations and having a "significant depth extent" (Clowes et al., 1995). Such low velocities in the Omineca Belt suggest that metamorphic-rock masses are also strongly permeated with granitic rocks. Along an imaginary line from 49°N/122°W to 52°N/121°W, the average upper/middle-crustal velocity has been reported to be 5.5 km/s to a depth of ~15 km in the Washington North Cascades; ~6.0 km/s to a depth of 6 km in the western British Columbia Coast Belt (and 6.4 km/s in the middle crust due to a more-mafic composition); ~6.25 km/s to a depth of 8 km in the east-central British Columbia Coast Belt and western Intermontane Belt, with a probable offset at the Fraser fault; and ~6.0 km/s in the Nicola horst area in the central zone of the Intermontane median-massif belt. Upper-crustal velocities increase over the Teslin orogen to a depth of 15-18 km, to 6.3-6.35 km/s, reflecting a greater abundance of mafic rocks (Zelt et al., 1992).

A vertical offset of ~4 km even at deep crustal and upper-mantle levels (including an offset on the Moho from 36 km on the east to 32 km on the west) is noted at the Okanagan Valley system of faults and fractures. It is associated not with a boundary of physiographic belts but with the western boundary of the Teslin "suture", where the Teslin orogen is squeezed into a narrow zone just ~10 km wide. Vertical structural offsets are attributed to the Okanagan Valley fault system (actually, the aligned Teslin and Kutcho faults) and to the steep, trans-crustal (e.g., Jones and Gough, 1995) Fraser fault.

Unlike the upper and middle crust, the lower crust in the Canadian Cordillera is interpreted to be more uniform, from the western Omineca Belt to the eastern Coast Belt. However, thickness of the lower crust in the Intermontane Belt is unclear: it has been modeled as a thin layer wedging out to the west (Cook et al., 1992) or a 10-km-thick layer between the Okanagan Valley and Fraser fault zones (Clowes et al., 1995). A 3-4-km rise in the Moho seems to lie beneath the Intermontane Belt, compared to the flanking orogens (also Cook, 1995b-c).

The suggestion that Lithoprobe seismic data were not very successful at detecting the boundaries between Cordilleran tectonic zones largely arises from misidentifying which geologic boundaries are the most important. Despite the poor ability of seismic data to detect steep discontinuities, in the southern Canadian Cordillera lateral changes in the crust's seismic properties do correspond to major geologic boundaries at the surface (e.g., Kootenay Lake fracture system). Many other surface fault-related geologic boundaries, assumed to be significant in the terrane-collage model (e.g., Slocan Lake fault), are actually of second-order significance to the velocity structure of the crust. Combination of results from geologic mapping and crustal seismic studies suggests which fault systems, having both surface geologic and deep seismic expression, are the most significant. Steep discontinuities are a fundamental property of continental crust in cratons and mobile megabelts, and their study provides the key to understanding a region's tectonic evolution.

Modern physiography and the geomorphological history of a region mimic its tectonic zonation only loosely, as numerous other factors also affect the topographic surface. Processes of erosion and accumulation modify the exposed surface and mask tectonic boundaries. Mismatches between topographic and tectonic boundaries can lead to confusion and misunderstanding in the tectonic analysis and interpretation of geophysical data. Cook's (1995c, p. 1805) statement that in the Canadian Cordillera "...there is little correspondence between major surface features, such as the regional geotectonic belts, with variations in structure of the Moho" is a result of misidentifying the main boundaries.

Magnetic and gravity data are more suitable than seismic surveys for detection of steep discontinuities (Nettleton, 1971; Hinze, ed., 1985). No magnetic anomalies are sourced at depths below the Curie isotherm, but above it a rock body can carry many different magnetizations. Besides, the magnetic field is dipolar and has a non-vertical inclination and a declination away from geographic north. As a result, even simple rock-body geometries may be associated with very complex magnetic-anomaly patterns that are hard to interpret. Gravity anomalies avoid these problems, and the relationships between anomalies and rock bodies are much simpler. Still, as with magnetic data, these relationships are not unique.

Interpretations can be constrained by using various types of geophysical data together, processing them in different ways, and above all by considering all the available geological constraints (see Geological Survey of Canada, 1990; Lyatsky, 1996; Lyatsky et al., 1999). Only direct geologic observations can provide unambiguous information about rocks and rock bodies.

Attempts to interpret various geophysical data have been made by Lithoprobe, as reflected in a range of geophysical surveys and final reports (e.g., Cook, ed., 1995). But integration of these interpretations with each other and with geological information has been less than full. Geophysical modeling, rather than geological considerations, was the main guide for these studies of the crust and lithosphere. As well, the interpretations of both geophysical and geological data were fitted to a pre-conceived idea of the Cordillera as a collage of accreted terranes, which provided a comprehensive but arbitrary and factually incorrect basis for interpreting the data. In the overviews of Lithoprobe results in the southern Canadian Cordillera, regional tectonics is very oversimplified and reduced substantially to non-indigenous, far-field, extra-regional influences. Mixing together dissimilar phenomena, Clowes et al. (1995, p. 1486) regarded the Cordillera as an orogen where "five major morphological belts and the offshore tectonic plates... define the transect region". According to Cook (1995c, p. 1815), "the North American craton 'bulldozed' into the accreted terranes" during plate convergence since the Jurassic; according to Hyndman (1995), plate convergence is continuing at the Canadian Pacific continental margin to this day. As new geological information becomes available in the Cordillera, outdated ultra-mobilistic templates, essentially unchanged since the 1960s and 1970s, are contradicted by new facts and are turning into an intellectual straightjacket.

From paleomagnetic studies of many Jurassic and Cretaceous volcanic and plutonic rocks, largely in the Intermontane Belt, huge translations and rotations of assumed terranes and their fragments have been postulated (Monger and Irving, 1980; Symons, 1983; Marquis and Globerman, 1988; Irving et al., 1995; Wynne et al., 1995). Terrane displacements have been modeled to be many hundreds or thousands of kilometers. For the eastern Intermontane Belt, for example, northward translation of ~1,100 km and clockwise rotation of ~60° have been suggested to have occurred since the mid-Cretaceous. Even greater translations, on the order of 3,000 km, have been proposed for the Coast Belt.

Critical reviews have shown that these interpretations of translation from paleomagnetic data contain a variety of methodological errors (e.g., Butler et al., 1989), and the displacements are overestimated (also Dickinson and Butler, 1998). Some of the translation models have been shown to contradict geologic facts such as the geometry and kinematics of major faults (Monger and Price, 1996) and sediment-provenance determinations (Mahoney et al., 1999). At any rate, it is clear that without geologic control such reconstructions can be very misleading.

Due to shortage of reliable constraints, also very speculative are many tectonic conclusions and models drawn from the variations in heat flow on the ground surface and the assumed thermal state of deeper crustal levels (e.g., Lewis et al., 1992; Lowe and Ranalli, 1993; Hyndman and Lewis, 1995; Marquis et al., 1995). If used with proper caution, surface measurements of heat flow can provide important information about the physical conditions of the crust and patterns of fluid circulation. However, models of the thermal state of deep crust, and tectonic speculations derived from them, are inevitably very unreliable. In Germany, for example, deep continental drilling has revealed large unforeseen variations in the thermal state of the upper crust and fluid-flow patterns (Clauser et al., 1997; Emmermann and Lauterjung, 1997).

Gough (1986) speculated about mantle-upflow tectonics under the Intermontane and Omineca belts in the Cordilleran hinterland. Lewis et al. (1992) proposed that at deep crustal levels, the Intermontane and Omineca belts can be joined into a single heat-flow province (though no intimate similarity is indicated by seismic data, as was noted by Clowes et al., 1995). Sometimes, the modeled thermal boundaries and transition zones in the Cordilleran crust coincide with discontinuities inferred from other geophysical data, but results obtained with different geophysical techniques are all too often in disagreement, depending in no small part on specific assumptions and modeling procedures (e.g., cp. Majorowicz et al., 1993; Cook, 1995c). Clowes et al. (1995, p. 1509) noted that the seismic-velocity crustal structure of the southern Canadian Cordillera corresponds poorly to estimates of temperature at depth: in particular, "...there appears to be no distinct correlation between the seismic velocity structure at depth to the 450° isotherm", which is the temperature conventionally associated with the brittle-ductile transition.

Generally in the southern Canadian Cordillera, heat-flow values tend to increase where the crustal thickness increases; the same tendency is apparent in radiogenic heat production (Lewis et al., 1992; Hyndman and Lewis, 1995). Suggesting indigenously sourced heating, the highest heat-flow values (>110 mW/m^2) and heat production have been reported for heavily granitized and thick crust in the Coast and Omineca belts (see also Cook, 1995c). Lower heat-flow values (~70 mW/m^2 or less) have been measured in the Intermontane median-massif belt south of the Skeena Arch. The lowest values, as low as <50 mW/m^2, have been recorded on the western and eastern sides of the central zone of the Coast Belt orogen, as well as in this orogen's southern part near the Canada-U.S. border (Hyndman and Lewis, 1995, their Fig. 1; also Lewis et al., 1992). This distribution indicates some sort of modern thermal segmentation within the Coast Belt and in the whole of the Cordillera, perhaps along megabelt-parallel and transverse faults.

Crustal zones with various electrical conductivities can be depicted with magnetotelluric methods (e.g., Jones and Gough, 1995), but the problem of non-uniqueness occurs in these interpretations too. It is particularly hard to interpret an electrical-conductivity model of the crust in terms of its rock composition and structure. For example, Kurtz et al. (1986, 1990) modeled a mid-crustal zone of high conductivity in the continental-margin region in British Columbia, and interpreted it as a product of active subduction of an oceanic plate. But just as easily, it could be a zone of structural shearing or a front of metamorphism (see discussion in Lyatsky, 1996). Increased electrical conductivity has been detected at depths of 12-15.4 km in the Intermontane Belt. It may be related to the base of the upper crust (Jones and Gough, 1995), but it may also correspond to the top of the crystalline basement of the median massif. In a seismic velocity model of this massif (Lithoprobe Line 1), a low-velocity zone at that depth may be assigned to the upper crust, or to the middle crust as is done sometimes (cp. Zelt et al., 1992).

Along the southern Cordilleran Lithoprobe transect, Clowes et al. (1995) noted a number of areas where the velocity models do not match the crustal structure inferred with other geophysical methods. A partial explanation could be that, because different techniques depict the distribution of different rock properties, and exact correlation should not be expected. Even coincidence of some bodies defined with different geophysical methods may be incidental and lacking geologic cause (Cook and Jones, 1995). Clowes et al. (1995) noted that, contrary to Marquis and Hyndman (1992), variations in resistivity do not necessarily correspond to changes in seismic velocity. Inferences about the distribution of various properties of rocks and melts, obtained from a wide range of geophysical, geochemical and geological methods, should all be combined in a single regional geological analysis, permitting to define mappable and remotely sensed geological bodies and their boundaries in an inhomogeneous rock mass.

A lot of effort has been made to understand the causes of gravity anomalies and rock-density distribution in the lithosphere of the Canadian Cordillera. Gravity anomaly maps with different reductions (Stacey, 1973; Geological Survey of Canada, 1990; Sobczak and Halpenny, 1990; Lyatsky, 1991, 1993, 1996) have been used to relate regional anomalies to the composition and geometry of rock bodies and the position of their boundaries. Perhaps because so many causes can affect the anomalies, no simple, straightforward relationship is apparent between Moho relief and gravity anomalies tens and even hundreds of kilometers in wavelength, and these relationships vary from region to region (Sweeney et al., 1991).

Topography of the mountains does not everywhere have a mirror image in the Moho relief (cp. Sweeney et al., 1991; Clowes et al., 1995; Cook et al., 1995; Miller et al., 1997). Sweeney et al. (1991, their Figs. 2.28-2.29) presented gravity-based profiles across the southern Canadian Cordillera which show the Intermontane Belt has a comparatively shallow and irregular Moho, but the correlation of Moho-depth variations with the Intermontane Belt is very approximate. The Intermontane plateaus generally have a comparatively low isostatic-gravity-anomaly relief. Very loosely, this picture corresponds to crustal seismic reflection patterns, which indicate slight Moho depressions under the Coast and Omineca mountains and a 3-4-km rollover centered beneath the Intermontane median-massif belt (Clowes et al., 1995; Cook, 1995b). Of course, neither gravity not seismic data can give an exact indication of the Moho relief, because the sources of gravity anomalies and seismic reflectivity may not coincide and have different geologic nature, but the tendency of the crust to change its thickness from one area to another is apparent. It seems unduly pessimistic to suggest (Clowes et al., 1995, p. 1507) that along the southern Cordilleran Lithoprobe transect "generally, there is little correlation between crustal thickness and Bouguer gravity values..." These authors noted that significant variations in crustal and perhaps mantle densities exist across the southern Canadian Cordillera. This conclusion is consistent with big lateral variations in the bulk rock composition of exposed crustal levels between and within the Cordilleran tectonic belts.

In the Intermontane Belt, with its shallower Moho and 4-5-km-thick Triassic and Jurassic volcanic successions at supracrustal levels, Bouguer gravity anomaly values are on the order of -120 mGal, considerably higher than the -150 to -200 mGal values in the Coast and Omineca belts. There is no gravity indication of a gradually westward-deepening top of the crystalline basement, and there is no gradual shallowing of the Moho from the craton on one side of the Cordillera to the continental margin on the other.

Though the lower, middle and upper levels of Cordilleran crust have dissimilar seismic properties (Clowes et al., 1995), to assume they are structurally detached is unfounded. Rheological division into the brittle upper and ductile lower continental crust, across a mid-crustal transition zone, is commonplace worldwide. But through geologic time, the vertical position of these rheologic levels and transition zones changed in response to variations in heating, water abundance and vertical movements of crustal blocks.

Structural discontinuities that stop being active are sometimes healed, especially in ductile lower crust. Metamorphism homogenizes older compositional and structural variations, and creates new ones.

Orogenic deformational structures in the lower crust and upper mantle are hard to image geophysically and impossible to map directly, and anyway they are commonly obliterated by subsequent reworking. In the brittle upper crust, by contact, geological and geophysical observations are easier, and remnants of previous tectonism are more decipherable. To assume the structural styles in the upper and lower crust are similar is to ignore these major differences in physical conditions and rheologic behavior.

Thickness and depth of the upper, middle and lower crust, as well as of the levels of mantle lithosphere, vary in time and space. On average, the brittle upper crust continues to depth of 10-15 km. There the rock behavior becomes semi-ductile, and metamorphism reaches amphibolite grade. The brittle-ductile transition is usually thought to lie around the 450°C isotherm, but variations in rock composition, wetness and so on also affect its position. From 20-25 km depth to the base of the crust, metamorphism may be as high as granulite-grade, and rocks are ductile and in places partly molten (e.g., Greenwood et al., 1991; Holbrook et al., 1991). Generally, density of rocks in the crust increases stepwise with depth, reaching the maximum at granulite metamorphic grades; seismic velocities are distributed similarly. From laboratory experiments and petrological studies, the conditions at depth are generally as follows: at mid-crust, pressures >300 MPa (>3 kbar) and temperatures of 450°-650°C; in the lower crust, pressures >600 MPa (>6 kbar) and temperatures >650°C (see also Rudnick and Fountain, 1995).

Changes in stress regime, induced by indigenous or external influences, cause changes in the physical conditions of the crust; the state of the crust, in turn, affects the stress patterns. Some estimates suggest that compressional horizontal stresses of only 100 MPa (1 kbar) may considerably distort the relationships between isostatic gravity anomalies and topography (Stephenson and Lambeck, 1985). In the conventional model of Cordilleran tectonics (Cook et al., 1992; Cook and Varsek, 1994; Clowes et al., 1995), ongoing subduction is occurring at the continental margin. Since such a process could easily produce horizontal stresses of 100 MPa (1 kbar) or more, the model bears an internal contradiction between the postulated permanence of deep crustal structure and the postulated active processes that could alter it. More compellingly, from geological evidence, the crust in this megabelt was tectonically reworked in the Antler orogeny and in the numerous subsequent orogenies in the Mesozoic and Cenozoic.

Failure to recognize whole-crust reworking and steep trans-crustal and trans-lithospheric faults serves the incorrect idea that tectonic deformation in the Cordillera concentrated in the upper and middle crust. Early in the Lithoprobe program, in the Insular Belt, terranes such as Wrangellia were thought to be tablets no deeper than ~15 km (see review in Lyatsky, 1996), but now Cordilleran terranes are generally interpreted as being rooted deeper (Clowes et al., 1995). Before that, in the 1970s, terranes had been regarded by geologists as whole-crust blocks attached to the leading edge of the continent (cp. Monger et al., 1982).

General recognition that steep discontinuities are fundamental features of continental crust predates plate and terrane tectonics. A great many high-angle faults have been mapped all across the Cordillera, but in the papers describing the Lithoprobe interpretations such faults are conspicuous only by their absence.

Many faults that are steep at the surface remain steep deep in the crust. They are usually invisible in seismic sections. Some faults are listric-shaped, and they flatten out at depth. However, to assume that most steep faults flatten out is groundless. Besides, listric faults are also hard to depict seismically, because their steep shallow parts are not imaged and to link a steep surface fault with this or that low-angle reflection at depth can be speculative and subjective.

Seismically reflective discontinuities in the crust may be of many types. Returning seismic signal is detected if it was reflected back to the surface rather than scattered sideways. The lower the angle of incidence on a discontinuity, the harder it is to recognize a returning signal (if it returns at all) as coming from a particular reflector. With near-vertical-incidence reflection surveys, steep discontinuities are often impossible to recognize because they return no reflected signal to the surface. Refraction surveys rely on subhorizontally traveling signal that hits subvertical discontinuities at a high angle, but such surveys have a low resolution because they are conducted in long profiles along which crustal seismic properties are largely averaged. Coherent noise in reflection seismic sections, such as off-line arrivals, diffractions, multiples and so on, insidiously contaminates the data, particularly in the highly deformed mobile megabelts.

Sometimes in crustal reflection sections, steep discontinuities appear as blank or chaotic zones lacking coherent reflections but containing trains of diffractions; across these zones, the reflection character changes (cp. Hajnal et al., 1996). But the overwhelming predominance in seismic sections of low-angle reflections gives only a partial picture of the structure of the crust, resulting in a false impression that all the discontinuities are subhorizontal or dipping at low angles.

Such a picture is geologically unrealistic. Profound knowledge of the real geology of the region of study is indispensable in interpreting geophysical data, seismic or any other. In the Canadian Cordillera, mismatches between gravity and seismic interpretations (Clowes et al., 1995) are a red flag not to accept any interpretation *a priori.* An agreement between the interpretations of different types of data greatly strengthens the resulting integrated inferences - but above all, the main guide to interpretation should be the observed facts about the region's geology.

Despite a general inability to image steep discontinuities, deep seismic reflection data provide useful images of low-angle ones, permitting to interpret their geometry and position. Even in the lower crust, modern data-analysis techniques permit to locate a reflector vertically to within several hundred or thousand meters (e.g., Clowes et al., 1995). Many hypotheses have been put forward to explain the lower crust's common reflectivity (e.g., Meissner, 1989; Pakiser and Mooney, eds., 1989). Chadwick et al. (1989), for example, suggested that partial melting, metamorphic differentiation and recrystallization could create many lateral and vertical inhomogeneities in the lower crust. Acoustic-impedance contrasts could be due to lithologic variations and molten-rock or intrusive-rock lenses, zones of creep, shearing and ductile flow.

Combined seismic and petrological considerations (Rudnick and Fountain, 1995) show that many complexities exist in relating seismic-reflection or -velocity properties to lithology. The lower crust is inhomogeneous in many of its properties: there are alternating bodies of mafic and felsic rocks, sometimes melt, metamorphic facies, emplaced upper-mantle material, anastomosing networks of faults and shears, etc. This rocks mass is metamorphosed and ductile, and exists under high temperature and pressure. In many areas worldwide, uplift and unroofing caused lower-crustal and upper-mantle rocks to be exposed at the surface, and although cooling and stress relaxation have altered their physical properties, these rocks give a keyhole view of the compositional complexity at depth. Tonalitic gneiss, amphibolite, mafic granulite and anorthosite are the principal lithologies in the lower crust, with secondary intermediate to felsic granulite. On the whole, the continental lower crust is more mafic than the upper, due to emplacement of mafic-ultramafic mantle-derived material and the tendency to differentiate the felsic crustal components into the upper crust.

Petrological and radiometric studies suggest that rock-making processes in the Earth produced similar results from the Archean till at least the Tertiary (no younger metamorphic rocks are exposed), which indicates these processes themselves were essentially similar through time. But any one metamorphic event could have reworked any available protoliths, regardless of their age and prior history, as is suggested by the known Precambrian radiometric inheritance in Phanerozoic granitoid rocks in many parts of the North American Cordillera.

In most continental regions, the average P-wave seismic velocity in the lower crust is usually around 6.8 to 7 km/s. The age of protoliths cannot be determined geophysically, nor in many instances geochemically, if older geochemical signatures were obliterated by reworking. On the other hand, history of the protoliths can be inferred from broadly-based regional geological analysis. Dynamic and kinematic (and generally geologic) processes in the lower crust and upper mantle need to be considered, because these endogenic processes were largely responsible for the geologic and geomorphologic manifestations observed at the surface.

From studies of xenoliths, Rudnick (1992) found that metapelites, i.e. metamorphosed shales whose protoliths formed at the surface, can compose up to 10% of deep high-velocity crustal layers in post-Archean terrains (also Rudnick and Fountain, 1995). This requires subsidence and burial in excess of 20-25 km. Similar crustal movements could have provided protoliths for voluminous intermediate and felsic deep-crustal granulites. Paragneisses make up ~30% of Archean terrains in present-day cratons, and a large proportion of granites in these terrains were generated by anatexis of metasedimentary rocks. Peraluminous palingenic magmas, resulting from melting of Al-rich and K-rich sedimentary protoliths, can yield S-type granites conventionally associated with continental crust. The origin of I-type granites, generally thought to form where oceanic-crust components are present at depth, is less clear; their protoliths could be intermediate to mafic (basaltic-andesitic gabbro to diorite) igneous rocks (Chadwick et al., 1989).

In regions which, like the Coast and Omineca orogenic belts, underwent large subsidence and uplift, pre-orogenic layering is overprinted completely by deep-seated processes. If rocks reach ductile conditions, especially, new fabrics may be imposed easily, while pre-existing discontinuities such as unconformities and faults may be obliterated. Yet, old lithospheric weakness zones may still be preserved in such transformations, as they channel and localize new strain. Control on crustal faults by weakness zones at subcrustal levels, in the upper mantle, would also favor long-time preservation. For these reasons, some major continental weakness zones have survived for billions of years, with their segments reactivated repeatedly at different times, in different modes.

But it is implausible that, in a mobile megabelt, any level of the crust can be preserved for billions of years without reworking. This holds true, in particular, for the Intermontane median-massif belt, where Cook (1995b-c) considered the lower crust and the Moho to be very ancient features, and assumed the "basement ramp on the west side of the Rocky Mountain Trench" to be old and stationary. It was proposed also that "the Moho and lower crust could be a zone of regional detachment" (Cook, 1995b, p. 1525) and that deformed rocks in the middle and lower crust "are detached from, and transported above, the underlying basement" (op. cit., p. 1528, also Cook and Varsek, 1994). It was presumed that some structural variations along the Moho surface "must be accommodated by regional structural detachment in the lower crust". Strangely, it was argued that offsets recognized in the reflection Moho bear "no evidence to suggest that they are related to high-angle faults" (Cook, 1995b, p. 1529). The lower crust was modeled as a thin wedge gradually thinning westward beneath the western Omineca, Intermontane and eastern Coast belts (also Cook et al., 1992; Varsek et al., 1993). This basement wedge was assumed to be a direct continuation of the North American cratonic basement beneath the accreted terranes (Cook, 1995b-c).

Not all Lithoprobe interpretations follow this scheme. Clowes et al. (1995) modeled a rather uniform, ~10-km-thick lower-crustal slab between the Okanagan Valley and Fraser fault systems. Marquis et al. (1995) proposed the Cordilleran basement includes not just the lower but also the middle crust, leaving only 10-12 km for the presumed accreted terranes. Varsek et al. (1993) showed the large faults near the Coast-Intermontane belt boundary - Fraser, Pasayten, Yalakom - as listric (profile 18, their Fig. 2); no fault offsets were shown at the Moho, and most faults were shown as stopping at the top of the lower crust. Yet, from magnetotelluric data, Jones and Gough (1995) interpreted the Fraser fault as trans-crustal and steep.

The volcano-sedimentary cover in the Intermontane Belt is Devonian and younger. It is weakly metamorphosed. No exposures of the underlying basement are known (Gabrielse et al., 1991b), but xenoliths in intrusive granitic rocks along seismic profile 11 contain zircons which seem to be Proterozoic (Armstrong and Ghosh, 1990; pers. comm. with R.L. Armstrong in Cook, 1995b, p. 1529). From this information, Cook (1995b-c) concluded that subsequent tectonic processes did not "severely overprint" the basement. Proterozoic ages of zircons from some xenoliths indeed suggest the presence of ancient ex-cratonic crustal remnants in the Intermontane Belt's basement, but no simple correlation can be made between these remnants and the modern North American cratonic basement. Metamorphism at deep levels in the Intermontane-Belt sedimentary cover suggests the underlying basement did not escape reworking completely. Still, it is encouraging to see a revival of the old idea that the Cordilleran mobile megabelt developed on a basement which had at one time belonged to a westward extension of a proto-North American craton. Reworked, orogenized remnants of this ancient craton are now noted in many parts of the Cordillera in Canada and the U.S. alike. In particular, semi-reworked ex-cratonic crustal fragments underlie the median massifs.

Armstrong and Ghosh (1990) suggested that in southern British Columbia the initial $^{87}Sr/^{86}Sr=0.704$ line in rocks of the volcano-sedimentary cover shifted westward during Triassic to Eocene time. They believed this shift reflects an influence of the continental crust on the oceanic-derived accreted exotic terranes. Many of the terranes postulated in the Cordillera since the 1970s are indeed conventionally regarded as oceanic-crust entities. There is an opinion among some geochemists and petrologists that from analysis of rocks it is possible to precisely classify terranes in terms of their original tectonic setting (mid-ocean ridge, seamount, abyssal oceanic plateau, island arc, etc.). This opinion is much too optimistic, as it exceeds the real capabilities of existing scientific methods. Rocks formed in various settings commonly have similar petrological and geochemical characteristics. Despite these limitations, Souther (1991) tried to describe many assumed far-traveled terranes quite definitively as oceanic: for example, the Slide Mountain terrane, which supposedly includes ophiolites and mid-ocean ridge and seamount volcanic rocks; the Quesnel terrane, with Mesozoic island-arc volcanism; the Cache Creek terrane, with oceanic basalts and ultramafics; the Stikine terrane, with island-arc affinities and superimposed oceanic volcanism. More cautiously, Woodsworth et al. (1991) reported they were unable to prove the existing tectonic speculations unequivocally from their petrological and geochemical data.

All too often, specialists operate in a pre-fabricated conceptual framework that the Cordillera is largely made up of terranes, and into that framework they fit their own findings. The major tiers in that framework are (a) more or less continuous subduction of "Pacific" oceanic lithosphere from early to middle Mesozoic to the present; and (b) accretion to the margin of the continent, in an accidental collage, of exotic, non-native terranes and stacking of these terranes as thrust tablets in the upper and middle continental crust of the Cordillera west of the miogeosyncline. The latter point means the presumed terrane collage does not represent whole-lithosphere, or even whole-crust, tectonics in this region. However, evidence discussed in this book indicates that local subduction zones and vertical movements and minor lateral shuffling of in-situ blocks, are a more realistic explanation for the geological, geophysical and geochemical observations.

Traditionally, the Canadian Cordillera is described in terms of five geomorphological belts (Douglas, ed., 1970). The recent version of this scheme is rather sophisticated (Gabrielse et al., 1991a), in that geomorphologic belts are associated with essentially coincident tectonic belts separated by fault boundaries (op. cit., their Fig. 2.4). Broadly speaking, the Rocky Mountain and Omineca belts correspond to the miogeosyncline in Kay's (1951) old designation, and the Intermontane, Coast and Insular belts correspond roughly to the eugeosyncline. The Intermontane and Insular belts continue into the Cordilleran regions of the U.S. (Plafker and Berg, eds., 1994), and the Cascadia subduction zone is presumed to contain a distinctive composite belt - an accretionary wedge located mostly offshore (cp. Hyndman et al., 1990).

No consensus exists about the boundaries of the Intermontane Belt. In the west, it could be a system of the Yalakom, Fraser and Hozameen faults, but the Fraser fault's southern continuation (the Straight Creek fault in the U.S.) does not follow this belt boundary. Sometimes the Intermontane Belt's western boundary in Canada is placed at the Pasayten fault, which may or may not be connected with the Pinchi fault farther north. Different authors in Gabrielse and Yorath (eds., 1991) showed the fault linkages differently. Similarly uncertain is the eastern boundary of the Intermontane Belt. The Teslin fault seems to be connected with the Kutcho fault, but links with the Thibert fault are also possible; perhaps all three are parts of the same system related to the Teslin "suture". Whether the Pinchi or Isaac Lake faults are offset segments of this system is also unclear (Gabrielse, 1985). In the south, the Okanagan Valley fault is commonly regarded as the Intermontane Belt's eastern boundary. However, in many cases fault correlations are very uncertain due to limited accessibility, blanket of Quaternary deposits, or just great number of ways to link the many faults in the Cordillera.

Of the geophysical methods, short-wavelength magnetic anomalies have been found to be particularly useful for delineation of some faults in the Canadian Cordillera (Cook et al., 1995). From variously processed magnetic data, these authors have added new fault correlations. They supposed, for example, that quite prominent might be: the Yalakom-Fraser-Pasayten fault system, highlighted in magnetic shadowgrams; the NNW-SSE-trending Cherry Creek fault in the middle of the Intermontane Belt, running north of and parallel to seismic profile 11; some NE-SW-trending lineaments in the Rocky Mountains. They also noted that their magnetic data highlight presumed fault boundaries of the Nicola structure.

This picture is at odds with Gabrielse's (1985) opinion that the Fraser-Straight Creek fault system, which was active in the Eocene, "clearly offsets" the Yalakom and Pasayten faults. The steep Fraser and Pasayten faults in Canada are well expressed in the topography, and they accommodate west-side-down and strike-slip displacements. Eocene activity on the Yalakom fault was proposed to be responsible for the upturn of migmatitic gneisses of presumed mid-crustal origin (Friedman and Armstrong, 1991). Topographic and potential-field lineaments suggest the Yalakom fault is linked, across the Coast Belt orogen, with the Kitkatla fault that crosses the Insular Belt (Lyatsky, 1991, 1993). An ultramafic body at Lillooet, near the junction of the Yalakom and Fraser faults, was evidently squeezed up from the mantle along a deep fault zone (Wright et al., 1982). The Hozameen fault, which is subparallel to the Pasayten fault, is associated with high-grade metamorphic rocks and Alaskan-type ultramafic bodies of Permian to Middle Jurassic age, which also suggests this fault penetrates deep into the crust (Fig. 65).

The fault system on the east side of the Intermontane Belt is also very complex. The biggest is the Teslin fault and its apparent NNW continuation the Semenof fault. The Kutcho and Thibert faults are also aligned with and could belong to the same system. In the upper Fraser River area, a system of Isaac Lake, Matthew, McLeod Lake and Willow faults may form another segment of the Intermontane-Omineca belt boundary. Not all parts of this system were active simultaneously, and their geologic and topographic expressions vary. Yet, they are all probably rooted in the same zone of crust-mantle weakness which separates the Cordilleran miogeosynclinal zone from the coeval but strikingly different eugeosyncline (Kay, 1951; Monger, 1977).

These parallel fault pairs (Teslin-Semenof, Thibert-Kutcho, Matthew-McLeod Lake) commonly contain discontinuous, usually serpentinized, Alaskan-type ultramafic bodies, as well as the rare, less altered ultramafic rocks of Triassic age (Irvine, 1976; Nixon and Hammack, 1991; Woodsworth et al., 1991). Concentric zoning typical in Alaskan-type mafic-ultramafic plutons, with dunite or peridotite in the core and gabbro to diorite on the rims (see also Patton et al., 1994), also appears in the Triassic Polaris suite of mafic-ultramafic complexes in British Columbia. Due to magma fractionation and differentiation, Polaris mafic-ultramafic rocks are associated with many batholiths of intermediate composition in the Quesnel and Stikine terranes. Woodsworth et al. (1991) also pointed out a possible genetic relationship of these intrusions with Middle-Late Triassic volcanism. Such similarity in magmatism in the supposedly-dissimilar Quesnel and Stikine terranes suggests they have long belonged to the same tectonic system in the Intermontane median-massif belt.

In the south, the eastern boundary of the Intermontane Belt is also ambiguous. The Cenozoic extensional faults in that region cannot be a boundary between the much older Omineca and Intermontane tectonic zones. Conventionally, the boundary is placed at

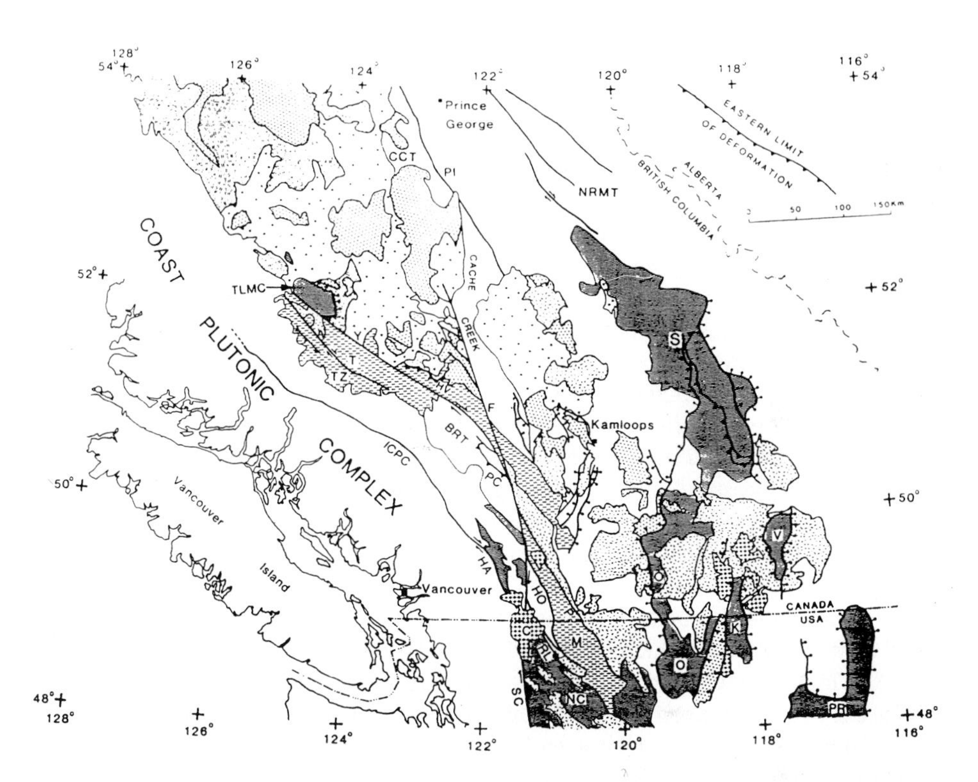

Figure 65. Main faults and metamorphic core complexes in the southern eugeosynclinal Coast Belt orogen and Intermontane median-massif belt, as identified by geologic mapping (simplified from Friedman and Armstrong, 1988). CCT - Cache Creek assemblage; F - Fraser fault; HA - Harrison Lake fault; HO - Hozameen fault; HV - Hungry Valley fault; ICPC - Intra-Coast Plutonic Complex fault; K - Kettle Complex; M - Methow Basin; NC - North Cascades; NRMT - Northern Rocky Mountain Trench fault system; O - Okanagan complex; P - Pasayten fault; PC - Phair Creek fault; PI - Pinchi fault; PR - Priest River complex; RL - Ross Lake fault; S - Shuswap complex; SC - Straight Creek fault; T - Tyaughton Basin; TLMC - Takla Lake metamorphic complex; TZ - Tchaikazan fault; V - Valhalla complex; Y - Yalakom fault.

the Okanagan Valley fault. Cook et al. (1995) noted two "obvious" straight gravity gradient zones associated with the Fraser fault in the west and the Okanagan Valley fault in the east. Jones et al. (1992) noted an abrupt change in crustal electrical resistivity at the Fraser fault, which they interpreted as a steep trans-crustal discontinuity (also Jones and Gough, 1995; contrary to Varsek et al., 1993). Low electric resistivity under the main part of the Omineca Belt is also in contrast to the crustal resistivity in the hinterland (Jones et al., 1992), but the exact location of the transition is unclear.

The Okanagan Valley fault has been recognized as one of the many shallow, gently dipping faults accommodating Early Tertiary extension that also produced the metamorphic core complexes (Parrish et al., 1988). On the west side of the N-S-trending Kettle-Manson metamorphic-core zone, the Okanagan Valley fault has been mapped to run N-S for ~300 km, with festoon surface-trace shapes common in thrust faults, from northern Washington state to east-central British Columbia (Parrish, 1995). These festoons are local, and they are arranged along a clear N-S trend. Complex structural patterns along the proposed deep-rooted Omineca-Intermontane belt boundary are common in the area south of the Teslin-Thibert-Louis Creek fault system. In places, steep faults offset the thrust sheets; elsewhere thrust sheets overlap the high-angle faults (e.g., complex relationships of the Pleasant Valley thrust with the Willow, Matthew and Isaac Lake faults have been mapped in the Cariboo Mountains; Struik et al., 1991, their Fig. 17.36). To the south, near the Shuswap and Adams lakes, several thrust sheets are thought to continue westward, creating a protrusion of the Omineca-Belt rocks towards the Intermontane median-massif belt. The mapped thrust splays seem to mask a deep, steep zone of crustal weakness which serves as the Intermontane-Omineca Belt boundary. As discussed in the previous chapters, joint examination of geological and geophysical evidence shows that the most fundamental crustal structural discontinuity in this area is not any of the thrust faults but the steep Kootenay Lake fracture zone.

The Intermontane median-massif belt is a whole-crust or whole-lithosphere entity containing a partly reworked ex-cratonic basement and a volcano-sedimentary cover. Its basement has for a long time been tectonically separated from that of the Interior Platform of the North American craton, and its volcano-sedimentary cover is sharply different from the formations in the adjacent orogenic zones in the Omineca and Coast belts.

Insights into geologic history of the Intermontane median massif(s) from analysis of its post-basement structural-formational étages

In our analysis, the three first-order tectonic zones in the Cordilleran interior - Omineca, Intermontane and Coast - evolved essentially *in situ* and in close connection with one another. Their common basement and mutual influences can be traced back to Precambrian time. Although tectonic manifestations in these zones differed, the timing of some tectonic stages was shared. Because only parts of these stages correspond to preserved tectonic étages in the rock record, because the extent of reworking differs from zone to zone, and because the collection and rock composition of étages even in the same tectonic zone varies from block to block, only large-scale analysis can unravel the tectonic evolution of the whole region. To concentrate exclusively on specific areas, types of structure, lithologies or geophysical models (as is often done in terrane-tectonic reconstructions) is not enough. Differences in étages from area to area are not reliable evidence for mutually exotic nature of adjacent belts and blocks.

The base of the Intermontane Belt's volcano-sedimentary cover is not exposed, and the oldest known rocks are Lower Devonian carbonates in the Iskut River area on the belt's western side. Several rock assemblages interpretable as structural-formational, tectonic étages can be recognized in the Lower Devonian through Tertiary stratigraphic section.

The Devonian and older étages are studied poorly due to lack of adequate exposure, and the tectonic regimes during that time are difficult to discern. Rocks of Mississippian to Permian age, recognized as the Cache Creek terrane, have been described mainly in two areas, north of the Stikine Arch and south of the Skeena Arch (Monger, 1977; Monger et al., 1991; Gabrielse et al., 1991a-b). The former is an allochthonous thrust sheet displaced from the Teslin orogenic zone. The latter could be an extension of that orogen into the interior of the Intermontane Belt, separating it into similar but distinct blocks commonly regarded as the Stikine and Quesnel terranes. The Quesnel terrane is thought to be an upper part of a more complex tectonic unit, which included the Slide Mountain thrust sheets (now displaced eastward into western Omineca Belt) as a lower part of a single stratigraphic succession. If so, Slide Mountain-type rock assemblages could lie beneath the exposed Mesozoic succession of the Quesnel terrane. Monger and Ross (1971) suggested that absence of Permian fusilinaceans similar to those on the craton to the east reflects the large distance from which the presumably exotic terranes arrived. This conclusion is weak, because biofacies assemblages in mobile megabelts commonly change sharply along and across the tectonic grain. Besides, a major role in the global dispersal of marine life forms is played by sea currents (e.g., Newton, 1988).

Permian clastic sequences containing pebble and boulder conglomerates have been studied in the Harper Ranch assemblage of the Quesnel terrane. Fragments of granodiorite and andesite plutonic rocks and quartz-mica schistose metamorphic rocks suggest an uplifted proximal provenance area (probably, Omineca Belt or Kootenay Arc). Thrusting of the Slide Mountain assemblage over the miogeosynclinal successions is evidence of late Paleozoic tectonic activity at the boundary zone between the miogeosynclinal and eugeosynclinal parts of the Cordillera. Later rejuvenation of this zone was accompanied by another, more-local thrusting of Quesnellian rocks over the Slide Mountain thrust sheets by Late Triassic time. North-northwest from Vernon and west of Adams Lake, an almost straight line separates Slide Mountain sheets from those of probable Quesnellian origin. This line represents part of the long fault system separating the Intermontane and Omineca belts.

In the Omineca Belt, the late Paleozoic Antler orogeny was manifested by differential movements of zones and blocks, taking advantage of ancient crustal weakness zones as old as Early Proterozoic which continue from the North American craton. Late Paleozoic magmatism and metamorphism, accumulation of coarse conglomerates (Earn Formation), appearance of the Carboniferous Prophet and Permian Ishbel foredeeps - all these events were expressions of orogenic activity in the miogeosynclinal region. Signs of Antler orogenic activity exist also in lower parts of Cordilleran-hinterland median massifs (e.g., Yukon-Tanana). The Prophet/Ishbel foredeep received sediments from both sides, east and west, including the Omineca highlands and Intermontane upland. Thus, both western miogeosyncline and eastern part of the median massif were affected by orogenic uplift. Rejuvenation of the Omineca-Intermontane boundary zone is marked by Carboniferous volcanics exposed on the western side of the foredeep, along the modern Selkirk, Cassiar, Cariboo and Omineca mountains.

The end of the Antler orogeny was followed by a period of relative tectonic quiescence, accompanied by rapid erosion and peneplanation. Triassic formations, of the next étage, overlie the Paleozoic rock assemblages unconformably in the area east of the Okanagan Lake. This unconformity is angular, separating dissimilar tectonic étages. Distribution of Triassic rocks above the beveled Paleozoic rocks indicates a westward tilt. The basal Triassic volcanic and clastic unit is rather thin, but the overlying volcanic succession (particularly the Nicola Group) have a wedge-like general form (Souther, 1991). The distribution of lithofacies also indicates subsidence of the eastern part of the Intermontane Belt. Souther (1991) speculated that Late Triassic volcanic and plutonic rocks developed above an east-directed subduction zone, but in the absence of corroborating evidence this idea is no more than conjectural.

A broad spectrum of Late Triassic plutonic rock types in the Intermontane Belt, from andesites to Alaskan-type ultramafics, could have resulted from cooling of magmas under rather quiescent tectonic conditions. Plutonic rocks are distributed along faults, but those faults are extensional (not compressional or strike-slip; see also Woodsworth et al., 1991). On the western and northern flanks of the Jurassic-Cretaceous Bowser Basin, granitic batholiths are predominantly I-type. These rocks are not necessarily related to oceanic crust, as teleorogenic magmatism is not unexpected in a median-massif area close to a major orogenic zone such as the Coast Belt.

Plutonic facies of the Late Triassic magmatism are well correlated with its sub-volcanic and volcanic facies. A belt of volcanoes has been traced to run NNE-SSW, oblique to the overall NNW-SSE elongation of the Intermontane Belt. Unlike the Late Triassic plutons, whose tectonic affinity is debated, these coeval volcanic rocks are more clearly related to continental lithosphere. Souther (1991) argued that the Late Triassic Nicola Group and its stratigraphic equivalents are of oceanic island-arc type, but evidence such as abundant volcaniclastic rocks (sandstones, breccias), subaerial lava flows (in places with granitic cobbles), calc-alkaline composition (including K-rich shoshonite) indicates a continental setting. The Nicola Group's volcanogenic formations have a cumulative thickness of >3,500 m, with extensive fields of subaerial volcanic and volcaniclastic rocks. Besides the main volcanic belt, there are two other, discontinuous belts related to the Intermontane-Coast and Intermontane-Omineca belt boundaries. The thickest volcanic successions and the most abundant comagmatic granitic batholiths are mapped along the NE-SW-trending Stikine Arch.

Tectonically, this was perhaps the time of considerable E-W crustal extension across the Intermontane median-massif belt. The distribution of plutons and volcanic centers followed a network of N-S-trending faults (Preto, 1977). Locally, folding occurred along major faults, and N-S structures re-deformed the Paleozoic formations (Gabrielse et al., 1991b). Steep foliation is common in the Late Triassic tectonic étage. At major faults, such as those in the Intermontane Belt's boundary zones, penetrative foliation is accompanied by the high (by this belt's standards), greenschist grades of metamorphism (e.g., in the Mt. Lytton complex in the west and in the Louis Creek fault zone in the east; Gabrielse et al., 1991b; Greenwood et al., 1991).

The subduction-related tectonism, assumed previously for the Nicola-assemblage time, is now undergoing revision. Island-arc volcanism is hard to tell apart from volcanism in a continental arc. Ore compositions, sensitive to tectonic regime, in this assemblage point to links with continental crust and do not require subduction (cp. Titley, 1987; Sillitoe, 1987). Though I-type granites may be derived from oceanic lithosphere as well as continental mantle, their geochemical signatures and Mo, Sn and W concentrations point to continental lithosphere as a source (McMillan, 1991a). Au is common in both the Intermontane and Omineca belts, but not in the Coast Belt (Panteleyev, 1991; Ray and Webster, 1991).

By the end of the Late Triassic-earliest Jurassic tectonic stage, the entire Canadian Cordillera east of the Coast Belt was probably a lowland, too low to shed clastic material to the few areas of shallow-shelf sedimentation in the Yukon and in the incipient Bowser Basin. Only a slight unconformity separates the widespread Lower Jurassic rock sequences from the Upper Triassic ones. The latest Triassic Norian and the earliest Jurassic Hettangian were the time of tectonic quiescence in most of the region.

The Lower Jurassic rocks are conventionally assigned largely to the Hazelton Group. It is overlain by the Middle to Late Jurassic Bowser Lake Group (Tipper and Richards, 1976; Evenchick, 1991, 1992). Widespread rocks of that age record shallow marine environments in a broad region from the western Omineca Belt to the Queen Charlotte Islands in the Insular Belt (Tipper and Richards, 1976; Haggart, 1993). Several local provenance areas marked the uplifted blocks. The Hazelton rock assemblage is typically volcanogenic, with intercalated sedimentary units. This is broadly reminiscent of the paleoenvironments and lithostratigraphic patterns of the Triassic tectonic étage, but throughout the Jurassic differential block movements were more apparent. In the Intermontane Belt, similarities between the Nicola and Hazelton assemblages include the presence of rocks resulting from granitic plutonism and comagmatic volcanism. In many places, the igneous rocks of these étages are lithologically hardly distinguishable, and even their radiometric dates are commonly uncertain. Some tectonic features of the Jurassic étage, such as the NE-SW-trending Skeena and Stikine arches and the Bowser Basin between them, may be partly inherited from the preceding tectonic stages (cp. Read, 1983); the causative transverse crustal structures date back to Early Proterozoic time.

Souther (1991) noted a progressive eastward shift of the magmatic front in the Quesnel terrane, i.e. in the southern Intermontane Belt, from the Early Triassic to Middle Jurassic. This progression has not actually been proved, and exceptions from this trend exist. After a hiatus near the Triassic-Jurassic boundary, Early Jurassic magmatism began in the Rossland area near the western fault boundary of the Kootenay Arc. These Sinemurian to Pliensbachian mafic volcanics, basaltic flows and pyroclastic debris have shoshonitic characteristics. The geochemical spectrum from low-K tholeiites to alkaline derivatives may be a product of some sort of island arc, but its crustal background was continental.

The Sinemurian to Bajocian Hazelton Group and its rough stratigraphic equivalents, present over the middle and northern parts of the Intermontane Belt, are volcano-sedimentary, with highly variable lithofacies. The recorded thickness of this group is estimated to be large, ~3,400 m. Subaerial lava flows up to 700 m thick are found in the continuous, NNW-SSE-oriented Hazelton Trough occupying much of the southern and central Intermontane Belt south of the Stikine Arch. These volcanic rocks range from basaltic derivatives to andesites and dacite-rhyolites. Associated with them are comagmatic granitic sub-volcanic bodies and plutons (Souther, 1991, his Fig. 14.6).

Coarse sedimentary rocks of the Hazelton Group contain fragments of older formations, including granitic plutons of Triassic age. Uplifts that supplied the debris were located in parts of the Coast Belt, western Omineca Belt, and the Skeena and Stikine arches. The Early Proterozoic NE-SW-trending faults, continuing from crustal weakness zones on the craton, were reactivated, segmenting the Intermontane median-massif belt into southern, central and northern parts separated by transcurrent arches. Early Middle Jurassic uplifts along the Skeena Arch cut the old Hazelton Trough, and since that time predominantly sedimentary deposits filled the subsiding area between the two rising arches. North of the Stikine Arch, the elongated NNW-SSE Whitehorse Trough west of the Intermontane-Omineca belt boundary may have been a huge half-graben, whose fill-thickness distribution suggests the northern part of the belt was tilted to the east.

Between the Stikine and Skeena arches, the Bowser Basin is rather isometric in map view, but its structure is complex (Tipper and Richards, 1976; Gabrielse and Yorath, eds., 1991). Whereas Triassic magmatism had been strong in the Stikine Arch, the Skeena Arch was it locus in the Jurassic. Both these zones were inverted upward in the Jurassic, shedding abundant clasts to the north and south; the Bowser Basin between them was mostly filled with finer clastic material. An apparent wedging-out of this basin to the north is marked by conglomeratic units along the Stikine Arch. This suggests the northern part of the basin was shallower than it southern part, which was downdropped against the Skeena Arch. Some conglomerate and sandstone lithofacies rim the basin on the east, where uplift of blocks in the Omineca Belt provided metamorphic-rock fragments. Marine intrusions in the Bowser Basin continued till the Late Jurassic.

South of the Skeena Arch, sedimentation took place in part of the Nechako Basin and (since the Sinemurian) the Tyaughton Basin. In the north, the Whitehorse Trough Basin developed by subsidence of an elongated NNW-SSE-trending zone. It lay on the western side of the southern Intermontane Belt. Since Middle Jurassic Aalenian-Bajocian time, depocenters in the Intermontane Belt seem to have lain mostly along its boundary with the Coast Belt. Rocks accumulating along the Intermontane-Omineca belt boundary were largely volcanic, with intercalated sedimentary units (at least, in a segment between the Stikine Arch and Kamloops Lake). The small Sustut Basin developed east of the Bowser basin, along the Intermontane-Omineca belt boundary.

These arches, basins, and plutonic and volcanic features are conventionally interpreted as "components of one complex island-arc terrane" (Monger et al., 1991, p. 305). This explanation is speculative, and unrealistically simplistic. Two large sedimentary basins in the central zone of the Intermontane median-massif belt, Bowser and Nechako, developed in the Jurassic on subsiding blocks. Areas on these blocks' rims were relatively uplifted, especially around the Bowser Basin; these uplifted areas were characterized by higher magmatism. Souther (1991) proposed these volcanic areas contain basaltic seamounts, but Jurassic volcanism there was bimodal, basaltic and rhyolitic, with predominantly andesitic composition (Ash et al., 1996). Jurassic granitic plutonism is represented by numerous bodies, including batholithic ones, scattered over much of the Intermontane, Omineca and Coast belts. Faults controlled the distribution of these plutons along the Intermontane-Omineca and Intermontane-Coast belt boundaries, along the Skeena and Stikine arches, and on the rim of the Nechako Basin. Some batholiths are slightly older, Late Triassic-Early Jurassic, and thus could have been emplaced during the time of comparative tectonic quiescence, when other tectonic manifestations were subdued. Plutons of the Early Jurassic suite are typically polyphase, multicomponent bodies derived from mid-crustal granodioritic magma chambers. Many alkaline plutons contain porphyry-type economic Cu, Ag and Mo mineralization. To the north, the plutons are more felsic and smaller, and their largest concentration is along the Skeena Arch.

Throughout the Jurassic, plutonic magmatism continued at a decreasing rate, with variable compositions and styles and with overlap between pulses in space and time (Woodsworth et al., 1991). This scenario is at odds with the ideas about accretion of far-traveled terranes. Irving et al. (1980) and Tipper (1984) argued, from paleomagnetic and paleontological studies, for large translations of many crustal fragments, in chaotic order, along the continental margin from the south. Monger et al. (1982) preferred a model where the fragments amalgamated into larger Intermontane and Insular superterranes before docking to North America; the Intermontane superterrane presumably docked in the Middle Jurassic (but before the deposition of the Bowser Lake Group). Collision of this superterrane with the continent supposedly resulted in the uplift of the Omineca Belt. In yet another model, Wernicke and Klepacki (1988) presumed the Stikine "block" traveled a much shorter distance, from the Oregon-Washington margin. Van der Heyden (1992) favored an earlier time of amalgamation and collision - before the Middle Jurassic - and thought the Intermontane and Insular superterranes had been joined before they were accreted to North America. At present, the idea of large displacement is in question for many blocks in the Intermontane Belt (Monger and Price, 1996) and elsewhere in the Cordillera (Dickinson and Butler, 1998; Mahoney et al., 1999).

The presumed distances of travel of the Intermontane superterrane and its constituent terranes - Quesnel, Stikine, Cache Creek, Slide Mountain - have decreased in the last decades from thousands of kilometers to hundreds, in the longitudinal and latitudinal directions alike. Harms (1986) proposed that shortening of the Slide Mountain terrane (in the Sylvester allochthon) involved a large part of the early Mesozoic Pacific oceanic crust. Struik (1987), however, was more cautious: in his palinspastic restoration of a Triassic cross-section across the eastern Intermontane Belt and western Omineca Belt, shortening was still large, ~200%, but only slightly exceeding 500 km in absolute terms. Width of the original Slide Mountain basin was thought to be 184 km, of the Triassic Quesnel island arc as much as 120 km.

The notions of Cache Creek, Sylvester, Teslin, Anvil and other presumed intra-Cordilleran oceans have never been clearly formalized or correlated. But nobody calls ocean the modern Sea of Japan, which separates the Japanese island arc from the Asian mainland. Neither are the seas of the Mediterranean basin called oceans. If the Cache Creek and Sylvester basins were marginal seas, they could partly be floored by oceanic crust and partly by continental crust (by analogy, for example, with the seas of Japan and Okhotsk). To regard an island-arc paleogeographic setting as evidence of oceanic tectonic setting is commonly misleading.

Miller (1987, 1989) cast doubt on the idea of genetic incompatibility of paleogeographic islands at North America's western continental margin with the continent itself. On the contrary, results of paleontological analysis are consistent with genetic links. Newton (1988) criticized the use of paleontological evidence to establish exoticism. McMillan (1991a, p. 13), who examined Cordilleran mineral-deposit settings, noted that Cu-Zn massive sulfides hosted in marine sediments (Besshi-type deposits), polymetallic sulfides which accumulated in volcanogenic sequences (Kuroko-type deposits) and massive cupriferous pyrite ores (Cyprus-type deposits) could have formed in a variety of tectonic settings, "with or without generation of oceanic crust".

Modern, and probably ancient, island chains, arcuate or otherwise, fall into two main categories: those formed on oceanic crust, and those whose crustal background is continental. The first group is exemplified by some archipelagoes in the southern Pacific;

the second by many islands in the northwestern Pacific Ocean and the Mediterranean Sea. Typical oceanic volcanism, such as that in Hawaii, is also recognized in many other south Pacific island systems. Transitional cases also exist, where both oceanic-crust and continental-crust characteristics are present (e.g., the Kurile-Japan or Antilles arcs). Island chains can arise on broad continental shelves (e.g., Novaya Zemlya in the Russian Arctic). On continental-type or transitional arcuate island chains, volcanics commonly have non-oceanic affinities (e.g., Permian volcanics on the northern Novaya Zemlya). Island arcs built on extensive, if modified, continental-crust pedestals may have mixed characteristics that could confuse the definition of tectonic setting.

Thus, assignments of rock assemblages to island-arc, forearc or backarc settings are inherently ambiguous, because these notions are poorly formalized and it is often unclear what kind of arc is suggested. The inboard extent of backarc tectonic domains has so far escaped definition. Hutchinson (1980) noted that Besshi-type Cu-Zn deposits are found in forearcs of both oceanic-type and intermediate-type paleogeographic island arcs, and that Kuroko-type polymetallic ores can form in arcs of both these types. An extensional tectonic regime in any sort of backarc basin may be conducive to the development of Besshi-type deposits hosted in marine sediments. Cyprus-type deposits can be attributed to either sea-floor ridges or arc settings (McMillan, 1991a).

Most Cordilleran Triassic and Jurassic plutono-volcanic suites are generally thought to have formed from granitic magmas. Jurassic bimodal magmatism produced a broad spectrum of rock compositions from felsic (granite, dacite, rhyolite) to mafic (gabbro, basalt), consistent with the expected igneous products in mediterranean-sea island archipelagoes preserved in mobile-megabelt interiors (Trümpy, 1960; Aubouin, 1965). These features have been described variously as geosynclinal or eugeosynclinal archipelagoes, geanticlines, or island arcs. At one time they were regarded as results of structural inversion, later as consequences of subduction. Also noted was their coincidence with the central zones in the Alps, the Appalachians and the Cordillera (e.g., Kent et al., eds., 1969). Median massifs in mobile-megabelt hinterlands often served as pedestals for such positive features, both tectonically and geomorphologically. Tectonic tilting or upwarping of a median massif may create island chains along the upthrown edge or axis; extensive volcanism would contribute to the construction of such chains.

Not all island chains are arcs, as their shapes are not arcuate. Geological data usually do not permit to determine the shape of an ancient chain whose remnants are incorporated into a mobile megabelt. Where such reconstructions are made, the inferred chains tend to be not arcuate but straight. Even modern island chains of arcuate shape are fairly rare, whereas straight and rectilinear patterns seem common. Straight chains often form along major fault systems dissecting continental crust or separating continental crust from oceanic; island archipelagoes formed along them lack a direct genetic relationship to plate motions and interactions or to terrane accretion.

The Late Triassic and Jurassic island chains developed in the Cordilleran megabelt's central zone over median massifs. They reveal no direct structural or magmatic links to oceanic-lithosphere plates. The abundant volcanic and plutonic bodies in the Intermontane Belt are obviously associated with faults, which were created or rejuvenated easily when the tectonic regime sporadically changed. The resulting distribution of plutonic suites, volcanic centers, island chains and mineral concentrations was determined by linear patterns of deep-seated extensional faults. In McMillan's (1991a, p. 6) description, individual terranes in the Intermontane Belt, as elsewhere, "can be produced by plate tectonic movements *or by fault offsets*" (italics here and below ours); "terrane-boundary faults... *tapped deeply into the crust*"; and "*deep fractures* with related zones of high heat flow, *mantle hot spots*, or other thermal events" (op. cit, p. 21) created favorable conditions for volcanism, plutonism and formation of mineral deposits.

Granitic magmatism in the Intermontane median-massif belt lasted through Jurassic time. It was indigenous rather than teleorogenic, as no nearby orogeny is recorded at least for the Early Jurassic Hazelton time interval. Genesis of granitic magmas largely involves partial melting and metamorphism at lower-crustal levels. Anatexis, palingenesis, migmatization, supply of water and CO_2 from the mantle - all these processes contribute to the production of granitic magmas. Dehydration, indicated by granulite-grade metamorphism which is typical in the lower crust (Rudnick, 1992, 1995), may be offset partly by wet, S-type, peraluminous or K-rich granitic magmas. Biotite or hornblende granites produced from mafic or intermediate as well as oceanic-lithosphere source rocks fall in the I-type category. S-type granitic magmas tend to be moderately wet and rather cool, with limited mobility; before solidifying, they commonly migrate to depths of 12-16 km in the continental crust. The wetter and hotter I-type granitic magmas are more fluid and able to reach shallower crustal levels before they freeze (McMillan, 1991a).

Buoyancy causes the light granitic material to rise, commonly along conduits provided by deep faults. The rising granitic magmas are contaminated by material absorbed into it from the surrounding rocks, and deep-sourced magma may thus acquire upper-crustal contaminants. This complicates the genetic interpretation of geochemical and petrological data. Due to these ambiguities, many alkaline magmatic provinces have been ascribed variously to backarcs in active island-arc settings, passive continental margins, or cratonic interiors. To interpret the Canadian Cordilleran eugeosynclinal belts as accreted-terrane regions on a shaky basis such as this seems adventurous. There are also many unexplained local peculiarities: for example, British Columbia's alkalic-porphyry-deposit province is unusual in the Pacific Rim, and Fox (1989) whimsically called it one of "schizophrenic cousins of real porphyry coppers". McMillan (1991a) noted that, on a global scale, British Columbia Cu-Au porphyry deposits have properties "uncommon" for oceanic island-arc settings.

Calc-alkaline and alkaline plutono-volcanic belts in the Late Triassic Nicola assemblage have been described by different investigators in terms of different tectonic settings (convergent or divergent plate margins, island-arc or continental-margin rifts; e.g., Sawkins, 1990; Höy, 1991). Common for the Jurassic plutono-volcanic episodes in the Intermontane Belt is a spatial differentiation of (overlapping) calc-alkaline and alkaline zones. Also common for the Jurassic magmatism is a wide distribution of porphyry as well as mesothermal and epithermal vein mineral deposits; such deposits elsewhere in the world are indigenous. Favorable for their formation is the longevity and relative stability of median massifs, which permitted fractionation of magmas at depths.

In the Intermontane Belt, indigenous, non-teleorogenic, genesis of many economic deposits of Cu, Au, Ag and other metals is inferred when the corresponding magmatism and mineralization are not well represented in the surrounding orogenic zones. The Jurassic was probably the main period of development of precious-metal epithermal ores in the Intermontane Belt (Panteleyev, 1991). By contrast, Au and Cu skarn deposits are less common, as their genesis was associated with teleorogenic plutons. Sn and W are extracted in large amount from skarns in many parts of the Pacific Rim, but in the Intermontane Belt they are less abundant. In Meinert's (1989) metallogenic scheme, the Sn, Cu, Zn, Pb, Mo and W occurrences in the Cordillera evolved in regions with mostly continental crust. Not terrane tectonics, but presence (or absence) of silicic continental crust strongly influenced the chemical composition of ores. Base-metal (Cu, Zn, Pb, Mo) occurrences worldwide are commonly located in continental-crust regions; in the Intermontane Belt, deposits of Mo, Sn and W also indicate a continental-crust background (e.g., McMillan, 1991a; Ray and Webster, 1991).

Typically in the Intermontane Belt, as in median massifs in other mobile megabelts, porphyry systems are related to plutons thought to have been at depths of just 2 to 4 km (Sutherland Brown, 1976). Fluid I-type magmas could have reached these shallow depths before solidifying. Predominance of I-type magmas in the Jurassic structural-formational étage (Woodsworth et al., 1991) and known shallow depths of pluton emplacement give a measure of the amount of erosion since ~190 Ma. Derived from the continental mantle or subducted oceanic lithosphere, I-type granites are not definitive indicators of tectonic setting. However, their emplacement in areas with low initial $^{87}Sr/^{86}Sr$, not exceeding 0.704 (Armstrong, 1988), suggests the semi-reworked basement of the Intermontane median-massif belt was heterogenous. But the geologic record of the Canadian Cordilleran interior contains no proof of exotic-terrane accretion.

Compared to the Late Triassic étage, Jurassic volcanism was more localized, and Jurassic granites generally lack foliation (Gabrielse, 1996). Block movements accounting for the rise of the Skeena and Stikine arches and the subsidence of the Bowser and Nechako basins gave the Jurassic tectonic stage and étage its distinctive look. A predominantly extensional tectonic regime facilitated the Early to Middle Jurassic plutono-volcanic magmatism. Magmatism also continued later, into the Early Cretaceous, but its manifestations changed: by the middle of the Jurassic, magmatism decreased considerably, while block-fault movements continued to account for local and regional uplift and subsidence.

The base of the Bowser Lake Group is a regional marker in the Early Jurassic to Early Cretaceous structural-formational étage. The 3,500-m-thick marine and non-marine sedimentary-rock assemblage, predominantly coarse-grained, with only minor volcanic layers, accumulated in subsided areas between the Skeena, Stikine and Omineca uplifts. These raised areas provided clastic material for the Bowser Basin (Eisbacher, 1974; Evenchick, 1991, 1992; Bassett and Kleinspehn, 1997). South of the Skeena Arch, the Nechako Basin developed in similar paleoenvironmental conditions (Fig. 60; Tipper and Richards, 1976). The sea transgressed from the west and covered a broad shelf, where islands were formed by intermittently uplifted blocks. Their erosion, combined with sea-level fluctuations and marine abrasion of the sea floor, created many local disconformities and unconformities; in some areas Early Cretaceous formations directly overlie the Late Triassic volcanics. Exhumation of some Early Jurassic granitic plutons, whose fragments are found in younger Jurassic conglomerates, requires uplift of ~2 km, especially near faults on the eastern and western boundaries of the Intermontane Belt. Since the Late Jurassic-Early Cretaceous, the elongated and partly disrupted Sustut Basin accentuated the Intermontane-Omineca boundary zone. Thick deltaic assemblages in the Intermontane Belt's basins indicate differential uplifts of the surrounding blocks. These blocks movements embraced larger areas than had the similar movements of the Late Triassic-earliest Jurassic tectonic stage: they included parts of the Omineca Belt and were more persistent. Since late Oxfordian time (Poulton, 1989, 1991), metamorphic clasts from the Omineca orogenic zone were shed not just into the Bowser Basin to the west but also into the Fernie Basin in the southern Rocky Mountain Belt to the east (Fig. 66).

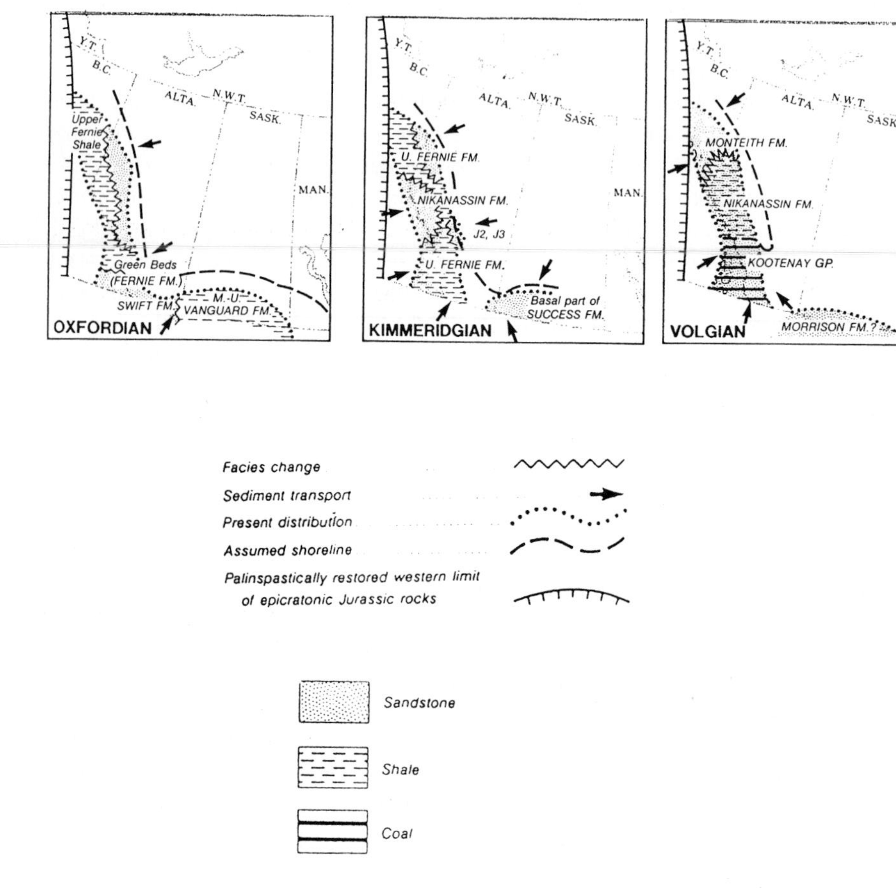

Figure 66. Distribution of Jurassic lithofacies in the eastern Cordillera and Western Canada Sedimentary Province, from geologic mapping and drillhole data (modified from Poulton, 1989). The NNW-SSE-trending elongated depositional area near the Alberta-British Columbia border represents the Fernie Basin.

Appearance of the Fernie Basin marks the creation of an embryonic foredeep in front of the rising Omineca orogenic zone. Mountain ranges grew intermittently, shifting from the Intermontane-Omineca boundary eastward. Jurassic stratigraphic sections (beginning with the Bathonian-Callovian so-called Grey beds) in southeastern British Columbia record the initiation of this mountain-building episode (Poulton, 1991, his Figs. 8.58, 8.60). The Jurassic succession is depositionally thin (up to ~400 m) in southeastern British Columbia, and even thinner in southwestern Alberta. In the thickest, western sections, it is stratigraphically rather continuous and exhibits an upward increase in sandy-material content over shale (mostly since Late Jurassic Kimmeridgian-early Tithonian time). The sandy upper part contains many seams of coal throughout the section, without significant intraformational disconformities or facies changes. This indicates the southern Canadian Omineca Belt, unlike that belt's central part facing the Bowser Basin to the west, underwent a much smaller uplift. Variations in lithofacies and thicknesses (from hundreds to tens of meters) in basin-edge areas lying next to the miogeosynclinal Omineca tectonic zone attest to the activity of major "Cordilleran" and transcurrent faults reaching upsection from the Precambrian basement. Some of these faults trend NE-SW and have been traced from the cratonic Canadian Shield and the interior Platform (Lyatsky et al., 1999).

The active fault networks of Jurassic time were complex, partly re-using and partly superimposed on the older networks of faults. Together, they defined the structural and tectonic physiognomy of the Jurassic-Early Cretaceous tectonic étage. The Hazelton time interval, for instance, marked the tectonic movements and associated tectonomagmatic activity along the prominent Pinchi fault inherited from the Teslin orogen, and many block-bounding faults are found in the transcurrent Skeena and Stikine whose origins predate the Cordilleran mobile megabelt. Syn-sedimentary faults have been identified in the Bowser and Nechako basins, which have been called "successor basins" (Eisbacher, 1974). The elongated Sustut Basin was apparently related to rejuvenation of faults (Cassiar, Hanson, Saunders and others) in the Intermontane-Omineca belt boundary zone. The NNW-SSE and NE-SW faults controlled magmatism and basin formation during this tectonic stage, whereas many of the controlling faults in the Late Triassic had been N-S.

Magmatism was apparently uncommon in the Intermontane Belt from ~155 to 130 Ma (Late Jurassic Kimmeridgian to Early Cretaceous Valanginian), but some granitic plutons and Mo accumulations did form along NNW-SSE- and N-S-trending faults (Woodsworth et al., 1991). Relative (but incomplete) tectono-magmatic quiescence in the Intermontane Belt lasted till the Late Cretaceous Coniacian. Souther (1991), linking the volcanism in the Canadian Cordillera to Pacific plate-reconstruction models, reported island-arc volcanic events in the Tyaughton-Methow Trough along the Intermontane-Coast belt boundary in the south. This volcanism was presumably related to accretion of the exotic Insular superterrane to the North American continent in the mid-Cretaceous. However, plate-interaction models proposed by various authors differ (Engebretson et al., 1985; Stock and Molnar, 1988; Atwater, 1989; Babcock et al., 1992, 1994), and the nature of mid-Cretaceous magmatism can also be interpreted variously. In the Intermontane Belt, these igneous manifestations were rather limited, although a broad but discontinuous magmatic belt did develop across the northern Intermontane and Omineca tectonic zones. This transcurrent belt is similar in style and rock types to another belt, which crosses the Yukon-Tanana massif (Foster et al., 1994).

In central parts of the Intermontane Belt, away from large fault zones, stratal rocks of the Jurassic-Early Cretaceous tectonic étage are moderately folded. Open concentric folds are common. Mode of folding and development of cleavage depend on the position and shape of individual blocks and their boundaries. Folding in this tectonic étage might have evolved diachronously (e.g., Gabrielse et al., 1991b), but Monger et al. (1982) nonetheless ascribed these mid-Jurassic folds to a supposed accretion of the Intermontane superterrane to the Omineca continental margin. Subordination of folding to the pattern of fault-bounded blocks is generally common for median massifs. Vertical movements and tilts of blocks are recognized by their structural and sedimentary manifestations in the volcano-sedimentary cover, where relatively intense folds and faults mark block boundaries.

In the Jurassic-Early Cretaceous étage in the Intermontane Belt, metamorphic grades vary from zeolite to subgreenschist. Aside from contact metamorphism in aureoles around large granitic plutons and dynamic metamorphism at major fault zones, these rocks have not experienced significant regional metamorphism that would result from deep burial. The prehnite-pumpellyite metamorphic facies seems to be the norm, indicating a subsidence of just 2.5 to 5 km (Greenwood et al., 1991). This amount of subsequent erosion matched the considerations of the depth of pluton emplacement.

Gough (1986) had proposed that convective upflow in the upper mantle may be occurring beneath the Cordilleran Basin and Range province in the U.S. and the Intermontane and Omineca belts in Canada. From a regional electromagnetic study of the Canadian Cordillera, Jones et al. (1992) proposed a partially molten crust under the Omineca Belt even at a rather shallow depth of slightly more than 20 km. If local convection cells indeed exist in the upper mantle, junctions of their downgoing limbs could account for localized subsidence or downwarped furrows in the crust, in cratons and mobile megabelts alike. While plate-boundary processes are restricted to zones where lithospheric plates rub shoulders, mantle dynamics takes place in intraplate regions as well. At any rate, plutono-metamorphic rocks of the Omineca and Coast orogenic belts are not necessarily related to former or present subduction. Large vertical movements of intracontinental crustal blocks, differing in their timing, amplitude, surface expression and resulting structural peculiarities are more easily explained by intraplate tectonic activity (Krohe, 1996; Lyatsky, 1996).

8 - LESSONS FROM CORDILLERAN GEOLOGY FOR THE METHODOLOGY OF REGIONAL TECTONIC ANALYSIS OF MOBILE MEGABELTS

Hypotheses, models and rock reality

The only reliable information about the geologic past is obtainable from observations of rocks and rock bodies and their interrelationships. For this reason, all geologic (including tectonic) hypotheses have traditionally rested on rock-based observed facts. This logic has been distorted in the last several decades, as hypotheses about the geology of continental regions have relied on interpretations of linear magnetic anomalies in oceanic regions far away. The idea that these stripes simply record the history of sea-floor spreading and hence plate movements (Vine and Matthews, 1963; Heirtzler et al., 1968) rests on three assumptions: (a) that sources of these magnetic anomalies lie in the basaltic layer of the oceanic upper crust; (b) that these anomalies and anomaly sources have not been altered since their formation; and (c) that oceanic-lithosphere plates are rigid and internally undeformed. Another common assumption in tectonics is that continents are essentially inert, responding passively to influences from plate boundaries.

Modern studies show that lithospheric plates are not rigid, that oceanic-floor basalts and magnetic sources in them are commonly altered, and anyway these basalts do not account for all the magnetic anomalies in oceanic regions (e.g., Ballu et al., 1998; Lawrence et al., 1998; Tivey and Tucholke, 1998). As the supposed tectonic silver bullets and golden keys lose their luster under detailed scrutiny, inevitably we must return to the old practice of gathering and examining rock samples, mapping rock bodies, and interpreting this information based on the fundamental principles of geology. Crucially, it must be remembered that continents are not inert, that they have indigenous heat sources within their own lithosphere, and that they are capable of self-development independent from external (plate-boundary and deep-mantle) influences.

The Permian faunal assemblages in the Canadian Cordillera, previously assigned to a then-exotic Tethyan paleontologic province (Monger and Ross, 1971), are no longer considered exotic to this region or regarded as evidence for accreted terranes (Nelson and Nelson, 1985; Newton, 1988). Reliability of paleomagnetic speculations (e.g., Irving and Monger, 1987) is also questioned (e.g., Butler et al., 1989; Monger and Price, 1996; Dickinson and Butler, 1998; Mahoney et al., 1999). Uncertainties and errors in plate-motion reconstructions are acknowledged and examined (e.g., Stock and Molnar,

1988), and the assumption of pervasive crustal-scale low-angle thrusting is no more than a prejudice (Lyatsky, 1993, 1996).

No modern subduction regime exists at the Pacific continental margin of Canada (Lyatsky, 1996). No hypothesized ancestral North American continental margin or corresponding crustal ramp in the eastern Cordilleran miogeosyncline (Cook et al., 1991a) is actually apparent in the seismic profiles or from geological information. The Cordilleran mobile megabelt began as an Early Proterozoic rift in the pre-existing continental crust. This deep-rooted zone controlled the position of two-sided basins in the Proterozoic (Struik, 1987) and Paleozoic (Henderson et al., 1993; Richards et al., 1993). During the Phanerozoic, the Cordilleran miogeosyncline in Canada underwent recurrent orogenic tectonism: in the middle to late Paleozoic, Middle Jurassic, mid-Cretaceous and in the Late Cretaceous to Early Tertiary. Each episode of large-scale crustal reorganization was marked by deep burial and unroofing, metamorphism, magmatism, deformation, crustal compression and extension, and large vertical block movements.

Crustal movements along steep concurrent and transcurrent faults produced the block patterns which are mappable geologically at the surface. Extrapolation of these patterns deep into the crust is less well constrained, because steep crustal boundaries have mostly been overlooked in the interpretations of deep seismic data. Vaguely, Clowes et al. (1995, p. 1509) noted that the principal seismically defined zones they interpreted in the southern Canadian Cordillera do not consistently match the geologically mapped belts, although "in broad terms" this zoning is "consistent with the evolutionary history of the region".

That the Canadian Cordillera consists of crustal blocks has been generally accepted regardless of whether these blocks are treated as accreted terranes or not (Douglas, ed., 1970; Price and Douglas, eds., 1972; Gabrielse and Yorath, eds., 1991). The Intermontane Belt has always been treated as relatively stable, with subdued vertical movements and other tectonic manifestations. The Omineca and Coast belts were much more orogenized, and had much greater vertical tectonic movements. Rapid vertical movements up to 30 km in amplitude require large, deep faults at the boundaries of these blocks, to accommodate the episodic subsidence and uplift.

Clowes et al. (1995) modeled in the Canadian Cordilleran crust subhorizontal seismic domains in which the velocity gradually increases downward until encountering a sharp break. From the modeled configuration of these domains, these authors obtained a picture of layered crust whose layering is only slightly irregular. Other authors (cp. Marquis and Hyndman, 1992; Cook and Jones, 1995; Marquis et al., 1995) argued that the nature and position of geophysically defined crustal layers may vary from area to area in an assumed Cordilleran collage of accreted terranes.

Steep boundaries defining the crustal-block patterns may be obscured to some degree by ductile flowage and subhorizontal shearing at depth. But fundamental, deep-rooted lithospheric boundaries survive tectonic reworking and manifest themselves in many successive tectonic stages. Subhorizontal seismic reflections can be from a great variety of sources: unconformities, magmatic bodies, structures, etc. To discriminate between these countless possibilities from geophysical data alone is usually impossible.

Mechanical properties of crystalline continental crust are defined largely by the properties of quartz (in contrast, olivine properties are normally used for the oceanic crust and for the mantle). However, using laboratory-determined quartz properties has not been successful for estimating the behavior of the lower continental crust. Rheological discontinuities are common there, in cratonic and orogenic regions alike. Anyway, to divide the crust into a vertical succession of units based on their physical properties is a physical exercise rather than geological. Besides, these physical boundaries can shift in fluctuating tectonic conditions and may not represent the inherited structure of the crust.

High-grade metamorphism and ductile creep in the lower and middle crust may greatly affect the seismic reflection pattern. The extent of ductile flow depends on the temperature and pressure fluctuations and on the predominant composition of rocks (Kirby and Kronenberg, 1987). A pressure or temperature perturbation can alter the reflection signature of a particular block in an unpredictable way: differences between blocks may be enhanced or smoothed. To reduce the uncertainties and pitfalls, interpretation of geophysical data should be driven by the history of the region recorded in its rocks.

The influence of deep high-angle faults on the overlying low-angle thrust sheets has been studied in Europe for over a century, particularly in the Alps (e.g., Butler, 1989), and European geologists have routinely described the effects of steep basement faults (see summary in Coward and Dietrich, 1989). Among North American Cordilleran geologists, inheritance from the basement is often underestimated, while patterns of low-angle faults mapped at shallow levels of the crust attract almost all the attention. Yet, fault-bounded, semi-reworked basement blocks are common in mobile megabelts on all continents (e.g., in the internal zones of the Eurasian megabelts; Lyatsky, 1967,

1978). In the North American Cordillera, the huge Yukon-Tanana and Intermontane median massifs defined some of the main characteristics of megabelt interior.

On the other hand, underestimation of the principal significance of geologic constraints derived from actual rock properties and rock-body interrelationships leads to gross errors. Without evidence, Cook (1995a-c) and Clowes et al. (1998) equated the geophysically modeled lower crust of the Omineca and Intermontane belts with the cratonic basement continuing from the Precambrian basement of the modern North American craton. As a thin sliver, this basement was shown as reaching the eastern Coast Belt (also Cook et al., 1992; Varsek et al., 1993). The idea was that a fundamentally unchanged Proterozoic craton was affected by subduction of ~13,000 km of paleo-Pacific oceanic lithosphere, from which pieces were obducted onto the continent as accreted terranes (also Clowes, 1996). Such a scenario reduces tectonics to just mechanical slicing of the crust and lithosphere, ignoring the multi-aspect richness of actual tectonic reworking. Indeed, other authors (working, for example, from electromagnetic data; Jones and Gough, 1995) have expressed skepticism about this speculative model. From seismic characteristics of the lower crust, Burianyk and Kanasewich (1995) did not continue the supposed cratonic lower crust westward beyond the Omineca miogeosyncline.

This is not to say remnants of ex-cratonic basement are not present under the Intermontane median-massif belt. These remnants could have originally been linked with the North American or Western craton, but they are semi-reworked, surrounded by very reworked crust in orogenic zones, and not in direct continuity with modern cratonic regions. The Intermontane Belt has been relatively stable since Jurassic time: Jurassic and later magmatism in it was teleorogenic, block movements were moderate, and the thrust deformation (notably, in the so-called "Skeena foldbelt" of Evenchick, 1991, 1992) was shallow and rootless, induced by the Coast Belt orogen.

The Intermontane median-massif belt is rimmed by the very mobile Omineca and Coast Belt orogens, where the whole crust has been heavily metamorphosed, magmatized and affected by great vertical tectonic movements. In these orogenic zones, regional metamorphism and deformation indicate the rocks were buried, more than once, to 15-30 km depth and then returned to the surface - in the Late Jurassic, Cretaceous and Tertiary (Parrish, 1983; Parrish et al., 1988; Rusmore and Woodsworth, 1991). Differential vertical movements of such amplitude require steep faults at the boundaries of these belts and of crustal blocks within them.

Clowes et al. (1995, 1998) modeled the crustal seismic velocity structure of the Canadian Cordillera as a fairly rather regular layered arrangement, although even the concept of accreted terranes (to which these authors subscribe) requires differences between the terranes' crustal nature and structure. Marquis et al. (1995) believed Cook et al. (1992; cp. also Cook, 1995b-c) had assigned a too-thick portion of the crust to the presumed accreted terranes in the Intermontane Belt: in their own estimate, the cumulative thickness of terrane tablets spans only the upper crust. Yet, Clowes et al. (1995) acknowledged the terranes should have "deep roots".

Despite all its inadequacy, the terrane-tectonics concept still drives the interpretation of the Lithoprobe deep seismic data. Only the Omineca Belt is presumed to be native to North America, whereas the rest of the Canadian Cordillera to the west is assumed to be made up predominantly of exotic terranes. The terranes are variously supposed to be whole-crust blocks (Monger, 1989, 1991, 1993) or thin tablets in the upper crust (Marquis and Hyndman, 1992). Most terranes were supposedly obducted and thrusted over pre-existing Cordilleran crustal entities, but others were wedged into the tectonically delaminated miogeosyncline. A result of this assumed wedging, in the presence of an assumed Proterozoic crustal ramp, was the formation of west-vergent thrust faults (Price, 1986). Still, Monger et al. (1994) acknowledged that estimates of crustal shortening in the terrane-made part of the Cordillera, "based simply on the asymmetry of the magnetic stripes in the Pacific Ocean", are only "an approximation". In fact, oceanic magnetic stripes should not be used as a major tool for reconstructing continental tectonics at all.

Cook (1995b) noted that the geometry of the reflection Moho in the southern Canadian Cordillera is rarely correlative with the distribution of tectonic belts, even though no continuous reflection band is apparent at the Moho level beneath the Cordillera (also Cook et al., 1992). Truncation of modern steep crustal faults by the modern reflection Moho, whose age is thus post-Early Tertiary, is recognized in the COCORP seismic profiles in the Cordillera just south of the Canada-U.S. border (Potter et al., 1986). Subvertical offsets of the Moho on faults, about 2 km in amplitude, is recognized just north of the Canada-U.S. border (Cook, 1995b). Tectonic, lithologic and rheologic origins of the reflection Moho are all strong possibilities, and to choose one of them without full consideration of others is indefensible. Post-Eocene thermal relaxation (Parrish, 1995) in the southern Omineca and Intermontane belts was conducive to creation of a new Moho. High-angle displacements of the Moho are probably associated with block tectonics: vertical displacements of several kilometers are not surprising in a post-orogenic regional regime characterized by extension of the crust, with proved block movements at the surface.

The oft-mentioned progressive crustal thickening of the Omineca orogenic zone during the Mesozoic and Cenozoic (e.g., Parrish, 1995) is a myth. Crustal thickening resulted from several discrete orogenic cycles, punctuated by periods of extension and crustal thinning. Just west of the Selkirk fan structure, the Monashee complex has been correlated with the Early Proterozoic basement of the North American craton (Parrish, 1995), but in reality these rocks could be no more than an ancient craton's strongly reworked remnants. The Middle Jurassic Kuskanax pluton has produced a zircon age of ~2,000 Ma (Parrish and Wheeler, 1983), but it is not clear if the ancient zircons were native or transported as clasts from elsewhere.

Parrish (1995) showed an inverted distribution of metamorphic grades in the rock assemblage conventionally referred to as the Selkirk allochthon (Brown and Read, 1983; Brown et al., 1986). Parrish (1995) rejected the possible effects of periodic decompression, and assumed a downward heat transfer from the eastward-translated allochthonous Selkirk rock mass. In the hypothesis of Brown et al. (1986, 1992) and Brown and Carr (1990), this allochthon was displaced >80 km from the Intermontane Belt, as was also the Monashee complex. In a new re-interpretation, Gibson et al. (1998, p. 193) stated that their new data "refute previous interpretations", and in their new "thermo-tectonic model" an assumed "substantial northeasterly-directed shear strain... led to lateral transfer of rocks preserving evidence of strongly diachronous (>=18.3 Ma) and apparent inverted metamorphism". If so, the inverted metamorphic gradient is secondary, induced by a young, Middle Miocene, deformation episode.

Confusingly, Johnston and Williams (1998, p. 194) argued for an assumed significance of "westward underthrusting of North American basement and its sedimentary cover beneath a burgeoning hinterland" of the Cordillera. They speculated that "with continued westward underthrusting during the Mesozoic, eventually a gravitationally unstable crustal edifice formed that may have been on the scale of the modern Himalayan front", and that "the crystalline rocks of the Monashee complex were then exhumed along the... GLSBZ [Greenbush Lake shear band zone] and other major Eocene normal faults (e.g., Columbia River fault)". But the GLSBZ is not mapped in this region as a grandiose feature. Even the Gwillim Creek fault and the "Monashee décollement" appear in seismic sections as discontinuous and relatively shallow (Kanasewich et al., 1994): the "Monashee" reflections flatten out at a depth of 16 km and disperse.

In a very unrealistic oversimplification, Clowes (1996, p. 112) believed that "the visible portion of the southern Canadian Cordillera is a result of plate divergence from the Mesoproterozoic to late Paleozoic, and plate convergence from the late Paleozoic to the Present" (also Cook, 1995c, p. 1821-1822). Besides confusing passive continental margins and divergent plate boundaries, this statement is factually incorrect. Visible rocks in the Canadian Cordillera record a multi-stage, multi-cycle tectonic history in a mobile megabelt within a continent, driven by indigenous tectonic causes.

Application of the guiding principles of tectonics in regional analysis

Mathematics is an abstract discipline, dealing with formalized procedures that relate quantities and permit to predict mathematical structures and undefined parameters. Physics and chemistry deal with experiments, where the outcome can be changed by modifying the selection and specification of a limited number of parameters that define the system under investigation. The most meaningful and general physical and chemical relationships are formulated as laws of nature. For all relevant systems, these laws are inviolate. Tectonics, as part of geology, deals with endogenic phenomena in the rock-made outer shell of the Earth. The systems it considers are very complex, and the number of parameters is extremely large. Such systems cannot usually be reproduced in the laboratory (in part also because geologic processes are very slow) or simulated with numerical procedures.

Geology lacks precise laws (of the kind that exist in chemistry and physics). However, from the centuries-long practice of field work and theoretical evaluation of observations, geologists have been able to formulate empirical generalizations, or principles, whose general adequacy has been tested by observing the real geologic world. For people with exact-science mindsets, such imprecision and empiricism can be annoying. All too often, people leaning towards exact parametrization, experiments and models are prone to reduce complex systems to the smallest possible number of assumed parameters. In the process, they often miss the emergent and very real complexity of geologic systems, and their simplistic models produce unrealistic and unusable results. Incorrect or over-simplified assumptions rarely produce results that are helpful for understanding the real world.

The crust and lithosphere are always under tremendous stresses. Changes in the overlying load and thermal conditions cause the stress to fluctuate. Lateral tectonic stresses, acting in compression or extension, may be indigenous (e.g., due to thermal changes or magma intrusion) or external (e.g., teleorogenic or subduction-induced). Thermal conditions in the lithosphere, affected radioactive decay of certain minerals, supply of heat from deeper mantle, fluid circulation in the lithosphere, etc., are notoriously difficult to predict (Clauser et al., 1997; Emmermann and Lauterjung, 1997). This inevitable shortcoming limits the realism and usefulness of many numerical models of lithosphere evolution (Lyatsky and Haggart, 1993; Lyatsky, 1994a-b, 1996).

Physical models of crust and lithosphere that ignore the unglamorous but fundamental empirical tectonic principles cannot fail to be misleading. One of the fundamental tectonic principles is that the crust and lithosphere are structured as blocks; these blocks are bounded by steep fault zones. To ignore this block structure would result in an unrealistic model.

A common example of tectonic postulate derived from an assumed model rather than real geologic observations is the concept of Wilson Cycle. It assumes that great ocean basins, such as the Atlantic, have existed several times, opening and closing repeatedly. In Wilson's (1966) reasoning, compressional orogenic regions on both sides of the Atlantic are explained by the existence of a precursor ocean (called Iapetus) that was squeezed out of existence by the colliding continental masses in the Paleozoic. As these continents split apart again in the Mesozoic, roughly along the line of presumed Iapetus suture, the modern Atlantic Ocean was created. The supposed suture is marked by ophiolitic formations, which are thought to be all that remains of the Iapetus oceanic crust. Difficulties with this concept have been recognized all along, arising from mechanical, dynamic, structural and petrogenic considerations. The ease with which the assumed extent and evolution of the Iapetus ocean, and the paleo-position of its flanking continents, vary from one set of authors to another (e.g., Mac Niocaill et al., 1997, and references therein) attests to these reconstructions arbitrary nature and unreliability.

In the Cordillera, in the description of Monger et al. (1994, p. 373), "western and northern parts of the Belt-Purcell basin may have been rifted from North America at [Windermere] time and transported to another part of the globe (Sears and Price, 1978)"; early Paleozoic rocks "record a third interval of extension, separation and possible removal of parts of the Late Proterozoic continental terrace wedge" (p. 373); and yet, after being pulled away, Precambrian cratonic rocks such as those in the Monashee metamorphic core complex were "then pushed back during Mesozoic compression" (p. 377). Such an improbable scenario ignores the principle that alternating burial and inversion, extension and compression, and deep reworking of the lithosphere are normal in zones of intense tectonic activity. Harms (1986) suggested that the Sylvester Allochthon accommodated the closure of an oceanic basin up to half the width of the modern Pacific (cp. also Gabrielse et al., 1991a-b). Nelson (1993), instead, regarded the Sylvester Allochthon as a late Paleozoic marginal basin near a continent, coupled with an island arc. Presence in the Yukon-Tanana median massif of siliciclastic debris and continent-derived metasedimentary rocks affiliated with the tectonic platform to the east, and of abundant orthogneisses of continental affinity, is a clear indication of continental links (see in Plafker and Berg, eds., 1994). The Devonian to Mississippian granitoids (now metamorphosed into augen gneiss) contain ancient, probably basement-derived, grains dated as Precambrian (Aleinikoff et al., 1986; Mortensen, 1992). The partly correlative Slide Mountain terrane has been related to a marginal basin (e.g., Thompson, 1998) and interpreted as parautochthonous (Wheeler and McFeely, comps., 1991).

Despite these facts, Hansen and Dusel-Bacon (1998, p. 228) speculated about a hypothetical Anvil ocean which supposedly appeared due to splitting of the western North American continent in the Paleozoic, "prior to Permian subduction" when this ocean was squeezed into a suture as these separated fragments came together again. Speculations about late Paleozoic - Anvil, Cache Creek, Teslin - multiply the number of supposed ocean opening-and-closure cycles in the Cordillera, with associated rifts and subduction zones, beyond any plausibility, requiring some kind of hectic motion of big and small continental masses, pulling away and colliding again for unclear reasons. These scenarios overlook Ernst's (1988) warning that geochemical signatures in the western U.S. do not support ultra-mobilistic ideas that regard many of the Cordilleran continental-crust blocks as far-traveled terranes. In Canada, isotopic similarities have indeed been noted between the Slide Mountain terrane and the Cordilleran miogeosyncline to the east (Roback et al., 1994).

A growing volume of geological facts calls for a revision of earlier terrane postulates. The well-studied Cache Creek assemblage is not a mélange nor a very disrupted accretionary prism scraped off the ocean floor. The assumed structural juxtaposition there of rocks from different paleogeographic provinces (based on Tethyan-vs.-Boreal fusilinid faunas; Monger and Ross, 1971; Ross and Ross, 1983) has not been confirmed (Nelson and Nelson, 1985; Newton, 1988). The suppositions about opening and closure of Cordilleran oceans are weakened by the observation that the Cache Creek assemblage is surrounded by blocks of continental crust (Nelson and Mihalynuk, 1993; Mihalynuk et al., 1994) which are parts of the Intermontane and Yukon-Tanana median massifs. Discussions of whether this or that block is a terrane, superterrane or subterrane, parautochthonous or allochthonous, are sterile sophistry and word games that do little to elucidate the real geologic facts. What matters is that the Cache Creek assemblage lies among continental-crust blocks with close geologic links.

In a self-contradiction, Hansen and Dusel-Bacon (1998, p. 224) stated that the "Slide Mountain-Seventymile assemblages represent Anvil oceanic strata... although they have some isotopic similarities to North American marginal basin strata and North America" (cp. also Nelson, 1993; Roback et al., 1994; Smith and Lambert, 1995). Big Late Triassic to Early Jurassic granitoid bodies, which intruded the Seventymile assemblage, were classified by Hansen and Dusel-Bacon (1998, p. 212) as "...arc granitoids within an oceanic or marginal-basin setting" and correlated with "the Stikinia arc".

But granitoids are an attribute of continental rather than oceanic lithosphere. In the Stikine area and to the south, the Late Triassic-Early Jurassic granitic suites have no relation to an "oceanic setting". The entire sedimentary cover of the Intermontane median-massif belt affected by that magmatic episode overlies blocks of continental crust.

The subsequent Middle-Late Jurassic, mid-Cretaceous and Late Cretaceous-Early Tertiary orogenies also occurred in a continental setting, requiring no Wilson-Cycle perturbations and transportation of continental fragments to "another part of the globe" to make room for an assumed "oceanic setting". As was discussed in previous chapters, the Cache Creek assemblage could have formed in a mediterranean marine setting with some oceanic crust, in a narrow rift that was subsequently inverted and closed, but not in any wide ocean.

Speculations about previous existence of a now-closed ocean are usually set off by presence of mafic-ultramafic rock assemblages in a region's geologic record. But mafic-ultramafic complexes differ and fall into various categories (Wright et al., 1982; Moskaleva, 1989; Patton et al., 1994), and not all of them are ophiolitic complexes of the same origin (Jenner et al., 1991). Besides, great difficulties arise in correlating ophiolites to the assumed (and still undrilled) deep structure of the crust in modern oceans. The classic Alpine ophiolites are correlated with Mediterranean-type orogenic tectonic settings, associated with marginal seas underlain by transitional crust whose similarities to oceanic crust proper are incomplete. Ophiolites in Newfoundland were interpreted by early workers as representing oceanic crust of the pelagic Atlantic, but subsequent studies point to marginal-basin or island-arc affinities (Coish et al., 1982; Swinden et al., 1989). Links to Mediterranean-type subduction, local and discontinuous, seem more probable. Jenner et al. (1991, p. 1636) found "no compelling evidence to suggest that any of these [Newfoundland] ophiolites represents a fragment of the Cambrian-Ordovician major oceanic basin (i.e., Iapetus)". These authors added, broadly: "We doubt that there are any fragments of the initial Iapetus major ocean preserved in the Appalachians". The concept of Wilson Cycles is just that, a concept, driven by temporarily fashionable models postulated without verification by geologic facts. Confronted with reality, such concepts are often shown to be unrealistic and fade away. But principles of tectonics that receive factual confirmation endure.

Shaky foundations of terrane-tectonic speculations

Terrane tectonics was born of the practice to reconstruct plate motions in the geologic past from assumed age correlation of oceanic lithosphere with magnetic-anomalies. Such reconstructions assume that oceanic lithosphere is stiff at all its levels and internally undeformed, and that magnetic anomalies reflect the age belts of lithospheric material accreted panel-like at spreading centers and remaining essentially unreworked. These approximations are too simplistic.

The complexity of a system increases with the system's size and scope, and lithospheric plates are enormous and complex indeed. Simple principles of kinematics and dynamics can in some cases be used to describe uncomplicated systems with few elements and components, such as atoms bouncing around in a small space. To apply the same level of simplification to lithospheric plates is unrealistic. No one plate is the same as any other: plates differ in size, history, composition, structure. Smaller parts of the plates also differ from one another: there are no two identical continents, cratons, mobile megabelts, orogens, median massifs, oil fields or mineral occurrences.

Mass production on a conveyor belt works in an artificial, highly standardized modern industrial environment. In nature, it is harder to imagine, especially as it is supposed to have continued for hundreds of millions of years. Such constancy and regularity contradict the very principles of tectonics. On continents, where rocks can be observed directly, no process of comparable steadiness has been inferred: episodicity, cyclicity and change are integral parts of a region's geologic evolution. In oceanic crust, no comparably extensive geological observations are possible, leaving more room for oversimplification, inference and speculation. At any rate, the assumed stiffness of plates has evidently been exaggerated. Even oceanic crust is now known to be susceptible to considerable internal deformation: warping, faulting of various types. Magmatism at spreading centers is not regular or steady, but sporadic and episodic. Many fractures that conduct magma to the ocean floor run at an angle to the magnetic anomalies. The assumption that magnetic anomaly stripes reflect the age when the corresponding parts of the plate were first formed does not allow for subsequent alteration, deformation, magmatic intrusion or other effects of exogenic sea-water action and endogenic tectonic reworking.

In the current geophysical models, modern tectonic regimes at the North American Pacific margin are just two: convergence and subduction in some areas, strike-slip movements elsewhere. The reality is considerably more complex: plate fragments offshore British Columbia are too broken-up and not rigid enough to permit exact reconstructions, plate triple junctions are too diffuse to pinpoint exactly, and even the subduction off Oregon and Washington has many very unusual manifestations (Lyatsky, 1996).

Riddihough and Hyndman (1976, 1991) uncritically used the standard assumptions about plate motions and interactions to interpret the tectonic regime at the Canadian Pacific continental margin, and Monger et al. (1994) uncritically applied such assumptions in their description of the Canadian Cordilleran interior. Monger et al. (1994) identified four principal tectonic regimes in the history of this megabelt: (a) regime of passive continental margin along the Omineca Belt from ~1,500 to 170 Ma; (b) regime of terrane accretion from the Omineca to Coast belts from 170 to 60 Ma; (c) regime of terrane accretion in the Insular Belt from 60 to 40 Ma; and (d) modern regime of plate convergence and strike-slip plate interactions at the modern Pacific margin since 40 Ma. Age estimates for regime changes in Mesozoic and Cenozoic time were derived from studies of magnetic anomalies offshore. Even more simplistically, Clowes (1996) noted only two regimes in the southern Canadian Cordillera: plate divergence from the Middle Proterozoic to late Paleozoic, and plate convergence thereafter.

Lyatsky (1996) showed that, with regards to the northeast Pacific at least, this scheme needs fundamental revision, because no evidence exists to support the idea of ongoing subduction off British Columbia as this model requires. Combined analysis of seismic, potential-field, geological and geochemical data negates the explanation of the observed facts in terms of ongoing subduction there. Identification of major subvertical faults made it possible to delineate crustal blocks. Instead of a subduction megathrust, the huge Mesozoic-Cenozoic, Olympic-Wallowa zone of crustal weakness was traced from the Cordilleran interior in the conterminous U.S. through British Columbia into southern Alaska. Drillholes and outcrop mapping put constraints on speculations about accretionary prisms, crustal thrust stacks, terrane amalgamation and presumed subduction-related magmatism and deformation. Analysis of the available facts resulted in an internally consistent concept of that region's evolution in terms of crustal development by restructuring and vertical movements.

Monger et al. (1994) supposed that the Middle Proterozoic to Middle Jurassic (1,500 to 170 Ma) crustal regime in the eastern Cordillera was one of predominant extension. However, geologic evidence of several orogenic episodes, accompanied by deformation, magmatism and metamorphism of various types indicates a variable tectonic regime, with alternating periods of compression and extension. The Middle Jurassic to Paleocene, according to Monger et al. (1994), saw an "active margin in which the dominant features of Cordilleran crust were established in a contractional and transpressive regime". But three distinct Mesozoic to Early Tertiary orogenic cycles resulted in deep burial, inversion, extension and uplift in orogenic zones, particularly the miogeosynclinal Omineca and eugeosynclinal Coast belts. As before, tectonic regimes varied, including numerous periods of extension and compression.

The Cordilleran crust west of the Omineca Belt is supposed to be made up of exotic, accreted terranes. They were delaminated from the subducted oceanic lithosphere and thrusted flake-like onto the leading edge of the North American continent (or craton; some authors use these terms interchangeably). The terranes are of oceanic-crust origin, although some may bear pieces of continental derivation. Before accretion, terranes and subterranes were somehow amalgamated into composite superterranes, Intermontane and Insular, which finally docked to the continent. The Intermontane superterrane was amalgamated by the end of the Triassic and accreted in the Jurassic; the Insular superterrane amalgamated by the end of the Jurassic and docked in the Cretaceous. This hypothetical scenario, in several permutations and variants, continues to strongly influence the perceptions of Cordilleran evolution.

Terranes, in this scenario, are thin, imbricated, stacked thrust sheets. Presumed thickness of the terrane stack is 10-12 km (Marquis et al., 1995), or 10-20 km (Monger et al., 1994), or more (Clowes et al., 1995). The exotic nature of terranes is still accepted based on dissimilarities of Permian and early Mesozoic faunas, which are interpreted as evidence of origin in different biogeographic provinces, as well as on the much-questioned paleomagnetic data. Intra-oceanic affinities of stratal rocks in the Intermontane Belt are now convincingly rejected. The attribution of Cache Creek rocks to an ophiolitic formation, mélange or disrupted accretionary prism is revised, and the associated blueschists are no longer ascribed to any specific tectonic setting. Teleorogenic magmatism in the median massif(s) is well correlated with that in the flanking Coast Belt and Omineca orogenic zones. Structures in the volcano-sedimentary cover are complex and variously oriented, they are not simply products of nappe tectonics. Long-lived close links between the orogenic zones and a median massif(s) are rooted in the common ancient basement, which differs from one Cordilleran tectonic zone to another in its degree of reworking.

The Yukon-Tanana massif during its entire history had demonstrable close connections with the conterminous Proterozoic cratonic mass to the east (McClelland et al., 1991; Mortensen, 1992). Its polygonal shape is clearly outlined by steep faults, which strongly influenced the regional structural configurations in northern British Columbia, Yukon and Alaska. In the Canadian Cordillera to the south and in the Russian Far East to the west, other massifs have since the Proterozoic acted as rigid crustal elements, determining the structural patterns of Paleozoic and Mesozoic orogenic zones. The multifaceted and complex history of these regions does not eliminate the partial inheritance of their tectonic grain from pre-megabelt crustal entities whose remnants are preserved in these and other crustal blocks.

Impressed by the tectono-metamorphic, -magmatic and -deformational complexities, Hansen and Dusel-Bacon (1998) interpreted the Yukon-Tanana massif as a "composite terrane", in which the amalgamated smaller terranes are stacked in structurally detached upper and lower plates. These authors acknowledged, nonetheless, that post-Jurassic thrusting, if any, produced only a thin upper-crustal nappe unit over the intact continental-crust basement. The hypothetical Anvil ocean, if it ever existed at all, left behind only a few blueschist rock assemblages, which were attributed to subduction of unknown but assumed-large extent (Hansen et al., 1991).

Hansen and Dusel-Bacon (1998, p. 227) noted that the supposed ocean-derived crust in the upper structural plate in the Yukon-Tanana massif represents "very little volume" (contrary to the estimates of Coney, 1981). According to Hansen and Dusel-Bacon (1998), many of the assumed upper-plate rocks were separated from parautochthonous and autochthonous lower plate of the North American crust in the Late Triassic-Early Jurassic, and pushed over the Yukon-Tanana massif which is volumetrically dominated by North American crust. To the south, the Intermontane median-massif belt also contains a diverse record of the Cordilleran mobile megabelt's complex history. In Monger's (1989, 1993) conventional description, it is a composite superterrane consisting of the amalgamated large Stikine and Quesnel terranes, the Cache Creek terrane between them, and the Slide Mountain terrane on the east. The Cache Creek assemblage, regarded as a squeezed remnant of a hypothetical ocean similar to the Anvil ocean to the north, contains several dismembered rock packages presumed to be ophiolitic, as well as several outcrops of confirmed blueschists.

The volcano-sedimentary cover over the probable crystalline basement in the Stikine block contains a mapped succession of middle-late Paleozoic to Jurassic and Early Cretaceous stratal rocks with a cumulative thickness of ~10 km. Information about the rocks beneath is indirect, obtained from geochemistry of magmatic rocks and xenoliths. The similar volcano-sedimentary cover in the Quesnel block comprises stratal rocks of Late Devonian to Early Cretaceous age. The Slide Mountain assemblage was previously interpreted as far-traveled and intra-oceanic, and related to the Sylvester allochthon (Harms, 1986). Today it is regarded as a product of marginal-basin setting, with strong connections to the Omineca miogeosyncline (Roback et al., 1994; Thompson, 1998).

Presence of ultramafic rocks does not prove an ophiolitic origin (Wright et al., 1982; Patton et al., 1994), and presence of ophiolites in a regional geologic record does not require a major ocean such as Iapetus or Pacific (Jenner et al., 1991). Even if blueschist rocks do indicate subduction (which is not everywhere certain), that subduction could be only local. In the interior of the Canadian Cordillera, blueschist paragneisses are found in a N-S-trending zone between the Stikine and Quesnel blocks in the Intermontane median-massif belt. Northern continuations of this zone are also known (the so-called northern Cache Creek and Bridge River terranes). The age ranges of blueschist-bearing lenses in these rocks are narrow and similar (based on mica minerals, the metamorphism ended at ~230 Ma, in the Late Triassic). In southern British Columbia, a growing volume of evidence puts all the numerous small terranes east of the presumed Bridge River suture into a single Stikine block of the Intermontane Belt. Glaucophane with jadeite and quartz is common in the northern blueschist lenses, epidote and garnet in southern ones, and lawsonite in the southernmost Tyaughton areas (Gabrielse and Yorath, eds., 1991; Monger et al., 1994). These differences indicate different grades of metamorphism and burial, as well as variations in hydration/dehydration metasomatism (Wood, 1979a-b).

To rely on presumed existence of oceanic-crust terranes and sutures resulting from squeezing of hypothetical oceans may be misleading in tectonic analysis, as former oceans can easily be confused with local marginal or intracontinental basins enclosed by continental-crust blocks (cp. Nelson and Mihalynuk, 1993; Mihalynuk et al., 1994). Mediterranean-type seas are entirely possible in mobile-megabelt zones, where ophiolite-bearing or transitional-crust depressions are local and discontinuous. Ephemeral, local subduction zones may develop around such depressions. Examples of this kind of depressions and zones are known in the Appalachians (Jenner et al., 1991) and in the Eurasian mobile megabelt, which stretches from southern European to southeast Asian orogenic areas. The Anvil and Cache Creek depressions in the Cordillera could have also of this type.

Real pre-existing continental-crust structures vs. preconceived assumptions in regional tectonic analysis of mobile megabelts

A lesson from Eurasia

In the current literature, a lot of discussion of tectonic and geodynamic phenomena lacks proper formalization of the notions being discussed. Interactions between lithosphere and asthenosphere, crust and mantle, and different rheological layers are discussed loosely, without specifications of what exactly is alluded to. The overall assumed scheme is very simple: currents in the mantle move the lithospheric plates; plates are destroyed at subduction zones but created at spreading centers; magnetic anomaly lineations in oceanic crust accurately record the time of plate formation. But this assumed simplicity in oceanic regions helps little in understanding the real continental tectonics. Kearey and Vine (1990, p. 89) stressed the "necessity to consider the nature and magnitude of all the [interacting] forces", and noted several caveats: "the depth to which mantle convection extends is still controversial" (op. cit., p. 220), "the nature of convective flow in the mantle is still problematic" (p. 218), "the mechanism behind the motion of plates is still controversial" (p. 214), and "the various forces acting on plates are disputable" (p. 223). The very foundations of the Vine and Matthews (1963) model are now in doubt, because the oceanic lithosphere is known to be significantly deformed and the sources of magnetic stripes are uncertain and complex (e.g., Yañez and LaBreque, 1997; Ballu et al., 1998; Lawrence et al., 1998; Mitchell et al., 1998; Tivey and Tucholke, 1998).

The outdated, simplistic 1960s-vintage model of plate dynamics overlooks a lot of now-known real-world complexity, but its simplicity continues attract many workers (Froidevaux and Nataf, 1981; Ma et al., 1984). In the mid-1980s, some Western geologists applied the model-driven methodology of tectonic analysis to new regions that became accessible to them in Asia (e.g., Chen and Dickinson, 1986). In 1989-1993, a large project was conducted by the International Union of Geological Sciences (IUGS) and UNESCO to reconstruct the geologic history of a paleo-Asian ocean (e.g., Coleman, 1990, 1994). The main goal was formulated, from the beginning, to be a study of amalgamation of Asia from a variety of terranes, assuming in advance the existence of the terranes and the now-closed ocean. Instead of field mapping, the effort was focused on "geodynamic interpretations", which are inevitably speculative. The long-recognized Eurasian mobile megabelt (Fig. 67) was artificially divided into terranes and superterranes supposedly containing ocean-floor and island-arc assemblages. Successive terrane accretion in the Paleozoic was assumed. Other assumptions concerned presumed post-collisional phenomena, appearance of huge granite-bearing provinces in the Trans-Baikal region, and development of extensive sedimentary basins.

More realistically, a great number of crustal blocks have long been recognized in the Eurasian mobile megabelt all across Asia, from Turkey and Iran through Central Asia to China and Korea. This megabelt lies between several cratons: Russian and Siberian in the north, Yangtze and Sino-Korean in the south. Blocks within this megabelt, as well as the flanking cratons, indicate a great extent of Precambrian crust before the Eurasian and Western Pacific mobile megabelts were formed. Presence of ex-cratonic crustal blocks beneath the Tarim Basin and central Kazakhstan is well established; so is the sharp lateral heterogeneity of the lithosphere in Kyrgyzstan (Vinnik and Saipbekova, 1984; Gilder et al., 1991). Pre-megabelt basement blocks largely define the general structural pattern of this entire region (Ren et al., 1996) and control the position of many generations of major faults, rifts, linear orogenic zones and zones of folding. The variety of structural-formational characteristics of individual orogens is not a result of their belonging to disparate exotic terranes that arrived from elsewhere on the globe.

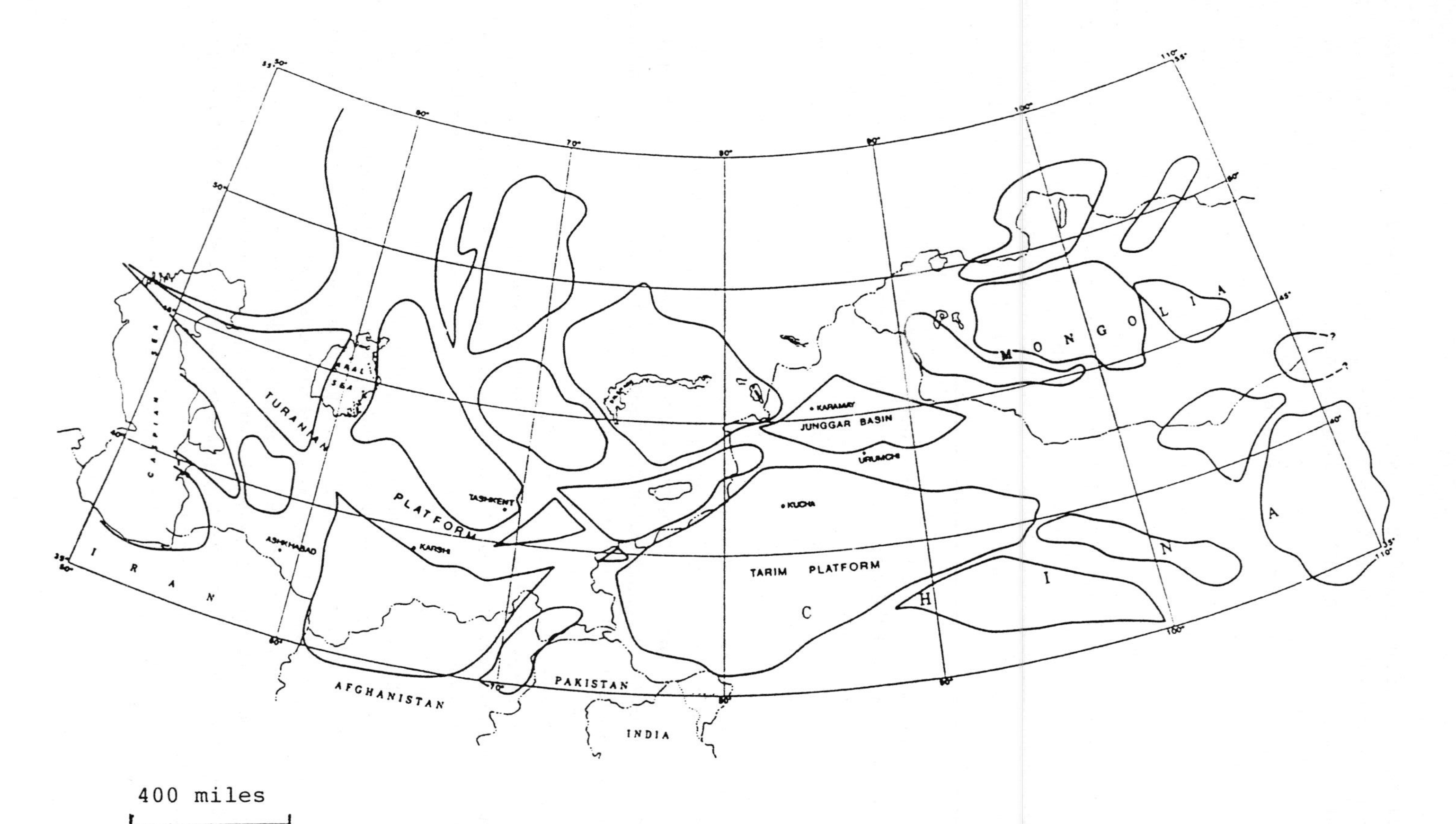

Figure 67. Precambrian crustal blocks, separated by more-orogenized zones, within the Eurasian mobile megabelt in Asia (modified from Lyatsky et al., 1990).

Rather, it is a consequence of these orogens' position relative to big and small continental-crust blocks (which some workers call microcontinents; Coleman, ed., 1994).

The tectonic history of the Eurasian continent is complex, but very differently from the complexities artificially assumed in a terrane-amalgamation scenarios. Coleman (1994, p. 35-36) admitted that evidence of former subduction in the region is "scant" and that "the [paleo-Asian ocean's] final closure was passive, perhaps marked by rotation and strike-slip motion of the assembled terranes". In short, this entire scenario is no more than assumed.

Strike-slip displacements are rarely well constrained. All too often, their estimates are based on notoriously unreliable paleomagnetic data. In the North American Cordillera, Beck and Nosun (1972), Irving et al. (1995), Wynne et al. (1995) and others inferred terrane translations and displacements along the Pacific margin to have been thousands of kilometers since the mid-Cretaceous. Price and Carmichael (1986), Butler et al. (1989), Monger and Price (1996) and Dickinson and Butler (1998) showed that estimates such as these often arise from pitfalls in paleomagnetic methods and are contradicted by the real geologic structure of this mobile megabelt. The same is probably true in the world's other megabelts.

In the Eurasian mobile megabelt, the long-lived zones of crustal weakness that determined its general smooth, sigmoid structural pattern are quite apparent. The shape of big orogenic ranges and platforms reflects a framework of pre-existing zones of crustal weakness as well as several generations of younger faults. A very old, middle-late Paleoproterozoic, WNW-ESE trend controls complex rock assemblages of epicontinental sediments and volcanics, locally with spilite basaltic suites (Brezhnev and Shvanov, 1980; also in Gilder et al., 1991) in the North Qilian, Quruktagh and Central Tien Shan areas. The same trend demarcates the southern boundary of the Turanian Platform in the Trans-Caspian region (Lyatsky et al., 1990). Transcurrent to that prominent trend is a series of generally N-S major inter-regional fault zones: the Uralo-Oman lineament, Talasso-Fergana fault and others (Amursky, 1976). Major sedimentary basins correspond to blocks delimited by big crustal faults.

An innovative idea to come out of the paleo-Asian ocean project was the one about soft subduction. Coleman (ed., 1994, and 1994) outlined it only vaguely. He noted that not one but many basins between numerous island-arc systems existed in the Paleozoic in the Eurasian megabelt, although he stood by his perceived "clear picture of widespread accretion" (Coleman, 1994, p. 35). In these assumed tectonic conditions, soft subduction was invoked to explain the lack of evidence for frontal subduction. "Passive collisions" without underthrusting of subducting plates were considered to have been common in Asia (except for a few diamond- and coesite-bearing zones of very high-grade metamorphism and zones of blueschist metamorphism - in northern Kazakhstan, Altai, East Sayan and central China). Neither diamonds not blueschists are exact indicators of tectonic setting (Snyder et al., 1997). Abundance in a sedimentary basin of molasse or flysch deposits does not necessarily require hard (frontal) or soft (passive) accretion or subduction. To Coleman (ed., 1994, and 1994), frontal collision is that of India with Eurasia in the Himalayas, or that of the Arabian plate with Eurasia in Zagros, although the latter is presumed to show some signs of "passivity" or "collisional softness".

Ren (1991) and Ren et al. (1996) tried to explain these notions from a purely physical, kinematic viewpoint: what counts is not subduction of oceanic lithosphere but collision between continental-lithosphere entities. Weak collision between microcontinents is called soft; collision between large continental masses is regarded as hard. In a collisional orogen, these authors explained, there are actually two orogenies: one due to oceanic-crust subduction, the other due to collision of continental-crust blocks. Thus, in a mature orogen it should be possible to find rock assemblages related to (a) oceanic-crust subduction, (b) continental-crust collision, and (c) post-collisional crustal extension and differential block movements. In this scheme, hard collision should induce large-scale mountain uplift and significant compensatory-depression subsidence with molasse sedimentation, whereas soft collision should induce only local mountain uplift and limited molassic basins.

This generalization was derived from the considerations of the tectonics of China (particularly of the long-lived, ~1,000 to 100 Ma, Qiling-Dabie system of orogenic belts separating the Sino-Korean and Yangtze platforms; see also Gao et al., 1998). Some of its points may be debatable, but it reveals enough regional complexity to illustrate that reduction of the geologic evolution of the Central Asian and Chinese segment of the Eurasian mobile megabelt to just the evolution of a paleo-Asian ocean is an error. Pre-megabelt cratons extended all over the Eurasian continent, and their basement has influenced the subsequent tectonic history for a long time till the present. Orogenic processes were diachronous: the Middle Triassic-Early Jurassic (240 to 190 Ma) Indosinian and the Late Jurassic-Early Cretaceous (140 to 120 Ma) Yanshanian orogenies were particularly strong in China, whereas the Late Jurassic-Cretaceous and Tertiary orogenies, Cimmerian and Alpine, were the strongest in regions to the west (Markovsky, ed.-in-chief, 1972; *Atlas of Paleogeography of China*, 1985; Ma, chief comp., 1986). But despite this heavy overprint, pre-existing continental-crust structural trends are found all over the Eurasian mobile megabelt.

False certainties in regional tectonic analysis

The temptation to reduce tectonic analysis to a search for a silver bullet that would resolve all the questions simply is geologically indefensible. Similar phenomenological manifestations may be produced by various causes, while on the other hand not every cause produces an expected range of diagnostic manifestations. The fundamental principles of tectonics, established by trial and error in a long practice of field mapping and analytical efforts, must take precedence over facile, assumption-driven "modeling" that all too often masquerades as sophistication and leads to shallow and factually unjustified conclusions.

For example, not all tholeiitic basalts are necessarily MORBs (mid-ocean-ridge basalts), and they cannot simply be interpreted in terms of oceanic crust, seamounts, oceanic plateaus and so on (contrary to Duncan, 1982). Not all depleted magmas were formed in oceanic plates, not all ophiolite-like formations were created in pelagic oceans (Jenner et al., 1991), and not all closely-spaced dikes are oceanic-crust sheeted-dike complexes (Lyatsky, 1996). Sills of amphibolite-grade mafic and ultramafic rocks in the Omineca-Belt miogeosynclinal formations (Pell and Simony, 1987) in the proximity of the Monashee complex have been interpreted as derived from continental mantle (Sevigny, 1988). The nature and origin of geologic phenomena such as these can be understood only in the context of bigger analysis of the regional tectonic setting, relying on the empirical basic principles of tectonics.

One of these fundamental principles is that continental crust and lithosphere have a block structure. Bounded by big steep faults, these blocks are recognizable by analysis of structural-formational étages and geophysical data in cratonic and orogenic areas (Lyatsky, 1996; Lyatsky et al., 1999). These blocks are expressed in long-lived lateral tectonic units of mobile megabelts - orogenic zones and median massifs. Old, pre-megabelt structural trends are known in megabelts in Eurasia, the Circum-Pacific region and elsewhere.

Poor ability of modern seismic methods to detect steep physical discontinuities results in only a partial picture of crustal structure. Unaware of the fundamental principles of tectonics, physics-minded geophysicists tend to rely too confidently on these partial images when they interpret the structure and evolution of the crust. Uncritical reliance on ideas of terrane accretion, and forgetting that continental crust and lithosphere are not inert but capable of self-development, leads to over-confident interpretation of low-angle seismic events as evidence of past terrane collision and thrust stacking.

Low-angle seismic events, even those confirmed to be primary reflections, can be caused by rock discontinuities of countless types: metamorphic fronts, brittle-ductile transitions, unconformities, shear zones, faults and so on. To assume all reflection geometries are created structurally is incorrect. Equally incorrect is to assume that all faults, even those which are steep at the surface, must flatten out at depth. These errors lead to interpretations of local faults as major while missing the true major faults, and to groundless postulation of trans-megabelt crustal detachments.

By postulating the Canadian Cordillera west of the Omineca miogeosyncline consists of accreted oceanic-crust-derived terranes (Hyndman et al., 1990; Clowes et al., 1992; Cook et al., 1992; Monger et al., 1994), the Lithoprobe program fails to consider possible alternatives. Regional tectonic analysis shows that accretion, if any, had only limited significance in Cordilleran tectonics. Even the famous Franciscan ophiolitic complex in western California contains clastic material of native origin (Tagami and Dumitru, 1996). The Coast Range ophiolite overlies the Franciscan complex structurally, and is overlain depositionally by the Great Valley sedimentary basin coeval with the Franciscan complex. The Precambrian Salinia block parallel to the San Andreas fault is oriented obliquely to the Franciscan-Great Valley rock assemblage (Wakabayashi, 1992). Detailed studies show that blocks in coastal and Baja California are paleomagnetically coherent with the interior of the North American continent (Dickinson and Butler, 1998).

In contrast to the typical circum-Pacific subduction zones, with Benioff zones reaching 600-700 km depth, subduction zones in mediterranean settings are much smaller and more ephemeral, shallow and local. In the Paleozoic and Mesozoic, the Eurasian and Cordilleran mobile megabelts evidently contained such local subduction zones. Frequent fluctuations between compression and extension seem common in such settings (Mercier et al., 1987). Soft collision and soft subduction, unfortunately oxymoronic as these terms may be, could be useful notions when describing intra-megabelt settings. The Anvil and Cache Creek rock assemblages, with their mantle-derived mafic and ultramafic rocks, may owe their origin to splitting and subsequent reconstruction of median massifs. No oceanic setting is required. Besides local subduction, some mafic-ultramafic complexes tend to develop along deep faults, which provide upward conduits for mantle-derive magmas impregnating the continental crust. It is too simplistic to ascribe all mafic-ultramafic rock complexes in a megabelt to off-scraping or obduction from converging oceanic crust (as was done in the Cordillera by Monger et al., 1982, 1994). The possibility of intrusion along faults and crustal weak spots (Wright et al., 1982; Patton et al., 1994) or minor local subduction zones should also be considered.

Not hypothetical low-angle faults, assumed by Lithoprobe to be the dominant characteristic of the crust (Clowes et al., 1992, 1995, 1998; Cook et al., 1992; Varsek et al., 1993), but real high-angle faults determined the general crustal structure of the Cordillera. As any mobile megabelt, this one contains a number of principal lateral tectonic units - orogens and median massifs. These units and their fault boundaries defined the principal tectonic and metallogenic zonation of this economically important province (McCartney, 1965; Sutherland Brown, 1976).

Expressions of Cordilleran tectonic zones in metallogenic zonation

The geographic distribution of petroleum, coal and metal-ore deposits all across the Cordillera is distinctly regular (Figs. 68-73). Oil and gas fields are essentially restricted to the eastern Rocky Mountain fold-and-thrust belt, where they are trapped in deformed rocks of the cratonic cover. Coal fields are common there and in the semi-reworked Intermontane Insular belts. Metals are concentrated in zones containing certain structural features (anticlines, faults) and rocks assemblages (sedimentary, volcanic, plutonic). Stratiform polymetallic deposits of Zn, Pb etc. and skarn deposits of W, Mo and Su tend to be found in miogeosynclinal rocks - in the Kootenay Arc, Cassiar platform and Nisling terrane. The Intermontane median-massif belt hosts volcanogenic massive sulfide (VMS) deposits of Cu and Zn, and polymetallic porphyry ores of Cu, Mo and other base metals. Mafic-ultramafic complexes in the Intermontane Belt are rich in Cu, Ni, Cr and platinum-group elements, as well as non-ore deposits of asbestos, jade and so on. Au is particularly common in the Coast Belt orogen, which is ~80% made of granitic rocks.

Distribution of metals and ore types and ages in the Cordillera shows strong links with the major rock types, structures and tectonic regimes. For many decades, long before the emergence of plate-tectonic and terrane-tectonic concepts, empirical principles of regional metallogeny successfully guided the mineral exploration. Later, attempts were made to link metallogeny to presumed terranes and pre-, syn- and post-accretionary stages of terrane evolution. If the terranes formed in different parts of the globe and remained apart till the Mesozoic, their pre-accretionary metallogenic characteristics should be different and not correlatable. Such an approach restricts the normal metallogenic reasoning only to the time after the assumed amalgamation and accretion of terranes, and it randomizes the metallogeny of and hence exploration strategy for older deposits. In the modern thinking, terranes are assumed to be thrust slices confined essentially to the top part of the crust. Being disparate, they did not form common geologic and metallogenic systems till their amalgamation, and even after that they were displaced by strike-slip movements from their original positions in the Cordillera. Derived from oceanic crust, these essentially mafic terranes had to undergo transformation into juvenile continental crust (Condie and Chomiak, 1996).

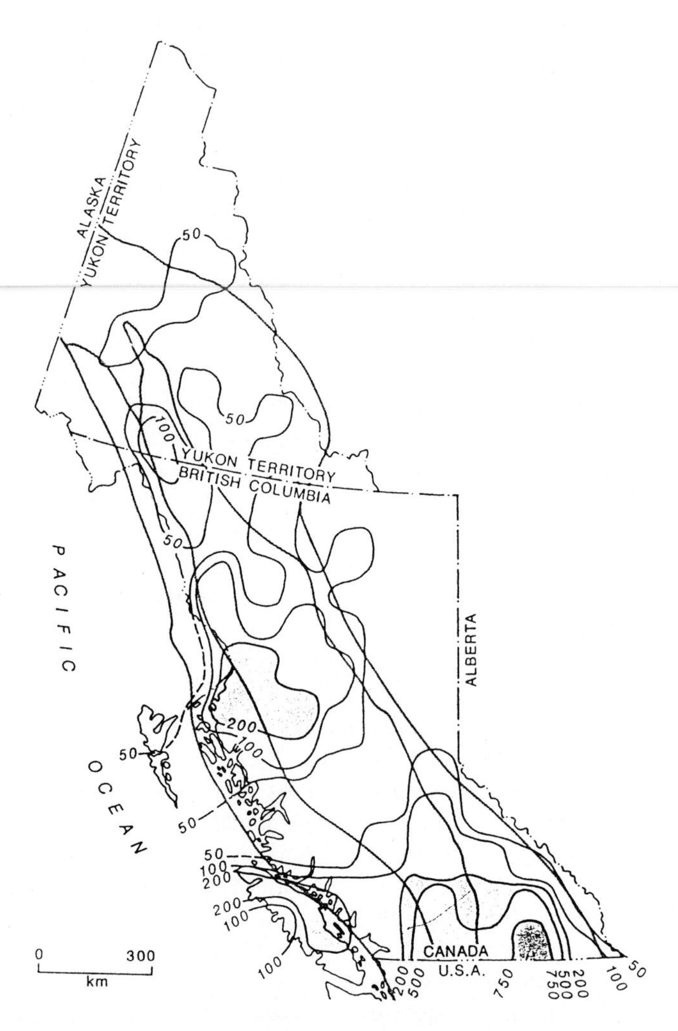

Figure 68. Distribution of mineral occurrences in the Canadian Cordillera (modified from Dawson et al., 1991). Contours are of the number of known mineral occurrences in each two-by-one-degree 1:250,000 map area.

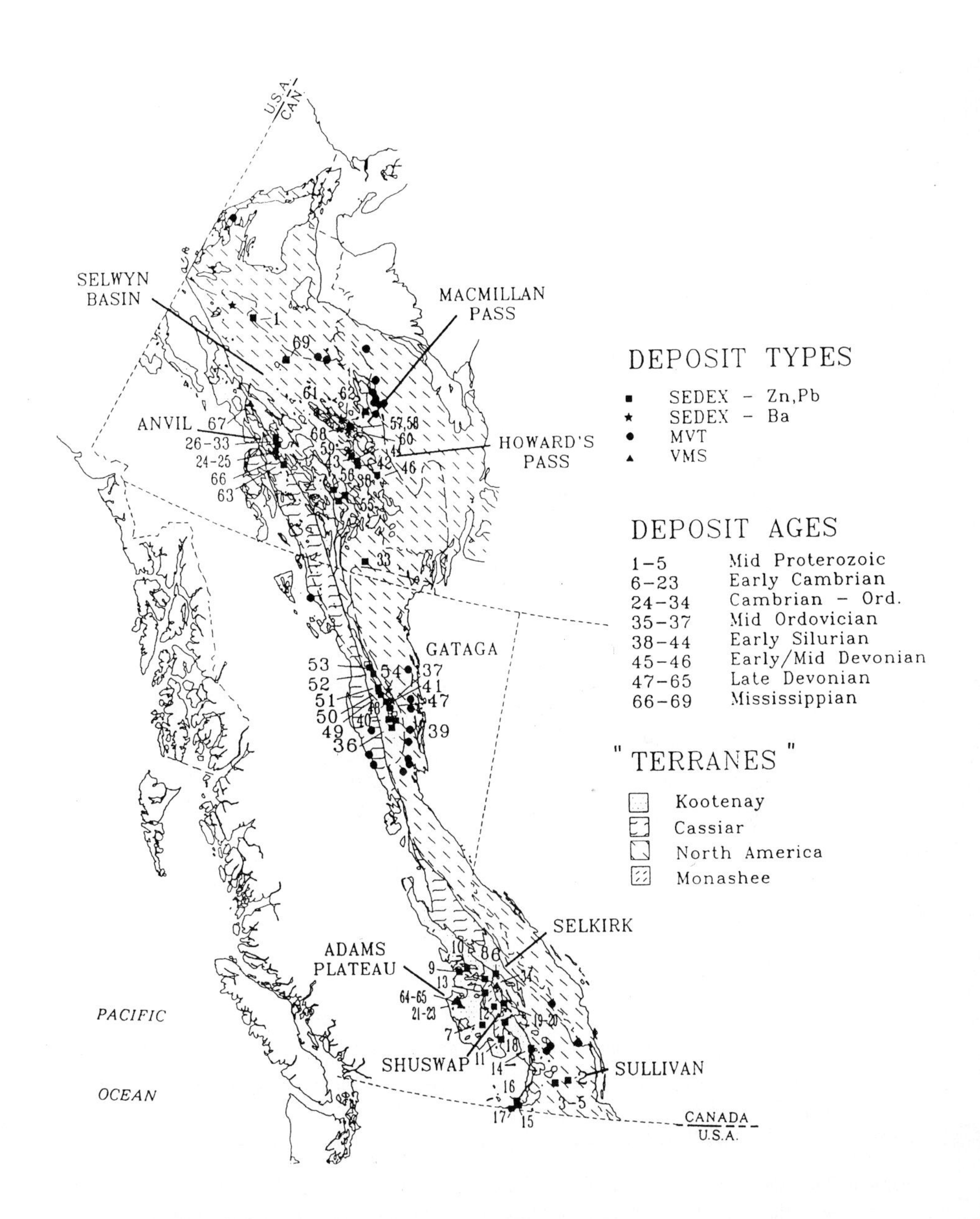

Figure 69. Distribution of sediment-hosted exhalative (SEDEX) mineral deposits in the Canadian Cordillera (modified from MacIntyre, 1991). Other abbreviations of mineral-deposit types: MVT - Mississippi-Valley-type; VMS - volcanogenic massive sulfide.

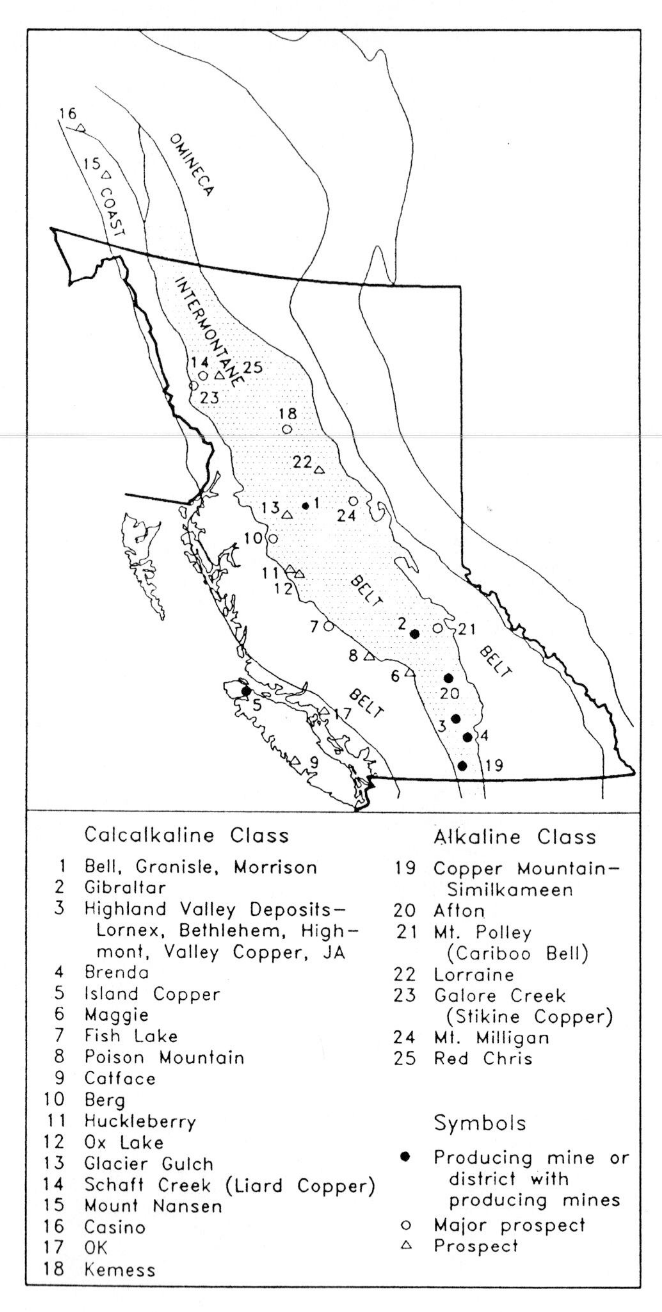

Figure 70. Distribution of porphyry deposits in the Canadian Cordillera (modified from McMillan, 1991b). The heavy outline shows the borders of British Columbia.

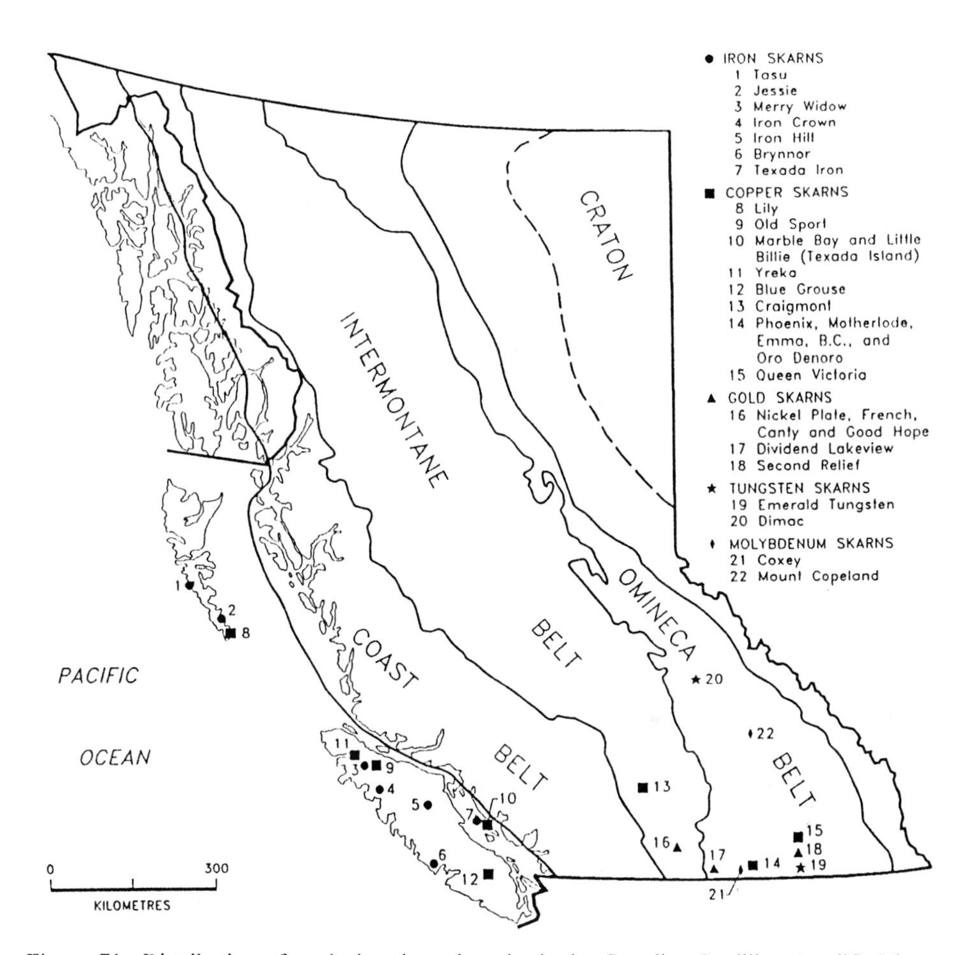

Figure 71. Distribution of producing skarn deposits in the Canadian Cordillera (modified from Ray and Webster, 1991). The heavy outline shows British Columbia. The dashed line separates the rootless Rocky Mountain fold-and-thrust belt from the undisturbed part of the North American craton.

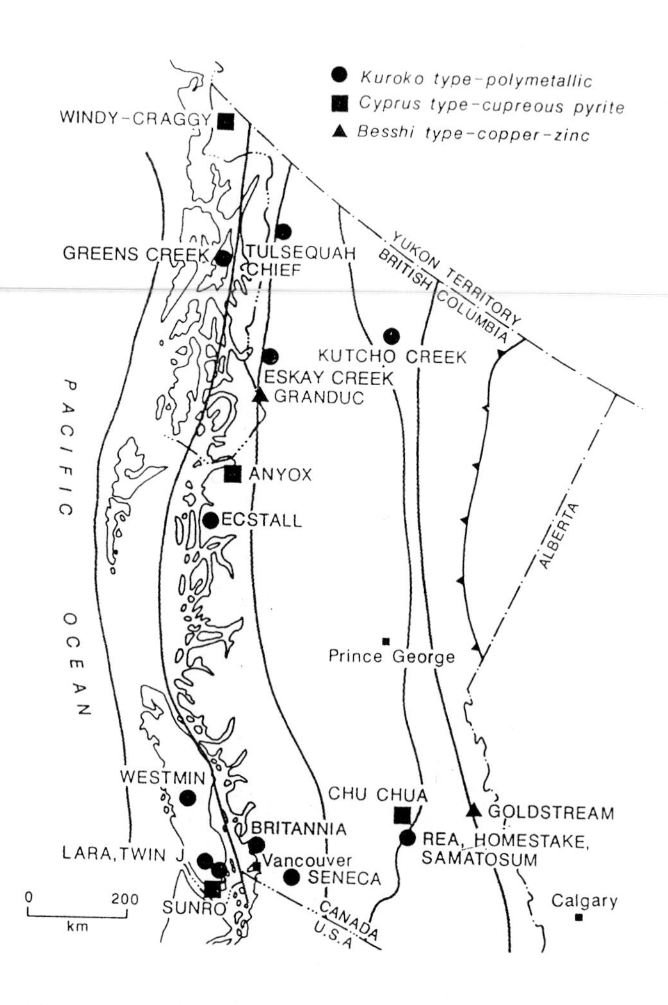

Figure 72. Distribution of volcanogenic massive sulfide (VMS) deposits in the Canadian Cordillera (modified from Dawson et al., 1991).

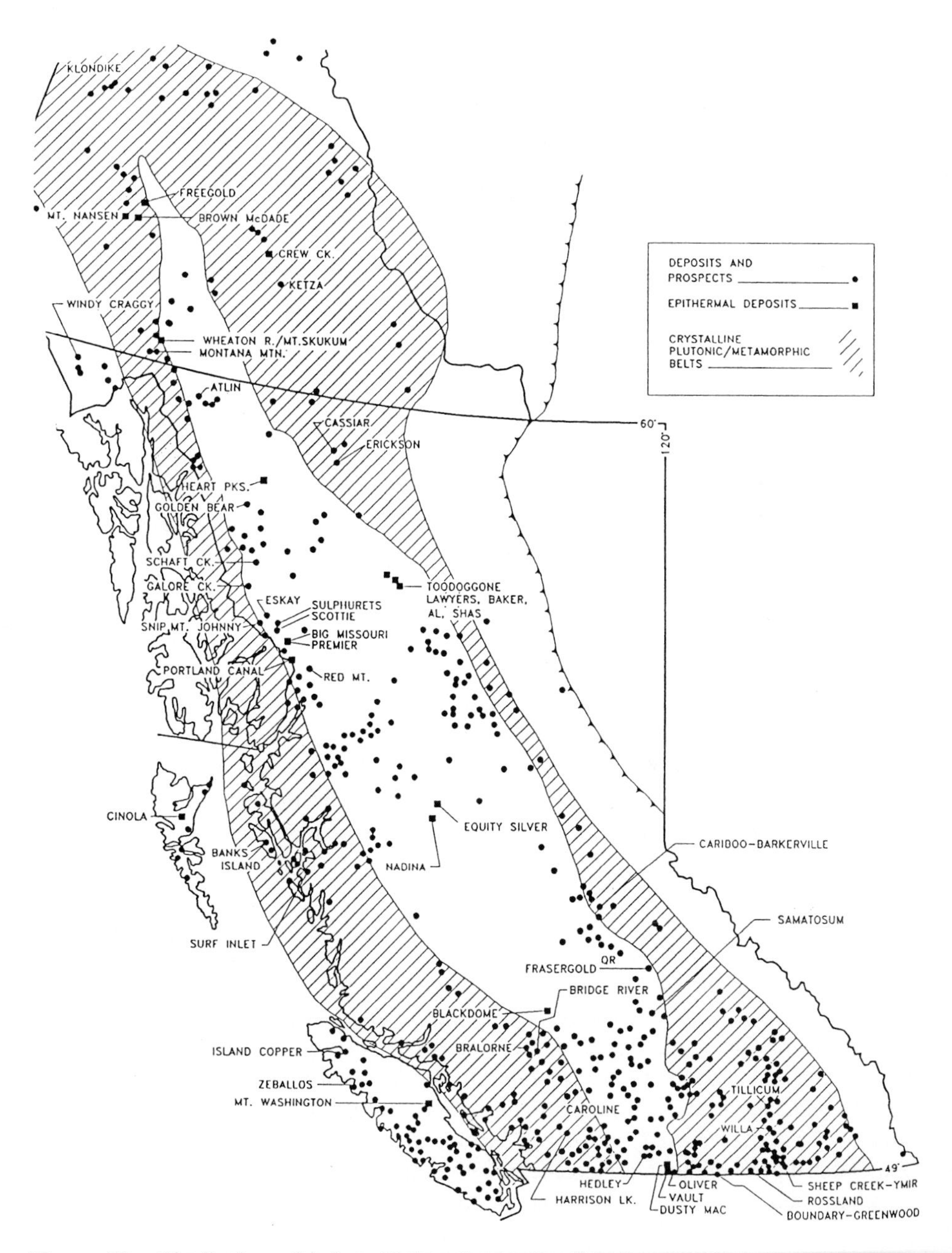

Figure 73. Distribution of lode gold deposits in the Canadian Cordillera (modified from Panteleyev, 1991). The transverse Skeena Arch in central British Columbia is clearly expressed.

How exactly a relatively simple chemical system of the oceanic lithosphere can change into a very complex continental-lithosphere system is unclear. Condie and Chomiak (1996, p. 101-102) speculated that "although some Mesozoic oceanic terranes began to evolve into continental crust before accretion to North America, most of the transition occurred during and shortly after accretion... The accreted Mesozoic crust has not yet evolved into mature continental crust..." The changes to oceanic-crust terranes needed to make them part of continental crust would be fundamental and, depending on the geologic peculiarities different from region to region. Yet, thermo-metallogenic zonation in large segments of the circum-Pacific mobile megabelts is remarkably consistent, particularly in western North America. This zonation corresponds not to speculatively and arbitrarily defined terranes, but rather to lithospheric belts well distinguished as miogeosynclines, median massifs and orogenic zones of various ages.

Porphyry deposits provide much of the world's Cu and Mo (Sutherland Brown, ed., 1976; Titley, ed., 1982). Generally, worldwide Cu-Mo prophyry deposits are associated with intermediate to felsic hypabyssal intrusions common in semi-stable median massifs. Continuous plutono-volcanic systems created favorable conditions for various ores, depending on the composition of intruding magmas and country rocks. Contact metamorphism in aureoles, particularly around alkalic intrusions, may have a significant impact on ore characteristics (Dawson et al., 1991).

Episodes of maximum granite-granodiorite magmatism in the Canadian Cordilleran interior are related to the Mesozoic orogenies. Owing to teleorogenic magmatism, the Intermontane Belt was affected along with the Omineca and Coast Belt orogenic zones. Association of the genesis of Cu, Mo, Sn, W, Ni, Co, Au and Ag (Sinclair, 1986; Dawson et al., 1991; McMillan, 1991a-b) with specific tectono-magmatic episodes is quite obvious. Mo-bearing magmas are usually derived from melting of sedimentary rocks that contained continent-derived detritus. In the subduction-accretion hypothesis of Cordilleran evolution, these magmas were produced when sedimentary prisms at Mesozoic subduction zones were placed at depths of some kilometers. On the other hand, lower-level rocks of the sedimentary cover of the Intermontane median-massif belt could also provide the Mo if buried deep enough. Such an event of deep block subsidence seems to have occurred in the Bowser and Methow Basin areas at the end of the Jurassic, marked by emplacement of the large Endako deposit at ~143 Ma (near the Jurassic-Cretaceous boundary; Anderson et al., 1998). Polymetallic porphyry deposits tend to be concentrated in zones rimming the deepest-buried parts of the Bowser and Methow-Tyaughton basins, as well as inboard along the middle zones of the Stikine and Quesnel parts of the Intermontane Belt. (A rootless fold-and-thrust zone induced by the Coast Belt orogen seems to distort this regular arrangement in the northern part of the Intermontane Belt.) Late Miocene is the age of the youngest known Mo-bearing deposits in the Coast Belt.

A common practice is to relate Mo-bearing deposits, including the world-class Endako one, to volcanic-arc magmatism. However, no clear choice has been made whether the supposed tectonic setting was island-arc or continental-arc. Anderson et al. (1998, p. 206) believed the Endako area contains Andean-type magmatic suites formed in continental-margin arcs. Dawson et al. (1991, p. 752), by contrast, favored a pre-accretionary setting predating terrane amalgamation with the North American continent.

Another economically important type of mineral deposits in the Canadian Cordillera is numerous ore bodies of massive sulfide (Höy, 1991). Many of them are large, and Windy Craggy is a giant. The VMS deposits, as the name implies, have a clear association with volcanogenic rocks. They are located in orogenic zones outside the Intermontane Belt (Hutchinson, 1980; Franklin et al., 1981; Lydon, 1984, 1988). Their paragenesis varies, but most of them fall into two classes: rich in Cu and Zn, or rich in Zn, Pb and Cu. Cupriferous ores are hosted in mafic volcanic and stratal sedimentary rocks, and polymetallic ores in felsic volcanic rocks and breccias. The former might have accumulated in deep-sea environments (e.g., trenches in subduction zones), the latter require abundant continental-crust contamination. Cyprus-type Cu ores in the Mediterranean and Besshi-type Cu-Zn ores in southeast Asia may reflect local and shallow subduction settings. Kuroko-type (Japan) polymetallic ores distinctly require big blocks of continental crust, remobilized in orogenic zones and sourcing felsic magmas and Zn-Pb-Ag concentrations (Panteleyev, 1991). Kuroko-type deposits in the Canadian Cordillera are found on both eastern and western rims of the Intermontane Belt, especially at its southern and northern ends (Kutcho and Eskay Creek in the north, Samatosum and Seneca in the south). Close spatial association of Kuroko-type deposits with big VMS deposits of other types complicates assignments of tectonic settings. Even "black smokers" can be found not only in pelagic oceanic regions but also in semi-enclosed basins at continental margins.

The miogeosynclinal zone of the Cordillera contains W-rich skarn deposits, and Au skarns are particularly common in the Kootenay Arc's stratal rock assemblages. Stratiform SEDEX (sediment-hosted exhalative) deposits, including the huge Sullivan accumulation of polymetallic ores, lie in the Omineca Belt, as do barite-bearing deposits.

Metallogenic zonation in the Canadian Cordillera ignores the assumed terrane boundaries. Instead, metallogenic belts are correlated with the real lateral tectonic units: miogeosynclinal and eugeosynclinal orogenic zones as well as median massifs. Systems of metallic elements are very sensitive to the types of crust and lithosphere (e.g., Bilibin, 1955) as well as to the tectonic regimes. The hypothetical thin terrane tablets are incapable of self-development, whereas big and self-developing crustal and lithospheric blocks serve as reservoirs of compatible and incompatible, lithophile and chalcophile elements. Their distribution defined the metallogenic zonation, accounting for its correspondence with real lateral tectonic units but not with assumed terranes. Prominent zones of crustal weakness, such as the Skeena and Stikine arches, served as conduits for mineral-bearing fluids, which explains the location of many deposits in such zones.

Wheeler and Gabrielse (coords., 1972) discriminated five belts in the Canadian Cordillera, from Rocky Mountain to Insular. Because the former is a rootless fold-and-thrust stack, only four belts correspond to the deep grain of the Cordilleran mobile megabelt. Monger et al. (1972) put this picture together with the concept of accreted terranes, and for the next two decades the definition of geological and morphological belts was subordinated to the seemingly-triumphant terrane definitions (Monger et al., 1982, 1991, 1994; Gabrielse et al., 1991a). But this triumph has turned out to be short-lived.

The revision of terrane boundaries, nature and times of accretion is under way (e.g., van der Heyden, 1992; Thompson, 1998). Field mapping shows the Kootenay-Cassiar "terranes" are rather native (Klepacki and Wheeler, 1985), as is the Quesnel terrane. Magmatic rock assemblages and trends in the Coast Belt do not fit this hypothesis (Woodsworth et al., 1991), and many terranes in the Cordilleran interior are being redefined (Leitch et al., 1991; Currie and Parrish, 1997).

In the reasoning of Gabrielse et al. (1991a), the appearance of four Cordilleran belts with contrasting geological and geophysical properties was a secondary phenomenon, following the accretion of terranes which ended in the Paleocene (Monger et al., 1994). But the principal metallogenic zonation corresponds to the belts and not to the assumed terranes. These belts are whole-crust entities with a long history, reflecting the *in situ* evolution of the main lateral tectonic units in the Cordilleran mobile megabelt.

When the prevailing fashion was to define ever more and smaller terranes, Monger and Berg (1987) defined three terranes at the Intermontane-Coast belt boundary: Bridge River, Cadwallader and Methow. The Bridge River terrane was thought to be oceanic, containing mafic-ultramafic complexes (Schiarizza et al., 1989), and the Cadwallader terrane was ascribed to an island arc. The Methow terrane was defined as made up of sedimentary rocks overlapping the rocks of the Cadwallader terrane and Intermontane Belt. The lower stratigraphic levels in the Bridge River and Cadwallader terranes may be Permian or older (Leitch et al., 1991), and it was logical to correlate the Bridge River formations with those assigned to the Cache Creek terrane (also Mortimer et al., 1990). Also logically, it was suggested that the Cadwallader and Methow terranes are former parts of a single basin linked to the heavily faulted Stikine block in the Intermontane median-massif belt (Friedman et al., 1998). Middle to late Mesozoic sedimentary rocks of the Methow-Tyaughton Basin contain clasts shed from the Intermontane Belt's Quesnel block. Such intimate links between the presumably exotic terranes contradict the conventional hypothesis of a terrane collage.

Soft subduction of some sort may have occurred in the Cordilleran interior in the Paleozoic. This is quite different from large-scale Benioff-Wadati subduction zones, but it is more consistent with the rock evidence from the Cordillera. Such an evolution would account for broad fields of elevated mineral potential, with a smeared distribution of antimonite, cinnabar, scheelite, sphalerite and other important deposits (see, e.g., Woodsworth, 1977). In the Late Cretaceous to Paleocene, in response to upper-crust brittle faulting, mesothermal and epithermal Au mineralization was superimposed over the southern Intermontane-Coast belt boundary area (Bridge River Camp area). Two main episodes of Au mineralization are recognized in the Canadian Cordillera: Late Triassic-Jurassic and Early Tertiary. They occurred during the post-Nevadan and post-Laramian crustal extension (Barr, 1980). The Bridge River Camp area is thus untypical: the principal Bralorne-Pioneer gold field there is related to extensive (to a depth of >2,000 m) veins dated as 91.4 to 85.7 Ma (Leitch et al., 1991).

Many other vein Au deposits in the Canadian Cordillera were associated with the late and post-Laramian regional extension (in the Stikine and Quesnel blocks; Dawson et al., 1991) and with thermo-tectonic activity induced from the Coast Belt orogenic zone (Woodsworth et al., 1977; Sinclair et al., 1978). Au deposits in the Bridge River Camp area become more epithermal in a direction away from the Coast Belt, to the east, where the oldest and largest deposits such as Bralorne-Pioneer are less eroded. General crustal tilting away from the still-rising Coast Belt orogen is evident even in modern topography (Mathews, 1986). In the Skeena Arch area, this overall trend is distorted by the NE-SW zone of crustal weakness. Mineralogically important epithermal vein systems can be correlated with local Cretaceous and Tertiary stress regimes, including the 95-50-Ma Skeena mineralogenic belt (roughly parallel to the Skeena Arch) and the Laramide porphyry belt (from 75 to 45 Ma) in the U.S. (MacIntyre and Diakow, 1989).

9 - PRACTICAL UTILITY OF ROCK EVIDENCE AND GEOPHYSICAL STUDIES IN RESTORING REGIONAL TECTONIC HISTORY IN MOBILE MEGABELTS

Diversity of crustal deformation types in the Cordilleran mobile megabelt

Tectonic processes generate various manifestations in rocks. Heating and cooling can lead to expansion and contraction, vertical uplift and subsidence of blocks, folding, faulting and transformation of rocks by metamorphism and magmatism. Structures created by tectonic processes depend on the composition and physical state of the rocks, particularly on whether the rocks are brittle or ductile (Wegmann, 1930; van Bemmelen, 1967). Current interpretations of deep reflection seismic profiles in the U.S. and Canadian Cordillera (Allmendinger et al., 1986, 1987; Hauge et al., 1987; Hauser et al., 1987; Potter et al., 1986, 1987; Clowes et al., 1992, 1995, 1998; Cook et al., 1992; Varsek et al., 1993; Cook et al., 1995a-c) miss two crucial points: these data (a) generally do not image steep (usually, >60°) faults and (b) are unable to determine the geologic nature of refections.

Historically, tectonics originated as a specific field of geological knowledge with the studies of deformation of stratal rocks. Regional buckling, folding, crumbling and faulting came to be recognized as tectonic phenomena. In cratonic platforms, big flat-lying stratal bodies may also be regarded as tectonic: their very existence, as well as lithofacies content, thickness, wedge-outs and structural deformation, reflect movements of crustal blocks on which these cover units lie (see Lyatsky et al., 1999). Cratonic platforms' basement contains Precambrian rock. Laterally, these old basement units and structures may continue away from the platforms into the basement of neighboring mobile megabelts on the craton's flanks.

Non-cratonic platforms exist also. If the platform's basement is incompletely cratonized and has an age similar to that of adjacent late Precambrian-Phanerozoic mobile megabelts, the platform is regarded as young. An example is the epi-Variscan Turanian platform in central Asia. It formed when that part of the Eurasian mobile megabelt ceased its orogenic development, but this young platform remains part of the huge Eurasian megabelt rather than of any new craton.

Whereas old (Archean-Proterozoic) platforms are constituent lateral tectonic units of cratons, young, post-cratonic platforms on Phanerozoic or Neoproterozoic basement are tectonic units of mobile megabelts. In some phenomenological aspects - magmatic, deformational - median massifs resemble young tectonic platforms. The difference is that massifs develop owing to incomplete orogenic reworking of their crust during orogenic pulses, whereas young platforms form by platform-like cover deposition on top of megabelt crust whose orogenic development had stopped. The Cordillera contains several big median massifs, but only one big young platform (on the North Slope of Alaska).

Median massifs, orogenic zones and young non-cratonic platforms are distinguished by their dissimilar intensity and persistence of orogenic processes. The most visible geologic manifestations of these processes are commonly deformational. Stratal rocks are folded and faulted especially strongly in orogens (hence the old word *foldbelt*). Axial parts of orogenic zones contain rocks that had been lowered to >20 km depth, where they underwent ductile deformation and high-grade metamorphism. On an orogenic zone's flanks, outward-vergent fold-and-thrust belts formed predominantly in a brittle regime. Deformation in thick volcano-sedimentary cover of median massifs may also be strong compared to that in cratonic platforms, especially near the massifs' boundaries with orogens, but generally this deformation is considerably gentler than in orogens, as the degree of crustal reworking in massifs is lower. The least deformed, metamorphosed and magmatized are young platforms, because they lie in those parts of a megabelt where orogenic manifestations ceased.

Model-driven interpretation of Cordilleran tectonics reduces the continental crust to being a passive subject of formative influences supplied from the continental margin: the Cordilleran crust west of the miogeosyncline was "created" by terrane accretion, thrust stacking of accreted terranes, and subduction-related magma emplacement (Monger et al., 1994). In fact, the tectonic fabric of the Cordillera owes its existence to a combination of inherited pre-megabelt crustal features and megabelt-age tectonic reworking. A crucial thing to remember is that continental crust is not inert, but rather a self-developing geodynamical system.

Existence throughout the Cordilleran mobile megabelt of basement-related protoliths, basement-sourced magmas, basement-involving ductile deformation points to considerable geologic inheritance from pre-megabelt time. The extent of orogenic reworking varies within the megabelt from one zone to another. Zones of intense faulting can be impressively large, but structural manifestations are just one of the four main manifestations of tectonism (the others are sedimentological, magmatic and metamorphic). Some of the largest fault zones, even those associated with mafic-ultramafic rock complexes, were short-lived and not responsible for first-order crustal-unit boundaries in the

Cordillera. On the other hand, less visible but very ancient zones of crustal weakness inherited from pre-megabelt time expressed themselves repeatedly as transcurrent structural lineaments and zones crossing the principal Cordilleran tectonic units. Their reactivation was triggered by tectonism in the megabelt. The orogenic zones, for their part, developed through several cycles, each involving compression and extension, burial and uplift, heating and cooling.

Monger et al. (1994) wrongly restricted regional tectonic extension in the Canadian Cordilleran interior to only a single, Eocene episode; Cook (1995a-c) followed this hypothesis. Yet, crustal extension occurred in the Omineca miogeosynclinal orogenic zone each time an orogenic cycle drew to a close. Both the Omineca and Coast Belt orogens are polycyclic and structurally bilateral. Their flanks facing the more-rigid and less-reworked Intermontane median-massif belt demonstrate the strong influence of this belt on the deformation styles in the Cordillera.

Contractional and extensional structures in the Canadian Cordilleran interior did not develop just once in the Mesozoic and early Cenozoic, as the prevailing tectonic model assumes. Rather, they developed several times, in several cycles, throughout the Phanerozoic.

Shortcomings in many poorly substantiated numerical models, which grossly oversimplify tectonics, are now abundantly clear. One set of such models, made to predict crustal regimes and rheologies, has recently come under critical review after it signally failed to predict the temperatures in the German superdeep KTB drillhole, forcing the drilling to stop before reaching the intended total depth (Clauser et al., 1997; Emmermann and Lauterjung, 1997). There, heat production and circulation in an epi-Variscan young platform was greatly underestimated, and temperature-dependent rock properties (including acoustic ones) were predicted incorrectly. The KTB hole is thought to have reached the zone of brittle-ductile transition. Remnant of older such transitions, like older Mohos, formed in different temperature and pressure conditions, may be preserved in some localities. In the Canadian Cordillera, such old transition zones are common in many areas, including the miogeosynclinal Omineca orogenic zone (e.g., in the Shuswap field of metamorphic core complexes), eugeosynclinal Coast Belt orogenic zone (e.g., in the Cascade field of metamorphic core complexes) and Intermontane median-massif belt (e.g., in the Nicola structure).

Existence in the continental crust of anastomosing systems of mylonitic zones, shears and detachments was proved by the Arizona-1 deep well in the southwestern U.S.

Most faults there are low-angle (<30°) or moderately dipping (30° to 60°). Many disparate and controversial models have been proposed about how these faults might be connected (e.g., Beratan, ed., 1996). Some authors believe the low-angle fabric predates the steeper fabric, others favor the opposite. Both scenarios are possible, and which set of structures came first can vary from area to area. At any rate, presence of crustal blocks bounded by steep faults and diapir-like structures of overheated lower-crustal material is supported by field observations and results of isotopic, potential-field, seismic and paleomagnetic studies (e.g., Lister and Davis, 1989; Livaccari et al., 1995; Brocher et al., 1998).

For Eocene time, Monger et al. (1994) divided most of the Cordillera near the Canada-U.S. border into two tectonic domains: the domain of crustal extension in the east and the domain of transtension in the west. Phenomenological criteria for this division were not explained, and a gradation between these postulated domains was pointed out. Metamorphic core complexes of Early Tertiary age are present in both these domains. South of the Canada-U.S. border, the COCORP deep seismic data were reported as showing east-dipping reflections cutting west-dipping ones (Potter et al., 1986). North of the border, west-dipping reflections have been interpreted to indicate a sliced crustal structure (with a notable exception of the assumed-major, east-dipping Slocan Lake fault, which is assumed to cut the principal structural fabric; Cook et al., 1992). In both these interpretations, the geologically proved high-angle faults are absent.

Nonetheless, steep faults and the crustal blocks they bound determined the structure of the Cordilleran mobile megabelt, including its three main interior lateral tectonic units - the Omineca, Intermontane and Coast belts. All three are long-lived, with Precambrian continental-crust basements, complex magmatic, metamorphic and deformational histories, and repeated vertical movements of these entire belts and their constituent blocks. Kanasewich et al. (1994) emphasized that trans-crustal low-angle projections of discontinuous ("floating") reflections, such as those arbitrarily designated "Monashee" and "Gwillim Creek" (Cook et al., 1992), are not substantiated by other seismic (chiefly refraction) data. Assumption-based speculations about the origins of seismic images without the benefit of rigorous geological confirmation can only confuse the regional tectonic analysis.

Traces of big faults at the ground surface, and senses of motion determined for them by field mapping, are much more reliable indicators of the nature of these faults. Steep faults tend to be straight regardless of topography, whereas low-angle ones usually have sinuous traces if the topography is not flat. The main geologic belts of the Canadian Cordilleran interior (Omineca, Intermontane, Coast) are strongly linear, oriented NNW-

SSE. Although their deep crustal boundaries are somewhat masked by surficial geomorphological processes and orogen-flanking fold-and-thrust zones, they are controlled by deep-seated faults which separate the lateral tectonic units.

These high-angle faults determined the position of fundamental lithospheric discontinuities, which are manifested in the zonation of metamorphic and magmatic phenomena, deformation styles and metallogenic peculiarities. In the lower crust, ductile flowage of rocks may distort the downward projection of boundary faults to some degree, but confirmed persistence of many major continental crustal boundaries for billions of years points to some overall stability of their deep lithospheric roots. In the Cordilleran mobile megabelt in Canada, major NNW-SSE Late Proterozoic and Phanerozoic discontinuities define the position of principal lateral tectonic units, including the Omineca and Coast Belt orogens and their axial zones (Wheeler and McFeely, comps., 1991; Woodsworth et al., 1991). Contrasting geochemical reservoirs on both sides of the $^{87}Sr/^{86}Sr$ initial-ratio isopleths of 0.704 and 0.705 are correlatable, with some offsets, to fundamental mobile-megabelt crustal units. Near the Canada-U.S. border, this boundary is located in the middle of the Intermontane Belt and has a more N-S trend. In a 50-km-wide zone, the 0.704 isopleth remained stable for a long time from the Middle Jurassic through Cretaceous to Paleocene-Eocene time (Armstrong and Ghosh, 1990). Its position and orientation remained rather steady during the time when the terrane hypothesis (Monger, 1993; Monger et al., 1994; Cook, 1995a-c) calls for the main episodes of terrane-accretion tectonism.

Attention to the basement of mobile megabelts is traditional among European and Russian geologists. In North America, Rodgers (1987, 1995) recommended caution in the interpretations of the basement in the Appalachians and Cordillera. In the Lithoprobe practice, the geophysically modeled lower crust is equated with the Precambrian basement of the modern North American craton (Clowes et al., 1992, 1995, 1998; Cook, 1995a-c). This equation is incorrect, because the Cordilleran basement belonged not to the modern craton (which did not yet exist) but to one of its predecessors that had a different configuration. At present, the ex-cratonic Cordilleran basement is much too reworked, even in the median massifs, to be considered cratonic. Besides, seismic-data-based subdivision of the crust into layers with certain acoustic properties may not be correlative with vertical tectonic units, and seismic data are unable to discriminate between rocks by age. Old rocks probably do lie at the lower levels of Cordilleran crust, but they are reworked in the megabelt (especially in orogenic zones) and therefore not cratonic. Where ancient rocks are identified in the Cordillera on the ground surface, crustal-scale thrusting is far from the only mechanism to bring them there. Early Proterozoic and even Archean rocks are common in the metamorphic core-complex areas (e.g., Doughty et al., 1998), and xenoliths with Proterozoic ages are encountered often in magmatic rock bodies across the Intermontane and Yukon-Tanana median massifs as well as in the Omineca and Coast Belt orogens.

Cook (1995b-c) modeled the assumed westward basement extension as far as the eastern Coast Belt to be a thin lower-crustal tongue. From geological, refraction seismic and electromagnetic viewpoints, this model is unrealistic (Kanasewich et al., 1994; Jones and Gough, 1995). Parrish et al. (1988) and Parrish (1995) noted that the Monashee décollement, where it is geologically mappable, does not coincide with the basement-cover contact but rather is a low-angle fault zone cutting through both the basement and the cover (also Journeay, 1986). The Monashee décollement was at one time interpreted as part of a master fault underlying and detaching the Laramian-age Rocky Mountain fold-and-thrust belt (Brown and Read, 1983). Carr (1992) continued the Rocky Mountain fold-and-thrust detachment into the Valhalla-complex area. Doughty et al. (1998) speculated that the Spokane dome mylonite zone may be a southern continuation of the same regional décollement. In a Lithoprobe-like supposition, Evenchick (1991) proposed this basal detachment underlies the Intermontane and Coast belts.

In the same logic, thin-skinned (above-basement) shortening was supplemented, west of the Rocky Mountain Trench in the north and Purcell Trench in the south, by imbrication of the crystalline basement and its involvement in the shortening (cp. Brown and Carr, 1990). The assumed Selkirk allochthon, thought to have mid-crustal rock affinities, was interpreted as translated ~80 km east from its original position somewhere in the Intermontane Belt (e.g., Brown et al., 1992). In the prevailing hypothetical scenario, the Selkirk allochthon is considered to have been involved in terrane collision, obduction and backthrusting in a supposed former continental-margin zone which is buried in the crust at a depth of ~25 km. The assumed Jurassic collision "deformed and thickened the continental margin" and "initiated the development of [the] Omineca Belt" which became "the metamorphic and plutonic hinterland of the Rocky Mountain Foreland Thrust and Fold Belt" (Brown and Carr, 1990).

The miogeosynclinal Omineca orogenic zone was in fact initiated long before the Jurassic and even the Mesozoic. The large mid-Cretaceous magmatic episode there has petrological and geochemical signatures suggesting not subduction but an intracontinental tectonic background (Driver et al., 1998). How the long-lived, poly-orogenic Omineca Belt can be a "metamorphic and plutonic hinterland" to the very short-lived and shallow Rocky Mountain fold-and-thrust belt is unclear. More accurately, and restrictively, only the Laramian-stage Omineca orogen could have been a hinterland to the Laramian Rocky Mountain deformed belt.

Metamorphic and plutonic history of Selkirk-fan rocks contains several distinct episodes (Greenwood et al., 1991; Woodsworth et al., 1991). Middle Jurassic (ca. 170 Ma) burial metamorphism is predominant on the fan's west-verging flank, and mid-Cretaceous (ca. 100 Ma) metamorphism on its east-verging flank (Marchildon et al., 1998). The latter is correlative in time with an episode of intrusive magmatism. A Late Jurassic-Early Cretaceous intrusion episode on the west flank of the fan took place at ~150 to 140 Ma. Marchildon et al. (1998) also noted that the distribution of the rock record of metamorphic pulses on the western flank of the Selkirk fan varies spatially.

Normal throw is recognized by geologic mapping at the Southern Rocky Mountain Trench and Purcell Trench faults, where previous thrust faults (Dogtooth, Purcell) are cut by normal faults of Tertiary age. Mapping of metamorphic grades suggests that on both sides of the Purcell thrust the metamorphic conditions may be rather similar (Gal and Ghent, 1990). Between the western Rocky Mountains and the northern Selkirk and Monashee mountains, however, west-side-up throw must be ~6 to 7 km in amplitude on a post-metamorphic fault (Greenwood et al., 1991).

The end of the Laramian metamorphic and magmatic episode in many parts of the Cordillera was apparently associated with extensional deformation. It is recognized in ductile shearing in the metamorphic core complexes, and elsewhere in grabens, half-grabens, steep brittle faults and veins. In the Valhalla core-complex area, the young Ladybird leucocratic granitic rocks formed in the Paleocene-Eocene below the ductile Valkyr shear zone; above this shear zone, these rocks are absent. In the Rocky Mountain Belt, the imbricated, brittle fold-and-thrust deformation ended in the Oligocene (Constenius, 1996). That belt is essentially amagmatic and not metamorphosed. The regional thrust detachment at the base of this zone of shallow deformation cannot be traced seismically beyond the Purcell Trench. There, coherent reflections that could be correlated with the top of the crystalline basement, which dip gently westward from the Interior Platform in Alberta and Montana, deepen more steeply and dissipate (Bally et al., 1966; Lyatsky et al., 1999). Seismic and structural fabrics change across the Kootenay Lake fracture zone; geometric correlations of reflections across this fundamental crustal boundary are questionable.

On the other hand, based on geologic observations of various exposed crustal levels, stratigraphic, metamorphic-grade and deformation-style correlations for the top ~20 km of the crust are possible across the entire northern Selkirk and Monashee mountains. From U-Pb geochronologic dating, the regional Middle Jurassic (~160 Ma) metamorphic peak in the southern Omineca Belt is represented widely. In many areas, it was overprinted by two younger thermal events, in the mid-Cretaceous (Marchildon et al., 1998) and Tertiary. A strong regional thermo-tectonic event is also recorded in a 70-Ma geochronologic peak (Crowley et al., 1998).

It is possible that crustal thickening due to shortening causes heating at crustal and subcrustal levels. As the accumulated heat breaks through the crust which acts like a thermal cap, vivid tectonic manifestations result at the surface (cp. Cooper, 1990). It is therefore natural for thermal relaxation and extension to come at the end of an orogenic event.

Benefits of compiling and critically reviewing geological and geophysical data to provide rational constraints on combined interpretations

Sedimentary, metamorphic, magmatic and deformational aspects of regional tectonics are essential in reconstructing the endogenic reworking of the crust during its evolution. In observed stratal rock assemblages, suites of cogenetic magmatic rocks, complex structural-formational étages, regional metamorphic domains and panels (including brittle-ductile transitions) one finds the rock-based information from which the geologic history is read. Deformation of rock bodies (by folding, faulting, buckling), analyzed in a specific geological discipline called structural geology, permits to understand the past tectonic regimes as well as the distribution and variations of regional stresses in space and time, in a reliable time frame obtained from observed properties of rocks. Gravity anomalies and topographic relief help understand the nature and extent of vertical tectonic movements (upward and downward motion of blocks and warping of various wavelengths). Insights from deep crustal seismic profiling are rather geometrical, but still very important for determining the distribution of physical properties and discontinuities in the rock mass.

Cook (1995b) noted that the geometry of the supposedly Proterozoic-age reflection Moho in the southern Canadian Cordillera is rarely correlative with the distribution of tectonic belts. If true, this would be strange - but it is not true. No continuous reflection band is apparent at the Moho level across the Cordillera (see also Cook et al., 1991a-b, 1992; Clowes et al., 1995). On the other hand, truncation of steep crustal faults by the reflection Moho is recognized there (Potter et al., 1986). The young (Cenozoic?) reflection Moho may be ascribed to diverse causes such as subhorizontal layered intrusions, metamorphic fronts or structural features.

Clowes et al. (1995) have noted that the lowest average seismic velocities of the lower crust and the deepest Moho in the eastern Canadian Cordillera are found in the eastern part of the Omineca Belt. In that area, the ex-cratonic basement seems to continue as far as the eastern strands of the Kootenay Lake fracture zone, but Lithoprobe publications associate the apparent change in crustal structure there with the assumed-major Slocan Lake fault. To justify the existence of such a trans-crustal fault, Cook (1995c, his Fig. 7b) offered a complicated interpretation of Lithoprobe Line 9. In its eastern part, the steep Hall Lake fault turns sharply low-angle to run under the western Purcell Anticlinorium. There and in the Kootenay Arc area, it is shown as being a subhorizontal undulating surface at ~5 s two-way traveltime (~14-15 km depth). Beneath, other undulating thrust surfaces are interpreted to lie above the flat top of the crystalline basement. The presumed-major, east-dipping Slocan Lake fault is shown as cutting all the thrusts and merging with the master detachment which coincides with the basement top. Now available with various processing, the data in Line 9 do not show this geometry. From its surface trace on the east side of the Valhalla metamorphic core complex, the relatively local Slocan Lake fault may correspond to an upper-crust reflection dipping ~30° to the east. It may be traced to ~10 km depth, where it loses definition at the bottom of the Nelson pluton. Yet, across a gap in the data, Cook (1995c) and Cook et al. (1992) continued this fault by correlating it with a "floating" reflector that lies at ~15 km depth and has no eastward continuity.

The Nelson pluton is one of a series of Middle Jurassic batholithic bodies, which may be connected at depth and have laccolith shapes (i.e., have flat bottoms; Simony and Carr, 1997). Such flat base surfaces could produce seismic reflections which should not be misinterpreted as low-angle faults. Only ductile shears with limited offset are expected at a rheologic boundary between a tonalitic pluton and country rocks.

Potential-field and seismic geophysical data can say nothing about the age and genesis of rocks. The assumed continuity of North American basement far into the Cordilleran interior, beyond the thin-skinned Rocky Mountain fold-and-thrust belt, cannot be justified with geophysical data alone. Burianyk and Kanasewich (1995) stated that, from recorded seismic characteristics, this cratonic basement does not continue beyond the western Boundary of the Omineca Belt. Kanasewich et al. (1994) found large disagreements between the interpretation of Lithoprobe seismic reflection (Cook et al., 1992) and refraction data in the same transect. They pointed out the discontinuous character of the "floating" reflections which Cook et al. (1992) had interpreted as a shear zone of whole-crust significance. Neither the Monashee not the Gwillim Creek shear zone has direct links with these "floating" seismic events. Even if correlations between upper-crust seismic reflections and surface fault traces are correct (which is not certain), the

Gwillim Creek and Monashee shear zones can be traced only to depths of ~15 km, in the upper and middle crust.

No steep crustal faults are clearly depicted in the Lithoprobe seismic data. The clearly steep Fraser fault (Jones and Gough, 1995), like many others, was misinterpreted as low-angle (Varsek et al., 1993). In Lithoprobe publications, the abundant steep faults mapped geologically at the surface are generally shown as flattening out at depth to link them with low-angle seismic events (e.g., Brown et al., 1992; Clowes et al., 1992; Cook et al., 1992). This unjustifiable bias creates a distorted picture of Cordilleran crustal structure. For instance, Johnson and Brown (1996, their Fig. 3) showed a crustal-scale Monashee detachment continuing for a distance of ~250 km across the Omineca miogeosyncline and reaching a depth of ~25 km near the Omineca-Intermontane belt boundary. However, Kanasewich et al. (1994, their Fig. 8) demonstrated that the "floating" seismic events, on which these interpretations rely (also Cook et al., 1992), are actually discontinuous and associated with minor changes in acoustic properties of rocks. This recognition, along with geological considerations, raises doubt about the overall correctness of Cook and Varsek's (1994) postulation from seismic data of "orogen-scale" décollements.

Evidence for both steep and low-angle faults in the Shuswap Lake area (Johnson, 1994) is not strange if the low-angle fault is associated with the dynamics of the Tertiary metamorphic core complexes. Outward-dipping low-angle normal faults on the flanks of rising domes, and shear zones of various sorts over such domes' tops, are not surprising. One low-angle fault, with seismically uncertain downward extent, is the Okanagan Valley fault west of the Okanagan metamorphic complex. In contrast, the long-lived steep faults separating the Omineca and Intermontane belts are among the fundamental zones of crustal weakness that controlled the division of the Cordilleran mobile megabelt into lateral tectonic units.

That crustal blocks recognized from multi-aspect (sedimentological, magmatic, metamorphic, deformational) rock-based analysis are often ignored in seismic interpretations illustrates the danger of excessive reliance on just one type of data. Sub-vertical crustal discontinuities, which are normally not seen in seismic images, are fundamental to the structure of the crust. They may also lack gravity and magnetic expression, if there are no associated lateral changes in rock density and magnetization. With these limitations in mind, the Lithoprobe and COCORP seismic data give invaluable if partial images of the deep structure of the Cordillera. Where reflection and refraction images differ, even those differences may be informative. The biggest discrepancies occur in areas where well-developed and incipient metamorphic core complexes disturb the crust. Some uncertainties also surround the definition of the crust-mantle boundary, including an ill-constrained transition zone a few kilometers thick.

To account for these discrepancies in the interpretations of reflection and refraction seismic data, Kanasewich et al. (1994, p. 2667) suggested these two geophysical methods may reveal different seismic characteristics and discontinuities in the rock mass: "It is possible that, in the middle and lower crust, the refraction data are sensing profound crustal velocity variations that have overprinted the structural fabric imaged by the reflection data". Similar mismatches between different seismic data sets have also been reported from another region of Tertiary extensional tectonics - the Basin and Range Province in the U.S. (Holbrook, 1990). The most probable explanation for such mismatches lies in the sensitivity of different seismic methods to different physical discontinuities. Anastomosing but local fault networks on top of metamorphic core complexes, for example, tend to be highly reflective, whereas refraction seismic data reveal overall seismic characteristics of the crust.

None of the Cordilleran metamorphic complexes is exhumed deeply enough to examine its roots. These complexes are disrupted by many intrusive igneous bodies of several generations and are emplaced into lower-grade country rocks. On their flanks, they are commonly rimmed by normal faults dipping outward - perhaps in a manner similar to rising igneous stocks. One such feature may be the so-called Nicola horst (plug) in the Intermontane median-massif belt, imaged in Lithoprobe Line 7 (Clowes et al., 1992, 1995; Cook et al., 1992; Varsek et al., 1993; Burianyk and Kanasewich, 1994; Cook, 1995b). The Moho beneath it is smooth and flat. The middle crust and the upper part of the lower crust are involved in a swell that pierces the upper-crustal horizons as well. A large granitic intrusion and some metamorphic rocks have been mapped there at the surface (Parrish and Monger, 1992). Around it, faults at different crustal levels dip outward. The Nicola plug may be an underdeveloped metamorphic core complex. The Nicola swell is ~100 km across at its base in the middle and lower crust and 15-20 km at the surface. Its structural relief is ~25 km, suggesting considerable displacement of remobilized crustal material. But, where available for observation, displacements on ductile shears have been found to be quite limited (in the Valhalla-complex area, Simony and Carr, 1997).

A prominent deep seismic reflection band separating the highly reflective lower crust from the underlying weakly reflective subcrustal intervals is fairly strong and continuous from the Monashee-Valhalla-core-complex area to the Intermontane-Coast belt boundary. This reflection band lies at a depth of ~35-36 km (11.5 s traveltime) in the east and ~32 km (10.5 s) in the west. This Moho surface has a smooth and gentle upwarp

~2 km in amplitude in the middle of the Intermontane median-massif belt. It is probable that the imaged Cordilleran Moho interface is young, as it is thought to be in the U.S. to the south (Potter et al., 1986). Clowes et al. (1992) and Cook et al. (1992) interpreted the Moho as very old, Proterozoic. The Cordilleran lowest crust in that model is equated with a thin and uniform layer of cratonic crust, continuing westward for hundreds of kilometers under the Cordillera from the Interior Platform without structural disruption (also Cook, 1995a-c). Above this assumed ancient sliver lies a trans-Cordilleran detachment. This unrealistic interpretation ignores all the metamorphic, magmatic and deformational evidence for repeated deep crustal reworking and mobilization in the Tertiary and earlier in the Phanerozoic.

As interpreted by Cook et al. (1991a-b, 1992), the Omineca Belt's crust contains a dramatic ramp in the cratonic basement that was preserved from Proterozoic time. The ramp flattens into the modern lower crust, and is overlain by the Vernon metamorphic-core antiform. Brown et al. (1992) speculated that crustal-scale shear zones with large displacements create a single system of major faults, combining the detachment at the base of the Rocky Mountain fold-and-thrust stack with the assumed décollements in the Cordilleran interior. But the reflections at the top of the crystalline basement of the Rocky Mountain Belt are seen in the Lithoprobe and other seismic data to steepen and dissipate somewhere beneath the Purcell Anticlinorium, and low-angle whole-crust discontinuities remain unproven. Clowes et al. (1992, p. 1823) speculated that the Vernon antiform appeared initially due to compression, even though metamorphic core complexes are conventionally linked to extensional crustal regimes in the Cordillera in the Tertiary (Beratan, ed., 1996). An antiform, by definition, has outward-dipping limbs. Yet, in the description of the Vernon antiform by Clowes et al. (1992), only its western boundary is considered, and treated as a listric fault. The Nicola plug or horst farther west was also considered to have formed during compression.

In more-mature metamorphic complexes, links with the lower crust may no longer be so direct. Necking may have separated these bodies from their lower-crustal source layer, analogous to the heads of rising granitic and salt diapirs. The rise of metamorphic complexes must be associated with extensive fracturing of the country rocks, both above and on the flanks. Seismic images of fractured zones depend strongly on fracture dips and orientations. Reflectors above the necked core-complex bodies generally dip away from body apices, whereas reflectors at the base dip towards the body centers. Reflectors related to necking may be symmetric or asymmetric, depending on the shape and motion history of the metamorphic body. In Line 9, the Valhalla complex appears as an asymmetric lens in the upper crust, to a maximum depth of ~18 km (profiles 4, 5). Outward-dipping faults, including the Slocan Lake fault on the east, lie mostly in the upper crust. At the bottom of this necked lens lies a family of inward-dipping seismic events.

The Cordilleran interior from Malton Gneiss in the east to Skagit Gneiss in the west contains metamorphic core complexes at different stages of their evolution, from initial swells to necked lenses, and relatively depressed crustal areas between them. The pattern of gneiss-granite domes and relative depressions between them is known in deeply eroded Precambrian terrains elsewhere (e.g., in Zimbabwe: McGregor, 1951; in the Canadian Shield: Langenberg, 1983). For the Cenozoic lower and middle crust in the Cordilleran interior, imaged remotely with geophysical methods, a similar picture is now emerging. Added to the networks of ancient and juvenile steep trans-crustal faults, this pattern of domes and sags makes the structure of the crust exceedingly complex and not subject to simplistic interpretations.

In the oversimplified and model-driven interpretation of Clowes et al. (1992, p. 1855), the Omineca, Intermontane and other Cordilleran belts to the west "...were emplaced in a compressional environment with primarily east-verging structures", while "...the lower crust and Moho have acted as a detachment zone during both compression and extension". Regional extensional structures in the Cordillera are ascribed in this scenario to only one, Eocene, episode of extension (Monger et al., 1994; Cook, 1995a-c). Rock-based evidence from geologic mapping, however, shows that such episodes were numerous. As is normal in mobile megabelts, each orogenic pulse - Antler, Nevadan, Columbian, Laramian - ended with an episode of crustal extension.

Brown et al. (1992), Cook et al. (1992) and Varsek et al. (1993) speculated that the Central Nicola antiform may be "structurally connected" to the metamorphic-core antiforms in the Omineca Belt (there is no evidence to suggest such connection), and that low-angle faults in the Cordilleran interior may be "kinematically linked" with those in the Rocky Mountains (there is no evidence for that either). Steep faults dividing the whole-crust or whole-lithosphere Cordilleran mobile megabelt seem to begin in the Rocky Mountain Trench. In the fold-and-thrust belt to the east, shallow and low-angle thrusts that die out towards the craton disturb the cratonic cover without involving the basement, and the steep faults in the basement are essentially of the same type, offset amplitude and history as those in the un-thrusted parts of the cratonic Interior Platform. Between the Rocky Mountain and Purcell trenches, basement is involved in the thrusting. In seismic sections, the major detachment at the base of the of the Rocky Mountain thrust stack dissipates on the west side of the Purcell Anticlinorium, and the steep Kootenay Lake fracture belt separates the eastern Cordilleran miogeosyncline from the Cordilleran internides. Steep faults mark the Omineca-Intermontane belt boundary.

Seismic reflections from the low-angle Okanagan Valley fault system, which dips westward at ~10°-15° (Johnson, 1990), are rather weak, whereas diffractions from steep faults in that area are common in Lithoprobe profiles 9 and 10 (cp. also profile 19).

Analysis of an anomalous and unexpected seismic event in a Lithoprobe broadside wide-angle survey has led Burianyk and Kanasewich (1997) to an idea that, beneath the Omineca miogeosyncline, the upper mantle is thermally layered (they improperly used the term *stratification* to describe this layering). A low-velocity zone, <=8.0 km/s, modeled at a depth of 59 to 62 km, was thought to have elevated temperatures. The base of modern lithosphere in that area was interpreted to lie at a depth of only ~50 km. The Moho was found to lie at 35-36 km, with normal sub-Moho velocities of ~8.0 km/s (also Kanasewich et al., 1994; Burianyk et al., 1997). Burianyk and Kanasewich (1995), in contrast to Cook (1995b-c), did not continue the seismically recognizable cratonic basement west of the Okanagan Lake. In any case, any old basement in the eastern Cordilleran mobile megabelt is tectonically very reworked. Archean to Early Proterozoic rocks in the metamorphic core complexes are just one of the many lines of evidence for this conclusion.

The eastern part of Lithoprobe profile 19 crosses the Monashee Complex and the Selkirk tectonic unit. In the hypothesis of Johnson and Brown (1996), the latest Laramian thrusting and extensional shearing could have occurred there almost synchronously, as the regional tectonic regime changed drastically from predominantly compressional to predominantly extensional. These shears supposedly formed in the Late Paleocene at a depth of ~15 km. Since then, the Monashee complex and the 10-20-km-thick Selkirk unit were elevated. The Monashee shear zone between them was arched upward, and in this shape it is now exposed in the Monashee Mountains. Dip-slip offsets have been estimated there to be 15 km on the low-angle normal Columbia River fault on the east, and 32 km on the low-angle normal Okanagan Valley fault on the west. High-angle faults have also been recognized there by mapping (Johnson, 1990). They are oriented N-S, parallel to the main structural trend of the Cenozoic rifting. Although these steep faults are treated as relatively insignificant in Johnson and Brown's (1996) interpretation, they are related to fundamental, deep-seated, high-angle fault systems which bound the main lateral tectonic units in the Cordilleran megabelt.

To the west, a family of N-S-trending steep faults is highlighted by considerable strike-slip offsets on some of them - including the Pasayten fault, which is conventionally regarded as Intermontane-Coast belt boundary. The long, steep Straight Creek-Fraser fault system runs in Washington state and southwestern British Columbia (Monger, 1985; Gabrielse and Yorath, eds., 1991; Monger et al., 1994). The Late Cretaceous-

Early Tertiary Ross Lake fault in the North Cascade Mountains contains well-studied segments where juxtaposition of the same protoliths regionally metamorphosed to different (but high: amphibolite, granulite) grades requires ~10 km of differential offset. Eocene dikes stitch these offset blocks together (Baldwin et al., 1997).

Types of crustal movements in the Cordilleran mobile megabelt and adjacent cratonic regions: further overview

Physiographically, the mountainous Cordillera is recognized in contrast to the great Plain regions in the middle of the North American continent. Rugged mountains are easy to set apart from flat Plains - especially in Canada, where the Rocky Mountains abruptly disrupt the flat prairie landscape. Although this topographic contrast is striking, for a tectonist who concerns himself with the structure of the entire crust or lithosphere the distinction is no so clear-cut.

The Rocky Mountain fold-and-thrust structural belt is underlain by the same, orogenically undisturbed crystalline basement as the Interior Platform. Likewise, in the U.S., the Colorado Plateau is tectonically similar to the Plains to the east. The tectonic opposite of a craton is not a mountain range but a mobile megabelt: Grenville-Appalachian in the east, Cordilleran in the west. The inland boundary of the Cordilleran mobile megabelt lies west of the Colorado Plateau in the U.S. (King, 1977) and along the major Rocky Mountain Trench fault in Canada.

Yet, some types of lithospheric tectonic movements that deform the crust on both sides of the craton-megabelt boundary are similar: vertical block displacements, warping and epeirogenic fluctuations. This partial tectonic similarity can lead to similar physiographic manifestations: mountains, for example, can form even on a cratonic background. In mobile megabelts and cratons alike, juvenile crustal movements are measurable: e.g., by geomorphological correlation of old topographic surfaces and river terraces. Preserved in recognizable topographic remnants, these fossil features are often called *neotectonic*. This distinguishes them from fossil tectonic features preserved in rock bodies (e.g., unconformity-bounded successions), as well as from features whose existence is identified from measurements of ongoing, current crustal movements (Lyatsky et al., 1999). The latter - contemporary movements - are conventionally attributed to *active tectonics* (McKenzie, 1972), but this term is unfortunate because not all contemporary movements are caused by currently-active tectonic processes.

In the Cordilleran mobile megabelt in Canada, extensive topographic plateaus dissected by deep river valleys are clearly apparent in the Intermontane median-massif belt (Holland, 1964; Bostock, 1970; Mathews, 1986, 1991). Even in the high mountain ranges of the Omineca Belt, old beveled topographic surfaces have been mapped. They look like subdued, slightly undulating, hilly plains, with gentle depressions and round summits (e.g., Nechako Uplift, Cariboo Plateau, Okanagan Highland, Purcell Mountains). Boundaries of big remnants of old beveled surfaces commonly coincide with steep faults controlling the Rocky Mountain, Purcell and Tintina topographic trenches. Glacial sculpting has modified these neotectonic patterns further.

The present-day continental divide between the Pacific and non-Pacific (Arctic and Atlantic) drainage systems runs NNW-SSE along the Inner Ridges of the Rocky Mountains, following roughly the axis of regional neotectonic warping (it is used to demarcate the Alberta-British Columbia provincial border and a county boundary in Montana). Some parts of the modern drainage pattern in the Cordillera (e.g., the Colorado River valley) show evidence of antecedent drainage along structural lineaments (e.g., King, 1977; Mathews, 1991). After the Early Tertiary thermal peak and regional peneplanation, regional uplift in the southern part of the Canadian Cordillera since the Middle Eocene was rapid enough to maintain the considerable elevation of the eastern Cordillera (Struik et al., 1997). Evidence of vertical block movements accommodated by steep faults, and extensional magmatism, are recognized in many parts of the miogeosyncline and neighboring pericratonic domains in western Canada and U.S., including the Northern and Southern Sweetgrass arches in Alberta and Montana (Peterson and Smith, 1986; Lipman, 1992).

The Eocene mafic alkalic suite includes dikes, sills, plugs and volcanic complexes that form buttes (Woodsworth et al., 1991). Their age is usually 52 to 47 Ma (Middle Eocene), which partly fits the reported timing of extension in the Rocky Mountains (48 to 25 Ma; Constenius, 1996) but not in the Omineca Belt (60 to 57 Ma; Parrish et al., 1988; Carr 1992). Constenius (1996) noted that, near the Idaho-Montana state border, Paleogene extensional magmatism predates the Tertiary horsts and grabens by 1 to 3 m.y. In the area where Idaho and Montana meet British Columbia, and at the junction of Idaho, Wyoming and Utah, magmatism was localized in some spots (e.g., Fowkes and Kishenehn grabens) surrounded by broad concentric areas (with radius of ~100 km) where no magmatism occurred. Such a concentration of mantle-derived magmatism indicates uneven heating of the Paleogene crust and lithosphere, with a few centers of maximum heating.

Gilbert (1890) was first to describe epeirogenic tectonic movements in the interior of the Cordilleran mobile megabelt, in Utah. He considered them to be the opposite of orogenic movements. This was correct for his particular area of study, but in general these types of crustal movements can occur simultaneously and affect the same area: even orogenic zones are subject to epeirogenic movements. In the eastern Canadian Cordillera, crustal warping has been manifested in specific types of magmatism and tectonic relief. Of the latter, the most interesting are topographic bulges (Fig. 74). In Canada, the biggest semi-circular bulge coincides with the Mackenzie Mountains, which follow the shape of the Mackenzie miogeosyncline (Stott and Aitken, eds., 1993). It is intriguing but unexplained. The smaller bulge in southwestern Alberta, straddling the Canada-U.S. border, is understood better (Lyatsky et al., 1999): it was imposed on a complex geologic background as a semi-circular upwarp protruding into the cratonic Interior Platform. The last major subsidence and submergence of the Interior Platform, of 50-60 m.y. duration, ended in the Paleocene, and uplift that resumed there is still continuing.

During the Laramian orogeny, many vertical movements in the eastern Cordillera are thought to have occurred fast, mainly during just 10 m.y., in the latest Paleocene and Eocene (King, 1977; Sloss, 1988; Parrish, 1995). The uplift was accompanied by massive erosion, and between 1 and 3 km of rocks were shed from the raised area (Kalkreuth and McMechan, 1984; Issler et al., 1990). Much of that clastic material blanketed the cratonic platforms, including the Williston Basin far to the east (Edwards et al., 1994). In the eastern Cordillera, including regions of previous Laramian thrusting, Tertiary extensional movements created a number of grabens and horsts. A chain of them stretches along the Rocky Mountains from Nevada to Utah to Idaho and western Montana into eastern British Columbia (Constenius, 1988, 1996).

The end of crustal shortening (thrusting) in the eastern Cordillera in the Early Tertiary was not accompanied by a tectonic collapse (Lyatsky et al., 1999). The horsts and grabens were formed by differential tectonic movements of crustal blocks, both up and down, rather than by passive downdropping of a segmented crust which lost its support below (as was envisioned by Dewey, 1988). Despite the great cratonward extent of Laramian shortening, east of the Rocky Mountain Trench its was fairly superficial and rootless, in a thin-skinned fashion. Since the Middle Eocene, differential block movements spread westward, into the hinterland of the Cordilleran mobile megabelt. They continued till the Early Miocene, ca. 20 Ma (Constenius, 1996) or near 18.3 Ma (Gibson et al., 1998).

At least two distinct episodes of normal faulting, on listric and steep planar faults, are apparent from examination of the Tertiary volcano-sedimentary graben fill (Eisbacher,

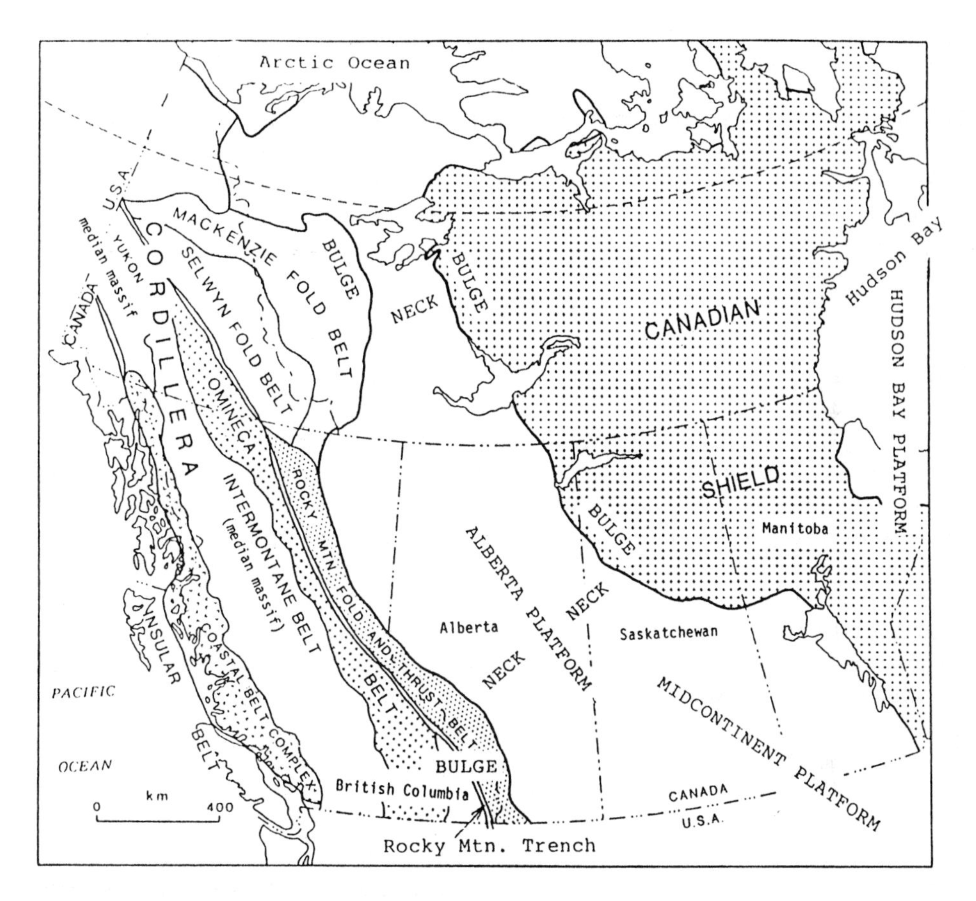

Figure 74. Bulges and necks in the outline of the Western Canada Sedimentary Province. Facing platformward protrusions of the Cordilleran mobile megabelt and the Canadian Shield (marked "bulge") create waist-like narrow zones in the sedimentary province (marked "neck"). These bulge-neck zones probably represent broad neotectonic warping of the continental crust.

1977; Lamerson, 1982). These faults originated variously at the base of thrust sheets or at the ground surface; in places they cut through the entire 6-8-km-thick thrust stack (Dahlstrom, 1970). Tertiary dikes and sills formed along some of these faults at various times but commonly in the Middle Eocene to Miocene (volcanics younger than 17 Ma are found in this region only in the Columbia River flood-basalt province; Reidel and Hooper, eds., 1989).

The change from Laramian compression to extension in the southern Canadian Cordilleran miogeosyncline was dated by Carr (1992) at 58±1 Ma (latest Paleocene). Syn-extensional sedimentation occurred in basins along major faults (e.g., the 150-km-long Kishenehn Basin). Most commonly, extension was accommodated by a multitude of small faults, but some very large faults, such as the one controlling the Southern Rocky Mountain Trench, are found as well.

These big block movements involved the crystalline basement. It has been reported that in seismic profiles across the Rocky Mountains the basement seems undeformed (Bally et al., 1966; Yoos et al., 1991), but in the inner parts of the eastern Cordilleran miogeosyncline the up and down block movements were very large. Supracrustal rocks in the metamorphic core complexes in the Omineca Belt were buried to depths of 20-30 km, and then brought back to the surface (Parrish, 1991a-b, 1995). Near the Moyie fault, depth to the ex-cratonic basement reaches ~20 km, implying considerable brittle or semi-brittle movements of the crystalline basement blocks during the Tertiary tectonism west of the axial zone of the Purcell Anticlinorium. Differential Tertiary block movements are also recognized in western U.S. (Peterson, ed., 1986; Sloss, ed., 1988), including a huge area of the Rocky Mountains (the miogeosyncline-platform tectonic domain of, e.g., Price, 1994). Rapid uplift and erosion of rocks next to the grabens created a source of sediments to fill numerous basins in the downgoing blocks.

In the Southern Rocky Mountain Trench in Canada, two distinct Paleocene and Miocene clastic successions contain boulders and cobbles of Omineca-miogeosyncline rocks, including quartzite (McMechan and Thompson, 1993; Constenius, 1996). In the Alberta Platform, gravel units also contain boulders and cobbles of quartzite that may be sourced from the Omineca Belt. Uplift in the Cordillera was rapid. Fission-track studies suggest rapid cooling in the Rocky Mountains around 45 Ma, in the late Middle Eocene (Constenius, 1996); in the Omineca Belt, cooling had occurred a few million years earlier (Carr, 1992). Mass movement of water-logged debris from the rugged mountain terrains could account for the wide distribution of pre-glacial Cenozoic gravels in the Alberta Plains (Edwards et al., 1994). As the rise of the Omineca Belt slowed, the Rocky Mountains' rise took over, providing another source for the accumulating Alberta Plains gravels for much of the Tertiary. After that uplift also slowed down in the Miocene, broad crustal upwarping involved the Rocky Mountains and the Plains of Alberta. A topographic scarp called Missouri Coteau forms the boundary of the elevated Alberta Plains with the smoother and lower-lying plains of Saskatchewan (Bostock, 1970).

The Alberta Plains today lie at an altitude of ~1,000 m above sea level, but contain remnants of erosional surfaces whose flat tops stand hundreds of meters higher: Cameron and Cypress Hills, Caribou and Birch Mountains, and so on. Some of them (e.g., Cypress Hills) have elevations of 1,300-1,500 m. On the other hand, the Plains are deeply cut by incised river valleys (e.g., Athabasca, North and South Saskatchewan rivers) forming spectacular bluffs many tens of meters high; some of these broad valleys are as much as 100-150 m deep. The scenic Alberta badlands are an example of fluvially carved terrain. The E-W profiles across Alberta (Shaw and Kellerhalls, 1982) show that east of the 50-km-wide Foothills, which are part of the Laramian fold-and-thrust belt, a western Plains belt 60-70 km wide forms a distinct topographic feature in the miogeosyncline-platform domain. Its uplift is also related to the juvenile Cordilleran mountain-building. Only farther east, across a topographic break (Misra et al., 1991), do plateaus in the shield-platform neotectonic domain lose a relationship to the Cordilleran mountain-building (Lyatsky et al., 1999).

It has been suggested (Shaw, 1963) that the general epeirogenic uplift of the North American continent was contemporaneous with neotectonic movements in the Appalachian and Cordilleran geomorphologic provinces. Epeirogenic movements occur on a very broad scale, involving whole continents or their large parts; the wavelength of epeirogenic undulations is thousands of kilometers. Such movements can easily embrace several continental geomorphologic and tectonic provinces at once. Warps like the one in southwestern Alberta are considerably more local, only hundreds of kilometers in wavelength. At a still lower level of the hierarchy of vertical continental crustal movements are movements of individual blocks (Lyatsky et al., 1999).

Due to the neotectonic uplift, most Cenozoic stratal rocks in the eastern Cordillera and western Interior Platform were eroded away. Up to 2 km of material were removed in the mountains (Kalkreuth and McMechan, 1984), and perhaps <1 km in the Plains (Nurkowski, 1984). Late-stage warping, since the late Early Oligocene, has been going on for ~30 m.y. (cp. Taylor et al., 1964). Purely cratonic warping was responsible for the round shape of the Williston Basin, which began to look more rectangular only in the Late Tertiary (Kent and Christopher, 1994) owing to block movements. Uplift of the fault-controlled Northern Sweetgrass Arch is expressed in a series of topographic highs running at an angle to the NNW-SSE structural trend of the Cordillera to the west.

Long-lived cratonic zones of crustal weakness defined the position of the Peace River Arch/Embayment and a series of steep NE-SW-trending faults crossing the Interior Platform (Edwards et al., 1998; Lyatsky et al., 1999). These faults, such as Great Slave Lake, MacDonald, Snowbird-Virgin River and others, evidently transect the Rocky Mountain Belt. They clearly correspond to some geologic and topographic features in the Intermontane median-massif belt, which contains the prominent transverse Skeena and Stikine arches. This is evidence that the basement of the Intermontane and Yukon-Tanana median massifs has ancient genetic links with the basement of the North American craton.

Round-shaped neotectonic upwarps may distort pre-existing structural and topographic patterns. Results of warping and block movements are usually clear in platforms and median massifs, where their effects are recorded in the volcano-sedimentary cover. With less clarity, they can be discerned in the geologic record of orogens as well. Abundant Early Tertiary quartz and calcite veins are known in the southern Canadian Cordillera, including the Rocky Mountains. Subvertical veins cut the Laramian faults and folds, but themselves are usually undeformed. These veins might be linked to the broad crustal upwarping that created the neotectonic structural domes.

Tertiary magmatism took place all over the Cordillera (Armstrong, 1988; Armstrong and Ward, 1991; Lipman, 1992), but in the eastern part of this megabelt it was comparatively subdued. In the Paleocene to Oligocene, from 58-55 to 35 Ma, it occurred mostly in the Intermontane Belt. In the Late Tertiary and Quaternary (17 Ma and after), flood basalts covered the Columbia River province in the Cordilleran interior. These basalts have been related variously to a continental arc, rifting or mantle plumes.

Although crust can be thickened in compression, the connection between warping and crustal thickness is by no means universal or direct. Existence in the Cordillera of metamorphic-core domes points to other causes of warping. In E-W transects across the southern Basin and Range province in the U.S., four whole-crust units are recognized: Colorado Corridor, metamorphic-core-complex belt, Transition Zone and Colorado Plateau. In the Colorado Plateau, the uplifted cratonic crust is unextended. Generally in the Basin and Range province, the crust is orogenic, subjected to Tertiary extension. By combining seismic reflection and refraction data, McCarthy and Parsons (1994) found the maximum crustal-thickness variations in that entire region to be 10-15 km.

But in the Cenozoic, areas belonging to the North American craton and the Cordilleran mobile megabelt in that region were involved in upwarping together. Extended and unextended crustal domains overlap the older crustal divisions and contrasts in composition and structure, and are themselves overlapped by younger neotectonic zonation.

Tertiary magmatism and structure in the eastern Cordillera in the U.S. are sometimes ascribed to assumed shallow subduction of oceanic plates from the Pacific (Stewart et al., 1977). From teleseismic data (Humphreys and Dueker, 1994), it has been shown that the modern subducted Juan de Fuca slab remains semi-coherent only for a few tens of kilometers inland from the Cascadia subduction zone, no farther east than the Oregon and Washington Cascades, and that the subducting slab is very steep. Inland continuity of a shallow subducting slab for ~1,000 km may be a tempting but speculative explanation for abundant (and ongoing) U.S. Cordilleran magmatism (Armstrong and Ward, 1991) and mountain building (Atwater, 1970, 1989). One alternative explanation is that the western U.S. is underlain by a broad mantle plume with several distinct apophysis-like upward protrusions (Blank et al., 1998). Yet, remarkably, many of the observed structural and geophysical variations of the shallow and deep lithosphere in the western U.S. (Humphreys and Dueker, 1994; Blank et al., 1998) and Canada are aligned NE-SW, on trend with the Precambrian tectonic grain of the North American craton.

The time-variant distribution of Cenozoic magmatism in the U.S. Cordillera (Cross and Pilger, 1978; Burke and McKee, 1979), including the northeastward progression of the eruptive loci, was not completely regular (Best, 1988; Blank et al., 1998). Hypothetical tectonic events inferred from interpretation of magnetic anomaly patterns and apparent hot-spot tracks in the Pacific (Atwater, 1970, 1989; Engebretson et al., 1985) were linked to the observed pattern of Cordilleran magmatism through an idea of roll-back subduction in response to hypothesized slowing-down of plate convergence. But these estimates rely on an assumption-based interpretation of what these magnetic stripes in the northeastern Pacific mean geologically (Lyatsky, 1996), and indeed these reconstructions do not produce the required closure of plate circuits, leaving impermissible gaps and overlaps (e.g., Stock and Molnar, 1988).

In the U.S. Rocky Mountains, no connection of mountain-building with either terrane docking or subduction has been compellingly proved by verifiable facts. In Canada, Late Tertiary time saw a cessation of subduction at the Pacific continental margin off Vancouver Island (Lyatsky, 1996). In the U.S., autochthonous whole-crust units of miogeosynclinal nature lie along the western edge of the Colorado Plateau, which is underlain by cratonic continental crust. Miogeosynclinal rocks were displaced to the east in thrust sheets during the Antler, Sevier and especially Laramian orogenies. King (1977) noted a common confusion between the geographic extent of the Rocky Mountains, which include all areas of rugged relief west of the plains, and the geologic extent of the huge shallow fold-and-thrust belt and eugeosynclinal rock successions.

Discriminating between these various classes of phenomena would help to understand the various tectonic movements recognized in modern megabelts. It is particularly important to separate regional deformation from results of movements (such as warping and epeirogenic motions) encompassing cratons and megabelts together; movements specific to orogenized and non-orogenized crust; and different episodes of movement. Orogenic zones may contain a broad spectrum of brittle and ductile structures, whereas brittle deformation predominates at the exposed crustal levels in median massifs. Varying degrees and extent of orogenization produce dissimilar sets of tectonic phenomena specific for individual regions. The Antler orogeny is characterized by little high-grade metamorphism in its Roberts Mountains tectonotype in the U.S., but metamorphic phenomena are well recognized in late Paleozoic orogenic areas in Canada. The Nevadan orogeny is the most widespread one in the Cordillera, in all its four main aspects (sedimentary, metamorphic, magmatic, deformational). Cretaceous orogenies (Columbian, Sevier), by contrast, are more localized and phenomenologically variable. The Laramian orogeny is widespread, and its products are extensive fold-and-thrust belts on the Cordilleran periphery and in the interiors.

Rock-based tectonic history of continental masses as a reliable basis for reconstructing the global tectonic evolution: examples from Eurasia and North America

Links and interactions of cratonic and mobile-megabelt structural features (e.g., transcurrent zones of crustal weakness) are established worldwide. These links put limits on speculations about randomly accreted terranes, and create a realistic framework for deciphering the long history of continental lithospheric masses. These masses are not inert but capable of self-development, having their own sources of heat.

Sadly, it has become commonplace to substitute exhaustive continental tectonic analysis based on studies of rocks and rock bodies with something facile - partial summaries of the data consistent with this or that subjectively preferred hypothetical tectonic scenario, derived from interpretation of irrelevant magnetic anomalies in oceanic areas far away on the basis of unproven or downright incorrect assumptions about the structure of oceanic lithosphere. Speculative efforts continue to be made to correlate the modern stress field of the entire North American continent to plate interactions (e.g., Zoback and Zoback, 1980, 1989). The same is done for large regions in Asia (e.g., Zang et al., 1992), even despite evidence that the great East Siberian rift system stretching for a distance of ~1,800 km and including the famous Baikal Rift was induced by intracontinental tectonic causes (Logatchev, 1993).

Classic examples of interactions of continental cratonic and mobile-megabelt tectonic units and processes come from pericratonic forelands and foredeeps (Lyatsky et al., 1999). North of the Caucasus Major orogenic zone, in southern Russia (Figs. 75-76) lies the big Stavropol basement uplift which separates two extensive oil-bearing foredeeps facing, correspondingly, the Black (Azov) Sea and north Caspian Sea. This Cis-Caucasian foredeep system can be regarded as a single if composite tectonic feature, in the sense that it developed mainly as a compensatory depression in front of the growing Mesozoic-Cenozoic (Cimmerian and Alpian) Caucasus Major. The Stavropol uplift was not a passive crustal block segmenting the foredeep system into constituent foredeeps and basins: its complex history, with variable amplitudes of differential uplift at various times relative to the depocenters on its flanks, involved irregular movements. Partly as a result, sedimentary settings and petroleum systems in the flanking basins differ (Raaben, 1978).

To the south, between Caucasus Major and Caucasus Minor, in the Trans-Caucasian countries of Georgia, Armenia and Azerbaijan, atop median massifs lies another zone of sedimentary basins, which contains the world-class south Caspian Baku oil fields. The Georgian crustal block, west of that country's capital city of Tbilisi, separates these basins. Precambrian massifs thus mark a broad transcurrent belt running from the young (non-cratonic) Scythian platform in the north far into the Trans-Caucasian segment of the Eurasian mobile megabelt in the south.

In a review of that region's petroleum geology, Patton (1993) assigned these intermontane basins to an Alpine island-arc setting, an inter-arc or intermontane rift, or the back of an active (in Armenia) magmatic arc. As in most of the Trans-Caucasus region, the basement of young volcano-sedimentary basins there is probably Paleozoic, probably with Precambrian rock bodies. During the Alpine orogeny, these formations became a basement for intermontane volcano-sedimentary basins developing over median massifs. A similar situation appears to exist to the east (in Iran) and to the west (in Turkey), where many median massifs have Paleozoic or Precambrian continental-crust affinities.

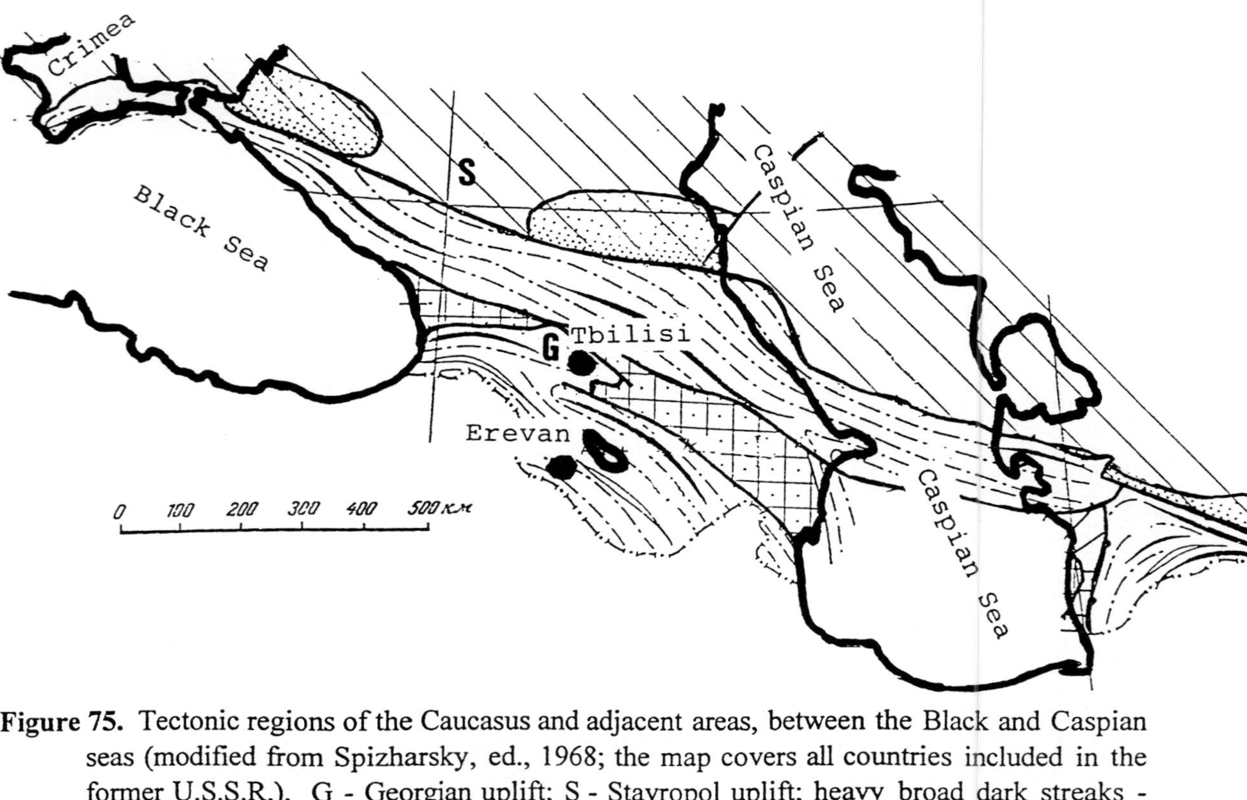

Figure 75. Tectonic regions of the Caucasus and adjacent areas, between the Black and Caspian seas (modified from Spizharsky, ed., 1968; the map covers all countries included in the former U.S.S.R.). G - Georgian uplift; S - Stavropol uplift; heavy broad dark streaks - anticlinoria; unfilled broad streaks - synclinoria; diagonally dashed area - platforms north of the Caucasus system; diagonally dashed and dotted areas - foredeep basins (separated by the Stavropol uplift).

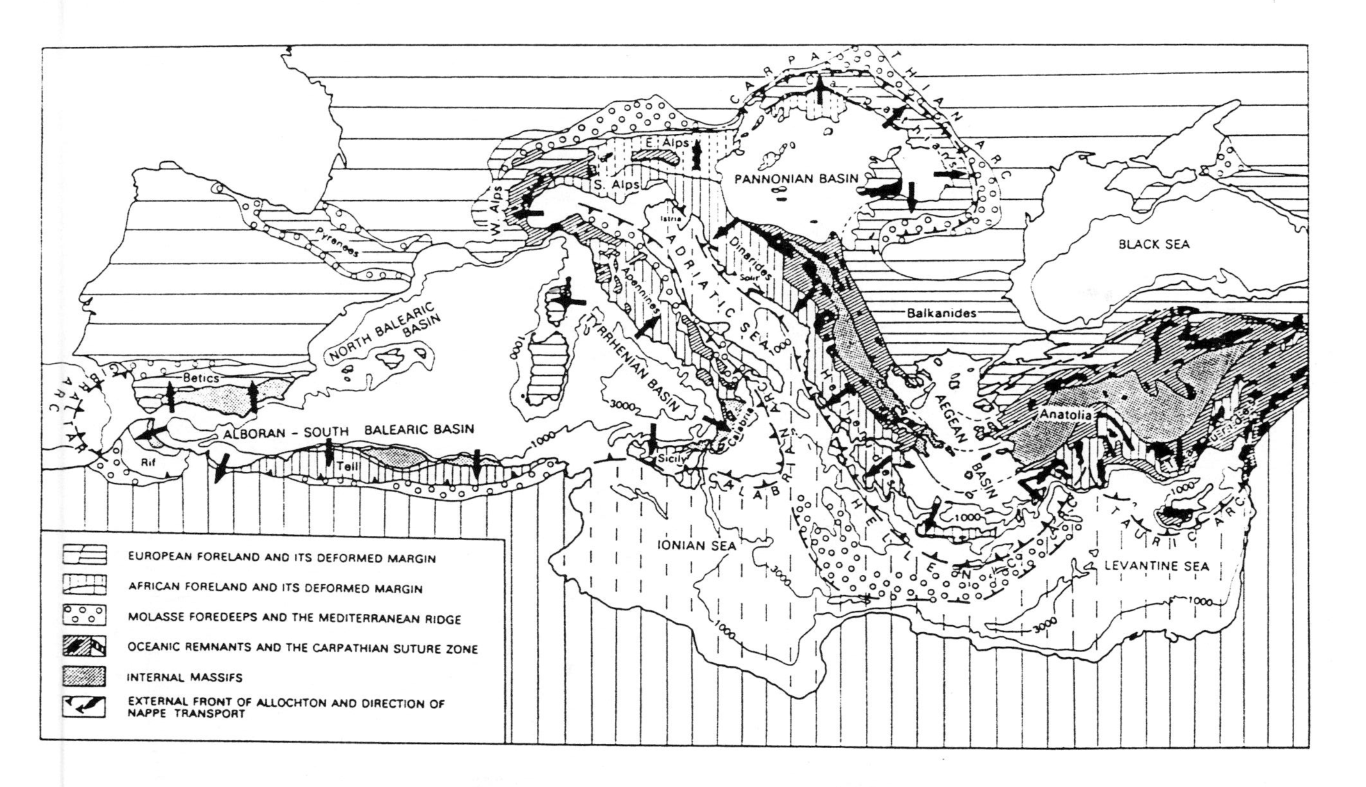

Figure 76. Major tectonic features of southern Europe and Mediterranean region (modified from Horváth, 1988), showing the Alpine mobile megabelt and the Pannonian Basin (median massif). Alpine suture zones and oceanic-crust remnants are assumed, not proved; the genetic equation of the Alpine molasse assemblage and some Mediterranean-floor sediments is also uncertain.

These considerations make it hard to explain the intra-continental tectonics in Eurasia by plate-boundary processes. Even the Eurasian plate's boundary in the Caucasus/Trans-Caucasus region is unclear and a matter of controversy (Brew et al., 1997). Interactions between plates, between continental-lithosphere and oceanic-lithosphere entities, between continental cratons and mobile megabelts, between median massifs and orogens, are not actions of active agents on passive objects. Lateral lithospheric tectonic units of this scale are not inert but active; they are tectonically alive internally, and they affect one another. Tectonic influences from neighboring active units commonly overlap with the indigenous, internally sourced tectonism.

But despite internal deformation, during certain periods of time some lateral tectonic units may act as coherent entities. In Asia, Yang et al. (1986) considered the early Paleozoic Caledonian Eastern Kunlun together with the Qilian foldbelt, and into the Western Kunlun they included the Tikelik uplift (a pre-late Paleozoic constituent of the Tarim cratonic block; Figs. 67, 77-78). The northeastern part of the original Tarim cratonic platform included the Quruktagh uplift containing a thick Sinian to Lower Devonian platformal cover; later it was incorporated into the Tien Shan, although it still remains largely unmetamorphosed. Huang et al. (1987) pointed out that the miogeosynclinal Southern Tien Shan and the eugeosynclinal Northern Tien Shan are separated by a median massif in central Tien Shan. The loci of burial and folding migrated along the flanks of this elevated massif in various directions.

Tectonic evolution of the middle part of the Eurasian continent occurred in a complex but integral system of lateral tectonic units - median massifs and orogens, as well as cratonic fragments - whose diverse interactions are not reducible to a simple stock of continental-margin processes (subduction, obduction, etc.). Even to discriminate subduction into hard and soft (Ren, 1991; Coleman, ed., 1994; Ren et al., 1996) is very insufficient to account for all the existing geologic complexity. No evidence exists to support the idea (Chen and Dickinson, 1986) that all sedimentary basins in western China are "compressional" whereas all basins in eastern China are "extensional". Even within the northern Tarim platform (basin), which lies on an ex-cratonic block, the nature and age of bounding faults changes from east to west: predominantly steep and normal in the east, steep and reverse in the Kucha area, and low-angle detachments in the Kelpin area (Lyatsky et al., 1990). The westernmost Karshi block is mostly influenced by orogenic processes in the neighboring Pamir. In the Hindukush-Pamir segment of the Eurasian mobile megabelt, convex-northward fault systems, such as those related to the Himalayan front and the North and South Pamir, interact complexly with the straighter, NW-SE-trending Karakorum and NE-SW-trending Middle Badakh faults. Two broad seismic zones are distinguished, north and south of latitude 36°-37°N (Zang et al., 1992, their Figs. 3a-c). The distribution of earthquake hypocenters varies in that

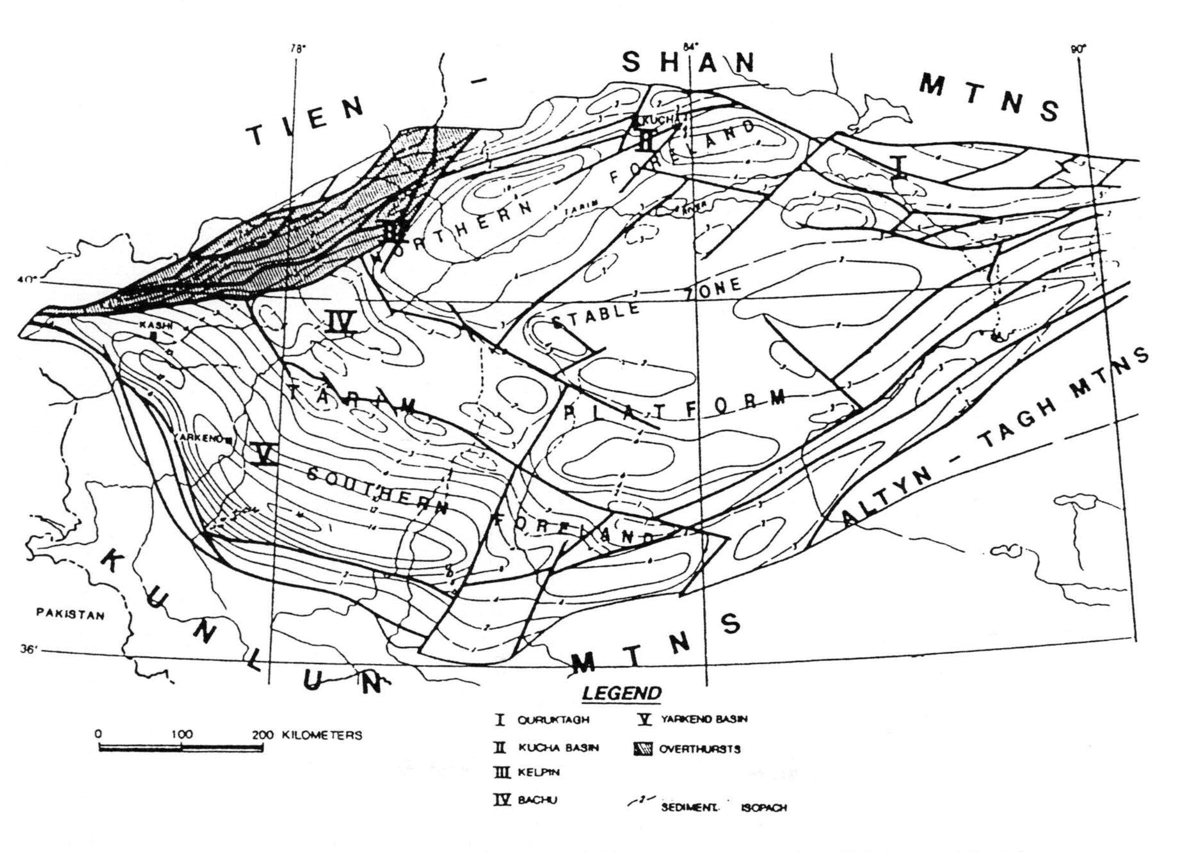

Figure 77. Structural pattern in the Tarim crustal block in northwestern China (modified from Lyatsky et al., 1990). Distribution of main ex-cratonic crustal blocks in the Eurasian mobile megabelt in Asia is shown in Fig. 67.

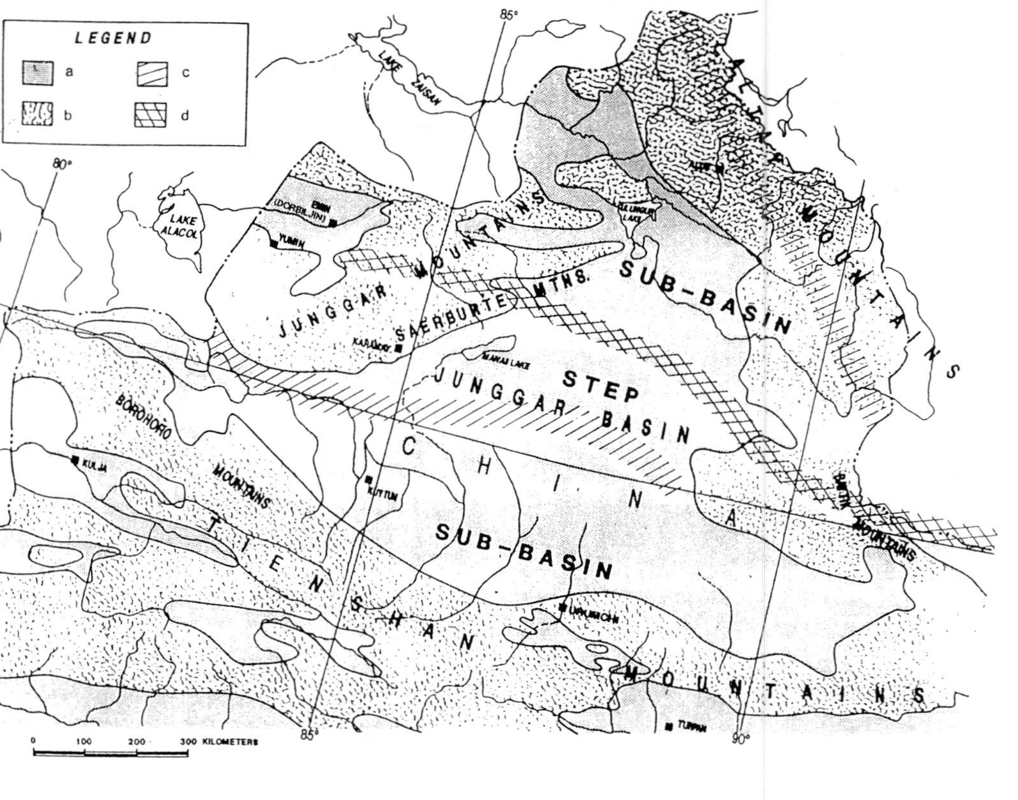

Figure 78. Principal tectonic zones in the Junggar Basin and adjacent areas in northwestern China (modified from Lyatsky et al., 1990). Distribution of main ex-cratonic crustal blocks in the Eurasian mobile megabelt in Asia is shown in Fig. 67. Notations: a - basin areas; b - mountain areas; c - gravity-anomaly zones; d - magnetic-anomaly zone.

region, with clusters in the top 70 km and at 100-200 km depth in the east, at 90-150 km and 180-260 km in the middle, and at 100-220 km in the west. Common earthquake magnitudes are 4 to 5. In the west, the focal zone of the northern seismic belt dips steeply to the north, away from India. In the east, its focal zone dips mainly to the south, towards India, in two distinct sections forming an S-shape. The nature of the northern seismic belt is still unclear, and Zang et al. (1992) proposed a new convergent boundary there.

The sea retreated from most of the middle of the Eurasian continent in the late Paleozoic, and the Variscan orogeny strongly affected southern Siberia and western China. Differential block and warp movements resulted in many sedimentary basins of thick terrigenous clastics, e.g., in Kazakhstan. In the western Tarim Platform, the new Bachu Basin had a new NW-SE to E-W structural trend, which corresponds to the Kashi-Yechang foredeep in front of the eastern Pamir and to the Talasso-Fergana fault. In the southern Tien Shan and western Kunlun, inverted Variscan orogens produced many impressive nappes, including those on the periphery of the Tarim Platform. Thrusting from the Kokshaal Range along China's northern border terminated the Permian development of the Bachu Basin. A new, narrow Late Permian foredeep overlapped the Bachu and Kucha basins. A series of horsts on the south side of the Tien Shan interacted complexly with foreland subsidence and growth of thrust-sheet stacks. In some areas, thrusting took place even over simultaneously evolving horsts (see Lyatsky et al., 1990).

Unlike in the southern Tien Shan, Variscan tectonism continued longer in the western Kunlun. General westward tilt of the Tarim Platform caused a new asymmetry of that platform's southern foreland. In the Yarkend area, in the west of the Tarim block, total thickness of sedimentary rocks reached 15-17 km, half of them Paleozoic. Western continuation of this large basin in the Afghan-Tajik Depression (whose modern successor is structurally divided) may be found in the Murgab Basin area. The giant Donmez-Davletabad hydrocarbon (largely gas) field lies on a big step in the crystalline basement straddling the Iran-Turkmenistan border (cp. Serber et al., 1997).

In Europe, links between structural histories of intracontinental regions and assumed formative plate-boundary processes are less evident. Their frequent invocation (e.g., Ziegler, 1982; Ziegler, ed., 1987) still leaves them unconvincing and far-fetched. Fault steps, horsts and grabens are common in the Paleozoic basement of the epi-Variscan, young Northwest European platform. Some of these structures are very young, including those in the famous Late Eocene-Early Oligocene Rhinegraben. Fundamental crustal zonation west of the Urals is apparent from the regional distribution of gravity and magnetic anomalies in the Russian craton, adjacent young platforms and orogenic provinces (e.g., Atakov et al., 1997).

Tectonic processes at margins of lithospheric plates - divergence, convergence, subduction, obduction etc. - important as they are in specific areas, do not overshadow indigenous intraplate tectonic processes that have a predominant role in shaping the continent. Tertiary evolution of the oceanic-crust North Atlantic certainly influenced the tectonic phenomena in Europe, Greenland and eastern North America. However, simultaneous evolution of the enormous Eurasian mobile megabelt, from Iberia through the Mediterranean region and all across Asia, had other, internal causes. The Pyrenean orogeny in the Late Eocene-Early Oligocene, produced presumably by local collisions, had time equivalents within this megabelt in areas far away: the Meso-Alpine orogeny (Trümpy, 1960) in the western and eastern Alps, northern Dinarides, south Crimea and Caucasus (Figs. 75-76). By the mid-Eocene, the eastern Alps and northern Dinarides became inactive (Becker, 1993), but the western Alps (e.g., Swiss Alps) experienced vigorous shortening in Neo-Alpine time. In the Apennines, and to a lesser extent in the Jura Mountains, such activity is still under way. Paleocene to Quaternary fold-and-thrust belts are superimposed on the Po River foredeep and onto the southeastern Cis-Carpathian foredeep in the Moesian Platform (in Bulgaria and Romania; Boyanov et al., 1989). Though specific local explanations may be devised for each of these events, their near-simultaneous occurrence all across the huge Eurasian megabelt suggests a common indigenous intra-continental causation.

Broad time coincidence of the opening and spreading of the North Atlantic Ocean with distant continental events, such as the inversion of some sedimentary basins in northern and central Europe or orogenies in western North America, is not proof that continental tectonics is driven by tectonics in distant and unrelated oceans. Ascribing the late Mesozoic and Cenozoic intra-continental tectonism in Europe to the assumed smaller and more-local plate boundaries (e.g., papers in Ziegler, ed., 1987) still requires tectonic influences to be sourced hundreds or even thousands kilometers away from the areas of action, and continues to assume the continent to be essentially inert. Yet, some presumed collisional manifestations in central and southern Europe were ascribed by Becker (1993) to interactions of not large but small continental-crust entities (Pannonian massif, Adriatic block, Corsica-Sardinia block and so on).

Foredeeps are strongly related to adjacent orogenic belts, being compensatory depressions in front of inverted and rising crustal zones (e.g., the Molasse Basin and Carpathian Foredeep in Europe, the Deep Basin in Alberta). Domal uplifts are also commonly recognized in Europe and North America. It is suggested they existed in the past as well (e.g., the inferred North Sea dome in the Jurassic; Underhill and Partington, 1993). At present, such domes are apparent from neotectonic remnants preserved in the modern topography. Warping is not restricted to cratons, as it also occurs in mobile megabelts (Lyatsky et al., 1999). It happens far inside continents, independently from any collisions at distant plate boundaries. The Central French, Rhenish and Bohemian domes lie some distance north and west of the Alpine deformation front, weakening the possibility of genetic links with that tectonic zone as well. Since the Late Cretaceous, sedimentation settings of the epi-Variscan cover in Germany and France were affected by general regional tilting. By the Late Oligocene, a single Vosges-Schwarzwald domal upwarp was disrupted by the formation of the Rhinegraben and Besse Graben. These and other rift-like features may be a result of thermo-tectonic evolution of warp structures, which do not need to be connected with far-field effects of the Alpine mobile-megabelt tectonism or with events in the even more remote North Atlantic oceanic lithosphere. Early Cenozoic upwarping which embraced the Scandinavian and Barents shelves may have had its own causes within the northwestern Eurasian continent (cp. Muir Wood, 1989).

Long-wavelength warps are difficult to distinguish in the Mediterranean region, which was recently affected by the powerful Alpine orogeny. But since the Miocene, the largest foredeeps in front of the Alps and the Carpathians have been involved in non-systematic and sharply diachronous uplifts. These differential warp and block movements, like those in the Interior and Midcontinent platforms in the heartlands interior of the North American continent (Lyatsky et al., 1999), have no connection to plate-boundary processes.

In North America, Laramian deformation is sometimes linked to the supposed behavior of the Pacific oceanic plates far to the west. It is assumed, for example, that the Farallon oceanic plate subducted at a very shallow angle, permitting the slab's influences to extent across most of the Cordillera (Atwater, 1970, 1989; Atwater and Severinghaus, 1989). In Europe, some deformation at that time is simply linked to supposed plate reorganizations in the North Atlantic (Becker, 1993). But rock-based evidence, such as that from the well-studied superdeep KTB drillhole area near the Bavarian-Czech border, do not support such speculations. In the KTB area, the NW-SE-trending Franconian lineament (fault) separates a shield of semi-cratonized epi-Variscan rocks from the Central German Platform. Permo-Triassic sedimentary cover is widespread from there to the northwest; Cretaceous rocks thicken from that area westward, into France. The Franconian fault has been active during its history more than once. Today it roughly separates the platformal upland from the higher-standing Bohemian Massif, which near the KTB site is dissected by the NE-SW-trending Tertiary Eger Graben containing Holocene volcanics as young as ~10,000 years. The probable cause of Cenozoic rifting in that whole part of Europe, from southeastern France through Germany's Schwarzwald (Black Forest) to western Czech Republic, was fairly proximal - the stress regime along the outer Alps (Letouzey, 1986). To invoke far-field influences from remote plate boundaries and oceans would be very far-fetched.

If tectonic interpretations are uncertain on land, they are extremely speculative in oceans, where rocks are unavailable for detailed observation. The age of oceanic lithosphere has for decades been derived from linear magnetic anomalies, supposedly created in a regular production-line fashion along the spreading centers. The real geologic world is certainly less regular and more diverse than the Vine-Matthews (1963) physical model of oceanic-lithosphere formation assumes. It is even unclear what drives the plate motions (e.g., Wilson, 1993). Old faults in oceanic crust are probably rejuvenated, and new ones are formed as even inactive faults often remain available for reactivation later. It is now recognized that spreading-center volcanism is quite irregular and sporadic, that spreading occurs at different centers with different rates and in different modes, and that oceanic crust does not survive immaculately unaltered for even a short time after its creation. Even the sources of magnetic lineations are still unclear (e.g., Ballu et al., 1998; Lawrence et al., 1998; Tivey and Tucholke, 1998). Many oceanic seismic profiles contain dipping reflections at deep oceanic-crustal levels (Eittreim et al., 1994). The origin of these reflections is uncertain: they may be due to discontinuities formed at the spreading centers with the crust, or they could indicate younger, secondary shears, dikes and other effects of reworking.

Besides, common magnetic polarity of different rocks is not proof of their common age: by themselves, geophysical anomalies say nothing about the rock age or genesis. Some linear anomalies in oceanic crust may reflect relatively young dikes emplaced along fault conduits following the structural grain parallel to original spreading centers; interpreting all source rocks as primary could easily lead to age miscorrelation. The unreliability of tectonic reconstructions based on oceanic magnetic stripes is compounded by the fact that large parts of the Atlantic bottom were formed in the Cretaceous during a long normal-polarity time interval when no geomagnetic reversals took place, leaving no record in the oceanic magnetic field. Given the uncertainties in reconstructing plate motions in the past, to use the assumed plate interactions as a framework for intra-continental tectonics is very unreliable. No model-based tectonic inferences can be accepted without local and regional, rock-based geological verification on land.

The idea of Wilson Cycle, involving recurrent opening and closing of oceans as continents separate and come together again, implies the Atlantic Ocean has existed in the Phanerozoic twice. Based on geological evidence from adjacent continental regions, that contention has come under criticism (e.g., Jenner et al., 1991). The idea of past existence of a huge supercontinent that included all the modern continents remains speculative. Strong disagreements in the literature about how the continents were arranged in such an entity (e.g., Mac Niocaill et al., 1997) reveal the lack of reliable, hard evidence. The known presence of continental-crust blocks in even pelagic areas of the North Atlantic makes the distribution of oceanic and continental crust irregular and complicates the reconstructions of a presumed former supercontinent from the modern continents the Atlantic Ocean separates.

The Variscan crystalline-crust basement in large areas of the North Atlantic is covered by sediments, making the ocean floor relatively smooth and masking the differences between individual blocks and warps. Still, downwarps ~50 km in diameter, as well as tilted blocks, are reflected in the ocean-floor morphology: water depth changes from ~900 m on the upward sides of some tilted blocks to ~2,500 m over Upper Jurassic depocenters. Granitic rocks recovered by drilling clearly indicate presence of continental crust in sampled areas, though the continental crust in these regions is evidently thin and easily fractured (e.g., Srivastava et al., 1990).

The off-handed and easy interpretations of many mafic and ultramafic rock complexes, even those lying far inside continents, as remnants of oceanic crust, are also suspect. Occurrence of some samples of serpentinized peridotite and gabbro peridotite with highly metamorphosed (amphibolite- and granulite-grade) rocks within hydrothermal and cataclastic zones complicates the definition of protoliths, to say nothing of original tectonic settings. Many mafic-ultramafic bodies have textures that suggest emplacement by intrusion, commonly along faults. A decision whether a particular mafic-ultramafic rock complex is ophiolitic or not can be very uncertain; even more unreliable are determinations of whether the rocks were formed in a break-up rift, at a spreading center, in a back-arc setting, and so forth. The only thing that can usually be stated with any certainty is that at present these rocks are found in a continental-crust setting.

In the north, the manner in which the opening of the Atlantic Ocean was hinged on the spherical globe is still unclear. Spreading between Greenland and northern Canada was very restricted in space and time, producing only a narrow oceanic basin. The narrow rift between Greenland and Baffin Island that split the northeastern part of the Canadian Shield did not extend beyond Baffin Bay and Ellesmere Island, although a window of oceanic crust is enclosed there on the sea floor.

Evolution of the Atlantic Ocean is poorly understood, and it appears to be far more complex than is sometimes assumed. West of the North Atlantic Ridge, granitoids have been recovered on Flemish Cap. Non-marine Jurassic sediments have been found at the base of sedimentary cover at Orphan Knoll. Mid-Cretaceous glauconite and fossil planktonic foraminifera indicate stable deep marine environments since that time (van Hinte et al., 1975; Aubry, 1995). Orphan Knoll and Flemish Cap are evidently foundered remnants of continental crust, now forming part of the North Atlantic ocean floor. The oldest sediments in the abyssal sedimentary basins on both sides of the North Atlantic Ridge are Middle-Late Jurassic, marking the timing of initial downwarping of the Variscan basement. Late Jurassic limestones are shallow-water, shelfal, suggesting water depths of only a few hundred meters. Late Jurassic and Early Cretaceous clay and chalk contain benthic foraminiferal assemblages. Only after a stratigraphic gap of ~50 m.y. did deep-oceanic environments become predominant.

The volcanic island of St. Helena stands 822 m above sea level some 800 km east of the Mid-Atlantic Ridge, where the ocean bottom is ~4,400 m deep and estimated to be 40 Ma old. This strange subaerial volcano appeared at 14.3 Ma, and at ~11 Ma its locus shifted sharply 12 km to the SW. Three distinct volcanic shields, lower, main and upper, indicate various localities and styles of principal edifices. The volcanism finally ended at ~7 Ma (Anderson and Weaver, 1991). The evolution of the St. Helena volcanic complex has no relation to a mid-ocean rift, and its instability and transience are at odds with the long life normally assumed for hot spots. (Questions about the reliability of plate-motion reconstructions based on the presumed-absolute hot-spot reference frame have also been discussed recently by Baksi, 1999).

Lessons from the comparatively well studied North Atlantic Ocean and its continental flanks point to great complexity in the thermo-tectonic regimes responsible for shaping the oceanic and continental lithosphere and inducing movements of the crust. To presume that continental tectonics is driven by the tectonics of oceans assumes the understanding of oceanic-region evolution to be much better than it actually is. As well, and fatally, this presumption fails to take into consideration the fundamental self-development capability of continental lithosphere.

Principal role of rock evidence in tectonic analysis

At plate boundaries with big subduction zones, as in much of the Pacific rim, interactions between the downgoing and overriding plates add to the indigenous complexity in the distribution of continental tectonic styles. More widespread is indigenous intracontinental tectonism driven by the lithosphere's own self-development, which creates tectonic manifestations in both cratons and mobile megabelts. Overprinting of tectonic patterns and interactions between tectonic domains are much too complex to be reducible to simple models based on just a few parameters. Differences between tectonic processes typical for continental and oceanic lithosphere, cratons and megabelts, remain despite the existence of lithospheric plates. Inter-plate kinematics must not be the sole or main problem to which regional tectonic analysis is reduced. True tectonic analysis, keeping in mind the self-development capability of continental lithosphere, is based on observations of rocks, and it accounts for indigenous as well as external tectonic influences.

On land, where rocks are available for direct observation, work on various continents shows that their evolution cannot be explained by simple interactions of continental-crust and oceanic-crust masses. In a self-developing continent, internal sources of energy are the principal cause of tectonism. This holds true for various parts of a continent, cratons and mobile megabelts alike.

With the benefit of modern technology and factual knowledge, a mature tectonic theory, combining regional and global rock-based observations into an internally coherent conceptual system, should more than ever be driven by results of detailed field mapping and laboratory analysis of rocks. The data for realistic, practical tectonic analysis of continental regions must come not from remote oceans but overwhelmingly from the continental regions themselves, because it is there that the rocks are available for examination. These observed rocks stand as guardians against groundless speculations, because they alone witnessed and recorded the truth.

REFERENCES

Acharya, H., 1992. Comparison of seismicity parameters in different subduction zones and its implications for the Cascadia subduction zone; Journal of Geophysical Research, v. 97, p. 8831-8842.

Aitken, J.D., 1993a. Tectonic framework; *in* D.F. Stott and J.D. Aitken (eds.), Sedimentary Cover of the Craton in Canada; Geological Society of America, The Geology of North America, v. D-1, p. 45-54.

Aitken, J.D., 1993b. Proterozoic sedimentary rocks; *in* D.F. Stott and J.D. Aitken (eds.), Sedimentary Cover of the Craton in Canada; Geological Society of America, The Geology of North America, v. D-1, p. 81-95.

Aitken, J.D., 1993c. Tectonic evolution and basin history; *in* D.F. Stott and J.D. Aitken (eds.), Sedimentary Cover of the Craton in Canada; Geological Society of America, The Geology of North America, v. D-1, p. 483-502.

Aitken, J.D. and McMechan, M.E., 1991. Middle Proterozoic assemblages; *in* H. Gabrielse and C.J. Yorath (eds.), Geology of the Cordilleran Orogen in Canada; Geological Society of America, The Geology of North America, v. G-2, p. 97-124.

Aleinikoff, J.N., Dusel-Bacon, C., and Foster, H.L., 1986. Geochronology of augen gneiss and related rocks, Yukon-Tanana terrane, east-central Alaska; Bulletin of the Geological Society of America, v. 97, p. 626-637.

Aleinikoff, J.N., Dusel-Bacon, C., Foster, H.L., and Nokleberg, W.J., 1987. Lead isotopic fingerprinting of tectonostratigraphic terranes, east-central Alaska; Canadian Journal of Earth Sciences, v. 24, p. 2089-2098.

Allmendinger, R.W., 1992. Thrust and fold tectonics of the western United States exclusive of the accreted terranes; *in* B.C. Burchfiel, P.W. Lipman, and M.L. Zoback (eds.), The Cordilleran Orogen: Conterminous U.S.; Geological Society of America, The Geology of North America, v. G-3, p. 583-607.

Allmendinger, R.W., Farmer, H., Hauser, E., Sharp, J., Von Tish, D., Oliver, J., and Kaufman, S., 1986. Phanerozoic tectonics of the Basin and Range-Colorado Plateau transition from COCORP data and geologic data: a review; *in* M. Barazangi and L.D. Brown (eds.), Reflection Seismology: The Continental Crust; American Geophysical Union, Geodynamics Series, v. 14, p. 257-268.

Allmendinger, R.W., Hauge, T.A., Hauser, E.C., Potter, C.J., Klemperer, S.L., Nelson, K.D., Knuepfer, P., and Oliver, J., 1987. Overview of the COCORP 40°N Transect, western United States: the fabric of an orogenic belt; Bulletin of the Geological Society of America, v. 98, p. 308-319.

Allmendinger, R.W. and Royse, Jr., F., 1995. Is the Sevier Desert reflection of west-central Utah a normal fault?: comment; Geology, v. 23, p. 669-670.

Amursky, G.I., 1976. Uralo-Oman lineament; Geotectonics, v. 10, p. 134-141.

Anders, M.H. and Christie-Blick, N., 1994. Is the Sevier Desert reflection of west-central Utah a normal fault?; Geology, v. 22, p. 771-774.

Anders, M.H., Christie-Blick, N., and Wills, S., 1995. Is the Sevier Desert reflection of west-central Utah a normal fault?: reply; Geology, v. 23, p. 670.

Anderson, D.L., 1995. Lithosphere, asthenosphere, and perisphere; Reviews of Geophysics, v. 33, p. 125-149.

Anderson, H.E. and Davis, D.W., 1995. Age and geological setting of the Aldridge Formation and Sullivan orebody: evidence from U-Pb geochronology of the Moyie sills, southeastern British Columbia (abs.); Geological Association of Canada/Mineralogical Association of Canada, Annual Meeting, v. 20, Victoria; Program and Abstracts, p. A-2.

Anderson, H.E. and Goodfellow, W.D., 1995. Petrogenesis of the Moyie sills, southeastern British Columbia: implications for the early tectonic setting of the Middle Proterozoic Purcell Basin (abs.); Geological Association of Canada/Mineralogical Association of Canada, Annual Meeting, v. 20, Victoria; Program and Abstracts, p. A-2.

Anderson, P.M. and Weaver, B.L., 1991. Geology, geochemistry, and origin of alkaline volcanic rocks from Saint Helena Island, South Atlantic ocean; The Compass, v. 68, p. 193-201.

Anderson, R.G., Whalen, J.B., Struik, L.C., and Villeneuve, M.E., 1998. Mesozoic to Eocene composite Endako Batholith redefined; *in* Slave-Northern Cordillera Lithospheric Evolution and Cordilleran Tectonics Workshop, Lithoprobe Report 64, p. 206-209.

Ando, C., Cook, F.A., Oliver, J., Brown, R.L., and Kaufman, S., 1983. Crustal geometry of the Appalachian orogen from seismic reflection studies; *in* R.D. Hatcher, Jr., H. Williams, and I. Zietz (eds.), Contributions to the Tectonics and Geophysics of Mountain Chains; Geological Society of America, Memoir 158, p. 83-101.

Andrew, K.P.E., Höy, T., and Simony, P., 1991. Geology of the Trail Map Area, Southeastern British Columbia; British Columbia Ministry of Energy, Mines and Petroleum Resources, Open File Report 1991-16.

Archibald, D.A., Glover, J.K., Price, R.A., Farrar, E., and Carmichael, D.M., 1983. Geochronology and tectonic implications of magmatism and metamorphism, southern Kootenay Arc and neighbouring regions, southeastern British Columbia; Part 1, Jurassic to mid-Cretaceous; Canadian Journal of Earth Sciences, v. 20, p. 1891-1913.

Archibald, D.A., Krogh, T.E., Armstrong, R.L., and Farrar, E., 1984. Geochronology and tectonic implications of magmatism and metamorphism, southern Kootenay Arc and neighboring regions, southeastern British Columbia; Part 2, Mid-Cretaceous to Eocene; Canadian Journal of Earth Sciences, v. 21, p. 567-584.

Armstrong, R.L., 1982. Cordilleran metamorphic core complexes; from Arizona to southern Canada; Annual Reviews of Earth and Planetary Sciences, v. 10, p. 129-154.

Armstrong, R.L., 1988. Mesozoic and early Cenozoic magmatic evolution of the Canadian Cordillera; *in* S.P. Clark, B.C. Burchfiel, and J. Suppe (eds.), Processes in Continental Lithospheric Deformation; Geological Society of America, Special Paper 218, p. 55-91.

Armstrong, R.L., 1991. The persistent myth of crustal growth; Australian Journal of Earth Sciences, v. 38, p. 613-630.

Armstrong, R.L. and Ghosh, D.K., 1990. Westward movement of the $^{87}Sr/^{86}Sr$=.704 line in southern British Columbia from Triassic to Eocene time: monitoring the tectonic overlap of accreted terranes on North America; *in* Southern Cordillera Transect Workshop, Lithoprobe Report 11, p. 53-56.

Armstrong, R.L., Parrish, R.R., van der Heyden, P., Scott, K., Runkle, D., and Brown, R.L., 1991. Early Proterozoic basement exposures in the southern Canadian Cordillera: core gneiss of Frenchman Cap, Unit I of the Grand Forks gneiss, and Vaseaux Formation; Canadian Journal of Earth Sciences, v. 28, p. 1169-1201.

Armstrong, R.L. and Ward, P., 1991. Evolving geographic patterns of Cenozoic magmatism in the North American Cordillera: temporal and spatial association of magmatism and metamorphic core complexes; Journal of Geophysical Research, v. 96, p. 13,201-13,224.

Ash, C., Friedman, R., Cordey, F., and Tipper, H., 1996. Stikine terrane stratigraphy along the northwestern margin of the Bowser Basin (abs.); *in* Slave-Northern Cordillera Lithospheric Evolution (SNORCLE) and Cordilleran Tectonics Workshop, Lithoprobe Report 50, p. 92.

Atakov, A.I., Belyaev, I.V., and Rzhevsky, N.N., 1997. New information about Navarin Basin deep structure based on the results of gravity and magnetometry data complex analysis (abs.); European Association of Geoscientists and Engineers, 59th Conference and Technical Exhibition, Geneva; Extended Abstracts, paper P189.

Atlas of Paleogeography of China, 1985. Compiled by Institute of Geology, Chinese Academy of Geological Sciences and Wuhan College of Geology; Cartographic Publishing House, Beijing.

Atwater, T., 1970. Implications of plate tectonics for the Cenozoic tectonic evolution of western North America; Bulletin of the Geological Society of America, v. 81, p. 3518-3536.

Atwater, T., 1989. Plate tectonic history of the northeast Pacific and western North America; *in* E.L. Winterer, D.M. Hussong, and R.W. Decker (eds.), The Eastern Pacific Ocean and Hawaii; Geological Society of America, The Geology of North America, v. N, p. 21-72.

Atwater, T. and Severinghaus, J., 1989. Tectonic maps of the northeast Pacific; *in* E.L. Winterer, D.M. Hussong, and R.W. Decker (eds.), The Eastern Pacific Ocean and Hawaii; Geological Society of America, The Geology of North America, v. N, p. 15-20.

Aubouin, J., 1965. Geosynclines; Developments in Geotectonics, v. 1, Elsevier, 335 p.

Aubry, M.-P., 1995. From chronology to stratigraphy: interpreting the Lower and Middle Eocene stratigraphic record in the Atlantic Ocean; *in* W.A. Berggren, D.V. Kent, M.-P. Aubry, and J. Hardenbol (eds.), Geochronology, Time Scales and Global Stratigraphic Correlation; Society of Economic Paleontologists and Mineralogists, Special Publication 54, p. 213-274.

Babcock, R.S., Burmester, R.F., Engebretson, D.C., Warnock, A.C., and Clark, K.P., 1992. A rifted margin origin for the Crescent basalts and related rocks in the northern Coast Range volcanic province, Washington and British Columbia; Journal of Geophysical Research, v. 97, p. 6799-6821.

Babcock, R.S., Suczek, C.A., and Engebretson, D.C., 1994. The Crescent "terrane", Olympic Peninsula and southern Vancouver Island; *in* R. Lasmanis and E.S. Cheney (eds.), Regional Geology of Washington State; Washington Division of Geology and Earth Resources, Bulletin 80, p. 141-157.

Baksi, A.K., 1999. Reevaluation of plate motion models based on hotspot tracks in the Atlantic and Indian oceans; Journal of Geology, v. 107, p. 13-26.

Baldwin, J.A., Whitney, D.L., and Hurlow, H.A., 1997. Metamorphic and structural evidence for significant vertical displacement along the Ross Lake fault zone, a major orogen-parallel shear zone in the Cordillera of western North America; Tectonics, v. 16, p. 662-681.

Ballu, V., Dubois, J., Deplus, C., Diament, M., and Bonvalot, S., 1998. Crustal structure of the Mid-Atlantic Ridge south of the Kane Fracture Zone from seafloor and sea surface gravity data; Journal of Geophysical Research, v. 103, p. 2615-2631.

Bally, A.W., 1975. A geodynamic scenario for hydrocarbon occurrences; 9th World Petroleum Congress, Tokyo, Proceedings, v. 2, p. 33-44.

Bally, A.W., 1989. Phanerozoic basins of North America; *in* A.W. Bally and A.R. Palmer (eds.), The Geology of North America - An Overview; Geological Society of America, The Geology of North America, v. A, p. 397-446.

Bally, A.W., Gordy, P.L., and Stewart, G.A., 1966. Structure, seismic data, and orogenic evolution of southern Canadian Rocky Mountains; Bulletin of Canadian Petroleum Geology, v. 14, p. 337-381.

Bally, A.W., Scotese, C.R., and Ross, M.L., 1989. North America; plate-tectonic setting and tectonic elements; *in* A.W. Bally and A.R. Palmer (eds.), The

Geology of North America - An Overview; Geological Society of America, The Geology of North America, v. A, p. 1-16.

Banda, E., Ranero, C.R., Danobeitia, J.J., and Rivero, A., 1992. Seismic boundaries of the eastern central Atlantic Mesozoic crust from multichannel seismic data; Bulletin of the Geological Society of America, v. 104, p. 1340-1349.

Barr, D.A., 1980. Gold in the Canadian Cordillera; Bulletin of the Canadian Institute of Mining and Metallurgy, v. 73, p. 59-76.

Barrell, J., 1914. The strength of the earth's crust; Geology, v. 22, p. 28-48.

Bassett, K.N. and Kleinspehn, K.L., 1997. Early to middle Cretaceous paleogeography of north-central British Columbia: stratigraphy and basin analysis of the Skeena Group; Canadian Journal of Earth Sciences, v. 34, p. 1644-1669.

Bates, R.L. and Jackson, J.A. (eds.), 1980. Glossary of Geology, second edition; American Geological Institute, 749 p.

Bates, R.L. and Jackson, J.A. (eds.), 1987. Glossary of Geology, third edition; American Geological Institute, 788 p.

Beaudoin, B.C., Fuis, G.S., Mooney, W.D., Nokleberg, W.J., and Christensen, N.I., 1992. Thin, low-velocity crust beneath the southern Yukon-Tanana Terrane, east central Alaska: results from Trans-Alaska Crustal Transect refraction/wide-angle reflection data; Journal of Geophysical Research, v. 97, p. 1921-1942.

Beaudoin, B., Roddick, J.C., and Sangster, D.F., 1991a. Eocene age for Ag-Pb-Zn-Au vein and replacement deposits of the Kokanee Range, southeastern British Columbia; Canadian Journal of Earth Sciences, v. 28, p. 3-14.

Beaudoin, G., Sangster, D.F., and Godwin, C.I., 1991b. Isotopic evidence for complex Pb sources in the Ag-Pb-Zn-Au veins of the Kokanee Range, southeastern British Columbia; Canadian Journal of Earth Sciences, v. 28, p. 418-431.

Beaudoin, G., Taylor, B.E., and Sangster, D.F., 1991c. Silver-lead-zinc veins, metamorphic core complexes, and hydrologic regimes during crustal extension; Geology, v. 19, p. 1217-1220.

Beck, M.E., Jr., and Nosun, L., 1972. Anomalous paleolatitudes in Cretaceous granitic rocks; Nature, v. 235, p. 11-13.

Becker, A., 1993. An attempt to define a "neotectonic period" for central and northern Europe; Geologische Rundschau, v. 82, p. 67-83.

Beikman, H.M., 1980. Geologic Map of Alaska, 2 sheets; U.S. Geological Survey.

Beloussov, V.V., 1962. Basic Problems in Geotectonics; McGraw-Hill, 809 p.

Beratan, K.K. (ed.), 1996. Reconstructing the History of Basin and Range Extension Using Sedimentology and Stratigraphy; Geological Society of America, Special Paper 303, 212 p.

Bercovici, D., 1993. A simple model of plate generation from mantle flow; Geophysical Journal International, v. 114, p. 635-650.

Berggren, W.A., Kent, D.V., Swisher III, C.C., and Aubry, M.-P., 1995. A revised Cenozoic geochronology and chronostratigraphy; *in* W.A. Berggren, D.V. Kent, M.-P. Aubry, and J. Hardenbol (eds.), Geochronology, Time Scales and Global Stratigraphic Correlation; SEPM (Society for Sedimentary Geology), Special Publication 54, p. 129-212.

Berman, R.G. and Armstrong, R.L., 1980. Geology of the Coquihalla Volcanic Complex, southwestern British Columbia; Canadian Journal of Earth Sciences, v. 17, p. 985-995.

Berry, M.J. and Forsyth, D.A., 1975. Structure of the Canadian Cordillera from seismic refraction and other data; Canadian Journal of Earth Sciences, v. 12, p. 182-208.

Berry, M.J., Jacoby, W.R., Niblett, E.R., and Stacey, R.A., 1971. A review of geophysical studies in the Canadian Cordillera; Canadian Journal of Earth Sciences, v. 8, p. 788-801.

Best, M.G., 1988. Early Miocene change in direction of least principal stress, southwestern United States; conflicting inferences from dikes and metamorphic core-detachment fault terranes; Tectonics, v. 7, p. 249-260.

Bevier, M.L., Armstrong, R.L., and Souther, J.G., 1979. Miocene peralkaline volcanism in west-central British Columbia - its temporal and plate-tectonics setting; Geology, v. 7, p. 389-392.

Bickford, M.E., 1988. The formation of continental crust: Part 1. A review of some principles; Part 2. An application to the Proterozoic evolution of southern North America; Bulletin of the Geological Society of America, v. 100, p. 1375-1391.

Bilibin, Y.A., 1955 (translated in 1967). Metallogenic provinces and metallogenic epochs; Geological Bulletin, Dept. of Geology, Queen's College Press, Flushing, NY, 35 p.

Blank, H.R., Butler, W.C., and Saltus, R.W., 1998. Neogene uplift and radial collapse of the Colorado Plateau - regional implications of gravity and aeromagnetic data; *in* J.D. Friedman and A.C. Huffman (coords.), Laccolith Complexes of Southeastern Utah: Time of Emplacement and Tectonic Setting - Workshop Proceedings; U.S. Geological Survey, Bulletin 2158, p. 9-32.

Bloodgood, M.A., Rees, C.J., and Lefebure, D.V., 1989. Geology and mineralization of the Atlin area, northwestern British Columbia; *in* Geological Fieldwork 1988; British Columbia Ministry of Energy, Mines and Petroleum Resources, Paper 1989-1, p. 311-322.

Bond, G.C. and Kominz, M.A., 1984. Construction of tectonic subsidence curves for the early Paleozoic miogeocline, southern Canadian Rocky Mountains: implications for subsidence mechanisms, age of breakup, and crustal thinning; Bulletin of the Geological Society of America, v. 95, p. 155-173.

Bostock, H.S., 1970. Physiographic subdivisions of Canada; *in* R.J.W. Douglas (ed.), Geology and Economic Minerals of Canada; Geological Survey of Canada, Economic Geology Report 1, p. 10-30.

Bott, M.H.P., 1993. Modelling the plate-driving mechanisms; Journal of the Geological Society (London), v. 150, p. 941-951.

Bowin, C., 1983. Depth of principal mass anomalies contributing to the Earth's geoidal undulations and gravity anomalies; Marine Geodesy, v. 7, p. 61-100.

Boyanov, I., Dabovski, H., Gocev, P., Harkovska, A., Kostadinov, V., Tzankov, Tz., and Zagorcev, I., 1989. A new view of the Alpine tectonic evolution of Bulgaria; Geologia Rhodopica, v. 1, p. 107-121.

Boyer, S.E., 1992. Geometric evidence for synchronous thrusting in the southern Alberta and northwest Montana thrust belts; *in* K.R. McClay (ed.), Thrust Tectonics; Chapman and Hall, p. 377-390.

Brandley, R.T., Krause, F.F., Varsek, J.L., Thurston, J., and Spratt, D.A., 1996. Implied basement-tectonic control on deposition of Lower Carboniferous carbonate ramp, southern Cordillera, Canada; Geology, v. 24, p. 467-470.

Brew, G.E., Litak, R.L., Seber, D., Barazangi, M., Sawaf, T., and Al-Imam, A., 1997. Summary of the geological evolution of Syria through geophysical interpretation: implications for hydrocarbon exploration; v. 16, p. 1473-1482.

Brewer, J.A., Cook, F.A., Brown, L.D., Oliver, J.E., Kaufman, S., and Albaugh, D.S., 1981. COCORP seismic reflection profiling across thrust faults; *in* K.R. McClay and N.J. Price (eds.), Thrust and Nappe Tectonics; Geological Society (London), Special Publication 9, p. 501-512.

Brezhnev, V.D. and Shvanov, V.N., 1980. Early troughs and formations in the western part of the southern Tien Shan; Geotectonics, v. 14, p. 289-296.

Brocher, T.M., Hunter, W.C., and Langenheim, V.E., 1998. Implications of seismic reflection and potential-field geophysical data on the structural framework of the Yucca Mountain-Crater Flat region, Nevada; Bulletin of the Geological Society of America, v. 110, p. 947-971.

Brosgé, W.P. and Dutro, J.T., Jr., 1973. Paleozoic rocks in northern and central Alaska; *in* M.G. Pitcher (ed.), Arctic Geology; American Association of Petroleum Geologists, Memoir 19, p. 361-375.

Brown, D.A. and Gunning, M.H., 1989. Geology of the Scud River area, northwestern British Columbia (104G5,6); *in* Geological Fieldwork 1988; British Columbia Ministry of Energy, Mines and Petroleum Resources, Paper 1989-1, p. 251-268.

Brown, E.H. and Walker, N.W., 1993. A magma-loading model for Barrovian metamorphism in the southeast Coast Plutonic Complex, British Columbia and Washington; Bulletin of the Geological Society of America, v. 105, p. 479-500.

Brown, R.L. and Carr, S.D., 1990. Lithospheric thickening and orogenic collapse within the Canadian Cordillera; *in* E.D. Ghent, P.S. Simony, and R.L. Brown, Metamorphism and Structure of Part of the Omineca Belt, southeastern British Columbia, IGCP Project 235-304, Field Trip #2 Guidebook.

Brown, R.L., Carr, S.D., Harvey, J.L., Williams, P.F., and de Keijszer, M., 1995. Structure and significance of the Teslin suture zone; *in* Slave-Northern Cordilleran Lithospheric Experiment (SNORCLE) Transect Meeting, Lithoprobe Report 44, p. 56-62.

Brown, R.L., Carr, S.D., Johnson, B.J., Coleman, V.J., Cook, F.A., and Varsek, J.L., 1992. The Monashee décollement of the southern Canadian Cordillera: a crustal-scale shear zone linking the Rocky Mountain foreland belt to lower crust beneath accreted terranes; *in* K. McClay (ed.), Thrust Tectonics; Chapman and Hall, London, p. 357-364.

Brown, R.L., Journeay, J.M., and Lane, L.S., 1991. Selkirk Allochthon; *in* H. Gabrielse and C.J. Yorath (eds.), Geology of the Cordilleran Orogen in Canada; Geological Society of America, The Geology of North America, v. G-2, p. 620-621.

Brown, R.L., Journeay, J.M., Lane, L.S., Murphy, D.C., and Rees, C.J., 1986. Obduction, back-folding and piggy-back thrusting in the metamorphic hinterland of the southeast Canadian Cordillera; Journal of Structural Geology, v. 8, p. 255-268.

Brown, R.L. and Read, P.B., 1983. Shuswap terrane of British Columbia: a Mesozoic 'core complex'; Geology, v. 11, p. 164-168.

Burchfiel, B.C., 1980. Plate tectonic and the continents: a review; *in* Continental Tectonics; National Academy of Sciences, Washington, DC, p. 15-25.

Burchfiel, B.C., Cowan, D.S., and Davis, G.A., 1992. Tectonic overview of the Cordilleran orogen in the western United States; *in* B.C. Burchfiel, P.W. Lipman, and M.L. Zoback (eds.), The Cordilleran Orogen: Conterminous U.S.; Geological Society of America, The Geology of North America, v. G-3, p. 407-479.

Burchfiel, B.C. and Davis, G.A., 1972. Structural framework and evolution of the southern part of the Cordilleran orogen, western United States; American Journal of Science, v. 272, p. 97-118.

Burchfiel, B.C. and Davis, G.A., 1975. Nature and controls of Cordilleran orogenesis, western United States: extensions of an earlier synthesis; American Journal of Science, v. 275-A, p. 363-396.

Burchfiel, B.C., Hodges, K.V., and Royden, L.H., 1987. Geology of the Panamint Valley-Saline Valley pull-apart system, California: palinspastic evidence for low-angle geometry of a Neogene range-bounding fault; Journal of Geophysical Research, v. 92, p. 10,422-10,426.

Burchfiel, B.C., Lipman, P.W., and Zoback, M.L. (eds.), 1992. The Cordilleran Orogen: Conterminous U.S.; Geological Society of America, The Geology of North America, v. G-3, 724 p.

Burianyk, M.J.A. and Kanasewich, E.R., 1995. Crustal velocity structure of the Omineca and Intermontane Belts, southeastern Canadian Cordillera; Journal of Geophysical Research, v. 100, p. 15,303-15,316.

Burianyk, M.J.A. and Kanasewich, E.R., 1997. Upper mantle structure in the southeastern Canadian Cordillera; Geophysical Research Letters, v. 24, p. 739-742.

Burianyk, M., Kanasewich, E.R., and Udey, N., 1997. Broadside wide-angle seismic studies and three-dimensional structure of the crust in the southeast Canadian Cordillera; Canadian Journal of Earth Sciences, v. 34, p. 1156-1166.

Burke D.B. and McKee, E.H., 1979. Mid-Cenozoic volcano-tectonic troughs in central Nevada; Bulletin of the Geological Society of America, v. 90, p. 181-184.

Burwash, R.A., Chacko, T., and Muehlenbachs, K., 1995. Tectonic interpretation of Kimiwan anomaly, northwestern Alberta; *in* Alberta Basement Transects Workshop, Lithoprobe Report 47, p. 340-349.

Burwash, R.A., Green, A.G., Jessop, A.M., and Kanasewich, E.R., 1993. Geophysical and petrophysical characteristics of the basement rocks of the western Canada sedimentary basin; *in* D.F. Stott and J.D. Aitken (eds.), Sedimentary Cover of the Craton in Canada; Geological Society of America, The Geology of North America, v. D-1, p. 55-77.

Butler, R.W.H., 1989. The geometry of crustal shortening in the Western Alps; *in* A.M.C. Sengor, Y. Yilmaz, A.I. Okay, and N. Gorur (eds.), Tectonic Evolution of the Tethyan Region; NATO Advanced Studies Institute, Series C, v. 259, p. 43-76.

Butler, R.F., Gehrels, G.E., McClelland, W.C., May, S.R., and Klepacki, D., 1989. Discordant paleomagnetic poles from the Canadian Coast Plutonic Complex: regional tilt rather than large-scale displacement?; Geology, v. 17, p. 691-694.

Caldwell, W.G.E. and Kauffman, E.G. (eds.), 1993. Evolution of the Western Interior Basin; Geological Association of Canada, Special Paper 39.

Campbell, R.B., Mountjoy, E.W., and Young, F.G., 1973. Geology of McBride Map-Area, British Columbia; Geological Survey of Canada, Paper 72-35.

Carr, S.D., 1992. Tectonic setting and U-Pb geochronology of the early Tertiary Ladybird leucogranite suite, southern Omineca Belt, British Columbia; Tectonics, v. 11, p. 258-278.

Carr, S.D., 1995. The southern Omineca Belt, British Columbia: new perspectives from the Lithoprobe geoscience program; Canadian Journal of Earth Sciences, v. 32, p. 1720-1739.

Carr, S.D., Parrish, R.R. and Brown, R.L., 1987. Eocene structural development of the Valhalla complex, southeastern British Columbia; Tectonics, v. 6, p. 175-196.

Cecile, M.P., Morrow, D.W., and Williams, G.K., 1997. Early Paleozoic (Cambrian to Early Devonian) tectonic framework, Canadian Cordillera; Bulletin of Canadian Petroleum Geology, v. 45, p. 54-74.

Cecile, M.P. and Norford, B.S., 1993. Ordovician and Silurian; *in* D.F. Stott and J.D. Aitken (eds.), Sedimentary Cover of the North American Craton in Canada; Geological Society of America, The Geology of North America, v. D-1, p. 125-149.

Chacko, T., King, R.W., Muehlenbachs, K., and Burwash, R.A., 1995. The Kimiwan isotope anomaly, a low ^{18}O zone in the Precambrian basement of Alberta: constraints on the timing of ^{18}O depletion from K-Ar and Rb-Sr data; *in* Alberta Basement Transects Workshop; Lithoprobe Report 47, p. 336-339.

Chadwick, R.A., Pharaoh, T.C., and Smith, N.J.P., 1989. Lower crustal heterogeneity beneath Britain from deep seismic reflection data; Journal of the Geological Society (London), v. 146, p. 617-630.

Chamot-Rooke, N., Jestin, F., de Voogd, B., and the Phèdre Working Group, 1993. Intraplate shortening in the central Indian Ocean determined from a 2100-km north-south deep seismic reflection profile; Geology, v. 21, p. 1043-1046.

Chandra, N.M. and Cumming, G.L., 1972. Seismic refraction studies in western Canada; Canadian Journal of Earth Sciences, v. 9, p. 1099-1109.

Chen Quanmao and Dickinson, W.R., 1986. Contrasting nature of petroliferous Mesozoic-Cenozoic basins in eastern and western China; Bulletin of the American Association of Petroleum Geologists, v. 70, p. 263-275.

Clauser, C., Giese, P., Huenges, E., Kohl, T., Lohman, H., Rybach, L., _afanda, K., Wilhelm, H., Windloff, K., and Zoth, G., 1997. The thermal regime of the crystalline continental crust: implications from KTB; Journal of Geophysical Research, v. 102, p. 18,417-18,441.

Clowes, R.M., 1996. Lithoprobe Phase IV: multidisciplinary studies of the evolution of a continent - a progress report; Geoscience Canada, v. 23, p. 109-123.

Clowes, R.M., Baird, D.J., and Dehler, S.A., 1997. Crustal structure of the Cascadia subduction zone, southwestern British Columbia, from potential field and seismic studies; Canadian Journal of Earth Sciences, v. 34, p. 317-335.

Clowes, R.M., Brandon, M.T., Green, A.G., Yorath, C.J., Sutherland Brown, A., Kanasewich, E.R., and Spencer, C., 1987. LITHOPROBE - southern Vancouver Island: Cenozoic subduction complex imaged by deep seismic reflections; Canadian Journal of Earth Sciences, v. 24, p. 31-51.

Clowes, R.M., Cook, F.A., Green, A.G., Keen, C.E., Ludden, J.N., Percival, J.A., Quinlan, G.M., and West, G.F., 1992. Lithoprobe: new perspectives on crustal evolution; Canadian Journal of Earth Sciences, v. 29, p. 1813-1864.

Clowes, R.M., Cook, F.A., and Ludden, J.N., 1998. Lithoprobe leads to new perspectives on continental evolution; GSA Today, v. 8, no 10, p. 2-7.

Clowes, R.M., Zelt, C.A., Amor, J.R., and Ellis, R.M., 1995. Lithospheric structure in the southern Canadian Cordillera from a network of seismic refraction lines; Canadian Journal of Earth Sciences, v. 32, p. 1485-1513.

Cogley, J.G., 1984. Continental margins and the extent and number of continents; Reviews of Geophysics, v. 22, p. 101-122.

Cohen, Y. and Achache, J., 1990. New global vector magnetic anomaly maps derived from Magsat data; Journal of Geophysical Research, v. 95, p. 10,783-10,800.

Cohen, Y. and Achache, J., 1994. Contributions of induced and remanent magnetization to long-wavelength oceanic magnetic anomalies; Journal of Geophysical Research, v. 99, p. 2943-2954.

Coish, R.A., Bailey, D., Fleming, S., and Poyner, R., 1982. REE geochemistry of rift and ocean ridge volcanics in the Vermont Appalachians (abs).; Geological Society of America, Abstracts with Programs, v. 14, p. 11.

Coleman, R.G., 1990. Can international cooperation help reconstruct the paleo-Asian ocean?; Episodes, v. 14, p. 184-185.

Coleman, R.G., 1994. Main scientific achievements: geodynamic evolution and main structures of the Palaeo-Asian Ocean; Geological Correlation, no. 22, p. 35-38.

Coleman, R.G. (ed.), 1994. Reconstruction of the Palaeo-Asian Ocean; VSP, Utrecht, 186 p.

Coles, R.L. and Currie, R.G., 1977. Magnetic anomalies and rock magnetizations in the southern Coast Mountains, British Columbia: possible relation to subduction; Canadian Journal of Earth Sciences, v. 14, p. 1753-1770.

Coles, R.L., Haines, G.V., and Hannaford, W., 1976. Large scale magnetic anomalies over western Canada and the Arctic: a discussion; Canadian Journal of Earth Sciences, v. 13, p. 790-802.

Collerson, K.D., Lewry, J.F., Bickford, M.E., and Van Schmus, W.R., 1990. Crustal evolution of the buried Precambrian of southern Saskatchewan: implications for diamond exploration; *in* L.S. Beck and C.T. Harper (eds.), Modern Exploration Techniques; Saskatchewan Geological Society, Special Publication 10, p. 150-165.

Colpron, M. and Price, R.A., 1995. Tectonic significance of the Kootenay terrane, southeastern Canadian Cordillera: An alternative model; Geology, v. 23, p. 25-28.

Colpron, M., Price, R.A., Archibald, D.A., and Carmichael, D.M., 1996. Middle Jurassic exhumation along the western flank of the Selkirk fan structure: thermobarometric and thermochronometric constraints from the Illecillewaet synclinorium, southeastern British Columbia; Bulletin of the Geological Society of America, v. 108, p. 1372-1392.

Colpron, M., Warren, M.J., and Price, R.A., 1998. Selkirk fan structure, southeastern Canadian Cordillera: tectonic wedging against an inherited basement ramp; Bulletin of the Geological Society of America, v. 110, p. 1060-1074.

Condie, K.C. and Chomiak, B., 1996. Continental accretion: contrasting Mesozoic and Early Proterozoic tectonic regimes in North America; Tectonophysics, v. 265, p. 101-126.

Coney, P.J., 1970. The geotectonic cycle and the new global tectonics; Bulletin of the Geological Society of America, v. 81, p. 739-747.

Coney, P.J., 1981. Accretionary tectonics in western North America; Arizona Geological Society Digest, v. 14, p. 23-37.

Coney, P.J., Jones, D.L., and Monger, J.W.H., 1980. Cordilleran suspect terranes; Nature, v. 288, p. 329-333.

Conrey, R.W., Sherrod, D.R., and Hooper, P.R., 1995. Changes in primitive magma composition along the Cascade arc in northern Oregon and southern Washington, U.S.A. (abs.); Geological Association of Canada/Mineralogical Association of Canada, Annual Meeting, v. 20, Victoria; Program and Abstracts, p. A-19.

Constenius, K., 1988. Structural configuration of the Kishenehn Basin delineated by geophysical methods, northwestern Montana and southeastern British Columbia; Mountain Geologist, v. 25, p. 13-28.

Constenius, K.N., 1996. Late Paleogene extensional collapse of the Cordilleran foreland fold and thrust belt; Bulletin of the Geological Society of America, v. 108, p. 20-39.

Cook, F.A., 1995a. The Southern Canadian Cordillera Transect of Lithoprobe: Introduction; Canadian Journal of Earth Sciences, v. 32, p. 1483-1484.

Cook, F.A., 1995b. The reflection Moho beneath the southern Canadian Cordillera; Canadian Journal of Earth Sciences, v. 32, p. 1520-1530.

Cook, F.A., 1995c. Lithospheric processes and products in the southern Canadian Cordillera: a Lithoprobe perspective; Canadian Journal of Earth Sciences, v. 32, p. 1803-1824.

Cook, F.A. (ed.), 1995. The Southern Canadian Cordillera Transect of Lithoprobe; Canadian Journal of Earth Sciences, special issue, v. 32, no. 10.

Cook, F.A., Brown, L., Kaufman, S., and Oliver, J., 1983. The COCORP seismic reflection traverse across the southern Appalachians; American Association of Petroleum Geologists, Studies in Geology 14, 61 p.

Cook, F.A. and Clark, E.A., 1990. Middle Proterozoic piggyback basin in the subsurface of northwestern Canada; Geology, v. 18, p. 662-664.

Cook, F.A., Green, A.G., Simony, P.S., Price, R.A., Parrish, R.R., Milkereit, B., Gordy, P.L., Brown, R.L., Coflin, K.C., and Patenaude, C., 1988. LITHOPROBE seismic reflection structure of the southern Canadian Cordillera: initial results; Tectonics, v. 7, p. 157-180.

Cook, F.A. and Jones, A.G., 1995. Seismic reflections and electrical conductivity: a case of Holmes' curious dog?; Geology, v. 23, p. 141-144.

Cook, F.A. and van der Velden, A., 1995. Three dimensional crustal structure of the Purcell anticlinorium in the Cordillera of southwestern Canada; Bulletin of the Geological Society of America, v. 107, p. 642-664.

Cook, F.A. and Varsek, J.L., 1994. Orogen-scale décollements; Reviews of Geophysics, v. 32, p. 37-60.

Cook, F.A., Varsek, J.L., and Clark, E.A., 1991a. Proterozoic craton to basin crustal transition in western Canada and its influence on the evolution of the Cordillera; Canadian Journal of Earth Sciences, v. 28, p. 1148-1158.

Cook, F.A., Varsek, J.L., and Clowes, R.M., 1991b. LITHOPROBE reflection transect of southwestern Canada: Mesozoic thrust and fold belt to mid-ocean ridge; *in* Continental Lithosphere: Deep Seismic Reflections; American Geophysical Union, Geodynamics Series 22, p. 247-255.

Cook, F.A., Varsek, J.L., Clowes, R.M., Kanasewich, E.R., Spencer, C.S., Parrish, R.R., Brown, R.L., Carr, S.D., Johnson, B.J., and Price, R.A., 1992. Lithoprobe crustal reflection cross section of the southern Canadian Cordillera. I. Foreland thrust and fold belt to Fraser River fault; Tectonics, v. 11, p. 12-35.

Cook, F.A., Varsek, J.L., and Thurston, J.B., 1995. Tectonic significance of gravity and magnetic variations along the Lithoprobe Southern Canadian Cordillera Transect; Canadian Journal of Earth Sciences, v. 32, p. 1584-1610.

Cooper, M.R., 1990. Tectonic cycles in southern Africa; Earth-Science Reviews, v. 28, p. 321-364.

Cooper, M., 1992. The analysis of fracture systems in subsurface thrust structures from the Foothills of the Canadian Rockies; *in* K.R. McClay (ed.), Thrust Tectonics; Chapman and Hall, p. 391-405.

Corbett, C.R. and Simony, P.S., 1984. The Champion Lakes fault in the Trail-Castlegar area of southern British Columbia; *in* Current Research, Part A; Geological Survey of Canada, Paper 84-1A, p. 103-104.

Cordey, F., Mortimer, N., DeWever, P., and Monger, J.W.H., 1987. Significance of Jurassic radiolarians from the Cache Creek terrane, British Columbia; Geology, v. 15, p. 1151-1154.

Coward, M.P. and Dietrich, D., 1989. Alpine tectonics - and overview; *in* M.P. Coward, D. Dietrich, and R.G. Park (eds.), Alpine Tectonics; Geological Society (London), Special Publication 45, p. 1-32.

Creaser, R.A., Heaman, L.M., and Erdmer, P., 1996. U-Pb zircon dating of eclogite from the Teslin tectonic zone: constraints for the age of high pressure metamorphism in the Yukon-Tanana terrane; *in* Slave-Northern Cordillera Lithospheric Evolution and Cordilleran Tectonics Workshop; Lithoprobe Report 50, p. 58-60.

Crittenden, M.D., Jr., Coney, P.J., and Davis, G.H. (eds.), 1980. Cordilleran Metamorphic Core Complexes; Geological Society of America, Memoir, 153, 490 p.

Cross, T.A. and Pilger, Jr., R.H., 1978. Tectonic controls on Late Cretaceous sedimentation, western Interior, USA; Nature, v. 274, p. 653-657.

Crowley, J.L., 1995. Geochronological constraints on Early Proterozoic tectonism and Cordilleran overprinting in the Monashee complex, British Columbia (abs.); Geological Association of Canada/Mineralogical Association of Canada, Annual Meeting, v. 20, Victoria; Program and Abstracts, p. A-21.

Crowley, J.L. and Brown, R.L., 1994. Tectonic links between the Clachnacudainn terrane and Selkirk allochthon, southern Omineca belt, Canadian Cordillera; Tectonics, v. 13, p. 1035-1051.

Crowley, J.L., Ghent, E.D., Carr, S.D., and Simony, P.S., 1998. U-Pb geochronologic evidence for Late Jurassic and Late Cretaceous thermal events in the northern Monashee and Selkirk mountains, Mica Creek area, B.C.; *in* Slave-Northern Cordillera Lithospheric Evolution and Cordilleran Tectonics Workshop, Lithoprobe Report 64, p. 243-244.

Cumming, G.L., Garland, G.D., and Vozoff, K., 1962. Seismological Measurements in Southern Alberta; Contract AF19 (604)-8470; Dept. of Physics, University of Alberta, Project Vela Uniform, Final Report.

Cumming, G.L. and Kanasewich, E.R., 1966. Crustal Structure in Western Canada; Contract AF19 (628)-2835, Project 8652; Dept. of Physics, University of Alberta, Project Vela Uniform, Final Report, 185 p.

Cumming, W.B., Clowes, R.M., and Ellis, R.M., 1979. Crustal structure from a seismic refraction profile across southern British Columbia; Canadian Journal of Earth Sciences, v. 16, p. 1024-1040.

Currie, L.D. and Parrish, R.R., 1997. Paleozoic and Mesozoic rocks of Stikinia exposed in northwestern British Columbia: implications for correlations in the northern Cordillera; Bulletin of the Geological Society of America, v. 109, p. 1402-1420.

Dahlstrom, C.D.A., 1969. Balanced cross-sections; Canadian Journal of Earth Sciences, v. 6, p. 743-757.

Dahlstrom, C.D.A., 1970. Structural geology in the eastern margin of the Canadian Rocky Mountains; Bulletin of Canadian Petroleum Geology, v. 18, p. 332-406.

Dana, J.D., 1873. On some results of the Earth's contraction from cooling, including a discussion of the origin of mountains and the nature of the Earth's interior; American Journal of Science, 3rd series, v. 5, p. 423-443; v. 6, p. 6-14, 104-115, 161-172.

Davis, G.A., Monger, J.W.H., and Burchfiel, B.C., 1978. Mesozoic construction of the Cordilleran "collage", central British Columbia to central California; *in*

D.G. Howell and K.A. McDougall (eds.), Mesozoic Paleogeography of the Western United States; SEPM (Society of Economic Paleontologists and Mineralogists), Pacific Coast Paleogeography Symposium, p. 1-32.

Dawson, K.M., Panteleyev, A., Sutherland Brown, A., and Woodsworth, G.J., 1991. Regional Metallogeny; *in* H. Gabrielse and C.J. Yorath (eds.), Geology of the Cordilleran Orogen in Canada; Geological Society of America, The Geology of North America, v. G-2, p. 707-768.

Day, S.J. and Matysek, P.F., 1989. Using the Regional Geochemical Survey Database: examples from the 1988 release (104B,F,G,K); *in* Geological Fieldwork 1988; British Columbia Ministry of Energy, Mines and Petroleum Resources, Paper 1989-1, p. 593-602.

De Keijzer, M. and Williams, P.F., 1996. Structural analysis of the Teslin tectonic zone, south-central Yukon; *in* Slave-Northern Cordillera Lithospheric Evolution and Cordilleran Tectonics Workshop, Lithoprobe Report 50, p. 45-53.

DeMets, C., Gordon, R.G., Argus, D.F., and Stein, S., 1990. Current plate motions; Geophysical Journal International, v. 101, p. 425-478.

Devlin, W.J. and Bond, G.C., 1988. The initiation of the early Paleozoic Cordilleran miogeocline: evidence from the uppermost Proterozoic - Lower Cambrian Hamill Group of southeastern British Columbia; Canadian Journal of Earth Sciences, v. 25, p. 371-412.

Dewey, J.F, 1988. Extensional collapse of orogens; Tectonics, v. 7, p. 1123-1139.

Dewey, J.F. and Burke, K.C.A., 1973. Tibetan, Variscan and Precambrian basement reactivation: products of continental collision; Journal of Geology, v. 81, p. 683-692.

Dewey, J.F. and Bird, J.M., 1970. Mountain belts and the new global tectonics; Journal of Geophysical Research, v. 75, p. 2625-2647.

Dickinson, W.R., 1976. Sedimentary basins developed during evolution of Mesozoic-Cenozoic arc-trench system in western North America; Canadian Journal of Earth Sciences, v. 13, p. 1268-1287.

Dickinson, W.R. and Butler, R.F., 1998. Coastal and Baja California paleomagnetism reconsidered; Bulletin of the Geological Society of America, v. 110, p. 1268-1280.

Dietz, R.S., 1961. Continent and ocean basin evolution by spreading of the sea floor; Nature, v. 190, p. 854-857.

Dietz, R.S. and Holden, J.C., 1967. Miogeoclines in space and time; Journal of Geology, v. 65, p. 566.

Dietz, R.S. and Holden, J.C., 1974. Collapsing continental rises: actualistic concept of geosynclines - a review; *in* R.H. Dott, Jr. and R.H. Shaver (eds.), Modern and Ancient Geosynclinal Sedimentation; SEPM (Society of Economic Paleontologists and Mineralogists), Special Publication 19, p. 14-25.

Dietz, R.S. and Sproll, W.P., 1970. Overlaps and underlaps in the North America to Africa continental drift fit; *in* The Geology of the East Atlantic Continental Margin: 1, General and Economic Papers; Report - Natural Environment Research Council, Institute of Geological Sciences, v. 70/13, p. 143-151.

Digel, S.G., Ghent, E.D., Carr, S.D., and Simony, P.S., 1998. Early Cretaceous kyanite-sillimanite metamorphism and Paleocene sillimanite overprint near Mount Cheadle, southeastern British Columbia: geometry, geochronology, and metamorphic implications; Canadian Journal of Earth Sciences, v. 35, p. 1070-1087.

Doglioni, C., 1990. The global tectonic pattern; Journal of Geodynamics, v. 12, p. 21-38.

Doglioni, C., 1993. Some remarks on the origin of foredeeps; Tectonophysics, v. 228, p. 1-20.

Dohr, G., 1989. Deep seismic - a tool in the recognition and interpretation of large geological elements, the starting point for deterministic basin modeling; Geologische Rundschau, v. 78, p. 21-48.

Doughty, P.T., Price, R.A., and Parrish, R.R., 1998. Geology and U-Pb geochronology of Archean basement and Proterozoic cover in the Priest River complex, northwestern United States, and their implications for Cordilleran structure and Precambrian continent reconstructions; Canadian Journal of Earth Sciences, v. 35, p. 39-54.

Douglas, B., 1986. Deformation history of an outlier of metasedimentary rocks, Coast Plutonic Complex, British Columbia, Canada; Canadian Journal of Earth Sciences, v. 23, p. 813-826.

Douglas, R.J.W. and Lebel, D., 1993. Geology and Structure Cross-Section, Cardinal River, Alberta; Geological Survey of Canada, Map 1828A, NTS 83C/15, scale 1:50,000.

Douglas, R.J.W. (ed.), 1970. Geology and Economic Minerals of Canada; Geological Survey of Canada, Economic Geology Report No. 1, 838 p.

Douglas, R.J.W., Gabrielse, H., Wheeler, J.O., Stott, D.F., and Belyea, H.R., 1970. Geology of Western Canada; *in* R.J.W. Douglas (ed.), Geology and Economic Minerals of Canada; Geological Survey of Canada, Economic Geology Report No. 1, p. 366-488.

Dover, J.H., 1994. Geology of part of east-central Alaska; *in* G. Plafker and H.C. Berg (eds.), The Geology of Alaska; Geological Society of America, The Geology of North America, v. G-1, p. 153-204.

Driver, L.A., Creaser, R., Chacko, T., and Erdmer, P., 1998. Petrogenesis of the Cretaceous Cassiar batholith, Yukon-B.C., Canada; *in* Slave-Northern Cordillera

Lithospheric Evolution and Cordilleran Tectonics Workshop, Lithoprobe Report 64, p. 158-170.

Duncan, R.A., 1982. A captured island chain in the Coast Range of Oregon and Washington; Journal of Geophysical Research, v. 87, p. 10,827-10,837.

Dunne, W.W., 1996. Revision of Alleghenian tectonics in Appalachians; American Journal of Science, v. 296, p. 549-575.

Durrheim, R.J. and Mooney, W.D., 1991. Archean and Proterozoic crustal evolution: evidence from crustal seismology; Geology, v. 19, p. 606-609.

Durrheim, R.J. and Mooney, W.D., 1994. Evolution of the Precambrian lithosphere: seismological and geochemical constraints; Journal of Geophysical Research, v. 99, p. 15,359-15,374.

Dusel-Bacon, C., 1994. Map and Table Showing metamorphic Rocks of Alaska, 2 sheets, scale 1:2,500,000; *in* G. Plafker and H.C. Berg (eds.), The Geology of Alaska; Geological Society of America, The Geology of North America, v. G-1, Plate 4.

Dusel-Bacon, C. and Aleinikoff, J.N., 1985. Petrology and tectonic significance of augen gneiss from a belt of Mississippian granitoids in the Yukon-Tanana terrane, east-central Alaska; Bulletin of the Geological Society of America, v. 96, p. 411-425.

Eaton, D.W.S. and Cook, F.A., 1988. LITHOPROBE seismic reflection imaging of Rocky Mountain structures east of Canal Flats, British Columbia; Canadian Journal of Earth Sciences, v. 25, p. 1339-1348.

Eaton, D.W.S. and Cook, F.A., 1990. Crustal structure of the Valhalla complex, British Columbia, from Lithoprobe seismic-reflection and potential-field data; Canadian Journal of Earth Sciences, v. 27, p. 1048-1060.

Edwards, D.E., Barclay, J.E., Gibson, D.W., Kvill, G.E., and Halton, E., 1994. Triassic strata of the Western Canada Sedimentary Basin; *in* G.D. Mossop and I. Shetsen (comps.), Geological Atlas of the Western Canada Sedimentary Basin; Canadian Society of Petroleum Geologists & Alberta Research Council, p. 259-275.

Edwards, D.J., Lyatsky, H.V., and Brown, R.J., 1998. Regional interpretation steep faults in the Alberta Basin from public-domain gravity and magnetic data: an update; Recorder (Canadian Society of Exploration Geophysicists), v. XXIII, no. 1, p. 15-24.

Eisbacher, G.H., 1974. Evolution of successor basins in the Canadian Cordillera; *in* R.H. Dott and R.H. Shaver (eds.), Modern and Ancient Geosynclinal Sedimentation; SEPM (Society of Economic Paleontologists and Mineralogists), Special Publication 19, p. 274-291.

Eisbacher, G.H., 1977. Mesozoic-Tertiary basin models for the Canadian Cordillera and their geological constraints; Canadian Journal of Earth Sciences, v. 14, p. 2414-2421.

Eittreim, S.L., Gnibidenko, H., Helsley, C.E., Sliter, R., Mann, D., and Ragozin, N., 1994. Oceanic crustal thickness and seismic character along a central Pacific transect; Journal of Geophysical Research, v. 99, p. 3139-3145.

Emmermann, R. and Lauterjung, J., 1997. The German Continental Deep Drilling Program KTB: overview and major results; Journal of Geophysical Research, v. 102, p. 18,179-18,201.

Engebretson, D.C., Cox, A., and Gordon, R.G., 1985. Relative Motions Between Oceanic and Continental Plates in the Pacific Basin; Geological Society of America, Special Paper 206, 59 p.

Erdmer, P., Ghent, E.D., Archibald, D.A., and Stout, M.Z., 1996. High pressure metamorphism in the Yukon-Tanana terrane - physical conditions, timing and tectonic implications; *in* Slave-Northern Cordillera Lithospheric Evolution and Cordilleran Tectonics Workshop; Lithoprobe Report 50, p. 61-68.

Erdmer, P., Ghent, E.D., Archibald, D.A., and Stout, M.Z., 1998. Paleozoic and Mesozoic high pressure metamorphism at the margin of ancestral North America in central Yukon; Bulletin of the Geological Society of America, v. 110, p. 615-629.

Erdmer, P. and Helmstaedt, H., 1983. Eclogite from central Yukon: a record of subduction at the western margin of ancient North America; Canadian Journal of Earth Sciences, v. 20, p. 1389-1408.

Erdmer, P., Thompson, R.I., and Daughtry, K.L., 1999. Pericratonic Paleozoic succession in Vernon and Ashcroft map areas, British Columbia; *in* Current Research, 1999-A; Geological Survey of Canada, p. 205-213.

Ernst, W.G., 1988. Metamorphic terranes, isotopic provinces, and implications for crustal growth of the western United States; Journal of Geophysical Research, v. 93, p. 7634-7642.

Evans, K.V. and Fischer, L.B., 1986. U-Th-Pb geochronology of two augen gneiss terranes, Idaho: new data and tectonic implications; Canadian Journal of Earth Sciences, v. 23, p. 1919-1927.

Evans, K.V. and Zartman, R.E., 1990. U-Th-Pb and Rb-Sr geochronology of middle Proterozoic granite and augen gneiss, salmon River Mountains, east-central Idaho; Bulletin of the Geological Society of America, v. 102, p. 63-73.

Evenchick, C.A., 1991. Geometry, evolution and tectonic framework of the Skeena fold belt, north central British Columbia; Tectonics, v. 10, p. 527-546.

Evenchick, C.A., 1992. The Skeena fold belt: a link between the Coast Plutonic Complex, the Omineca Belt and the Rock Mountain fold and thrust belt; *in* K. McClay (ed.), Thrust Tectonics; Chapman and Hall, p. 365-375.

Fermor, P., 1999. Aspects of the three-dimensional structure of the Alberta Foothills and Front Ranges; Bulletin of the Geological Society of America, v. 111, p. 317-346.

Fillipone, J.A. and Yin, A., 1994. Age and regional tectonic implications of Late Cretaceous thrusting and Eocene extension, Cabinet Mountains, northwest Montana and northern Idaho; Bulletin of the Geological Society of America, v. 106, p. 1017-1032.

Flint, R.F. and Skinner, B.J., 1977. Physical Geology, second edition; John Wiley & Sons, 571 p.

Forsyth, D.A., Berry, M.J., and Ellis, R.M., 1974. A refraction survey across the Canadian Cordillera at 54°N; Canadian Journal of Earth Sciences, v. 11, p. 533-548.

Forte, A.M., Peltier, W.R., Dziewonski, A.M., and Woodward, R.L., 1993. Dynamic surface topography: a new interpretation based upon mantle flow models derived from seismic tomography; Geophysical Research Letters, v. 20, p. 235-238.

Fossen, H., 1993. The role of extensional tectonics in the Caledonides of south Norway: reply; Journal of Structural Geology, v. 15, p. 1381-1383.

Foster, H.L., Keith, T.E., and Menzie, W.D., 1994. Geology of the Yukon-Tanana area of east-central Alaska; *in* G. Plafker and H.C. Berg (eds.), The Geology of Alaska; Geological Society of America, The Geology of North America, v. G-1, p. 205-240.

Fox, P., 1989. Alkaline copper-gold porphyries: schizophrenic cousins of real porphyry coppers; *in* Copper Gold Porphyry Workshop, Mineral Deposits Division, Geological Association of Canada, Vancouver.

Franklin, J.M., Lydon, J.W., and Sangster, D.M., 1981. Volcanic-associated massive sulfide deposits; Economic Geology, 75th Anniversary Volume, p. 485-627.

Friedman, G.M., 1998. Sedimentology and stratigraphy in the 1950s to mid-1980s: the story of a personal perspective; Episodes, v. 21, p. 172-177.

Friedman, G.M. and Sanders, J.E., 1991. Principles of Sedimentology; Oil & Gas Consultants International, Tulsa, 792 p.

Friedman, G.M., Sanders, J.E., and Kopaska-Merkel, D.C., 1992. Principles of Sedimentary Deposits: Stratigraphy and Sedimentology; Macmillan, 717 p.

Friedman, J.D. and Huffman, A.C. (coords.), 1998. Laccolith Complexes of Southeastern Utah: Time of Emplacement and Tectonic Setting - Workshop Proceedings; U.S. Geological Survey Bulletin 2158, 292 p.

Friedman, R.M. and Armstrong, R.L., 1988. Tatla Lake metamorphic complex: as Eocene metamorphic core complex on the southwestern edge of the Intermontane Belt of British Columbia; Tectonics, v. 7, p. 1141-1166.

Friedman, R.M. and Armstrong, R.L., 1991. Intermontane Belt; *in* H. Gabrielse and C.J. Yorath (eds.), Geology of the Cordilleran Orogen in Canada; Geological Society of America, The Geology of North America, v. G-2, p. 663-664.

Friedman, R.M., Schiarizza, P., Mustard, P., van der Heyden, P., Mortensen, J., Thompson, J., and Deyell, C., 1998. Middle and late Triassic arc volcanic rocks and associated intrusions in southern and central British Columbia: evidence for correlation of Stikine, Cadwallader and Methow terranes (abs.); *in* Slave-Northern Cordillera Lithospheric Evolution and Cordilleran Tectonics Workshop, Lithoprobe Report 64, p. 250.

Fritz, W.H., 1991. Cambrian assemblages; *in* H. Gabrielse and C.J. Yorath (eds.), Geology of the Cordilleran Orogen in Canada; Geological Society of America, The Geology of North America, v. G-2, p. 155-184.

Fritz, W.H., Cecile, M.P., Norford, B.S., Morrow, D., and Geldsetzer, H.H.J., 1991. Cambrian to Middle Devonian assemblages; *in* H. Gabrielse and C.J. Yorath (eds.), Geology of the Cordilleran Orogen in Canada; Geological Society of America, The Geology of North America, v. G-2, p. 151-218.

Froidevaux, C. and Nataf, H.C., 1981. Continental drift: what driving mechanism?; Geologische Rundschau, v. 70, p. 166-176.

Fyles, J.T. and Höy T., 1991. Kootenay Arc; *in* H. Gabrielse and C.J. Yorath (eds.), Geology of the Cordilleran Orogen in Canada; Geological Society of America, The Geology of North America, v. G-2, p. 621-625.

Gabrielse, H., 1985. Major dextral transcurrent displacements along the Northern Rocky Mountain Trench and related lineaments in north-central British Columbia; Bulletin of the Geological Society of America, v. 96, p. 1-14.

Gabrielse, H., 1991a. Fault-controlled basins; *in* H. Gabrielse and C.J. Yorath (eds.), Geology of the Cordilleran Orogen in Canada; Geological Society of America, The Geology of North America, v. G-2, p. 360-365.

Gabrielse, H., 1991b. Okanagan-Kootenay region; *in* H. Gabrielse and C.J. Yorath (eds.), Geology of the Cordilleran Orogen in Canada; Geological Society of America, The Geology of North America, v. G-2, p. 607-610.

Gabrielse, H., 1996. Early Jurassic tectonism in north-central British Columbia (abs.); *in* Slave-Northern Cordillera Lithospheric Evolution and Cordilleran Tectonics Workshop; Lithoprobe Report 50, p. 116.

Gabrielse, H. and Campbell, R.B., 1991. Upper Proterozoic assemblages; *in* H. Gabrielse and C.J. Yorath (eds.), Geology of the Cordilleran Orogen in Canada; Geological Society of America, The Geology of North America, v. G-2, p. 125-150.

Gabrielse, H., Monger, J.W.H., Wheeler, J.O., and Yorath, C.J., 1991a. Morphogeological belts, tectonic assemblages, and terranes; *in* H. Gabrielse and C.J. Yorath (eds.), Geology of the Cordilleran Orogen in Canada; Geological Society of America, The Geology of North America, v. G-2, p. 15-28.

Gabrielse, H., Monger, J.W.H., Tempelman-Kluit, D.J., and Woodsworth, G.J., 1991b. Intermontane Belt; *in* H. Gabrielse and C.J. Yorath (eds.), Geology of the Cordilleran Orogen in Canada; Geological Society of America, The Geology of North America, v. G-2, p. 591-603.

Gabrielse, H. and Yorath, C.J., 1991. Introduction; *in* H. Gabrielse and C.J. Yorath (eds.), Geology of the Cordilleran Orogen in Canada; Geological Society of America, The Geology of North America, v. G-2, p. 3-11.

Gabrielse, H. and Yorath, C.J. (eds.), 1991. Geology of the Cordilleran Orogen in Canada; Geological Society of America, The Geology of North America, v. G-2, 823 p.

Gal, L.P. and Ghent, E.D., 1990. Metamorphism of the Solitude Range, southwestern Rocky Mountains, British Columbia: comparison with adjacent Omineca Belt Rocks and tectonometamorphic implications for the Purcell Thrust; Canadian Journal of Earth Sciences, v. 27, p. 1511-1520.

Galster, R.W., Coombs, H.A., and Waldron, H.H., 1989. Engineering geology in Washington - an introduction; *in* R.W. Galster (ed.), Engineering Geology in Washington, v. I; Washington Division of Geology and Earth Resources, Bulletin 78, p. 3-12.

Gao, S., Zhang, B.-R., Jin, Z.-M., Kern, H., Luo, T.-C., and Zhao, Z.-D., 1998. How mafic is the lower continental crust?; Earth and Planetary Science Letters, v. 161, p. 101-117.

Garland, G.D. and Bower, M.E., 1959. Interpretation of aeromagnetic anomalies in Northeastern Alberta; World Petroleum Congress, Proceedings, v. 10, no. 18, p. 787-800.

Garland, G.D. and Tanner, J.G., 1957. Investigations of gravity and isostasy in the southern Canadian Cordillera; Publications of the Dominion Observatory, Ottawa, v. 19, p. 169-222.

Gary, M., McAfee, Jr., R., and Wolf, C.L. (eds.), 1972. Glossary of Geology; second edition; American Geological Institute, 805 p.

Geiger, H.D. and Cook, F.A., 1996. Directional filtering of 3D gravity models: lower crustal features revealed; *in* Slave-Northern Cordillera Lithospheric Evolution and Cordilleran Tectonics Workshop; Lithoprobe Report 50, p. 191-198.

Geldsetzer, H.H.J., James, N.P., and Tebbutt, G.E. (eds.), 1988. Reefs: Canada and Adjacent Areas; Canadian Society of Petroleum Geologists, Memoir 13, 775 p.

Geological Survey of Canada, 1990. Canadian Geophysical Atlas; 15 maps, scale 1:10,000,000.

Ghent, E.D., Knitter, C.C., Raeside, R.P., and Stout, M.Z., 1982. Geothermometry and geobarometry of pelitic rocks, upper kyanite and sillimanite zones, Mica Creek, British Columbia; Canadian Mineralogist, v. 21, p. 295-305.

Ghent, E.D., Nicholls, J., Simony, P.S., Sevigny, J.H., and Stout, M.Z., 1991. Hornblende geobarometry of the Nelson batholith, southeastern British Columbia: tectonic implications; Canadian Journal of Earth Sciences, v. 28, p. 1982-1991.

Ghosh, 1995. Nd-Sr isotopic constraints on the interactions of the Intermontane Superterrane with the western edge of North America in the southern Canadian Cordillera; Canadian Journal of Earth Sciences, v. 32, p. 1740-1758.

Gibson, D.W., 1993. Triassic; *in* D.F. Stott and J.D. Aitken (eds.), Sedimentary Cover of the Craton in Canada; Geological Society of America, The Geology of North America, v. D-1, p. 294-320.

Gibson, H.D., Brown, R.L., and Parrish, R.R., 1998. Deformation-induced inverted metamorphic field gradients, an example from southeastern Canadian Cordillera (abs.); *in* Slave-Northern Cordillera Lithospheric Evolution and Cordilleran Tectonics Workshop, Lithoprobe Report 64, p. 193.

Gilbert, G.K., 1890. Lake Bonneville; U.S. Geological Survey, Monograph, v. 1, 438 p.

Gilder, S.A., Keller, G.R., Luo, M., and Goodell, P.C., 1991. Timing and spatial distribution of rifting in China; Tectonophysics, v. 197, p. 225-243.

Goble, R.J., Treves, S.B., Ghazi, A.M., and Wampler, J.M., 1995. Geochemical differences between Proterozoic alkaline basalts in the Lewis thrust sheet, southwestern Alberta, Canada (abs.); American Geophysical Union Fall Meeting, San Francisco, Poster V11A-5.

Gordey, S.P., 1991. Devonian-Mississippian clastics of the Foreland and Omineca belts; *in* H. Gabrielse and C.J. Yorath (eds.), Geology of the Cordilleran Orogen in Canada; Geological Society of America, The Geology of North America, v. G-2, p. 230-242.

Gordey, S.P., Abbott, J.G., Tempelman-Kluit, D.J., and Gabrielse, H., 1987. "Antler" clastics in the Canadian Cordillera; Geology, v. 15, p. 103-107.

Gordey, S.P., Geldsetzer, H.H.J., Morrow, D.W., Bamber, E.W., Henderson, C.M., Richards, B.C., McGugan, A., Gibson, D.W., and Poulton, T.P., 1991. Upper Devonian to Middle Jurassic assemblages; *in* H. Gabrielse and C.J. Yorath (eds.), Geology of the Cordilleran Orogen in Canada; Geological Society of America, The Geology of North America, v. G-2, p. 219-327.

Gordy, P.L., Frey, F.R., and Norris, D.K., 1977. Geological Guide for the C.S.P.G. and 1977 Waterton-Glacier Park Field Conference; Canadian Society of Petroleum Geologists, Calgary.

Gossler, J. and Kind, R., 1996. Seismic evidence for very deep roots of continents; Earth and Planetary Science Letters, v. 138, p. 1-13.

Gough, I.D., 1986. Mantle upflow tectonics in the Canadian Cordillera; Journal of Geophysical Research, v. 91, p. 1909-1919.

Grant, A.C., 1987. Inversion tectonics on the continental margin east of Newfoundland; Geology, v. 15, p. 845-848.

Grant, S., Creaser, R., and Erdmer, P., 1996. Isotopic, geochemical, and kinematic studies of the Yukon-Tanana terrane in the Money klippe, SE Yukon; *in* Slave-

Northern Cordillera Lithospheric Evolution Transect and Cordilleran Tectonics Workshop, Lithoprobe Report 50, p. 27-31.

Greenwood, H.J., Woodsworth, G.J., Read, P.B., Ghent, E.D., and Evenchick, C.A., 1991. Metamorphism; *in* H. Gabrielse and C.J. Yorath (eds.), Geology of the Cordilleran Orogen in Canada; Geological Society of America, The Geology of North America, v. G-2, p. 533-570.

Gupta, J.C. and Jones, A.G., 1995. Electrical conductivity structure of the Purcell Anticlinorium in southeast British Columbia and northwest Montana; Canadian Journal of Earth Sciences, v. 32, p. 1564-1583.

Haggart, J.W., 1993. Latest Jurassic and Cretaceous paleogeography of the northern Insular Belt, British Columbia; *in* G.C. Dunne and K.A. McDougall (eds.), Mesozoic Paleogeography of the Western United States - II; SEPM (Society of Economic Paleontologists and Mineralogists), Pacific Section, Book 71, p. 463-475.

Hailwood, E.A., 1989. Magnetostratigraphy; Geological Society (London), Special Report 19, 84 p.

Haines, G.V. and Hannaford, W., 1972. Magnetic anomaly maps of British Columbia and the adjacent Pacific Ocean; Publications of the Earth Physics Branch, Ottawa, v. 42/7, p. 211-228.

Haines, G.V. and Hannaford, W., 1976. A Three-Component Aeromagnetic Survey of Saskatchewan, Alberta, Yukon and the District of Mackenzie; Earth Physics Branch, Ottawa, Geomagnetism Series 8, 34 p.

Hajnal, Z., Lucas, S., White, D., Lewry, J., Bezdan, S., Stauffer, M.R., and Thomas, M.D., 1996. Seismic reflection images of high-angle faults and linked detachments in the Trans-Hudson Orogen; Tectonics, v. 15, p. 427-439.

Hall, J., 1859. Paleontology: vol. III, containing descriptions and figures of the organic remains of the Lower Helderberg Group and the Oriskany Sandstone; New York Geological Survey, Natural History of New York, Part 6, 532 p.

Hamilton, W., 1980. Complexities of modern and ancient subduction systems; *in* Continental Tectonics; National Academy of Sciences, Washington, DC, p. 33-41.

Hannaford, W. and Haines, G.V., 1974. A three-component aeromagnetic survey of British Columbia and the adjacent Pacific Ocean; Publications of the Earth Physics Branch, Ottawa, v. 44/14, p. 323-379.

Hansen, V.L., 1992. P-T evolution of the Teslin suture zone and Cassiar tectonites, Canada: evidence of A- or B-type subduction; Journal of Metamorphic Geology, v. 10, p. 239-263.

Hansen, V.L. and Dusel-Bacon, C., 1998. Structural and kinematic evolution of the Yukon-Tanana upland tectonites, east-central Alaska: a record of late Paleozoic to Mesozoic crustal assembly; Bulletin of the Geological Society of America, v. 110, p. 211-230.

Hansen, V.L., Heizler, M.T., and Harrison, T.M., 1991. Mesozoic thermal evolution of the Yukon-Tanana composite terrane: new evidence from $^{40}Ar/^{39}Ar$ data; Tectonics, v. 10, p. 51-76.

Harland, W.B., 1992. Stratigraphic regulation and guidance: a critique of current tendencies in stratigraphic codes and guides; Bulletin of the Geological Society of America, v. 104, p. 1231-1235.

Harland, W.B., Armstrong, R.L., Cox, A.V., Craig, L.E., Smith, A.G., and Smith, D.G., 1990. A Geological Time Scale 1989; Cambridge University Press, 263 p.

Harms, T.A., 1986. Structural and Tectonic Analysis of the Sylvester Allochthon, northern British Columbia; Implications for Paleogeography and Accretion; Ph.D. thesis, University of Arizona, Tucson, 80 p.

Harms, T.A. and Stevens, R.A., 1996. A working hypothesis for the tectonostratigraphic affinity of the Stikine Ranges and a portion of the Dorsey terrane; *in* Slave-Northern Cordillera Lithospheric Evolution Transect and Cordilleran Tectonics Workshop, Lithoprobe Report 50, p. 93-95.

Harrison, J.A., Cressman, E.R., and Whipple, J.W., 1992. Geologic and Structure Maps of the Kalispell 1°x2° Quadrangle, Montana, and Alberta and British Columbia; U.S. Geological Survey, Miscellaneous Investigation Series, Map I-2267, scale 1:250,000.

Harrison, J.E., 1972. Precambrian Belt basin of northwestern United States: its geometry, sedimentation, and copper occurrences; Bulletin of the Geological Society of America, v. 83, p. 1215-1240.

Harvey, J.L., Brown, R.L., and Carr, S.D., 1996. Progress in structural mapping in the Teslin Suture Zone, Big salmon Range, Central Yukon Territory; *in* Slave-Northern Cordillera Lithospheric Evolution and Cordilleran Tectonics Workshop; Lithoprobe Report 50, p. 33-44.

Haug, E., 1900. Les géosynclinaux et les aires continentales. Contribution à l'étude des régressions et des transgressions marines; Bulletin de la Société Géologique de France, v. 28/3, p. 617-711.

Hauge, T.A., Allmendinger, R.W., Caruso, C., Hauser, E.C., Klemperer, S.L., Opdyke, S., Potter, C.J., Sanford, W., Brown, L., Kaufman, S., and Oliver, J., 1987. Crustal structure of western Nevada from COCORP deep seismic-reflection data; Bulletin of the Geological Society of America, v. 98, p. 320-329.

Hauser, E., Potter, C., Hauge, T., Burgess, S., Burtsch, S., Mutschler, J., Allmendinger, R., Brown, L., Kaufman, S., and Oliver, J., 1987. Crustal structure of eastern Nevada from COCORP deep reflection seismic data; Bulletin of the Geological Society of America, v. 98, p. 833-844.

Heirtzler, J.R., Dickson, G.O., Herron, E.M., Pitman, W.C. III, and Le Pichon, X., 1968. Marine magnetic anomalies, geomagnetic field reversals and motions of the ocean floor and continents; Journal of Geophysical Research, v. 73, p. 2119-2136.

Henderson, C.M., 1989. The Lower Absaroka Sequence: Upper Carboniferous and Permian; *in* B.D. Ricketts (ed.), Western Canada Sedimentary Basin, A Case Study; Canadian Society of Petroleum Geologists, Special Publication 30, p. 203-217.

Henderson, C.M., Bamber, E.W., Richards, B.C., Higgins, A.C., and McGugan, A., 1993. Permian; *in* D.F. Stott and J.D. Aitken (eds.), Sedimentary Cover of the Craton in Canada; Geological Society of America, The Geology of North America, v. D-1, p. 272-293.

Henderson, G.G.L., 1959. A summary of the regional structure and stratigraphy of the Rocky Mountain Trench; Transactions of the Canadian Institute of Mining and Metallurgy, v. 62, p. 156-161.

Hess, H.H., 1962. History of the ocean basins; *in* A.E.J. Engell, H.L. James, and B.F. Leonard (eds.), Petrologic Studies: A Volume in Honor of A.F. Buddington; Geological Society of America, p. 599-620.

Hickson, C.J., 1991. Volcano vent map and table; *in* C.A. Wood and J. Kienle (eds.), Volcanoes of North America; Cambridge University Press, p. 116-118.

Hinze, W.J. (ed.), 1985. The Utility of Regional Gravity and Magnetic Anomaly Maps; Society of Exploration Geophysicists.

Hoffman, P.F., 1988. United plates of America - Early Proterozoic assembly and growth of Laurentia; Annual Reviews of Earth and Planetary Sciences, v. 16, p. 543-603.

Hofmann, H.J., Mountjoy, E.W., and Teitz, M.W., 1985. Ediacaran fossils from the Miette Group, Rocky Mountains, British Columbia; Geology, v. 13, p. 819-821.

Holbrook, W.S., 1990. The crustal structure of the northwestern Basin and Range Province, Nevada, from wide-angle seismic data; Journal of Geophysical Research, v. 95, p. 21,843-21,869.

Holbrook, W.S., Reiter, E.C., Purdy, G.M., and Toksöz, M.N., 1992. Image of the Moho across the continent-ocean transition, U.S. east coast; Geology, v. 20, p. 203-206.

Holland, S.S., 1964. Landforms of British Columbia - a Physiographic Outline; British Columbia Department of Mines and Petroleum Resources, Bulletin 48, 138 p.

Hollister, L.S., Diebold, J., Smithson, S., Morozov, I., and Das, T., 1996. Preliminary seismic results from Accrete (abs.); *in* Slave-Northern Cordillera Lithospheric Evolution and Cordilleran Tectonics Workshop; Lithoprobe Report 50, p. 202.

Horváth, F., 1988. Neotectonic behavior of the Alpine-Mediterranean region; *in* L.H. Royden and F. Horváth (eds.), The Pannonian Basin: a Study in Basin Evolution; American Association of Petroleum Geologists, Memoir 45, p. 49-55.

Howell, D.G. (ed.), 1985. Tectonostratigraphic Terranes of the Circum-Pacific Region; Circum-Pacific Council for Energy and Mineral Resources, Earth Science Series, no. 1.

Höy, T., 1989. The age, chemistry, and tectonic setting of the Middle Proterozoic Moyie sills, Purcell Supergroup, southeastern British Columbia; Canadian Journal of Earth Sciences, v. 26, p. 2305-2317.

Höy, T., 1991. Volcanogenic massive sulfide deposits in British Columbia; *in* Ore Deposits, Tectonics and Metallogeny in the Canadian Cordillera; British Columbia Ministry of Energy, Mines and Petroleum Resources, Paper 1991-4, p. 89-123.

Höy, T. and Andrew, K., 1989. The Rossland Group, Nelson map area, southeastern British Columbia (82 F/6); *in* Geological Fieldwork 1988; British Columbia Ministry of Energy, Mines and Petroleum Resources, Paper 1989-1, p. 33-43.

Huang Jiqing and others (T.K. Huang, director), 1987. Geotectonic Evolution of China; Science Press Beijing & Springer-Verlag, 199 p.

Humphreys, E.D. and Dueker, K.G., 1994. Western U.S. upper mantle structure; Journal of Geophysical Research, v. 99, p. 9615-9634.

Hutchinson, R.W., 1980. Massive base metal sulphide deposits as guides to tectonic evolution; *in* D.W. Strangway (ed.), The Continental Crust and Its Mineral Deposits; Geological Association of Canada, Special Paper 20, p. 659-684.

Hutchison, W.W., 1982. Geology of the Prince Rupert-Skeena Map Area, British Columbia; Geological Survey of Canada, Memoir 394, 116 p.

Hyndman, R.D., 1995. The Lithoprobe corridor across the Vancouver Island continental margin: the structural and tectonic consequences of subduction; Canadian Journal of Earth Sciences, v. 32, p. 1777-1802.

Hyndman, R.D. and Lewis, T.J., 1995. Review: the thermal regime along the southern Canadian Cordillera Lithoprobe corridor; Canadian Journal of Earth Sciences, v. 32, p. 1611-1617.

Hyndman, R.D., Yorath, C.J., Clowes, R.M., and Davis, E.E., 1990. The northern Cascadia subduction zone at Vancouver Island: seismic structure and tectonic history; Canadian Journal of Earth Sciences, v. 27, p. 313-329.

Irvine, T.N., 1976. Alaskan-type ultramafic-gabbroic bodies in the Aiken Lake, McConnell Creek, and Toodoggone map-areas; *in* Current Research, part A; Geological Survey of Canada, Paper 76-1A, p. 149-152.

Irving, E. and Monger, J.W.H., 1987. Preliminary paleomagnetic results from the Permian Asitka Group, British Columbia; Canadian Journal of Earth Sciences, v. 24, p. 1490-1497.

Irving, E., Monger, J.W.H., and Yole, R.W., 1980. New paleomagnetic evidence for displaced terranes in British Columbia; *in* D.W. Strangway (eds.), The Continental Crust and Its Mineral Deposits; Geological Association of Canada, Special Paper 20, p. 441-456.

Irving, E., Thorkelson, D.J., Wheadon, P.M., and Enkin, R.J., 1995. Paleomagnetism of the Spences Bridge Group and northward displacement of the Intermontane Belt, British Columbia: a second look; Journal of Geophysical Research, v. 100, p. 6057-6071.

Irving, E. and Yole, R.W., 1972. Paleomagnetism and kinematic history of mafic and ultramafic rocks in fold mountain belts; Publications of the Earth Physics Branch; Energy, Mines and Resources, Ottawa, v. 42, p. 87-95.

Irwin, W.P., 1981. Tectonic accretion of the Klamath Mountains; *in* W.G. Ernst (ed.), The Geotectonic Development of California; Rubey Volume 1, p. 29-49.

Issler, D.R., Beaumont, C., Willett, S.D., Donelick, R.A., Mooers, J., and Grist, A., 1990. Preliminary evidence from apatite fission-track data concerning the thermal history of the Peace River Arch region, Western Canada Sedimentary Basin; Bulletin of Canadian Petroleum Geology, v. 38A, p. 250-269.

Jackson, J.A. (ed.), 1997. Glossary of Geology, fourth edition; American Geological Institute, 769 p.

Janecke, S.U., 1994. Sedimentation and paleogeography of an Eocene to Oligocene rift zone, Idaho and Montana; Bulletin of the Geological Society of America, v. 106, p. 1083-1095.

Jenner, G.A., Dunning, G.R., Malpas, J., Brown, M., and Brace, T., 1991. Bay of Islands and Little Port complexes, revisited: age, geochemical and isotopic evidence confirm suprasubduction-zone origin; Canadian Journal of Earth Sciences, v. 28, p. 1635-1652.

Jerzykiewicz, T. and Norris, D.K., 1992. Anatomy of the Laramide Foredeep and the structural style of the adjacent foreland thrust belt in Southern Alberta; Guidebook for Canadian Society of Petroleum Geologists Field Trip #3 for the American Association of Petroleum Geologists Annual Convention, Calgary, 88 p.

Jerzykiewicz, T. and Norris, D.K., 1994. Stratigraphy, structure and syntectonic sedimentation of the Campanian 'Belly River' clastic wedge in the southern Canadian Cordillera; Cretaceous Research, v. 15, p. 367-399.

Johnson, B.J., 1990. Geology Adjacent to the Western Margin of the Shuswap Metamorphic Complex (Parts of 82L, M); British Columbia Ministry of Energy, Mines and Petroleum Resources, Open File 1990-30.

Johnson, B.J., 1994. Structure and Tectonic Setting of the Okanagan Valley Fault System in the Shuswap Lake Area, Southern British Columbia; Ph.D. thesis, Dept. of Earth Sciences, Carleton University, Ottawa.

Johnson, B.J. and Brown, R.L., 1996. Crustal structure and early Tertiary extensional tectonics of the Omineca belt at 51°N latitude, southern Canadian Cordillera; Canadian Journal of Earth Sciences, v. 33, p. 1596-1611.

Johnston, D. and Williams, P., 1998. Westward underthrusting, extrusion and exhumation of crystalline rocks of the Monashee complex, southeastern B.C.; *in* Slave-Northern Cordillera Lithospheric Evolution and Cordilleran Tectonics Workshop, Lithoprobe Report 64, p. 194-195.

Jones, A.G. and Gough, I.D., 1995. Electromagnetic images of crustal structures in southern and central Canadian Cordillera; Canadian Journal of Earth Sciences, v. 32, p. 1541-1563.

Jones, A.G., Gough, D.I., Kurtz, R.D., DeLaurier, J.M., Boerner, D.M., Craven, J.A., Ellis, R.M., and McNeice, G.W., 1992. Electromagnetic images of regional structure in the southern Canadian Cordillera; Geophysical Research Letters, v. 12, p. 2373-2376.

Jordan, T.H., 1975. The continental tectosphere; Geophysics and Space Physics, v. 13, p. 1-12.

Journeay, J.M., 1986. Stratigraphy, Internal Strain and Thermotectonic Evolution of Northern Frenchman Cap Dome: an Exhumed Duplex Structure, Omineca Hinterland, S.E. Canadian Cordillera; Ph.D. thesis, Queen's University, Kingston, Ontario.

Kalkreuth, W. and McMechan, M.E., 1984. Regional pattern of thermal maturation as determined from coal-rank studies, Rocky Mountain Foothills and Front Ranges north of Grande Cache, Alberta - implications for petroleum exploration; Bulletin of Canadian Petroleum Geology, v. 32, p. 249-271.

Kanasewich, E.R., 1993. Deep seismic profiles; *in* D.F. Stott and J.D. Aitken (eds.), Sedimentary Cover of the Craton in Canada; Geological Society of America, The Geology of North America, v. D-1, p. 65-67.

Kanasewich, E.R., Burianyk, M.J.A., Ellis, R.M., Clowes, R.M., White, D.T., Côte, T., Forsyth, D.A., Luetgert, J.H., and Spence, G.D., 1994. Crustal velocity structure of the Omineca belt, southeastern Canadian Cordillera; Journal of Geophysical Research, v. 99, p. 2653-2670.

Kanasewich, E.R., Hajnal, Z., Green, A.G., Cumming, G.L., Mereu, R.F., Clowes, R.M., Morel-a-l'Huissier, P., Chiu, S., Macrides, C.G., Shahriar, M., and Congram, A.M., 1987. Seismic studies of the crust under Williston Basin; Canadian Journal of Earth Sciences, v. 24, p. 2160-2171.

Kay, M., 1951. North American Geosynclines; Geological Society of America, Memoir 48, 143 p.

Kearey, P. and Vine, F.J., 1990. Global Tectonics; Blackwell Science Publications, 302 p.

Kent, D.M. and Christopher, J.E., 1994. Geological history of the Williston Basin and Sweetgrass Arch; *in* G.D. Mossop and I. Shetsen (comps.), Geological Atlas of the Western Canada Sedimentary Basin; Canadian Society of Petroleum Geologists & Alberta Research Council, p. 421-430.

Kent, P.E., Satterthwaite, G.E., and Spencer, A.M. (eds.), 1969. Time and Place in Orogeny; Geological Society (London), Special Publication 3, 311 p.

King, P.B., 1969. Tectonic Map of North America, scale 1:5,000,000; U.S. Geological Survey.

King, P.B., 1977. The Evolution of North America; Princeton University Press, 197 p.

Kirby, S.H. and Kronenberg, A.K., 1987. Rheology of the lithosphere: selected topics; *in* U.S. National Report to the International Union of geodesy and Geophysics, Reviews o geophysics, v. 25, p. 1219-1244.

Klepacki, D.W. and Wheeler, J.O., 1985. Stratigraphic and structural relations of the Milford, Kaslo and Slocan Groups, Goat Range, Lardeau and Nelson map-area, British Columbia; *in* Current Research, Part A; Geological Survey of Canada, Paper 85-1A, p. 277-286.

Kober, L., 1923. Bau und Entstehung der Alpen; Bornträger, Berlin, 379 p.

Kober, L., 1951. Die Orogentheorie; Berlin.

Koch, N.G., 1973. Central Cordilleran region; *in* R.G. McCrossan (ed.), Future Petroleum Provinces of Canada; Canadian Society of Petroleum Geologists, Memoir 1, p. 37-72.

Kozlovsky, Ye.A. (ed.), 1984 (translated in 1987). The Superdeep Well of the Kola Peninsula; Springer-Verlag, 558 p.

Krohe, A., 1996. Variscan tectonics of central Europe: postaccretionary intraplate deformation of weak continental lithosphere; Tectonics, v. 15, p. 1364-1388.

Kurtz, R.D., DeLaurier, J.M., and Gupta, J.C., 1986. A magnetotelluric sounding across Vancouver Island sees the subducting Juan de Fuca plate; Nature, v. 321, p. 596-599.

Kurtz, R.D., DeLaurier, J.M., and Gupta, J.C., 1990. The electrical conductivity distribution beneath Vancouver Island: a region of active plate subduction; Journal of Geophysical Research, v. 95, p. 10,929-10,946.

Laffitte, R., Harland, W.B., Erbe, H.K., Blow, W.H., Haas, W., Hughes, N.F., Ramsbottom, W.H.C., Rat, P., Tinyant, H., and Ziegler, W., 1972. Some international agreement on essentials of stratigraphy; Geological Magazine, v. 109, p. 1-15.

Lamerson, P.R., 1982. The Fossil basin and its relationship to the Absaroka thrust system, Wyoming and Utah; *in* R.B. Powers (ed.), Geologic Studies of the Cordilleran Thrust Belt; Rocky Mountain Association of Geologists, p. 817-830.

Langenberg, C.W., 1983. Polyphase Deformation in the Canadian Shield of Northeastern Alberta; Alberta Geological Survey, Bulletin 45, 33 p.

Lawrence, R.M., Karson, J.A., and Hurst, S.D., 1998. Dike orientation, fault-block rotations, and the construction of slow-spreading oceanic crust at 22°40'N on the Mid-Atlantic Ridge; Journal of Geophysical Research, v. 103, p. 663-676.

Lawton, D.C., Spratt, D.A., and Hopkins, J.C., 1994. Tectonic wedging beneath the Rocky Mountain foreland basin, Alberta, Canada; Geology, v. 22, p. 519-522.

Lebel, D., Langenberg, W., and Mountjoy, E.W., 1996. Structure of the central Canadian Cordilleran fold-and-thrust belt, Athabasca-Brazeau area, Alberta: a large, complex intercutaneous wedge; Bulletin of Canadian Petroleum Geology, v. 44, p. 282-298.

Leech, G.B., 1967. The Rocky Mountain Trench; *in* The World Rift System; Geological Survey of Canada, Paper 66-14, p. 307-329.

Leitch, C.R.B., van der Heyden, P., Godwin, C.I., Armstrong, R.L., and Harakal, J.E., 1991. Geochronometry of the Bridge River Camp, southwestern British Columbia; Canadian Journal of Earth Sciences, v. 28, p. 195-208.

Letouzey, J., 1986. Cenozoic paleo-stress pattern in the Alpine Foreland and structural interpretation in a platform basin; Tectonophysics, v. 132, p. 215-231.

Lewis, T.J., Bentkowski, W.H., and Hyndman, R.D., 1992. Crustal temperatures near the Lithoprobe Southern Canadian Cordillera transect; Canadian Journal of Earth Sciences, v. 29, p. 1197-1214.

Lewry, J.F. and Stauffer, M.R. (eds.), 1990. The Early Proterozoic Trans-Hudson Orogen of North America; Geological Association of Canada, Special Paper 37.

Link, P.K., Christie-Blick, N., Devlin, W.J., Elston, D.P., Horodyski, R.J., Levy, M., Miller, J.M.G., Pearson, R.C., Prave, A., Stewart, J.H., Winston, D., Wright, L.A., and Wrucke, C.T., 1993. Middle and Late Proterozoic stratified rocks of the western U.S. Cordillera, Colorado Plateau, and Basin and Range Province; *in* J.C. Reed, Jr., M.E. Bickford, R.S. Houston, P.K. Link, D.W. Rankin, P.K. Sims, and W.R. Van Schmus (eds.), Precambrian: Conterminous U.S.; Geological Society of America, The Geology of North America, v. C-2, p. 463-595.

Lipman, P.W., 1992. Magmatism in the Cordilleran United States; progress and problems; *in* B.C. Burchfiel, P.W. Lipman, and M.L. Zoback (eds.), The Cordilleran Orogen: Conterminous U.S.; Geological Society of America, Geology of North America, v. G-3, p. 481-514.

Lister, G.S. and Davis, G.A., 1989. The origin of metamorphic core complexes and detachment faults formed during Tertiary continental extension in the northern Colorado River region, U.S.A.; Journal of Structural Geology, v. 11, p. 65-94.

Litak, R.K. and Hauser, E.C., 1992. The Bagdad reflection sequence as tabular mafic intrusions: evidence from seismic modeling of mapped exposures; Bulletin of the Geological Society of America, v. 104, p. 1315-1325.

Livaccari, R.L., Geissman, J.W., and Reynolds, S.J., 1995. Large-magnitude extensional deformation in the South Mountains metamorphic core complex, Arizona: evaluation with paleomagnetism; Bulletin of the Geological Society of America, v. 107, p. 877-894.

Logatchev, N.I., 1993. History and geodynamics of the Lake Baikal Rift in the context of the Eastern Siberia rift system: a review; Bulletin Centres Recherches Exploration-Production Elf Aquitaine, v. 17, p. 353-370.

Lowe, C. and Ranalli, G., 1993. Density, temperature, and rheological models for the southeastern Canadian Cordillera: implications for its geodynamic evolution; Canadian Journal of Earth Sciences, v. 30, p. 77-93.

Lyatsky, H.V., 1991. Regional geophysical constraints on crustal structure and geologic evolution of the Insular Belt, British Columbia; *in* G.J. Woodsworth (ed.), Evolution and Hydrocarbon Potential of the Queen Charlotte Basin, British Columbia; Geological Survey of Canada, Paper 90-10, p. 97-106.

Lyatsky, H.V., 1993. Basement-controlled structure and evolution of the Queen Charlotte Basin, west coast of Canada; Tectonophysics, v. 228, p. 123-140.

Lyatsky, H.V., 1994a. Formation of non-compressional sedimentary basins on continental crust: limitation on modern models; Journal of Petroleum Geology, v. 17, p. 301-316.

Lyatsky, H.V., 1994b. Book review of 'Foreland Basins and Fold Belts', American Association of Petroleum Geologists, Memoir 55; Journal of Petroleum Geology, v. 17, p. 247-248.

Lyatsky, H.V., 1996. Continental-Crust Structures on the Continental Margin of Western North America; Springer-Verlag, Lecture Notes in Earth Sciences 62, 352 p.

Lyatsky, H.V., Friedman, G.M., and Lyatsky, V.B., 1999. Principles of Practical Tectonic Analysis of Cratonic Regions, With particular Reference to Western North America; Springer-Verlag, Lecture Notes in Earth Sciences 84, 369 p.

Lyatsky, H.V. and Haggart, J.W., 1993. Petroleum exploration model for the Queen Charlotte Basin, offshore British Columbia; Canadian Journal of Earth Sciences, v. 30, p. 918-927.

Lyatsky, V.B., 1965. Towards a Common Geochronological Scale of Absolute Age of Geologic Formations; Proceedings of Meeting "General Regularities of Geologic Phenomena", U.S.S.R. Geographic Society, Leningrad (St. Petersburg), p. 177-181 (in Russian).

Lyatsky, V.B., 1967. Earthquakes and Geotectonics; U.S.S.R. Geographical Society, Leningrad (St. Petersburg), 22 p. (in Russian).

Lyatsky, V.B., 1978. The System Approach as a Methodological Basis for the Theory and Practice in the Studies of Shelves; U.S.S.R. Academy of Sciences, Institute of Zoology, Leningrad (St. Petersburg), 26 p. (in Russian).

Lyatsky, V.B., Fong, G., and Ha, T., 1990. Petroleum potential of Paleozoic sequences in sedimentary basins of north-western China; Carbonates and Evaporites, v. 5, p. 97-114.

Lydon, J.W., 1984. Volcanogenic massive sulphide deposits, Part I: a descriptive model; Geoscience Canada, v. 11, p. 195-202.

Lydon, J.W., 1988. Volcanogenic massive sulphide deposits, Part II: genetic models; Geoscience Canada, v. 15, p. 43-65.

Ma Xingyuan (chief comp.), 1986. Lithospheric Dynamics Map of China and Adjacent Seas, scale 1:4,000,000; Institute of Geology, State Seismological Bureau; Geological Publishing House, Beijing.

Ma Xingyuan, Liu Guodong, and Su Jian, 1984. The structure and dynamics of the continental lithosphere in north-northeast China; Annales Geophysicae, v. 2, p. 611-620.

Mac Niocaill, C., van der Pluijm, B.A., and Van der Voo, R., 1997. Ordovician paleogeography and the evolution of the Iapetus ocean; Geology, v. 25, p. 159-162.

Macdonald, K.C., Fox, P.J., Alexander, R.T., Pockalny, R., and Gente, P., 1996. Volcanic growth faults and the origin of Pacific abyssal hills; Nature, v. 380, p. 125-129.

Macdonald, K.C., Fox, P.J., Perram, L.J., Eisen, M.F., Haymon, R.M., Miller, S.P., Carbotte, S.M., Cormier, M.-H., and Shor, A.N., 1988. A new view of the mid-ocean ridge from the behaviour of ridge-axis discontinuities; Nature, v. 335, p. 217-225.

Macdonald, K.C., Scheirer, D.S., and Carbotte, S.M., 1991. Mid-ocean ridges: discontinuities, segments and giant cracks; Science, v. 253, p. 986-994.

MacIntyre, D.G., 1991. SEDEX - sedimentary-exhalative deposits; *in* W.J. McMillan (ed.), Ore Deposits, Tectonics and Metallogeny in the Canadian Cordillera; British Columbia Ministry of Energy, Mines and Petroleum Resources, Paper 1991-4, p. 25-70.

MacIntyre, D.G. and Diakow, L.J., 1998. Late Cretaceous to Early Tertiary tectonics, magmatism and mineral deposits, central British Columbia (abs.); *in* Slave-Northern Cordillera Lithospheric Evolution and Cordilleran Tectonics Workshop, Lithoprobe Report 64, p. 276-277.

MacIntyre, D.G., Desjardins, P., and Tercier, P., 1989. Jurassic stratigraphic relationships in the Babina and Telkwa Ranges (93L/10,11,14,15); *in* Geological Fieldwork 1988; British Columbia Ministry of Energy, Mines and Petroleum Resources, Paper 1089-1, p. 195-208.

Mahoney, J.B., Mustard, P.S., Haggart, J.W., Friedman, R.M., Fanning, C.M., and McNicoll, V.J., 1999. Archean zircons in Cretaceous strata of the western Canadian Cordillera: the "Baja B.C." hypothesis fails a "crucial test"; Geology, v. 27, p. 195-198.

Majorowicz, J.A., Gough, D.I., and Lewis, T.J., 1993. Correlation between the depth to the lower-crustal high conductivity layer and heat flow in the Canadian Cordillera; Tectonophysics, v. 225, p. 49-56.

Malpas, J., Moores, E., Panayiotou, A., and Xenophontos, C. (eds.), 1990. Ophiolites, Oceanic Crustal Analogues; Symposium "Troodos 1987 Cyprus", Proceedings.

Marchildon, N., Dipple, G.M., and Mortensen, J.K., 1998. New metamorphic and geochronological constraints on the tectonic evolution of the northern Selkirk Mountains, SE BC (abs.); *in* Slave-Northern Cordillera Lithospheric Evolution and Cordilleran Tectonics Workshop, Lithoprobe Report 64, p. 192.

Markovsky, A.P. (ed.-in-chief), 1972. Geologicheskaya Karta Evrazii (Geologic Map of Eurasia); U.S.S.R. Ministry of Geology, Moscow.

Marquis, G. and Globerman, B.R., 1988. Northward motion of the Whitehorse Trough: paleomagnetic evidence from the Upper Cretaceous Carmacks Group; Canadian Journal of Earth Sciences, v. 25, p. 2005-2016.

Marquis, G. and Hyndman, R.D., 1992. Geophysical support for aqueous fluids in the deep crust: seismic and electrical relationships; Geophysical Journal International, v. 110, p. 91-105.

Marquis, G., Jones, A.G., and Hyndman, R.D., 1995. Coincident conductive and reflective middle and lower crust in southern British Columbia; Geophysical Journal International, v. 120, p. 111-131.

Masters, J.A. (ed.), 1984. Elmworth - Case Study of a Deep Basin Gas Field; American Association of Petroleum Geologists, Memoir 38, 316 p.

Mathews, W.H. (comp.), 1986. Physiography of the Canadian Cordillera; Geological Survey of Canada, Map 1701A, scale 1:5,000,000.

Mathews, W.H., 1991. Physiographic evolution of the Canadian Cordillera; *in* H. Gabrielse and C.J. Yorath (eds.), Geology of the Cordilleran Orogen in Canada; Geological Society of America, The Geology of North America, v. G-2, p. 405-418.

McCarthy, J. and Parsons, T., 1994. Insights into the kinematic Cenozoic evolution of the Basin and Range-Colorado Plateau transition from coincident seismic refraction and reflection data; Bulletin of the Geological Society of America, v. 106, p. 747-759.

McCartney, W.D., 1965. Metallogeny of post-Precambrian geosynclines; *in* Some Guides to Mineral Exploration, Geological Survey of Canada, p. 33-42.

McClelland, W.C., Gehrels, G.E., Samson, S.D., and Patchett, P.J., 1991. Protolith relations of the Gravina belt and Yukon-Tanana terrane in central southeastern Alaska; Journal of Geology, v. 100, p. 107-123.

McDonough, M.R. and Simony, P.S., 1988. Structural evolution of basement gneisses and Hadrynian cover, Bulldog Creek area, Rocky Mountains, British Columbia; Canadian Journal of Earth Sciences, v. 25, p. 1687-1702.

McDonough, M.R. and Simony, P.S., 1989. The Valemount strain zone: a dextral oblique slip thrust system linking the Rocky Mountains and Omineca belts of the southeastern Canadian Cordillera; Geology, v. 17, p. 237-240.

McGregor, A.M., 1951. Some milestones in the Precambrian of Southern Rhodesia; Geological Society of South Africa, Transactions & Proceedings, v. 54, p. 27-71.

McGroder, M.F., 1991. Reconciliation of two-sided thrusting, burial metamorphism, and diachronous uplift in the Cascades of Washington and British Columbia; Bulletin of the Geological Society of America, v. 103, p. 189-209.

McKenzie, D., 1972. Active tectonics of the Mediterranean region; Geophysical Journal of the Royal Astronomical Society, v. 30, p. 109-185.

McMechan, M.E., 1990. Upper Proterozoic to Middle Cambrian history of the Peace River Arch: evidence from the Rocky Mountains; Bulletin of Canadian Petroleum Geology, v. 38A, p. 36-44.

McMechan, M.E., 1991. Purcell Anticlinorium; *in* H. Gabrielse and C.J. Yorath (eds.), Geology of the Cordilleran Orogen in Canada; Geological Society of America, The Geology of North America, v. G-2, p. 628-630.

McMechan, M.E., 1999. Geometry of the structural front in the Kakwa area, northern Foothills of Alberta; Bulletin of Canadian Petroleum Geology, v. 47, p. 31-42.

McMechan, M.E. and Price, R.A., 1982. Superimposed low-grade metamorphism in the Mount Fisher area, southeastern British Columbia - implications for the East Kootenay orogeny; Canadian Journal of Earth Sciences, v. 19, p. 476-489.

McMechan, R.D. and Price, 1984. Crustal extension and thinning in a foreland fold and thrust belt, southern Canadian Rockies (abs.); Geological Society of America, Abstracts with Programs, v. 16, p. 591.

McMechan, M.E. and Thompson, R.I., 1993. The Canadian Cordilleran fold-and-thrust belt south of 66°N and its influence on the Western Interior Basin; *in* W.G.E. Caldwell and E.G. Kauffman (eds.), Evolution of the Western Interior Basin; Geological Association of Canada, Special Paper 39, p. 73-90.

McMillan, W.J., 1991a. Overview of the tectonic evolution and setting of mineral deposits in the Canadian Cordillera; *in* Ore Deposits, Tectonics and Metallogeny in the Canadian Cordillera; British Columbia Ministry of Energy, Mines and Petroleum Resources, Paper 1991-4, p. 5-24.

McMillan, W.J., 1991b. Porphyry deposits in the Canadian Cordillera; *in* Ore Deposits, Tectonics and Metallogeny in the Canadian Cordillera; British Columbia Ministry of Energy, Mines and Petroleum Resources, Paper 1991-4, p. 253-276.

Meinert, L.D., 1989. Gold skarn deposits - geology and exploration criteria; *in* The Geology of Gold Deposits; The Perspectives in 1988; Economic Geology, Monograph 6, p. 537-552.

Meissner, R., 1989. Rupture, creep, lamellae and crocodiles: happenings in the continental crust; Terra Nova, v. 1, p. 17-28.

Mercier, J.L., Sorel, D., and Simeakis, K., 1987. Changes in the state of stress in the overriding plate of a subduction zone: the Aegean Arc from the Pliocene to the Present; Annales Tectonicae, v. 1, p. 20-39.

Merguerian, C. and Schweikert, R.A., 1987. Paleozoic gneissic granitoids in the Shoo Fly Complex, central Sierra Nevada, California; Bulletin of the Geological Society of America, v. 99, p. 699-717.

Mihalynuk, M.G., Nelson, J.L., and Diakow, L.J., 1994. Cache Creek terrane entrapment: oroclinal paradox within the Canadian Cordillera; Tectonics, v. 13, p. 575-595.

Miller, D.M., Nilsen, T.H., and Bilodeau, W.L., 1992. Late Cretaceous to early Eocene geologic evolution of the U.S. Cordillera; *in* B.C. Burchfiel, P.W. Lipman, and M.L. Zoback (eds.), The Cordilleran Orogen: Conterminous U.S.; Geological Society of America, The Geology of North America, v. G-3, p. 205-260.

Miller, E.L., Miller, M.M., Stevens, C.H., Wright, J.E., and Madrid, R., 1992. Late Paleozoic paleogeographic and tectonic evolution of the western U.S. Cordillera; *in* B.C. Burchfiel, P.W. Lipman, and M.L. Zoback (eds.), The Cordilleran Orogen: Conterminous U.S.; Geological Society of America, The Geology of North America, v. G-3, p. 57-106.

Miller, K.C., Keller, R.G., Gridley, J.M., Luetgert, J.H., Mooney, W.D., and Thybo, H., 1997. Crustal structure along the west flank of the Cascades, western Washington; Journal of Geophysical Research, v. 102, p. 17,857-17,873.

Miller, M.M., 1987. Dispersed remnants of a northeast Pacific fringing arc; Upper Paleozoic terranes of Permian McCloud affinity, western U.S.; Tectonics, v. 6, p. 807-830.

Miller, M.M., 1989. Intra-arc sedimentation and tectonism; late Paleozoic evolution of the eastern Klamath terrane, California; Bulletin of the Geological Society of America, v. 101, p. 170-187.

Miller, T.P., 1994. Pre-Cenozoic plutonic rocks in mainland Alaska; *in* G. Plafker and H.C. Berg, The Geology of Alaska; Geological Society of America, The Geology of North America, v. G-1, p. 535-554.

Misch, P., 1966. Tectonic evolution of the Northern Cascades of Washington State - a west-Cordilleran case history; *in* Tectonic History and Mineral Deposits of the

Western Cordillera; Canadian Institute of Mining and Metallurgy, Special Vol. 8, p. 101-148.

Misra, K.S., Slaney, V.R., Graham, D., and Harris, J., 1991. Mapping of basement and other tectonic features using Seasat and Thematic Mapper in hydrocarbon-producing areas of the Western Sedimentary Basin of Canada; Canadian Journal of Remote Sensing, v. 17, p. 137-151.

Mitchell, N., Escartin, J., and Allerton, S., 1998. Detachment faults at mid-ocean ridges garner interest; EOS, Transactions of the American Geophysical Union, v. 79, p. 127.

Moll-Stalcup, E.J., 1994. Latest Cretaceous and Cenozoic magmatism in mainland Alaska; *in* G. Plafker and H.C. Berg (eds.), The Geology of Alaska; Geological Society of America, The Geology of North America, v. G-1, p. 589-619.

Monger, J.W.H., 1977. Upper Paleozoic rocks of the western Canadian Cordillera and their bearing on Cordilleran evolution; Canadian Journal of Earth Sciences, v. 14, p. 1832-1859.

Monger, J.W.H., 1985. Structural evolution of the southwestern Intermontane Belt, Ashcroft and Hope map areas; Current Research, Part A; Geological Survey of Canada, Paper 85-1A, p. 349-358.

Monger, J.W.H., 1989. Overview of Cordilleran geology; *in* B.D. Ricketts (ed.), Western Canada Sedimentary Basin, A Case History; Canadian Society of Petroleum Geologists, Special Publication 30, p. 9-32.

Monger, J.W.H., 1991. Late Mesozoic to Recent evolution of the Georgia Strait-Puget Sound region, British Columbia and Washington; Washington Geology, v. 19, no. 4, p. 3-7.

Monger, J.W.H., 1993. Canadian Cordilleran tectonics: from geosynclines to crustal collage; Canadian Journal of Earth Sciences, v. 30, p. 209-231.

Monger, J.W.H. and Berg, H.C., 1987. Lithotectonic Terrane Map of Western Canada and Southeastern Alaska; U.S. Geological Survey, Miscellaneous Field Studies, Map MF-1874-B, scale 1:2,500,000, 12 p., 1 sheet.

Monger, J.W.H., Clowes, R.M., Cowan, D.S., Potter, C.J., Price, R.A., and Yorath, C.J., 1994. Continent-ocean transitions in western North America between latitudes 46 and 56 degrees: transects B1, B2, and B3; *in* R. Speed (ed.), Phanerozoic Evolution of North American Continent-Ocean Transitions; Geological Society of America, The Geology of North America, Transect Volume, p. 357-397.

Monger, J.W.H., Clowes, R.M., Price, R.A., Simony, P.S., Riddihough, R.P., and Woodsworth, G.J., 1985. Juan de Fuca plate to Alberta Plains; Geological Society of America, Continent-Ocean Transect B2.

Monger, J.W.H. and Irving, E., 1980. Northward displacement of north-central British Columbia; Nature, v. 285, p. 289-294.

Monger, J.W.H. and Price, R.A., 1996. Comment on "Paleomagnetism of the Upper Cretaceous strata of Mount Tatlow: evidence for 3000 km of northward displacement of the eastern Coast Belt, British Columbia" by P.J. Wynne et al., and on "Paleomagnetism of the Spences Bridge Group and northward displacement of the Intermontane Belt, British Columbia: a second look" by E. Irving et al.; Journal of Geophysical Research, v. 101, p. 13,793-13,799.

Monger, J.W.H., Price, R.A., and Tempelman-Kluit, D., 1982. Tectonic accretion and the origin of the two major metamorphic and plutonic welts in the Canadian Cordillera; Geology, v. 10, p. 70-75.

Monger, J.W.H. and Ross, C.A., 1971. Distribution of fusilinaceans in the Canadian Cordillera; Canadian Journal of Earth Sciences, v. 8, p. 259-278.

Monger, J.W.H., Souther, J.G., and Gabrielse, H., 1972. Evolution of the Canadian Cordillera: a plate-tectonic model; American Journal of Science, v. 272, p. 577-602.

Monger, J.W.H., Wheeler, J.O., Tipper, H.W., Gabrielse, H., Harms, T., Struik, L.C., Campbell, R.B., Dodds, C.J., Gehrels, G.E., and O'Brien, J., 1991. Cordilleran terranes; *in* H. Gabrielse and C.J. Yorath (eds.), Geology of the Cordilleran Orogen in Canada; Geological Society of America, The Geology of North America, v. G-2, p. 281-327.

Moore, T.E., Wallace, W.K., Bird, K.J., Karl, S.M., Mull, C.G., and Dillon, J.T., 1994. Geology of northern Alaska; *in* G. Plafker and H.C. Berg (eds.), The Geology of Alaska; Geological Society of America, The Geology of North America, v. G-1, p. 49-140.

Morgan, W.J., 1968. Rises, trenches, great faults and crustal blocks; Journal of Geophysical Research, v. 73, p. 1959-1982.

Morris, E., Detrick, R.S., Minshull, T.A., Mutter, J.C., White, R.S., Su, W., and Buhl, P., 1993. Seismic structure of oceanic crust in the western North Atlantic; Journal of Geophysical Research, v. 98, p. 13,879-13,903.

Morrow, D.W., 1991. The Silurian-Devonian sequence in the northern part of the Mackenzie Shelf, Northwest Territories; Geological Survey of Canada, Bulletin 413.

Mortensen, J.K., 1988. Geology of southwestern Dawson map area, Yukon Territory; *in* Current Research, Part E; Geological Survey of Canada, Paper 88-1E, p. 73-78.

Mortensen, J.K., 1992. Pre-Mid-Mesozoic tectonic evolution of the Yukon-Tanana Terrane, Yukon and Alaska; Tectonics, v. 11, p. 836-853.

Mortensen, J.K. and Jilson, G.A., 1985. Evolution of the Yukon-Tanana Terrane: evidence from southeastern Yukon Territory; Geology, v. 13, p. 806-810.

Mortimer, N., van der Heyden, P., Armstrong, R.L., and Harakal, J., 1990. U-Pb and K-Ar dates related to the timing of magmatism and deformation in the Cache

Creek terrane and Quesnellia, southern British Columbia; Canadian Journal of Earth Sciences, v. 27, p. 117-123.

Moskaleva, V.N., 1989. Petrological and Mineralogical Studies of Crystalline Rocks; VSEGEI, Leningrad (St. Petersburg), 152 p. (in Russian).

Mossop, G.D. and Shetsen, I. (comps.), 1994. Geological Atlas of the Western Canada Sedimentary Basin; Canadian Society of Petroleum Geologists & Alberta Research Council.

Mountjoy, E.W., 1980. Some questions about the development of Upper Devonian carbonate buildups (reefs), Western Canada; Bulletin of Canadian Petroleum Geology, v. 28, p. 315-340.

Mountjoy, E.W., 1992. Significance of rotated folds and thrust faults, Alberta Rocky Mountains; *in* S. Mitra and G.W. Fisher (eds.), Structural Geology of Fold and Thrust Belts; Johns Hopkins University Press, p. 207-223.

Muehlenbachs, K., Chacko, T., and Burwash, R.A., 1993. Oxygen isotope evidence for a metamorphic core complex in the Precambrian basement of Alberta (abs.); Geological Society of America, Abstracts with Program, p. A-80.

Muir Wood, R., 1989. Fifty million years of 'passive margin' deformation in north west Europe; *in* S. Gregersen and P.W. Basham (eds.), Earthquakes and North-Atlantic Passive margins: Neotectonics and Postglacial Rebound; NATO Advanced Studies Institute, Series C,: Mathematical and Physical Sciences, v. 266, p. 7-36.

Murphy, D.C., Walker, R.T., and Parrish, R.R., 1991. Gold Creek gneiss, southeastern Cariboo Mountains, British Columbia: age, geological setting, and regional implications; Canadian Journal of Earth Sciences, v. 28, p. 1217-1231.

National Academy of Sciences, 1980. Continental Tectonics; Washington, DC, 197 p.

Nelson, J., 1993. The Sylvester allochthon: Upper Paleozoic marginal-basin and island-arc terranes in northern British Columbia; Canadian Journal of Earth Sciences, v. 30, p. 631-643.

Nelson, J. and Mihalynuk, M.G., 1993. Cache Creek ocean: closure or enclosure?; Geology, v. 21, p. 173-176.

Nelson, S.J. and Nelson, E.R., 1985. Allochthonous Permian micro- and macrofauna, Kamloops area, British Columbia; Canadian Journal of Earth Sciences, v. 22, p. 442-451.

Nettleton, L.L., 1971. Elementary Gravity and Magnetics for Geologists and Seismologists; Society of Exploration Geophysicists, Monograph Series 1, 121 p.

Newton, C.R., 1988. Significance of the "Tethyan" fossils in the American Cordillera; Science, v. 242, p. 385-391.

Nixon, G.T. and Hammack, J.L., 1991. Metallogeny of ultramafic-mafic rocks in British Columbia with emphasis on the platinum-group elements; *in* W.J. McMillan (ed.), Ore Deposits, Tectonics and Metallogeny in the Canadian Cordillera;

British Columbia Ministry of Energy, Mines and Petroleum Resources, Paper 1991-4, p. 125-161.

Norford, B.S., 1969. Ordovician and Silurian stratigraphy of the southern Rocky Mountains; Geological Survey of Canada, Bulletin 176.

North American Commission on Stratigraphic Nomenclature, 1983. North American Stratigraphic Code; Bulletin of the American Association of Petroleum Geologists, v. 67, p. 841-875.

Nurkowski, J.R., 1984. Coal quality, coal rank variation and its relation to reconstructed overburden, Upper Cretaceous and Tertiary plains coals, Alberta, Canada; Bulletin of the American Association of Petroleum Geologists, v. 68, p. 285-295.

Okulitch, A.V., 1984. The role of the Shuswap Metamorphic Complex in Cordilleran tectonism: a review; Canadian Journal of Earth Sciences, v. 21, p. 1171-1193.

Oldow, J.S., Bally, A.W., Avé Lallemant, H.G., and Leeman, W.P., 1989. Phanerozoic evolution of the North American Cordillera; United States and Canada; *in* A.W. Bally and A.R. Palmer (eds.), The Geology of North America - An Overview; Geological Society of America, The Geology of North America, v. A, p. 139-232.

Oliver, D.H., Hansen, V.L., and Heizler, M.T., 1996. Variations in white mica $^{40}Ar/^{39}Ar$ cooling ages and calcite e-twin types: implications for subduction zone polarity (abs.); *in* Slave-Northern Cordillera Lithospheric Evolution and Cordilleran Tectonics Workshop, Lithoprobe Report 50, p. 32.

Oliver, J., Dobrin, M., Kaufman, S., Meyer, R., and Phinney, R., 1976. Continuous seismic reflection profiling of the deep basement, Hardeman County, Texas; Bulletin of the Geological Society of America, v. 87, p. 1537-1546.

Opdyke, N.D. and Channel, J.E.T., 1996. Magnetic Stratigraphy; Academic Press, 346 p.

Oreskes, N., Shrader-Frechette, K., and Belitz, K., 1994. Verification, validation, and confirmation of numerical models in the earth sciences; Science, v. 263, p. 641-646.

Osadetz, K.G., 1989. Basin analysis applied to petroleum geology in western Canada; *in* B.D. Ricketts (ed.), Western Canada Sedimentary Basin, A Case History; Canadian Society of Petroleum Geologists, Special Publication 30, p. 287-306.

Pakiser, L.C. and Mooney, W.D. (eds.), 1989. Geophysical Framework of the Continental United States; Geological Society of America, Memoir 172, 826 p.

Pakiser, L.C. and Zietz, I., 1965. Transcontinental crustal and upper-mantle structure; Reviews of Geophysics, v. 3, p. 505-520.

Panteleyev, A., 1991. Gold in the Canadian Cordillera - a focus on epithermal and deeper environments; *in* Ore Deposits, Tectonics and Metallogeny in the Canadian Cordillera; British Columbia Ministry of Energy, Mines and Petroleum Resources, Paper 1991-4, p. 163-212.

Parrish, R.R., 1983. Cenozoic thermal evolution and tectonics of the Coast Mountains of British Columbia. Part I. Fission track dating, apparent uplift rates, and patterns of uplift; Tectonics, v. 2, p. 601-631.

Parrish, R.R., 1991a. Precambrian basement rocks of the Canadian Cordillera; *in* H. Gabrielse and C.J. Yorath (eds.), Geology of the Cordilleran Orogen in Canada; Geological Society of America, The Geology of North America, v. G-2, p. 89-95.

Parrish, R.R., 1991b. Omineca Belt; *in* H. Gabrielse and C.J. Yorath (eds.), Geology of the Cordilleran Orogen in Canada; Geological Society of America, The Geology of North America, v. G-2, p. 661-663.

Parrish, R.R., 1995. Thermal evolution of the southeastern Canadian Cordillera; Canadian Journal of Earth Sciences, v. 32, p. 1618-1642.

Parrish, R.R. and Armstrong, R.L., 1987. The ca. 162 Ma Galena Bay stock and its relationship to the Columbia River fault zone, southeast British Columbia; *in* Radiogenic Age and Isotopic Studies: Report 1; Geological Survey of Canada, Paper 87-2, p. 25-32.

Parrish, R.R., Carr, S.D., and Parkinson, D.L., 1988. Eocene extensional tectonics and geochronology of the southern Omineca belt, British Columbia and Washington; Tectonics, v. 7, p. 181-212.

Parrish, R.R. and Monger, J.W.H., 1992. New dates from southwestern British Columbia; *in* Radiogenic Ages and Isotopic Studies: Report 5; Geological Survey of Canada, Paper 91-2, p. 87-108.

Parrish, R.R. and Wheeler, J.O., 1983. A U-Pb zircon age from the Kuskanax batholith, southeastern British Columbia; Canadian Journal of Earth Sciences, v. 20, p. 1751-1756.

Patton, D.K., 1993. *Samgori* field, Republic of Georgia: critical review of island-arc oil and gas; Journal of Petroleum Geology, v. 16, p. 153-168.

Patton, Jr., W.W., Box, S.E., and Grybeck, D.J., 1994. Ophiolites and other mafic-ultramafic complexes in Alaska; *in* G. Plafker and H.C. Berg (eds.), The Geology of Alaska; Geological Society of America, The Geology of North America, v. G-1, p. 671-686.

Pavoni, N., 1993. Pattern of mantle convection and Pangaea break-up, as revealed by the evolution of the African plate; Journal of the Geological Society (London), v. 150, p. 953-964.

Pell, J. and Simony, P.S., 1982. Hadrynian Kaza Group/Horsethief Creek Group correlations in the southern Cariboo Mountains, British Columbia; *in* Current Research, Part A; Geological Survey of Canada, Paper 82-1A, p. 305-308.

Pell, J.P. and Simony, P.S., 1987. New correlations of Hadrynian strata, south-central British Columbia; Canadian Journal of Earth Sciences, v. 24, p. 312-313.

Perry, W.J., Jr., Wardlaw, B.R., Bostick, N.H., and Maughan, E.K., 1983. Structure, burial history, and petroleum potential of frontal thrust belt and adjacent foreland, southwest Montana; Bulletin of the American Association of Petroleum Geologists, v. 67, p. 725-743.

Peterson, J.A. (ed.), 1986. Paleotectonics and Sedimentation in the Rocky Mountain Region, United States; American Association of Petroleum Geologists, Memoir 41.

Peterson, J.A. and Smith, D.L., 1986. Rocky Mountain paleogeography through geologic time; *in* J.A. Peterson (ed.), Paleotectonics and Sedimentation in the Rocky Mountain Region, United States; American Association of Petroleum Geologists, Memoir 41, p. 3-19.

Pettijohn, F.J., 1960. Some contributions of sedimentology to tectonic analysis; *in* International Geological Congress, Part 18, p. 446-454.

Pitman, W.C. III, and Heirtzler, J.R., 1966. Magnetic anomalies over the Pacific-Antarctic Ridge; Science, v. 154, p. 1164-1171.

Plafker, G. and Berg, H.C., 1994a. Introduction; *in* G. Plafker and H.C. Berg (eds.), The Geology of Alaska; Geological Society of America, The Geology of North America, v. G-1, p. 1-16.

Plafker, G. and Berg, H.C., 1994b. Overview of the geology and tectonic evolution of Alaska; *in* G. Plafker and H.C. Berg (eds.), The Geology of Alaska; Geological Society of America, The Geology of North America, v. G-1, p. 989-1021.

Plafker, G. and Berg, H.C. (eds.), 1994. The Geology of Alaska; Geological Society of America, The Geology of North America, v. G-1, 1055 p.

Plint, H.E. and Gordon, T.M., 1996. Structural evolution and rock types of the Slide Mountain and Yukon-Tanana terranes in the Campbell Range, southeastern Yukon; *in* Slave-Northern Cordillera Lithospheric Evolution and Cordilleran Tectonics Workshop; Lithoprobe Report 50, p. 76-90.

Podruski, J.A., Barclay, J.E., Hamblin, A.P., Lee, P.J., Osadetz, K.G., Proctor, R.M., and Taylor, G.C., 1988. Conventional Oil Resources of Western Canada (Light and Medium). Part 1: Resource Endowment; *in* Geological Survey of Canada, Paper 87-26, p. 1-125.

Poole, F.G., Stewart, J.H., Palmer, A.R., Sandberg, C.A., Madrid, R.J., Ross, R.J., Jr., Hintze, L.F., Miller, M.M., and Wrucke, C.T., 1992. Latest Precambrian to latest Devonian time; development of a continental margin; *in* B.C. Burchfiel, P.W. Lipman, and M.L. Zoback (eds.), The Cordilleran Orogen: Conterminous U.S.; Geological Society of America, The Geology of North America, v. G-3, p. 9-56.

Potter, C.J., Liu, C.-s., Huang, J., Zheng, L., Hauge, T.A., Hauser, E.C., Allmendinger, R.W., Oliver, J.E., Kaufman, S., and Brown, L., 1987. Crustal structure of north-central Nevada; results from COCORP deep seismic profiling; Bulletin of the Geological Society of America, v. 98, p. 330-337.

Potter, C.J., Sanford, W.E., Yoos, T., Prussen, E., Keach, R., II, Oliver, J.E., Kaufman, S., and Brown, L.D., 1986. COCORP deep seismic reflection traverse of the interior of the North American Cordillera, Washington and Idaho; Tectonics, v. 5, p. 1007-1027.

Poulton, T.P., 1989. Upper Absaroka to Lower Zuni; the transition to the Foreland Basin; *in* B.D. Ricketts (ed.), Western Canada Sedimentary Basin: A Case History; Canadian Society of Petroleum Geologists, Special Publication 30, p. 233-247.

Poulton, T.P., 1991. Lower and Middle Jurassic strata of the Foreland Belt; *in* H. Gabrielse and C.J. Yorath (eds.), Geology of the Cordilleran Orogen in Canada; Geological Society of America, The Geology of North America, v. G-2, p. 276-281.

Poulton, T.P. and Aitken, J.D., 1989. The Lower Jurassic phosphorites of southeastern British Columbia and terrane accretion to western North America; Canadian Journal of Earth Sciences, v. 26, p. 1612-1616.

Poulton, T.P., Braun, W.K., Brooke, M.M., and Davies, E.H., 1993. Jurassic; *in* D.F. Stott and J.D. Aitken (eds.), Sedimentary Cover of the Craton in Canada; Geological Society of America, The Geology of North America, v. D-1, p. 321-357.

Powell, W.G. and Ghent, E.D., 1996. Low-pressure metamorphism of the mafic volcanic rocks of the Rossland Group, southeastern British Columbia; Canadian Journal of Earth Sciences, v. 33, p. 1402-1409.

Preto, V.A., 1977. The Nicola Group: Mesozoic volcanism related to rifting in southern British Columbia; *in* W.R. Baragar, L.C. Coleman, and J.M. Hall (eds.), Volcanic Regimes in Canada; Geological Association of Canada, Special Paper 16, p. 39-57.

Price, R.A., 1972. Introduction; *in* R.A. Price and R.J.W. Douglas (eds.), Variations in Tectonic Styles in Canada; Geological Association of Canada, Special Paper 11, p. ix-x.

Price, R.A., 1981. The Cordilleran foreland fold and thrust belt in the southern Canadian Rocky Mountains; *in* K.R. McClay and N.J. Price (eds.), Thrust and Nappe Tectonics; Geological Society (London), Special Publication 9, p. 427-448.

Price, R.A., 1986. The southeastern Canadian Cordillera: thrust faulting, tectonic wedging, and delamination of the lithosphere; Journal of Structural Geology, v. 8, p. 239-254.

Price, R.A., 1994. Cordilleran tectonics and evolution of the Western Canada sedimentary basin; *in* G.D. Mossop and I. Shetsen (comps.), Geological Atlas of Western Canada; Canadian Society of Petroleum Geologists & Alberta Research Council, p. 13-24.

Price, R.A. and Carmichael, D.M., 1986. Geometric test for Late Cretaceous-Paleogene intracontinental transform faulting in the Canadian Cordillera; Geology, v. 14, p. 468-471.

Price, R.A. and Douglas, R.J.W. (eds.), 1972. Variations in Tectonic Styles in Canada; Geological Association of Canada, Special Paper 11, 688 p.

Price, R.A. and Fermor, P.R., 1985. Structure Section of the Cordilleran Foreland Thrust and Fold Belt West of Calgary, Alberta; Geological Survey of Canada, Paper 84-14.

Price, R.A., Monger, J.W.H., and Muller, J.E., 1981. Cordilleran cross-section - Calgary to Victoria; *in* R.I. Thompson and D.G. Cook (eds.), Field Guides to Geology and Mineral Deposits; Geological Association of Canada, Mineralogical Association of Canada, Canadian Geophysical Union, Joint Annual Meeting, Calgary, p. 261-292.

Raaben, V.F., 1978. Distribution of Oil and Gas in Regions of the World; Nauka Publishing, Moscow, 144 p. (in Russian).

Ranalli, G., Brown, R.L., and Bosdachin, R., 1989. A geodynamic model for extension in the Shuswap core complex, southeastern Canadian Cordillera; Canadian Journal of Earth Sciences, v. 26, p. 1647-1653.

Rankin, D.W., Chiarenzelli, J.R., Drake, Jr., A.A., Goldsmith, R., Hall, L.M., Hinze, W.J., Isachsen, Y.W., Lidiak, E.G., McLelland, J., Mosher, S., Ratcliffe, N.M., Secor, Jr., D.T., and Whitney, P.R., 1993. Proterozoic rocks east and southeast of the Grenville front; *in* J.C. Reed, Jr., M.E. Bickford, R.S. Houston, P.K. Link, D.W. Rankin, P.K. Sims, and W.R. Van Schmus (eds.), Precambrian: Conterminous U.S.; Geological Society of America, The Geology of North America, v. C-2, p. 335-461.

Rast, N., 1969. Orogenic belts and their parts; *in* P.E. Kent, G.E. Satterthwaite, and A.M. Spencer (eds.), Time and Place in Orogeny; Geological Society (London), Special Publication 3, p. 197-213.

Ray, G.E. and Webster, I.C.L., 1991. An overview of skarn deposits; *in* Ore Deposits, Tectonics and Metallogeny in the Canadian Cordillera; British Columbia Ministry of Energy, Mines and Petroleum Resources, Paper 1991-4, p. 213-252.

Read, P.B., 1983. Geology, Classy Creek (104J/2E) and Stikine Canyon (104J/1W); Geological Survey of Canada, Open File 940.

Read, P.B. and Brown, R.L., 1981. Columbia River fault zone: southeastern margin of the Shuswap and Monashee complexes, southern British Columbia; Canadian Journal of Earth Sciences, v. 18, p. 1127-1145.

Reed, J.C., Jr., Bickford, M.E., Houston, R.S., Link, P.K., Rankin, D.W., Sims, P.K., and Van Schmus, W.R. (eds.), 1993. Precambrian: Conterminous U.S.; Geological Society of America, The Geology of North America, v. C-2, 657 p.

Reidel, S.P. and Hooper, P.R. (eds.), 1989. Volcanism and Tectonism in the Columbia River Flood-Basalt Province; Geological Society of America, Special Paper 239, 386 p.

Ren Jinshun, 1991. Microcontinents, soft collision and polycyclic suturing - Dynamic characteristics of the tectonic evolution of continental Southeast Asia; Proceedings of the First International Symposium on Gondwana Dispersion and Asian

Accretion - Geological Evolution of Eastern Tethys, IGCP-321, Kunming, China, p. 216-219.

Ren Jinshun, Niu Baogui, and Liu Zhigang, 1996. Microcontinents, soft collision and polycyclic suturing; Continental Dynamics, v. 1, p. 1-9.

Richards, B.C., 1989. Upper Kaskaskia Sequence: uppermost Devonian and Lower Carboniferous; *in* B.D. Ricketts (ed.), Western Canada Sedimentary Basin, A Case Study; Canadian Society of Petroleum Geologists, Special Publication 30, p. 165-201.

Richards, B.C., Bamber, E.W., Higgins, A.C., and Utting, J., 1993. Carboniferous; *in* D.F. Stott and J.D. Aitken (eds.), Sedimentary Cover of the Craton in Canada; Geological Society of America, The Geology of North America, v. D-1, p. 202-271.

Richardson, R.M., 1992. Ridge forces, absolute plate motions, and the intraplate stress field; Journal of Geophysical Research, v. 97, p. 11,739-11,748.

Ricketts, B.D. (ed.), 1989. Western Canada Sedimentary Basin, A Case Study; Canadian Society of Petroleum Geologists, Special Publication 30.

Riddihough, R.P., 1979. Gravity and structure of an active margin - British Columbia and Washington; Canadian Journal of Earth Sciences, v. 16, p. 350-363.

Riddihough, R.P. and Hyndman, R.D., 1976. Canada's active margin - the case for subduction; Geoscience Canada, v. 3, p. 269-279.

Riddihough, R.P. and Hyndman, R.D., 1991. Modern plate tectonic regime of the continental margin of western Canada; *in* H. Gabrielse and C.J. Yorath (eds.), Geology of the Cordilleran Orogen in Canada; Geological Society of America, The Geology of North America, v. G-2, p. 437-455.

Roback, R.C., Sevigny, J.H., and Walker, N.W., 1994. Tectonic setting of the Slide Mountain terrane, southern British Columbia; Tectonics, v. 13, p. 1242-1258.

Roberts, R.J., 1949. Structure and stratigraphy of the Antler Peak quadrangle, north-central Nevada (abs.); Bulletin of the Geological Society of America, v. 60, p. 1917.

Roberts, R.J., 1964. Stratigraphy and Structure of the Antler Peak Quadrangle, Humbolt and Lander Counties, Nevada; U.S. Geological Survey, Professional Paper 459-A, 93 p.

Rodgers, J., 1987. Chains of basement uplifts within cratons marginal to orogenic belts; American Journal of Science, v. 287, p. 661-692.

Rodgers, J., 1995. Lines of basement uplifts within the external parts of orogenic belts; American Journal of Science, v. 295, p. 455-487.

Ross, C.A. and Ross, J.R.P., 1983. Late Paleozoic accreted terranes of western North America; *in* C.H. Stevens (ed.), Pre-Jurassic Rocks in Western North American Suspect Terranes; Society of Exploration Paleontologists and Mineralogists, Pacific Section, p. 7-22.

Ross, G.M., 1991. Tectonic setting of the Windermere Supergroup revisited; Geology, v. 19, p. 1125-1128.

Ross, G.M. and Harms, T.A., 1998. Detrital zircon geochronology of sequence 'C' grits, Dorsey Terrane (Thirtymile Range, Southern Yukon): provenance and stratigraphic correlation; *in* Radiogenic Age and Isotopic Studies: Report 11; Geological Survey of Canada, Current Research 1998-F, p. 107-115.

Rubin, C.M. and Saleeby, J.B., 1992. Thrust tectonics and Cretaceous intracontinental shortening in Southeast Alaska; *in* K.R. McClay (ed.), Thrust Tectonics; Chapman and Hall, p. 407-417.

Rudnick, R.L., 1992. Xenoliths - samples of the lower continental crust; *in* D.M. Fountain, R. Arculus, and R.W. Kay (eds.), Continental Lower Crust; Elsevier, p. 269-316.

Rudnick, R.L., 1995. Making continental crust; Nature, v. 378, p. 571-578.

Rudnick, R.L. and Fountain, D.M., 1995. Nature and composition of the continental crust: a lower crustal perspective; Reviews of Geophysics, v. 33, p. 267-309.

Rusmore, M.E. and Woodsworth, G.J., 1991. Coast Plutonic Complex: a mid-Cretaceous contractional orogen; Geology, v. 19, p. 941-944.

Ryan, B.D. and Blenkinsop, J., 1971. Geology and geochronology of Hell-Roaring Creek Stock, British Columbia; Canadian Journal of Earth Sciences, v. 8, p. 85-95.

Samson, S.D., Patchett, P.J., McClelland, W.C., and Gehrels, G.E., 1991. Nd isotopic characterization of metamorphic rocks in the Coast Mountains, Alaskan and Canadian Cordillera: ancient crust bounded by juvenile terranes; Tectonics, v. 10, p. 770-780.

Sawkins, M.J., 1990. Metal Deposits in Relation to Plate Tectonics, second edition; Springer-Verlag, 461 p.

Scammell, R.J., 1991. Structure and U-Pb geochronometry of the southern Scrip Range, southern Omineca Belt, British Columbia: a progress report; *in* Southern Canadian Cordillera Transect, Lithoprobe Report 18, p. 85-96.

Scammell, R.J., 1992. Composite deformation, intradeformational leucogranite magmatism and thermal history, northern Monashee Mountains, Omineca belt, B.C.; *in* Southern Canadian Cordillera Transect, Lithoprobe Report 24, p. 5-11.

Scammell, R.J., 1993. Mid-Cretaceous-Tertiary Thermotectonic History of Former Mid-Crustal Rocks, Southern Omineca Belt, Canadian Cordillera; Ph.D. thesis, Queens University, Kingston, Ontario.

Scammell, R.J. and Brown, R.L., 1990. Cover gneiss of the Monashee terrane: a record of synsedimentary rifting in the North American Cordillera; Canadian Journal of Earth Sciences, v. 27, p. 712-726.

Schaubs, P.M. and Carr, S.D., 1998. Geology of metasedimentary rocks and Late Cretaceous deformation history in the northern Valhalla complex, British Columbia; Canadian Journal of Earth Sciences, v. 35, p. 1018-1036.

Schiarizza, P., Gaba, R.G., Glover, J.K., and Garver, J.I., 1989. Geology and mineral occurrences in the Tyaughton Creek area; *in* British Columbia Ministry of Energy, Mines and Petroleum Resources, Geological Fieldwork 1988, Paper 1989-1, p. 115-130.

Sears, J.W. and Price, R.A., 1978. The Siberian connection - a case for Precambrian separation of the North American and Siberian platforms; Geology, v. 6, p. 267-270.

Seber, D., Vallvé, M., Sandvol, E., Steer, D., and Barazangi, M., 1997. Middle East tectonics: applications of geographic information systems (GIS); GSA Today, v. 7, no. 2, p. 2-6.

Sevigny, J.H., 1988. Geochemistry and Late Proterozoic amphibolites and ultramafic rocks, southeastern Canadian Cordillera; Canadian Journal of Earth Sciences, v. 25, p. 1323-1337.

Sevigny, J.H. and Parrish, R.R., 1993. Age and origin of Late Jurassic and Paleocene granitoids, Nelson batholith, southern British Columbia; Canadian Journal of Earth Sciences, v. 30, p. 2305-2314.

Shatsky, N.C. (ed.-in-chief), 1956. Tectonic Map of the U.S.S.R. and Adjacent Countries, scale 1:5,000,000; Gosgeoltechizdat Publishing, Moscow.

Shatsky, M., 1964. International Tectonic Map of Europe, scale 1:2,500,000; Subcommission on the Tectonic Map of the World.

Shaw, D.M., 1963. Composition of the Canadian Precambrian Shield; a progress report; Canadian Mineralogist, v. 7, p. 821.

Shaw, J. and Kellerhals, R., 1982. The Composition of Recent Alluvial Gravels in Alberta River Beds; Alberta Research Council, Bulletin 41, 151 p.

Silberling, N.J., 1973. Geologic events during Permian-Triassic time along the Pacific margin of the United States; *in* A. Logan and L.V. Hills (eds.), The Permian and Triassic systems and their mutual boundary; Canadian Society of Petroleum Geologists, Memoir 2, p. 345-362.

Silberling, N.J. and Jones, D.L. (eds.), 1984. Lithotectonic Terrane Maps of the North American Cordillera, scale 1:2,500,000; U.S. Geological Survey Open-File Report 84-523.

Silberling, N.J., Jones, D.L., Monger, J.W.H., Coney, P.J., Berg, H.C., and Plafker, G., 1994. Lithotectonic Terrane Map of Alaska and Adjacent Parts of Canada, scale 1:2,500,000; *in* G. Plafker and H.C. Berg (eds.), The Geology of Alaska; Geological Society of America, The Geology of North America, Plate 3.

Sillitoe, R.H., 1987. Copper, gold and subduction: a trans-Pacific perspective; Pacific Rim Congress, Gold Coast, Australia.

Silver, P.G. and Chan, W.W., 1988. Implications for continental structure and evolution from seismic anisotropy; Nature, v. 35, p. 34-39.

Simony, P.S., 1991. Northeastern Columbia Mountains; *in* H. Gabrielse and C.J. Yorath (eds.), Geology of the Cordilleran Orogen in Canada; Geological Society of America, The Geology of North America, v. G-2, p. 615-621.

Simony, P.S., 1995. Core complexes: lessons from little sheared margins of the Shuswap Complex (abs.); Geological Association of Canada/Mineralogical Association of Canada, Annual Meeting, Victoria; Final Program and Abstracts, p. A-98.

Simony, P.S. and Carr, S.D., 1997. Large lateral ramps in the Eocene Valkyr shear zone: extensional ductile faulting controlled by plutonism in southern British Columbia; Journal of Structural Geology, v. 19, p. 769-784.

Simony, P.S., Ghent, E.D., Craw, D., Mitchell, W., and Robbins, D.B., 1980. Structural and metamorphic evolution of northeast flank of Shuswap complex, southern Canoe River area, British Columbia; *in* M.D. Crittenden, Jr., P.J. Coney, and G.H. Davis (eds.), Cordilleran Metamorphic Core Complexes; Geological Society of America, Memoir 153, p. 445-461.

Sinclair, A.J., Wynne-Edwards, H.R., and Sutherland Brown, A., 1978. An analysis of distribution of mineral occurrences in British Columbia; British Columbia Ministry of Energy, Mines and Petroleum Resources, Bulletin 68, 125 p.

Sinclair, W.D., 1986. Molybdenum, tungsten and tin deposits and associated granitoid intrusions in the northern Canadian Cordillera and adjacent parts of Alaska; *in* J.A. Morin (ed.), Mineral Deposits of the Northern Cordillera; Canadian Institute of Mining and Metallurgy, Special Volume 37, p. 216-233.

Sloss, L.L., 1963. Sequences in the cratonic interior of North America; Bulletin of the Geological Society of America, v. 74, p. 93-114.

Sloss, L.L., 1988. Tectonic evolution of the craton in Phanerozoic time; *in* L.L. Sloss (ed.), Sedimentary Cover - North American Craton: U.S.; Geological Society of America, The Geology of North America, v. D-2, p. 25-51.

Sloss, L.L. (ed.), 1988. Sedimentary Cover - North American Craton: U.S.; Geological Society of America, The Geology of North America, v. D-2.

Sloss, L.L., Krumbein, W.C., and Dapples, E.C., 1949. Integrated facies analysis; *in* C.R. Longwell (chmn.), Sedimentary Facies in Geologic History; Geological Society of America, Memoir 39, p. 91-124.

Smith, A.D. and Lambert, R.S., 1995. Nd, Sr, and Pb isotopic evidence for contrasting origins of late Paleozoic volcanic rocks from the Slide Mountain and Cache

Creek terranes, south-central British Columbia; Canadian Journal of Earth Sciences, v. 32, p. 447-459.

Smith, D.K., Humphris, S.E., Tivey, M.A., and Cann, J.R., 1997. Viewing the morphology of the Mid-Atlantic Ridge from a new perspective; EOS, Transactions of the American Geophysical Union, v. 78, p. 265-269.

Smith, G.M. and Rubin, C.M., 1987. Devonian-Mississippian arc from the northern Sierras to the Seward Peninsula: record of a protracted and diachronous Antler-age orogeny (abs.); Geological Society of America, Abstracts with Programs, v. 19, p. 849.

Smith, M.T. and Gehrels, G.E., 1991. Detrital zircon geochronology of Upper Proterozoic to lower Paleozoic continental margin strata of the Kootenay Arc: implications for the early Paleozoic tectonic development of the eastern Canadian Cordillera; Canadian Journal of Earth Sciences, v. 28, p. 1271-1284.

Snyder, G.A., Taylor, L.A., Crozaz, G., Halliday, A.N., Beard, B.L., Sobolev, V.N., and Sobolev, N.V., 1997. The origins of Yakutian eclogite xenoliths; Journal of Petrology, v. 38, p. 85-113.

Sobczak, L.W. and Halpenny, J.F., 1990. Isostatic and Enhanced Isostatic Gravity Anomaly Maps of the Arctic; Geological Survey of Canada, Paper 89-16, 9 p.

Souther, J.G., 1991. Volcanic regimes; *in* H. Gabrielse and C.J. Yorath (eds.), Geology of the Cordilleran Orogen in Canada; Geological Society of America, The Geology of North America, v. G-2, p. 457-490.

Spear, F.S. and Parrish, R.R., 1996. Petrology and cooling rates of the Valhalla Complex, British Columbia, Canada; Journal of Petrology, v. 37, p. 733-765.

Speed, R.C., 1977. Island-arc and other paleogeographic terranes of late Paleozoic age in the western Great Basin; *in* J.H. Stewart, C.H. Stevens, and A.E. Fritsche (eds.), Paleozoic paleogeography of the western United States; Society of Economic Paleontologists and Mineralogists, Pacific Section, Paleogeography Symposium 1, Los Angeles, p. 349-362.

Speed, R.C., 1979. Collided Paleozoic microplate in the western United States; Journal of Geology, v. 87, p. 279-292.

Speed, R.C. and Sleep, N.H., 1982. Antler orogeny and foreland basin: a model; Bulletin of the Geological Society of America, v. 93, p. 815-828.

Spizharsky, T.N. (ed.), 1968. Geologic Structure of the U.S.S.R., v. II: Tectonics; Nedra Publishing, Moscow, 535 p. (in Russian).

Spizharsky, T.N., 1973. Overview Tectonic Maps of the U.S.S.R.; Nedra Publishing, Leningrad (St. Petersburg), 240 p. (in Russian).

Spizharsky, T.N. and Stelmak, N.K., 1977. Geologic bodies; *in* Paleontological Maps of the U.S.S.R., scale 1:5,000,000, Explanatory Notes, v. 1, p. 9-34 (in Russian).

Srivastava, S.P., Roest, W.R., Kovacs, L.C., Oakey, G., Levesque, S., Verhoef, J., and Macnab, R., 1990. Motion of Iberia since the Late Jurassic: results from detailed aeromagnetic measurements in the Newfoundland Basin; Tectonophysics, v. 184, p. 229-260.

Stacey, R.A., 1973. Gravity anomalies, crustal structure and plate tectonics in the Canadian Cordillera; Canadian Journal of Earth Sciences, v. 10, p. 615-628.

Staub, R., 1924. Der Bau der Alpen; Beitrage Geol. Karte Schweiz, v. 52, p. 2-72.

Stefanick, M. and Jurdy, D.M., 1992. Stress observations and driving force models for the South American plate; Journal of Geophysical Research, v. 97, p. 11,905-11,913.

Stephenson, R. and Lambeck, K., 1985. Isostatic response of the lithosphere with in-plane stress: application to central Australia; Journal of Geophysical Research, v. 90, p. 8581-8588.

Stephenson, R.A., Zelt, C.A., Ellis, R.M., Hajnal, Z., Morel-a-l'Huissier, P., Mereu, R.F., Northey, D.J., West, G.F., and Kanasewich, E.R., 1989. Crust and upper mantle structure and the origin of the Peace River Arch; Bulletin of Canadian Petroleum Geology, v. 37, p. 224-235.

Stevens, R.A., 1994. Structural and Tectonic Evolution of the Teslin Tectonic Zone, Yukon: a Doubly-Vergent Transpressive Shear Zone; Ph.D. thesis, University of Alberta, Edmonton, 130 p.

Stevens, R.A., 1996. Dorsey assemblage: pre-mid-Permian high temperature and pressure metamorphic rocks in the Dorsey Range, southern Yukon Territory; *in* Slave-Northern Cordillera Lithospheric Evolution Transect and Cordilleran Tectonics Workshop, Lithoprobe Report 50, p. 70-75.

Stevens, R.A. and Erdmer, P., 1996. Structural divergence and transpression in the Teslin tectonic zone, southern Yukon Territory; Tectonics, v. 15, p. 1342-1363.

Stevens, R.A., Erdmer, P., Creaser, R.A., and Grant, S.L. 1996. Mississippian assembly of the Nisutlin assemblage: evidence from primary contact relationships and Mississippian magmatism in the Teslin tectonic zone, part of the Yukon-Tanana terrane of south central Yukon; Canadian Journal of Earth Sciences, v. 33, p. 103-116.

Stevens, R.A. and Harms, T.A., 1995. Stratigraphy, structure and metamorphism in the Dorsey Range, southern Yukon Territory and northern British Columbia: investigations of the Dorsey Terrane, Part I; *in* Current Research 1995-A, Geological Survey of Canada, p. 117-127.

Stewart, J.H., MacMillan, J.R., Nichols, K.M., and Stevens, C.H., 1977. Deep water upper Paleozoic rocks in north-central Nevada - a study of the type area of the Havallah Formation; *in* J.H. Stewart, C.H. Stevens, and A.E. Fritscho (eds.), Paleozoic Paleogeography of the Western United States; Pacific Coast Paleogeography

Symposium I, Los Angeles; SEPM (Society of Economic Paleontologists and Mineralogists), p. 337-347.

Stille, H., 1924. Grundfragen der Vergleichenden Tektonik; Gebruder Bornträger, Berlin, 443 p.

Stille, H., 1941. Einführung in den Bau Amerikas; Bornträger, Berlin.

Stoakes, F.A. and Wendte, J.C., 1987. The Woodbend Group; *in* F.F. Krause and O.G. Burrowes (eds.), Devonian Lithofacies and Reservoir Styles in Alberta; 13th Canadian Society of Petroleum Geologists Core Conference and Display, Second International Symposium on the Devonian System, Canadian Society of Petroleum Geologists, p. 153-170.

Stock, J.M. and Molnar, P., 1988. Uncertainties and implications of the Late Cretaceous and Tertiary position of North America relative to the Farallon, Kula, and Pacific plates; Tectonics, v. 6, p. 1339-1384.

Stott, D.F. and Aitken, J.D. (eds.), 1993. Sedimentary Cover of the Craton in Canada; Geological Society of America, The Geology of North America, v. D-1, 826 p.

Stott, D.F., Caldwell, W.G.E., Cant, D.J., Christopher, J.E., Dixon, J., Koster, E.H., McNeil, D.H., and Simpson, F., 1993a. Cretaceous; *in* D.F. Stott and J.D. Aitken (eds.), Sedimentary Cover of the Craton in Canada; Geological Society of America, The Geology of North America, v. D-1, p. 358-438.

Stott, D.F., Dixon, J., Dietrich, J.R., McNeil, D.H., Russell, L.S., and Sweet, A.R., 1993b. Tertiary; *in* D.F. Stott and J.D. Aitken (eds.), Sedimentary Cover of the Craton in Canada; Geological Society of America, The Geology of North America, v. D-1, p. 439-465.

Stott, D.F., Yorath, C.J., and Dixon, J., 1991. The Foreland Belt; *in* H. Gabrielse and C.J. Yorath (eds.), Geology of the Cordilleran Orogen in Canada; Geological Society of America, The Geology of North America, v. G-2, p. 335-345.

Struik, L.C., 1987. The Ancient Western North American Margin: an Alpine Rift Model for the East-Central Canadian Cordillera; Geological Survey of Canada, Paper 87-15, 19 p.

Struik, L.C., 1991. Cariboo Mountains and Quesnel Highlands; *in* H. Gabrielse and C.J. Yorath (eds.), Geology of the Cordilleran Orogen in Canada; Geological Society of America, The Geology of North America, v. G-2, p. 632-634.

Struik, L.C., Currie, L.D., O'Sullivan, P.B., Kung, R.B., and Jackson, L.E., Jr., 1997. Tertiary drain patterns in central British Columbia: implications for Oligocene to Miocene tectonics (abs.); *in* Slave-Northern Cordillera Lithospheric Evolution (SNORCLE) Transect and Cordilleran Tectonics Workshop Meeting; Lithoprobe Report 56, p. 183.

Struik, L.C., Murphy, D.C., and Rees, C.J., 1991. Cariboo Mountains and Quesnel Highlands; *in* H. Gabrielse and C.J. Yorath (eds.), Geology of the Cordilleran

Orogen in Canada; Geological Society of America, The Geology of North America, v. G-2, p. 614-615.

Su, W.-j. and Dziewonski, A.M., 1995. Inner core anisotropy in three dimensions; Journal of Geophysical Research, v. 100, p. 9831-9852.

Su, W., Woodward, R.L., and Dziewonski, A.M., 1994. Degree 12 model of shear velocity heterogeneity in the mantle; Journal of Geophysical Research, v. 99, p. 6945-6980.

Sutherland Brown, A., 1976. Morphology and classification; *in* A. Sutherland Brown (ed.), Porphyry Deposits of the Canadian Cordillera; Canadian Institute of Mining and Metallurgy, Special Volume 15, p. 44-51.

Sutherland Brown, A. (ed.), 1976. Porphyry Deposits of the Canadian Cordillera; Canadian Institute of Mining and Metallurgy, Special Volume 15.

Sweeney, J.F., Stephenson, R.A., Currie, R.G., and DeLaurier, J.M., 1991. Crustal Geophysics; *in* H. Gabrielse and C.J. Yorath (eds.), Geology of the Cordilleran Orogen in Canada; Geological Society of America, The Geology of North America, v. G-2, p. 39-58.

Swinden, H.S., Jenner, G.A., Kean, B.F., and Evans, D.T.W., 1989. Volcanic rock geochemistry as a guide for massive sulphide exploration in central Newfoundland; *in* Report of Activities, Newfoundland Mineral Development Division, 89-1, p. 201-219.

Symons, D.T.A., 1983. New paleomagnetic data for the Triassic Guichon batholith of south-central British Columbia and their bearing on Terrane I tectonics; Canadian Journal of Earth Sciences, v. 20, p. 1340-1344.

Tagami, T. and Dumitru, T.A., 1996. Provenance and thermal history of the Franciscan accretionary complex: constraints from zircon fission track thermochronology; Journal of Geophysical Research, v. 101, p. 11,353-11,364.

Taylor, R.S., Mathews, W.H., and Kupsch, W.O., 1964. Tertiary; *in* R.G. McCrossan and R.P. Glaister (eds.), Geological History of Western Canada; Alberta Society of Petroleum Geologists, p. 190-200.

Teitz, M.W. and Mountjoy, E.W., 1989. The Late Proterozoic Yellowhead carbonate platform west of Jasper, Alberta; *in* H.H.J. Geldsetzer, N.P. James, and G.E. Tebbutt (eds.), Reefs, Canada and Adjacent Areas; Canadian Society of Petroleum Geologists, Memoir 13, p. 103-118.

Tempelman-Kluit, D.J., 1976. The Yukon crystalline terrane: enigma in the Canadian Cordillera; Bulletin of the Geological Society of America, v. 87, p. 1343-1357.

Tempelman-Kluit, D.J., 1979. Transported Cataclasite, Ophiolite and Granodiorite in Yukon: Evidence of Arc-Continent Collision; Geological Survey of Canada, Paper 79-14, 27 p.

Tempelman-Kluit, D.J., 1991. Nisling Terrane; *in* H. Gabrielse and C.J. Yorath (eds.), Geology of the Cordilleran Orogen in Canada; Geological Society of America, The Geology of North America, v. G-2, p. 605.

Tempelman-Kluit, D.J., Gabrielse, H., Evenchick, C.A., Mansy, J.L., Brown, R.L., Journeay, J.M., Lane, L.S., Struik, L.C., Murphy, D.C., Rees, C.J., Simony, P.S., Fyles, J.T., Höy, T., Gordey, S.P., Thompson, R.I., McMechan, M.E., and Harms, T.A., 1991. Omineca Belt; *in* H. Gabrielse and C.J. Yorath (eds.), Geology of the Cordilleran Orogen in Canada; Geological Society of America, The Geology of North America, v. G-2, p. 603-674.

Terry, J., 1977. Geology of the Nahlin ultramafic body, Atlin and Tulsequah map-areas, northwestern British Columbia; *in* Report of Activities, Part A; Geological Survey of Canada, Paper 77-1A, p. 263-266.

Thompson, R.I., 1989. Stratigraphy, Tectonic Evolution and Structural Analysis of the Halfway River Map Area (94 B), Northern Rocky Mountains, British Columbia; Geological Survey of Canada, Memoir 425, 119 p.

Thompson, R.I., 1998. NATMAP project proposed to study the Phanerozoic plate margin: highlights from Vernon map area (82L) (abs.); *in* Slave-Northern Cordillera Lithospheric Evolution and Cordilleran Tectonics Workshop; Lithoprobe Report 64, p. 201-202.

Thompson, R.I., Erdmer, P., Daughtry, K.L., Heaman, L., and Creaser, R.A., 1999. Tracking North American basement west of the Okanagan Valley: insights from the Vernon, Ashcroft and Hope map areas (abs.); Slave-Northern Cordillera Lithospheric Evolution and Cordilleran Tectonics Workshop; Lithoprobe Report 69, p. 195.

Thorstad, L.E. and Gabrielse, H., 1986. The Upper Triassic Kutcho Formation, Cassiar Mountains, North-Central British Columbia; Geological Survey of Canada, Paper 86-16, 53 p.

Tipper, H.W., 1984. The allochthonous Jurassic-Lower Cretaceous terranes of the Canadian Cordillera and their relation to correlative strata of the North American craton; *in* G.E.E. Westermann (ed.), Jurassic-Cretaceous Paleogeography of North America; Geological Association of Canada, Special Paper 27, p. 113-120.

Tipper, H.W. and Richards, T.A., 1976. Jurassic Stratigraphy and History of North-Central British Columbia; Geological Survey of Canada, Bulletin 270, 73 p.

Titley, S.R. (ed.), 1982. Advances in Geology of Porphyry Copper Deposits, Southwestern North America; University of Arizona Press, 560 p.

Titley, S.R., 1987. The crustal heritage of silver and gold ratios in Arizona; Bulletin of the Geological Society of America, v. 99, p. 814-826.

Tivey, M.A. and Tucholke, B.E., 1998. Magnetization of 0-29 Ma ocean crust on the Mid-Atlantic Ridge, 25°30' to 27°10'N; Journal of Geophysical Research, v. 103, p. 17,807-17,826.

Toomey, D.R., Purdy, G.M., Solomon, S.C., and Wilcock, W.S.D., 1985. The three-dimensional seismic velocity structure of the East Pacific Rise near latitude 9°30'N; Nature, v. 347, p. 639-645.

Trop, J.M. and Ridgway, K.D., 1997. Petrofacies and provenance of a Late Cretaceous suture zone thrust-top basin, Cantwell Basin, central Alaska Range; Journal of Sedimentary Research, v. 67, p. 469-485.

Trümpy, R., 1960. Paleotectonic evolution of the central and western Alps; Bulletin of the Geological Society of America, v. 71, p. 843-908.

Underhill, J.R. and Partington, M.A., 1993. Jurassic thermal doming and deflation in the North Sea; *in* J.R. Parker (ed.), Proceedings of the 4th Conference on the Petroleum Geology of Northwest Europe; Geological Society (London), p. 337-345.

Van Bemmelen, R.W., 1967. Stockwerktektonik *sensu lato*; *in* Étages Tectoniques; Institut de Géologie de l'Université de Neuchatel, p. 19-40.

Van der Heyden, P., 1992. A Middle Jurassic to Early Tertiary Andean-Sierran arc model for the Coast Belt of British Columbia; Tectonics, v. 11, p. 82-97.

Van der Velden, A.J. and Cook, F.A., 1994. Displacement of the Lewis thrust sheet in southwestern Canada: new evidence from seismic reflection data; Geology, v. 22, p. 819-822.

Van der Velden, A.J. and Cook, F.A., 1996. Structure and tectonic development of the southern Rocky Mountain trench; Tectonics, v. 15, p. 517-544.

Van Hinte, J.E., Adams, J.A.S., and Perry, D., 1975. K/Ar age of lower-upper Cretaceous boundary at Orphan Knoll (Labrador Sea); Canadian Journal of Earth Sciences, v. 12, p. 1484-1491.

Varsek, J.L., 1996. Structural wedges in the Cordilleran crust, southwestern Canada; Bulletin of Canadian Petroleum Geology, v. 44, p. 349-362.

Varsek, J.L., Cook, F.A., Clowes, R.M., Journeay, J.M., Parrish, R.R., Kanasewich, E.R., and Spencer, C., 1993. Lithoprobe crustal reflection survey of the southern Canadian Cordillera. 2. Coast Mountains Transect; Tectonics, v. 12, p. 334-360.

Vine, F.J. and Matthews, D.H., 1963. Magnetic anomalies southwest of Vancouver Island; Nature, v. 199, p. 947-949.

Vine, F.J. and Wilson, J.T., 1965. Magnetic anomalies over a young oceanic ridge off Vancouver Island; Science, v. 150, p. 485-489.

Vinnik, L.P. and Saipbekova, A.M., 1984. Structure of the lithosphere and the asthenosphere of the Tien Shan; Annales Geophysicae, v. 2, p. 621-626.

Wakabayashi, J., 1992. Nappes, tectonics of oblique plate convergence, and metamorphic evolution related to 140 million years of continuous subduction, Franciscan complex, California; Journal of Geology, v. 100, p. 19-40.

Walcott, R.I., 1967. The Bouguer Anomaly Map of Southwestern British Columbia; University of British Columbia, Institute of Earth Sciences, Scientific Report 15.

Walcott, R.I., 1968. The Gravity Field of Northern Saskatchewan and Northeastern Alberta; Gravity Map Series, Publications of the Dominion Observatory, Ottawa, No. 16 to 20.

Walker, J.D., 1988. Permian and Triassic rocks of the Mojave Desert and their implications for the timing and mechanisms of continental truncation; Tectonics, v. 7, p. 685-709.

Wanless, R.K., Stevens, R.D., Lachance, G.R., and Delabio, R.N., 1978. Age Determinations and Geological Studies: K-Ar Isotopic Ages, Report 13; Geological Survey of Canada, Paper 77-2, 60 p.

Wardlaw, B.R., Nestell, M.K., and Dutro, Jr., J.T., 1982. Biostratigraphy and structural setting of the Permian Coyote Butte Formation of central Oregon; Geology, v. 10, p. 13-16.

Wegmann, C.E., 1930. Über diapirismus (besonders im Grundbebirge); Bulletin Comm. Géol. Finlande, v. 92, p. 58-76.

Weller, J.M., 1958. Stratigraphic facies differentiation and nomenclature; Bulletin of the American Association of Petroleum Geologists, v. 42, p. 609-639.

Weller, J.M., 1960. Stratigraphic Principles and Practice; Harper & Row, 725 p.

Wernicke, B., 1985. Uniform-sense normal simple shear of the continental lithosphere; Canadian Journal of Earth Sciences, v. 22, p. 108-125.

Wernicke, B. and Klepacki, D.W., 1988. Escape hypothesis for the Stikine block; Geology, v. 16, p. 461-464.

Wheeler, J.O., 1970. Summary and discussion: structure of the southern Canadian Cordillera; *in* Geological Association of Canada, Special Paper 6, p. 155-166.

Wheeler, J.O., 1972. The Cordilleran structural province: introduction; *in* R.A. Price and R.J.W. Douglas (eds.), Variations in Tectonic Styles in Canada; Geological Association of Canada, Special Paper 11, p. 2-3.

Wheeler, J.O., Brookfield, A.J., Gabrielse, H., Monger, J.W.H., and Tipper, H.W., 1991. Terrane map of the Canadian Cordillera, scale 1:2,000,000; Geological Survey of Canada, Map 1713A.

Wheeler, J.O. and Gabrielse, H. (coords.), 1972. The Cordilleran structural province; *in* R.A. Price and R.J.W. Douglas (eds.), Variations in Tectonic Styles in Canada; Geological Association of Canada, Special Paper 11, p. 1-81.

Wheeler, J.O. and McFeely, P. (comps.), 1991. Tectonic Assemblage Map of the Canadian Cordillera and Adjacent Parts of the United States of America; Geological Survey of Canada, Map 1712A, scale 1:2,000,000.

White, R.S., 1988. The Earth's crust and lithosphere; Journal of Petrology, v. 29, special issue, p. 1-10.

White, R.S., Detrick, R.S., Mutter, J.C., Buhl, P., Minshull, T.A., and Morris, E., 1990. New seismic images of oceanic crustal structure; Geology, v. 18, p. 462-465.

White, W.H., 1959. Cordilleran tectonics in British Columbia; Bulletin of the American Association of Petroleum Geologists, v. 43, p. 60-100.

Wilson, J.T., 1966. Did the Atlantic close and then reopen?; Nature, v. 207, p. 343-347.

Wilson, M., 1993. Plate-moving mechanisms: constraints and controversies; Journal of the Geological Society (London), v. 150, p. 923-926.

Winston, D., 1986. Sedimentation and tectonics of the Middle Proterozoic Belt Basin and their influence on Phanerozoic compression and extension in western Montana and northern Idaho; *in* J.A. Peterson (ed.), Paleotectonics and Sedimentation in the Rocky Mountain Region, United States; American Association of Petroleum Geologists, Memoir 41, p. 87-118.

Wolfe, C.J. and Silver, P.G., 1998. Seismic anisotropy of oceanic upper mantle: shear wave splitting methodologies and observations; Journal of Geophysical Research, v. 103, p. 749-771.

Wood, R.M., 1979a. A re-evaluation of the blueschist facies. Part 1. The pressure-temperature hysteresis cycle; Geological Magazine, v. 116, p. 21-33.

Wood, R.M., 1979b. A re-evaluation of the blueschist facies. Part 2. The role of the mineral in metamorphism; Geological Magazine, v. 116, p. 191-201.

Woodsworth, G.J., 1977. Pemberton (92 J) Map Area, British Columbia; Geological Survey of Canada, Open File 482.

Woodsworth, G.J., 1991. Neogene to Recent volcanism along the east side of Hecate Strait, British Columbia; *in* G.J. Woodsworth (ed.), Evolution and Hydrocarbon Potential of the Queen Charlotte Basin, British Columbia; Geological Survey of Canada, Paper 90-10, p. 325-336.

Woodsworth, G.J., Anderson, R.G., and Armstrong, R.L., 1991. Plutonic regimes; *in* H. Gabrielse and C.J. Yorath (eds.), Geology of the Cordilleran Orogen in Canada; Geological Society of America, The Geology of North America, v. G-2, p. 491-531.

Woodward, N.B., Boyer, S.E., and Suppe, J., 1989. Balanced Geological Cross-Sections: An Essential Technique in Geological Research and Exploration; American Geophysical Union, Short Course in Geology, Volume 6, 132 p.

Workum, R.H. and Hedinger, A.S., 1987. Geology of the Devonian Fairholme Group Cline Channel, Alberta; Second International Symposium on the Devonian System, Field Excursion A6, 41 p.

Wright, R.L., Nagel, J., and McTaggart, K.C., 1982. Alpine ultramafic rocks of southwestern British Columbia; Canadian Journal of Earth Sciences, v. 19, p. 1156-1173.

Wu, P., 1991. Flexure of lithosphere beneath the Alberta foreland basin: evidence of an eastward stiffening continental lithosphere; Geophysical Research Letters, v. 18, p. 451-454.

Wyllie, P.J., 1971. The Dynamic Earth; John Wiley & Sons, 416 p.

Wynne, P.J., Irving, E., Maxson, J.A., and Kleispehn, K.L., 1995. Paleomagnetism of the Upper Cretaceous strata of Mount Tatlow: evidence for 3000 km of northward displacement of the eastern Coast Belt, British Columbia; Journal of Geophysical Research, v. 100, p. 6073-6091.

Yañez, G.A. and LaBreque, J.L., 1997. Age-dependent three-dimensional magnetic modeling of the North Pacific and North Atlantic oceanic crust at intermediate wavelengths; Journal of Geophysical Research, v. 102, p. 7947-7961.

Yang, J. and Hall, J.M., 1996. An intermediate-fast spreading rate of the Troodos type oceanic crust: a comparison to modern oceanic crusts; Continental Dynamics, v. 1, p. 70-80.

Yang Zunyi, Chen Yugi, and Wang Hongzhen, 1986. The Geology of China; Oxford Monographs on Geology and Geophysics, no. 3; Clarendon Press, 303 p.

Yoos, T.R., Potter, C.J., Thigpen, J.L., and Brown, L.D., 1991. The Cordilleran foreland thrust belt in northwestern Montana and northern Idaho from COCORP and industry seismic reflection data; Bulletin of the American Association of Petroleum Geologists, v. 75, p. 1089-1106.

Young, F.G., 1979. The lowermost Paleozoic McNaughton Formation and Equivalent Cariboo Group of Eastern British Columbia: Piedmont and Tidal Complex; Geological Survey of Canada, Bulletin 228, 60 p.

Young, G.M., 1982. The late Proterozoic Tindir Group, east-central Alaska: evolution of a continental margin; Bulletin of the Geological Society of America, v. 93, p. 759-783.

Zang, S.-x., Wu, Z.-l., Ning, J.-y., and Zheng, S.-h., 1992 (translated in 1993). The interaction of plates around China and its effects on the stress field in China: Part II, the influence of Indian plate; Chinese Journal of Geophysics, v. 35, p. 527-540.

Zelt, B.C., Ellis, R.M., and Clowes, R.M., 1993. Crustal velocity structure of the eastern Insular and southernmost Coast belts, Canadian Cordillera; Canadian Journal of Earth Sciences, v. 30, p. 1014-1027.

Zelt, B.C., Ellis, R.M., Clowes, R.M., Kanasewich. E.R., Asudeh, I., Luetgert, J.H., Hajnal, Z., Ikami, A., Spence, G.D., and Hyndman, R.D., 1992. Crust and upper mantle velocity structure of the Intermontane Belt, southern Canadian Cordillera; Canadian Journal of Earth Sciences, v. 29, p. 1530-1548.

Zelt, C.A. and White, D.J., 1995. Crustal structure and tectonics of the southeastern Canadian Cordillera; Journal of Geophysical Research, v. 100, p. 24,255-24,273.

Zen, E., 1988. Evidence for accreted terranes and the effect of metamorphism; American Journal of Science, v. 288-A, p. 1-15.

Zhamoida, A.I. (ed.), 1979. Stratigraphic Code of the U.S.S.R. (English edition); VSEGEI, Leningrad (St. Petersburg).

Ziegler, P.A., 1982. Geological Atlas of Western and Central Europe; Elsevier, 130 p.

Ziegler, P.A. (ed.), 1987. Compressional Intraplate Deformations in the Alpine Foreland; Tectonophysics, special issue, v. 137, 420 p.

Zoback, M.L. and Zoback, M.D., 1980. State of stress in the conterminous United States; Journal of Geophysical Research, v. 85, p. 6113-6156.

Zoback, M.L. and Zoback, M.D., 1989. Tectonic stress field of the continental United States; *in* L.C. Pakiser and W.D. Mooney (eds.), Geophysical Framework of the Continental United States; Geological Society of America, Memoir 172, p. 523-540.

Lecture Notes in Earth Sciences

For information about Vols. 1–19
please contact your bookseller or Springer-Verlag

Vol. 57: E. Lallier-Vergès, N.-P. Tribovillard, P. Bertrand (Eds.), Organic Matter Accumulation. VIII, 187 pages. 1995.

Vol. 58: G. Sarwar, G. M. Friedman, Post-Devonian Sediment Cover over New York State. VIII, 113 pages. 1995.

Vol. 59: A. C. Kibblewhite, C. Y. Wu,Wave Interactions As a Seismo-acoustic Source. XIX, 313 pages. 1996.

Vol. 60: A. Kleusberg, P. J. G. Teunissen (Eds.), GPS for Geodesy. VII, 407 pages. 1996.

Vol. 61: M. Breunig, Integration of Spatial Information for Geo-Information Systems. XI, 171 pages. 1996.

Vol. 62: H. V. Lyatsky, Continental-Crust Structures on the Continental Margin of Western North America. XIX, 352 pages. 1996.

Vol. 63: B. H. Jacobsen, K. Mosegaard, P. Sibani (Eds.), Inverse Methods. XVI, 341 pages, 1996.

Vol. 64: A. Armanini, M. Michiue (Eds.), Recent Developments on Debris Flows. X, 226 pages. 1997.

Vol. 65: F. Sansò, R. Rummel (Eds.), Geodetic Boundary Value Problems in View of the One Centimeter Geoid. XIX, 592 pages. 1997.

Vol. 66: H. Wilhelm, W. Zürn, H.-G. Wenzel (Eds.), Tidal Phenomena. VII, 398 pages. 1997.

Vol. 67: S. L. Webb, Silicate Melts. VIII. 74 pages. 1997.

Vol. 68: P. Stille, G. Shields, Radiogenetic Isotope Geochemistry of Sedimentary and Aquatic Systems. XI, 217 pages. 1997.

Vol. 69: S. P. Singal (Ed.), Acoustic Remote Sensing Applications. XIII, 585 pages. 1997.

Vol. 70: R. H. Charlier, C. P. De Meyer, Coastal Erosion – Response and Management. XVI, 343 pages. 1998.

Vol. 71: T. M. Will, Phase Equilibria in Metamorphic Rocks. XIV, 315 pages. 1998.

Vol. 72: J. C. Wasserman, E. V. Silva-Filho, R. Villas-Boas (Eds.), Environmental Geochemistry in the Tropics. XIV, 305 pages. 1998.

Vol. 73: Z. Martinec, Boundary-Value Problems for Gravimetric Determination of a Precise Geoid. XII, 223 pages. 1998.

Vol. 74: M. Beniston, J. L. Innes (Eds.), The Impacts of Climate Variability on Forests. XIV, 329 pages. 1998.

Vol. 75: H. Westphal, Carbonate Platform Slopes – A Record of Changing Conditions. XI, 197 pages. 1998.

Vol. 76: J. Trappe, Phanerozoic Phosphorite Depositional Systems. XII, 316 pages. 1998.

Vol. 77: C. Goltz, Fractal and Chaotic Properties of Earthquakes. XIII, 178 pages. 1998.

Vol. 78: S. Hergarten, H. J. Neugebauer (Eds.), Process Modelling and Landform Evolution. X, 305 pages. 1999.

Vol. 79: G. H. Dutton, A Hierarchical Coordinate System for Geoprocessing and Cartography. XVIII, 231 pages. 1999.

Vol. 80: S. A. Shapiro, P. Hubral, Elastic Waves in Random Media. XIV, 191 pages. 1999.

Vol. 81: Y. Song, G. Müller, Sediment-Water Interactions in Anoxic Freshwater Sediments. VI, 111 pages. 1999.

Vol. 82: T. M. Løseth, Submarine Massflow Sedimentation. IX, 156 pages. 1999.

Vol. 83: K. K. Roy, S. K. Verma, K. Mallick (Eds.), Deep Electromagnetic Exploration. X, 652 pages. 1999.

Vol. 84: H. V. Lyatsky, G. M. Friedman, V. B. Lyatsky. Principles of Practical Tectonic Analysis of Cratonic Regions. XX, 369 pages. 1999.

Vol. 85: C. Clauser, Thermal Signatures of Heat Transfer Processes in the Earth's Crust. X, 111 pages. 1999.

Vol. 86: H. V. Lyatsky, V. B. Lyatsky, The Cordilleran Miogeosyncline in North America. XX, 384 pages. 1999.